Physics at surfaces

PHYSICS AT SURFACES

ANDREW ZANGWILL
Georgia Institute of Technology

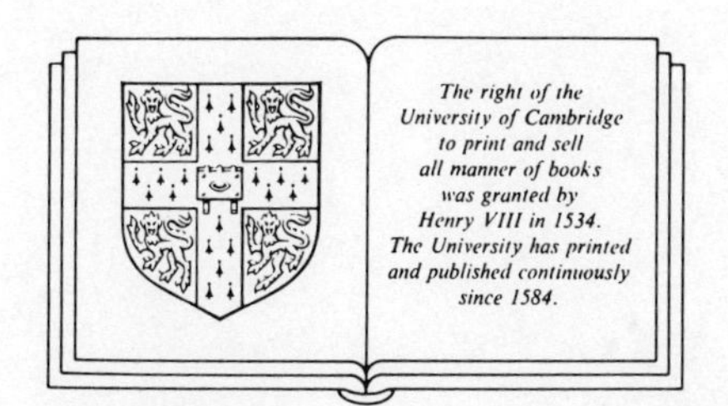

CAMBRIDGE UNIVERSITY PRESS
Cambridge
New York New Rochelle Melbourne Sydney

Published by the Press Syndicate of the University of Cambridge
The Pitt Building, Trumpington Street, Cambridge CB2 1RP
32 East 57th Street, New York, NY 10022, USA
10 Stamford Road, Oakleigh, Melbourne 3166, Australia

First published 1988

Printed in Great Britain at the University Press, Cambridge

British Library cataloguing in publication data

Zangwill, Andrew
Physics at surfaces.
1. Surfaces (Physics) 2. Solid state
physics
I. Title
530.4′1 QC176.8.S8

Library of Congress Cataloging in publication data

Zangwill, Andrew.
Physics at surfaces.

Bibliography: p.
Includes index.
1. Surfaces (Physics) 2. Surface chemistry.
I. Title.
QC173.4.S94Z36 1988 530.4′1 87-15126

ISBN 0 521 32147 6 hard covers
ISBN 0 521 34752 1 paperback

TM

To the memory of Ronald D. Parks

CONTENTS

PREFACE

> The surfaces of bodies are the field of very powerful forces of whose action we know but little. *Lord Rayleigh*

> The surface was invented by the devil. *Wolfgang Pauli*

The present volume is a graduate-level introduction to the physics of solid surfaces. It is designed for students of physics, physical chemistry and materials science who are comfortable with modern condensed matter science at the level of, say, *Solid State Physics* by Ashcroft & Mermin (1976) or *Principles of the Theory of Solids* by Ziman (1972). In the latter, Ziman points out that scientific knowledge passes from the laboratory to the classroom by a sequence of literary vehicles: original research papers, review articles, monographs and finally textbooks. I believe this book fits well into none of these categories. It is not a textbook – at least not in the traditional sense. The field of surface physics is simply not mature enough to support such an enterprise; too many results are untidy and too many loose ends remain. On the other hand, it is not a review or monograph either. My purpose is neither to set down an established wisdom nor to establish priority among claimants. Indeed, I steadfastly ignore who did what when – except when it is a matter of historical interest. Rather, my interest from the beginning has been to construct a coherent synthesis of an enormous range of material and to present the result in as heuristic and pedagogical a manner as possible. Consequently, I think it is useful to regard the account before you as a travelling companion – a tour guide if you will – through the world of surface physics. It possesses both the virtues and the faults of flesh-and-blood tour guides.

This book exists because Ron Parks wanted to learn something about surface physics. To that end, he asked me to prepare a graduate seminar course on the subject for the 1983–4 academic year at the Polytechnic Institute of New York. *Physics at Surfaces* is an expanded and refined version of lecture notes distributed to the attendees of that course. The notes were intended to fill what I perceived to be a yawning gap in the literature. At the undergraduate level, the slim volumes *Surface Physics* by Prutton (1983) and *Principles of Surface Chemistry* by Somorjai (1972) very ably meet the needs of their intended audiences. At the graduate level, *Chemistry in Two Dimensions: Surfaces* by Somorjai (1981) and *Introduction to the Theory of Solid Surfaces* by Garcia-Moliner & Flores (1979) provide unique perspectives from the point of view of very active researchers in the field. Remarkably, the subject matter presented in these books form two almost disjoint sets!

Physics at Surfaces is an attempt to bridge the gap between textbook treatments of condensed matter physics and the primary research literature of surface science. It was necessary, as usual, to choose between depth and breadth of presentation. In opting for the latter, one is challenged to provide unity to a subject which attracts scientists from sub-specialties as diverse as semiconductor device physics, critical phenomena, catalytic chemistry, electron spectroscopy and many-body physics. The choice of topics and logical organization reflect the concerns and prejudices of a condensed matter physicist with a healthy interest in chemical physics. Experiment and theory are intertwined whenever possible although there is little detailed discussion of technique. Explicit references are cited whenever theoretical formulae are quoted without derivation. For experimental detail, the reader should consult *Modern Techniques of Surface Science* by Woodruff & Delchar (1986) or *Solid State Physics: Surfaces* edited by Park & Lagally (1985).

A word about usage. The book is meant to be read as an organic whole. It is heavily self-referential in the sense that I continually revisit concepts and examples introduced in early chapters. It is not a reference work; one cannot look up the work function of Ge(111) or the sticking coefficient of N_2/Cu(100). In fact, the text contains no data tables (although I do provide an acronym table!). Instead, I stress trends which are presented visually in the figures. I wish to emphasize that careful study of the figures is of especial importance. This is so because the sheer volume of material covered and a desire to limit the cost of the book conspired to produce a rather terse prose style.

I owe a considerable debt to the Surface Physics Group at the University of Pennsylvania, *circa* 1976–80 (T. Gustafsson, E.W. Plummer, J.R. Schrieffer and P. Soven) for my initial introduction to this subject and to many members of the international surface science community for discussion and correspondence about their work since then. I am grateful to M. denBoer (CUNY), R. Bruinsma (UCLA), L. Roelofs (Haverford) and J. Tully (AT & T) who read and commented on selected chapters. Of course, any vagaries, misconceptions, or outright errors are entirely my responsibility. Special thanks go to Norton Lang (IBM) and Bill Gadzuk (NBS) for moral support and encouragement at all stages and to my wife Sonia for these and, equally importantly, for patience.

Atlanta
January 1987

A. Zangwill

ACRONYMS

AES	Auger electron spectroscopy
ATR	attenuated total reflection
CS	commensurate solid
DOS	density of states
EDC	energy distribution curve
EELS	electron energy loss spectroscopy
ESCA	electron spectroscopy for chemical analysis
ESD	electron stimulated desorption
ESDIAD	electron stimulated desorption ion angular distribution
FIM	field ion microscope
FWHM	full-width at half-maximum
FVM	Frank–Van der Merwe
HEIS	high energy ion scattering
IRAS	infrared absorption spectroscopy
IS	incommensurate solid
KT	Kosterlitz–Thouless
LDA	local density approximation
LDOS	local density of states
LEED	low energy electron diffraction
LEF	laser-excited fluorescence
LEIS	low energy ion scattering
LGW	Landau–Ginzburg–Wilson
MBE	molecular beam epitaxy
MBRS	molecular beam relaxation spectroscopy
MD	misfit dislocation
MEIS	medium energy ion scattering
MIGS	metal-induced gap states
MOCVD	metalorganic chemical vapor deposition

MPI	multi-photon ionization
NEXAFS	near-edge x-ray absorption fine structure
NFE	nearly-free electron
PSD	photon stimulated desorption
RHEED	reflection high energy electron diffraction
SBZ	surface Brillouin zone
SCLS	surface core level shift
SERS	surface-enhanced Raman scattering
SEXAFS	surface-extended x-ray absorption fine structure
SIMS	secondary ion mass spectroscopy
SK	Stranski–Krastanov
SOS	solid-on-solid
SP	surface polariton
STM	scanning tunnelling microscopy
TPD	temperature programmed desorption
UHV	ultra-high vacuum
UPS	ultraviolet photoelectron spectroscopy
VW	Volmer–Weber
XPD	x-ray photoelectron diffraction
XPS	x-ray photoelectron spectroscopy

0

HISTORICAL SKETCH

Physical phenomena explicitly associated with condensed matter surfaces have been studied since antiquity. Perhaps the oldest written record of experience in this area appears in Babylonian cuneiform dating from the time of Hammurabi (Tabor, 1980). A form of divination, known today as *lecanomancy*, involved an examination of the properties of oil poured into a bowl of water. The detailed behavior of the spreading oil film led the diviner, or *baru*, to prophesy the outcome of military campaigns and the course of illness.

In later years, many observers commented on the fact that choppy waves can be calmed by pouring oil into the sea. In particular, Pliny's account was known to Benjamin Franklin when he began his controlled experiments during one of his frequent visits to England. Franklin's apparatus consisted of a bamboo cane with a hollow upper joint for storage of the oil.

> At length being at Clapham, where there is, on the common, a large pond, which I observed one day to be very rough with the wind, I fetched out a cruet of oil, and dropped a little of it on the water. I saw it spread itself with surprising swiftness upon the surface... the oil, though not more than a tea spoonful, produced an instant calm over a space several yards square, which spread amazingly and extended itself gradually till it reached the lee side, making all that quarter of the pond, perhaps half an acre, as smooth as a looking-glass. (Seeger, 1973.)

Remarkably, Franklin did not perform the simple calculation which would have led him to conclude that the film thickness was only about one nanometer!

The firm establishment of modern methods of scientific analysis that occurred in the nineteenth century produced three notable results of importance to the future of surface science. First, in 1833, Michael Faraday directed his attention to a mysterious phenomenon observed ten years earlier by Dobereiner: the presence of platinum could induce the reaction of hydrogen and oxygen well below their nominal combustion temperature (Williams, 1965). In characteristic fashion, he designed a sequence of experiments which led him to propose a qualitative theory of catalytic action (a term coined in 1836 by Berzelius) which remains valid to this day.

A second critical discovery was made in 1874 by the future Nobel laureate Karl Ferdinand Braun (Susskind, 1980). During the course of electrical measurements of metallic sulfides, Braun noticed deviations from Ohm's law in the conduction of current through a sandwich of Cu and FeS. Only a few years later, he speculated that the cause of the unusual asymmetrical resistance (today called rectification) must reside in a thin surface layer at the interface.

Finally, in 1877, J. Willard Gibbs published (by subscription) the second part of his monumental memoir, 'The Equilibrium of Heterogeneous Substances' in the *Transactions of the Connecticut Academy*. This work, rightly considered one of the crowning achievements of nineteenth century science, established the mathematical foundations of thermodynamics and statistical mechanics (Rice, 1936). As part of this program, Gibbs completely described the thermodynamics of surface phases. Essentially all subsequent work in the field consists of elucidation of his rather difficult exposition.

Despite the impetus provided by these investigations, it was primarily Irving Langmuir's efforts in the early years of this century that led to the recognition of surface science as a significant research discipline (Rosenfeld, 1962). Langmuir received his doctorate under Nernst at Gottingen in 1906 for a problem involving the dissociation of various gases produced by a hot platinum wire. Three years later he joined the fledgling General Electric Research Laboratory and began a remarkable career of scientific achievement. Langmuir's early interest in gases at very low pressures near very hot metal surfaces soon bore fruit with his invention of the nitrogen-filled tungsten incandescent lamp.

At General Electric, Langmuir was free to pursue his broad scientific interests. Consequently, in addition to pioneering the experimental methods necessary for high vacuum studies, he introduced the concepts of the adsorption chemical bond, the surface adsorption lattice, the accommodation coefficient and adsorption precursors. He performed fundamental studies on the work function of metals, heterogeneous catalysis and adsorption kinetics, and provided a detailed model of thermionic emission. Most notably, of course, he and Katherine Blodgett explored the two-dimensional world of monomolecular films. In 1932, the Swedish Academy of Science rewarded Langmuir with its Nobel prize for 'outstanding discoveries and inventions within the field of surface chemistry'.

Two other Nobel prizes of the early twentieth century also have a direct bearing on the development of surface science, and surface physics in particular. The 1921 prize was awarded to Einstein for his explanation of the photoelectric effect and Clinton Davisson was co-recipient of the 1937 prize for his electron diffraction work with Lester Germer that confirmed the wave nature of quantum mechanical particles. Although Davisson and Germer were aware that they were probing the surface layer of their crystals, more than thirty years elapsed before photoemission spectroscopy and low energy electron diffraction became standard laboratory probes of surface electronic and geometrical structure, respectively.

The 1930s can be characterized as a period when a spurt of theoretical research defined a number of important directions for future work on the fundamentals of surface physics. The existence and properties of electron states localized at a crystal surface was explored by Tamm (1932), Maue (1935), Goodwin (1939) and Shockley (1939). In 1932, Lennard–Jones studied the nature of the physisorption precursor to dissociative chemisorption and soon thereafter, Gurney (1935) introduced the resonance level model of adsorbate electronic structure. The basic theory of a free metallic surface (which was to stand unchanged for over thirty years) was introduced at this same time (Bardeen, 1936). Fundamental studies of semiconductor surfaces quite naturally focused on the semiconductor/metal interface. Almost simultaneously, Mott (1938), Schottky (1939) and Davydov (1939) proposed theories of the rectifying junction.

Renewed interest in surfaces had to await the return of scientists from war-related research. In 1949, three papers appeared, each of which stimulated tremendous experimental activity. A sophisticated theory of crystal growth (Burton & Cabrera, 1949) motivated endeavors in that field while Cyril Stanley Smith's influential paper on 'Grains, Phases and Interfaces' (Smith, 1948) alerted much of the metallurgical community to the problems of surfaces. However, the most dramatic event by far was a discovery reported in *The New York Times* as 'a device called a transistor, which has several applications in radio where a vacuum tube ordinarily is employed' (Hoddeson, 1981). The invention of the point-contact transistor (Bardeen & Brattain, 1949) generated an unprecedented interest in the fundamental physics of surfaces, most particularly semiconductor surfaces. Fifteen years of intense research on surfaces and interfaces followed.

In the introduction to their classic monograph, *Semiconductor Surfaces*,

Many, Goldstein & Grover (1965) make an interesting distinction between a 'real' surface and a 'clean' surface. The former is obtained under ordinary laboratory procedures while the latter is prepared under 'carefully controlled conditions so as to ensure the absence of foreign matter'. Unfortunately, at the time, there did not exist any reliable experimental technique for the determination of the chemical composition of a 'clean' surface (Duke, 1984). This is not to say that a great deal of useful information was not obtained about practical rectifying junctions during this period. However, almost nothing was learned about atomically clean surfaces.

The true emergence of surface physics occurred in the late 1960s as a result of the coincidence of several events. The first of these was the realization that electron spectroscopy (Brundle, 1974) and Auger spectroscopy in particular (Harris, 1974) allows one to determine the chemical species present on a solid surface down to minute fractions of a monolayer. Second, technology associated with the space program permitted the commercial development of ultra-high vacuum chambers so that a sample could be *kept* clean for a substantial period of time. At last, controlled experiments could be performed on well-characterized solid surfaces and sensibly compared to theoretical expectations. Indeed, as a final ingredient, the development and availability of high-speed digital computers allowed sophisticated theoretical work to proceed far beyond the simple models of previous years.

The past decade bears witness to the evolution of surface physics out of its infancy. Experimental and theoretical progress has been truly striking. Nevertheless, in many cases, we lack the fundamental principles and unifying themes needed to guide a truly mature science. We are still in a groping phase. Accordingly, the chapters that follow should be regarded as a snapshot of this burgeoning field at a stage of development we might call adolescence.*

* I am indebted to Yves Chabal for this remark.

PART 1

CLEAN SURFACES

1

THERMODYNAMICS

Introduction

The basic tenets of classical thermodynamics derive from two centuries of observations. These experiments, performed almost exclusively for bulk matter, established that undisturbed macroscopic systems spontaneously approach *equilibrium* states that are characterized by a small number of thermodynamic variables. The logical consequences of this statement provide an essential underpinning to all other study of bulk condensed matter. By contrast, systematic study of solid surfaces is much more recent and a correspondingly smaller number of experimental observations are available. Therefore, we must inquire at the outset whether an independent thermodynamics of surfaces is required at the foundation of our subject. Fortunately, this question was thoroughly investigated by Gibbs (1948).

The essential features of bulk thermodynamics can be stated very succinctly (Callen, 1985). In equilibrium, a one-component system is characterized completely by the internal energy, U, which is a unique function of the entropy, volume and particle number of the system:

$$\begin{aligned} U &= U(S, V, N), \\ \mathrm{d}U &= \left.\frac{\partial U}{\partial S}\right|_{V,N} \mathrm{d}S + \left.\frac{\partial U}{\partial V}\right|_{S,N} \mathrm{d}V + \left.\frac{\partial U}{\partial N}\right|_{S,V} \mathrm{d}N, \\ \mathrm{d}U &= T\,\mathrm{d}S - P\,\mathrm{d}V + \mu N. \end{aligned} \tag{1.1}$$

These equations define the temperature, pressure and chemical potential of the bulk. The extensive property of the internal energy,

$$U(\lambda S, \lambda V, \lambda N) = \lambda U(S, V, N), \tag{1.2}$$

together with the combined first and second laws of (1.1), lead to the Euler equation,

$$U = TS - PV + \mu N. \tag{1.3}$$

Differentiating (1.3) and using (1.1) we arrive at a relation among the intensive variables, the Gibbs–Duhem equation:

$$S\,\mathrm{d}T - V\,\mathrm{d}P + N\,\mathrm{d}\mu = 0. \tag{1.4}$$

Surface tension and surface stress

How does this discussion change for a system with a free surface? We create a surface of area A from the infinite solid by a *cleavage* process. Since the bulk does not spontaneously cleave, the total energy of the system must increase by an amount proportional to A. The constant of proportionality, γ, is called the *surface tension*:

$$U = TS - PV + \mu N + \gamma A. \tag{1.5}$$

In equilibrium at any finite temperature and pressure, the semi-infinite solid coexists with its vapor. A plot of the particle density as a function of distance normal to the surface is shown in Fig. (1.1). Gibbs recognized that it is convenient to be able to ascribe definite amounts of the extensive variables to a given area of surface. Accordingly, the vertical lines in Fig. (1.1) indicate a partition of space into a bulk solid volume, a bulk vapor volume and a transition, or surface, volume. The remaining extensive quantities can be partitioned likewise:

$$\begin{aligned} S &= S_1 + S_2 + S_s, \\ V &= V_1 + V_2 + V_s, \\ N &= N_1 + N_2 + N_s. \end{aligned} \tag{1.6}$$

Fig. 1.1. Density of a one-component system as a function of distance from the surface.

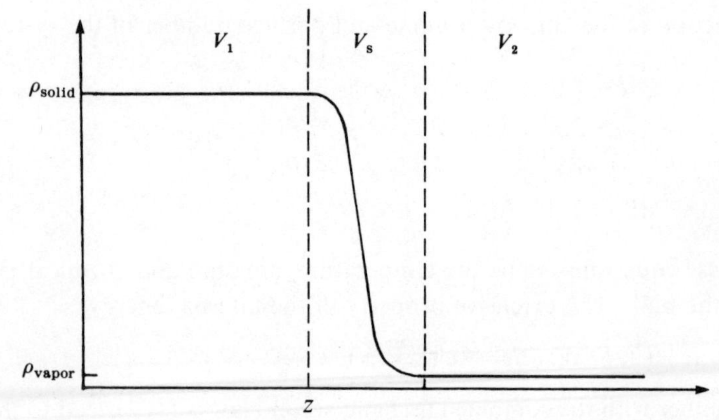

In these formulae, the bulk quantities are *defined* by

$$\left.\begin{aligned} S_i &= s_i V_i \\ N_i &= \rho_i V_i \end{aligned}\right\} \quad i = 1, 2, \tag{1.7}$$

where ρ_i and s_i characterize the uniform bulk phases. According to (1.6), once the surface volume is chosen, the other surface quantities are defined as *excesses.* Note that changes in the surface excess quantities are completely determined by changes in the bulk quantities:

$$\begin{aligned} \Delta S_s &= -\Delta S_1 - \Delta S_2, \\ \Delta V_s &= -\Delta V_1 - \Delta V_2, \\ \Delta N_s &= -\Delta N_1 - \Delta N_2. \end{aligned} \tag{1.8}$$

Evidently, there is nothing unique about the particular choice of the boundary positions illustrated in Fig. (1.1). Nevertheless, it will emerge that one always can choose a subset of the surface excesses that are perfectly well-defined quantities with values that are *independent* of any such conventional choices.

Now consider the effect of small variations in the area of the system, e.g., by *stretching.* We assume that the energy change associated with this process is described adequately by linear elasticity theory (Landau & Lifshitz, 1970). Accordingly, (1.1) should be replaced by

$$\begin{aligned} \mathrm{d}U &= \left.\frac{\partial U}{\partial S}\right|_{V,N,A} \mathrm{d}S + \left.\frac{\partial U}{\partial V}\right|_{S,N,A} \mathrm{d}V + \left.\frac{\partial U}{\partial N}\right|_{S,V,A} \mathrm{d}N \\ &\quad + A\sum_{i,j} \left.\frac{\partial U}{\partial \varepsilon_{ij}}\right|_{S,V,N} \mathrm{d}\varepsilon_{ij}, \\ \mathrm{d}U &= T\,\mathrm{d}S - P\,\mathrm{d}V + \mu\,\mathrm{d}N + A\sum_{i,j} \sigma_{ij}\,\mathrm{d}\varepsilon_{ij}, \end{aligned} \tag{1.9}$$

where σ_{ij} and ε_{ij} are components of the *surface* stress and strain tensors, respectively. These quantities are defined in direct analogy with the bulk. For example, consider any plane normal to the surface and label the normal to the plane as the direction j. σ_{ij} is the force/unit length which the atoms of the solid exert across the line of intersection of the plane with the surface in the i direction.

The corresponding Gibbs–Duhem equation for the *total* system follows from (1.5), (1.9) and the fact that $\mathrm{d}A/A = \sum \mathrm{d}\varepsilon_{ij}\delta_{ij}$:

$$A\,\mathrm{d}\gamma + S\,\mathrm{d}T - V\,\mathrm{d}P + N\,\mathrm{d}\mu + A\sum_{i,j}(\gamma\delta_{ij} - \sigma_{ij})\,\mathrm{d}\varepsilon_{ij} = 0. \tag{1.10}$$

However, the original Gibbs–Duhem relation (1.4) is still valid for each of the two bulk phases separately. Therefore, it can be used (twice) to reduce

(1.10) to a relationship among surface excess quantities only:

$$A\,d\gamma + S_s\,dT - V_s\,dP + N_s\,d\mu + A\sum_{ij}(\gamma\delta_{ij} - \sigma_{ij})d\varepsilon_{ij} = 0. \tag{1.11}$$

This is the *Gibbs adsorption equation*, a fundamental result of surface thermodynamics.

The proper interpretation of (1.11) requires some care. At first glance, it appears that there are five independent variables. γ, μ, P, T and ε. However, the two bulk phase Gibbs–Duhem relations reduce this number to three. For example, suppose we solve for $d\mu$ and dP in terms of dT and substitute them into (1.11):

$$A\,d\gamma + \left\{S_s - V_s\frac{(s_1\rho_2 - s_2\rho_1)}{\rho_2 - \rho_1} + N_s\left(\frac{s_1 - s_2}{\rho_2 - \rho_1}\right)\right\}dT$$

$$+ A\sum_{ij}(\gamma\delta_{ij} - \sigma_{ij})\,d\varepsilon_{ij} = 0. \tag{1.12}$$

The essential point is that (1.7) and (1.8) can be used to show that the quantity in brackets above is *independent* of the arbitrary boundary positions (cf. Fig. 1) which define N_s, V_s, and S_s. Consequently, with Gibbs, we can choose $V_s = N_s = 0$ with no loss of generality (Fig. 1.2). The adsorption equation then takes the simple form,

$$A\,d\gamma + S_s\,dT + A\sum_{i,j}(\gamma\delta_{ij} - \sigma_{ij})\,d\varepsilon_{ij} = 0 \tag{1.13}$$

from which it follows that

$$S_s = -A\left.\frac{\partial\gamma}{\partial T}\right|_{\varepsilon} \tag{1.14}$$

Fig. 1.2. The 'equal area' Gibbs convention, $V_s = N_s = 0$.

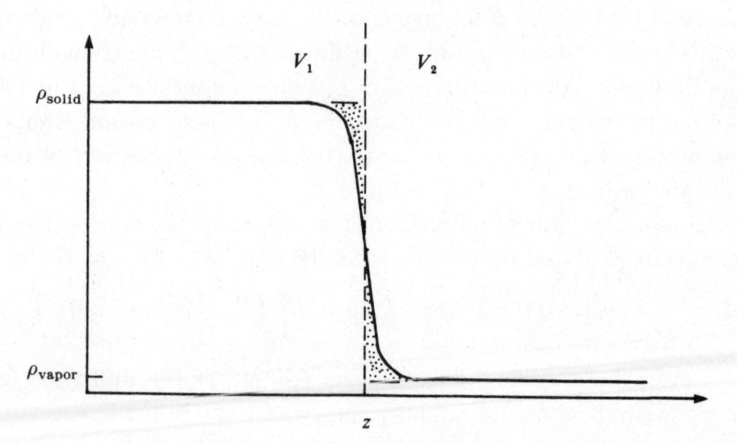

and

$$\sigma_{ij} = \gamma\delta_{ij} + \left.\frac{\partial\gamma}{\partial\varepsilon_{ij}}\right|_T . \tag{1.15}$$

Equation (1.15) shows that the surface tension and the surface stress are not identical in general. A special case occurs when γ is independent of small strains. This is true only when the system is free to rearrange itself in response to a perturbation, i.e., in a liquid. In a solid, non-zero surface stresses must be relieved in other ways. A detailed analysis (Herring, 1951a; Andreussi & Gurtin, 1977) shows that if $\partial\gamma/\partial\varepsilon < 0$, atomic dislocations and elastic buckling of the surface can be expected. A dramatic

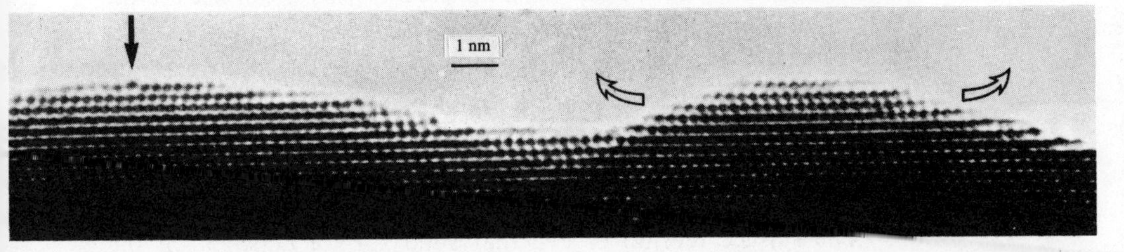

Fig. 1.3. A Au(111) surface buckled under surface stresses. Vertical arrow marks a surface dislocation (Marks, Heine & Smith, 1984).

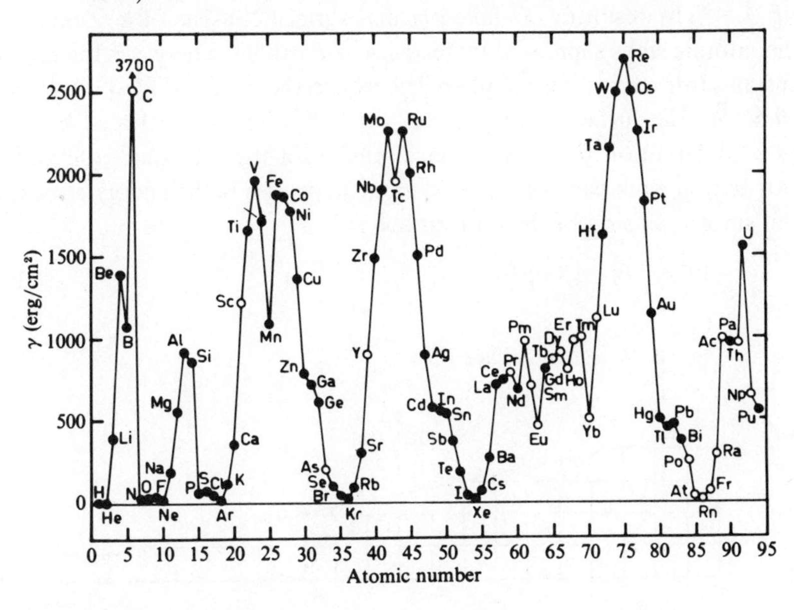

Fig. 1.4. Surface tension of the elements in the liquid phase (Schmit, 1974).

illustration of this phenomenon has been observed for a Au(111) surface using very high resolution electron microscopy (Fig. 1.3).

According to (1.5), the surface tension can be regarded as an excess *free* energy/unit area. This fact suggests a number of experiments that might be used to measure γ directly (Lindford, 1973). For example, in 1857, Faraday noticed that gold foils contracted as they were heated near their melting points. This effect, known as *creep*, occurs because of rapid atomic diffusion under the influence of surface forces. By opposing this creep with known external forces, the surface tension can be determined. These experiments are rather difficult to perform accurately. Nonetheless, we can obtain a crude order-of-magnitude estimate as follows. By definition, γ is the energy cost/unit area to cleave a crystal, i.e., to break surface bonds. Hence, we write $\gamma = E_{coh}\,(Z_s/Z)\,N_s$ where E_{coh} is the *bulk* cohesive energy, Z_s/Z is the fractional number of bonds broken (per surface atom) when the cleave occurs and N_s is the areal density of surface atoms. Putting in typical numbers ($E_{coh} \sim 3\,\mathrm{eV}$, $Z_s/Z \sim 0.25$, $N_s \sim 10^{15}$ atom/cm^2) we get $\gamma \sim 1200\,\mathrm{erg/cm^2}$. The variations in this number across the periodic table can be inferred from measured values of *liquid* surface tensions (Fig. 1.4) and simply reflect the variations in E_{coh} itself.

Anisotropy of γ

The surface tension of a planar solid surface depends on the crystallographic orientation of the sample. To see this, consider a two-dimensional solid which is very slightly misaligned from the [01] direction (Fig. 1.5). The resulting so-called *vicinal* surface consists of a number of monoatomic steps separated by terraces of width na, where a is the lattice constant. For n large, the small angle between the [01] and $[1n]$ directions is $\theta \cong 1/n$. The surface tension along the $[1n]$ direction, denoted by $\gamma(\theta)$, has a contribution from the surface tension of the (01) face, $\gamma(0)$, and a contribution from each of the individual steps. If β is the energy/step, the total surface tension of the $(1n)$ surface is

$$\gamma(\theta) = \gamma(0) + (\beta/a)|\theta|. \tag{1.16}$$

Fig. 1.5. A vicinal surface.

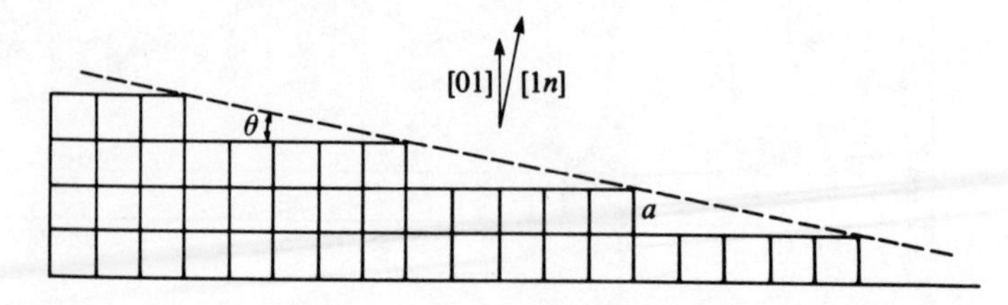

The change of sign in the second term reflects the fact that it always costs energy to produce steps on a flat surface. Notice that (1.16) implies that $\gamma(\theta)$ is a continuous function near $\theta=0$ but that it has a *discontinuous* derivative at that point, i.e., there is a cusp. More precisely,

$$\Delta\left(\frac{\mathrm{d}\gamma}{\mathrm{d}\theta}\right)_{\theta=0}=2\beta/a. \tag{1.17}$$

Now take θ to be a large angle. The density of steps will increase and a proper calculation of the surface tension must include the energy of interaction between steps. In this case, Landau (1965) has shown that $\gamma(\theta)$ has a cusp at every angle which corresponds to a rational Miller index! The sharpness of the cusp is a rapidly decreasing function of index:

$$\Delta\left(\frac{\mathrm{d}\gamma}{\mathrm{d}\theta}\right)\sim\frac{1}{n^4}. \tag{1.18}$$

Hence, a polar plot of the surface tension at $T=0$ has the form illustrated by the solid curve in Fig. 1.6.

The anisotropy of the surface tension determines the equilibrium shape of small crystals because a crystal will seek the shape that minimizes the

Fig. 1.6. Polar plot of the surface tension at $T=0$ (solid curve) and the Wulff construction of the equilibrium crystal shape (dashed curve) (Herring, 1951b).

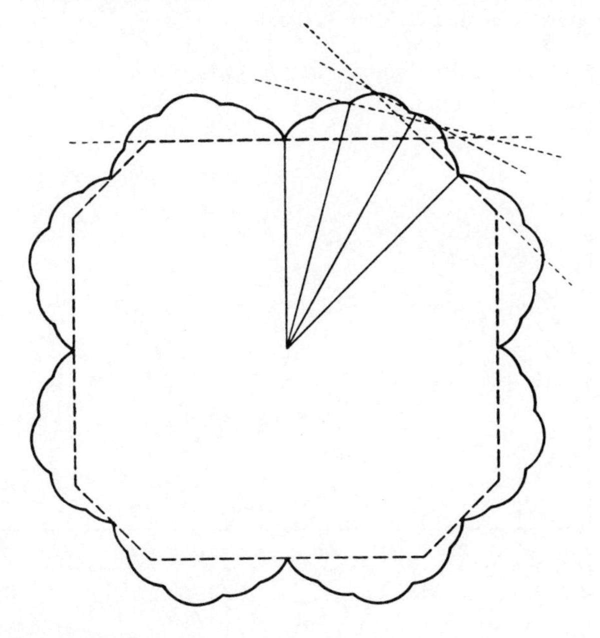

quantity

$$\oint \gamma(\theta)\,\mathrm{d}A, \tag{1.19}$$

subject to the constraint of fixed volume. This question amounts to a

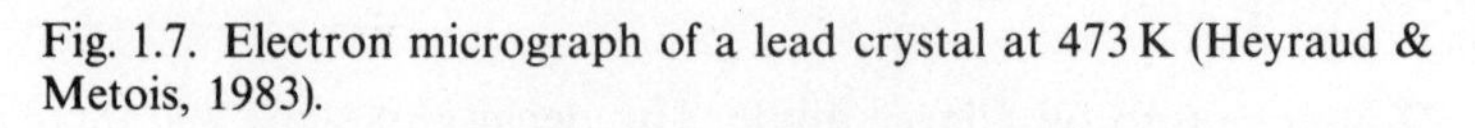

Fig. 1.7. Electron micrograph of a lead crystal at 473 K (Heyraud & Metois, 1983).

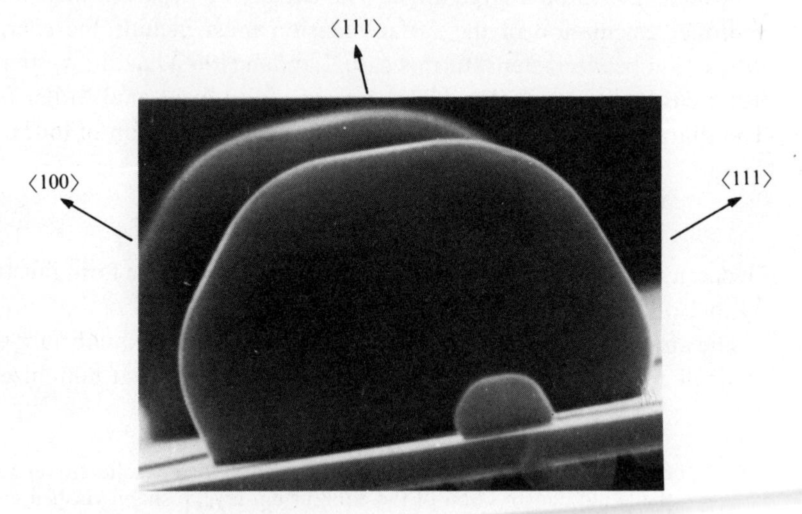

Fig. 1.8. Anisotropy of γ relative to ⟨111⟩ for lead as a function of temperature (Heyraud & Metois, 1983).

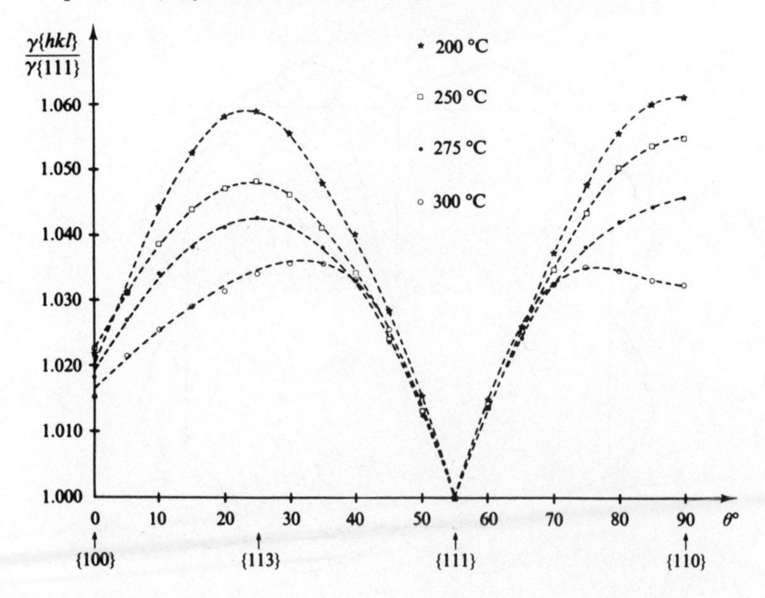

problem in affine geometry which first was solved by Wulff (1901). To find the equilibrium shape, draw a radius vector that intersects the polar plot at one point and makes a fixed angle with the horizontal. Construct the plane that is perpendicular to the vector at the point of intersection. Repeat this procedure for all angles. The interior envelope of the resulting family of planes is a convex figure whose shape is that of the equilibrium crystal (Fig. 1.6). If one tries to cleave a crystal along a direction which does not form part of this equilibrium boundary the crystal will spontaneously facet along those directions that do. It must be borne in mind that this construction is relevant only when the crystal is in true thermodynamic equilibrium. Unfortunately, crystal growth generally occurs under highly non-equilibrium conditions so that the equilibrium shape rarely is achieved; kinetic constraints restrict the necessary transport of mass along the surface. Nonetheless, some reliable data are available.

Electron microscopy has been used to study the shape of very small (diameter $\cong$ 10 microns) lead crystallites. The equilibrium shape (Fig. 1.7) is found to be a cubo-octahedron with a number of well-developed facets. The Wulff theorem then can be run in reverse to determine the anisotropy of the surface tension (Fig. 1.8). Note the 'blunting' of the cusp near [111] as the temperature is raised.

The roughening transition

At finite temperatures, the discussion of the previous section must be supplemented to include entropy effects. At very low temperature any given facet is microscopically flat with only a few thermally excited surface vacancies or defects (Fig. 1.9*a*). However, at higher temperature more and more energetic fluctuations in the local height of the surface can occur

Fig. 1.9. Surface morphology: (*a*) $T < T_r$; (*b*) $T > T_r$ (Muller-Krumbhaar, 1978).

leading to a *delocalized* interface with long wavelength variations in height (Fig. 1.9*b*). The step *free* energy, β, decreases with increasing temperature, blunting the Wulff cusps and causing the facets to shrink. At a certain *roughening temperature*, T_r, the facet disappears and only a smoothly rounded macroscopic morphology remains. The passage between these two extremes occurs via a phase transition.

The nature of the roughening transition can be appreciated quite simply with use of the so-called *solid-on-solid* (SOS) model. We view the crystal as a collection of interacting columns (one for each surface atom) and suppose that there is a finite energy cost J if nearest neighbor columns differ in height by one lattice constant. More generally, we take

$$\mathscr{H} = J \sum_{\langle i,j \rangle} |h_i - h_j|^2, \tag{1.20}$$

where the column heights, h_i, are restricted to integer values. Note that no overhangs are permitted and that at zero temperature all columns have the same height, i.e., the surface is flat at, say, $Z = 0$. The lowest energy excitations are monoatomic steps on the surface that form themselves into plateaus (Fig. 1.10). Therefore, a loop of length L bounding a plateau has energy JL/a, where a is the lattice constant. The number of possible loops of this length is equivalent to the number of self-avoiding random walks that return to the origin in L/a steps. If each column has z nearest neighbor columns, this number is $z^{L/a}$, to within a constant of order unity (Feller,

Fig. 1.10. Top view of a crystal surface in the SOS model (Schulz, 1985).

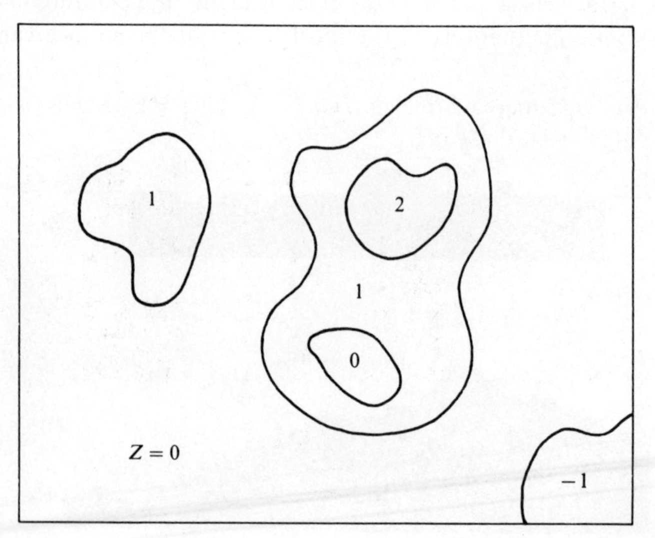

1968). Thus, the free energy of the system is:

$$F = U - TS = \frac{L}{a}(J - T\ln z). \tag{1.21}$$

Below the roughening temperature, $kT_r = J/\ln z$, $L = 0$ is favored whereas above this temperature, loops of arbitrarily large length must occur. Fig. 1.9(*b*) can be regarded as a side view of the system with a large number of concentric loops of (almost) infinite length.

Phase transitions generally are characterized by non-analytic behavior in the free energy function (Stanley, 1971). Interestingly, detailed studies of the roughening transition show that the required singularity at the transition point is extremely weak (Weeks, 1980):

$$F \sim \exp\left(-\frac{A}{|T - T_r|^{1/2}}\right). \tag{1.22}$$

Observe that the derivatives of this free energy, i.e., the usual thermodynamic observables, do not exhibit any unusual behavior as the transition occurs. However, as noted above, surface 'roughening' precisely corresponds to the disappearance of a crystal facet at T_r. Therefore, the transition can be detected by direct optical observation. A particularly attractive candidate in this regard is hexagonal close-packed ^{4}He coexisting with its own superfluid. The second sound mode of the superfluid provides a large thermal conductivity which facilitates equilibration of the sample. On purely dimensional grounds, a crude estimate of the roughening temperature can be found from $kT = a^2\gamma$, where a is the lattice constant. Using the measured* value of γ for ^{4}He, 0.2 erg/cm^2, we find $T_r \sim 1$ K. For the

Fig. 1.11. Optical holograms of a 2 mm ^{4}He crystal above and below the roughening temperature of the (1120) face (*B*) (Avron *et al.*, 1980).

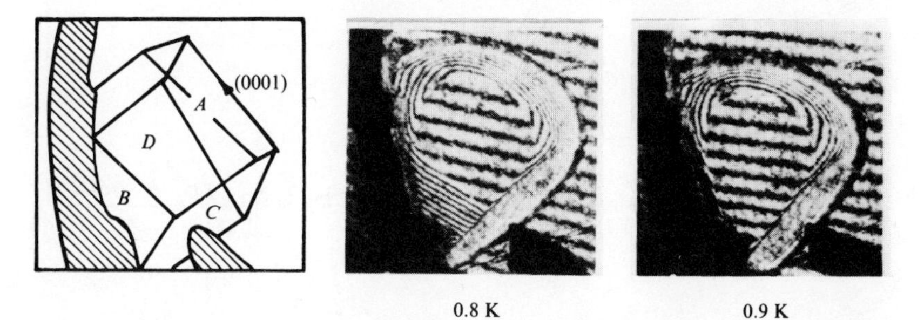

* More precisely, it is the surface *stiffness*, $\gamma + \gamma''$, that is obtained from measurements of surface curvature. γ'' is the second derivative of γ with respect to azimuthal angle (Herring, 1951a).

(1120) face, the phase transition actually occurs at about 0.85 K (Fig. 1.11).

The existence of a roughening transition has profound implications for crystal growth from the melt. Above T_r, the growth rate simply is proportional to the difference in chemical potential across the liquid–solid interface. But, for temperatures $T < T_r$, growth involves the nucleation of two-dimensional facet terraces – an activated process which depends on the step energy β. As a result, the behavior of $\beta(T)$ can be inferred directly from careful observations of the velocity of a growing crystal's solidification front (Fig. 1.12).

The roughening transition is but one example of a generic class of two-dimensional phase transformations that first were analyzed systematically by Kosterlitz & Thouless (1973). A number of other examples will appear in succeeding chapters.

Fig. 1.12. Experimental values of the temperature dependence of the terrace step energy β/a of the (0001) surface of ^{4}He near its roughening temperature ($T_r = 1.28$ K) (Gallet, Nozières, Balibar & Rolley, 1986).

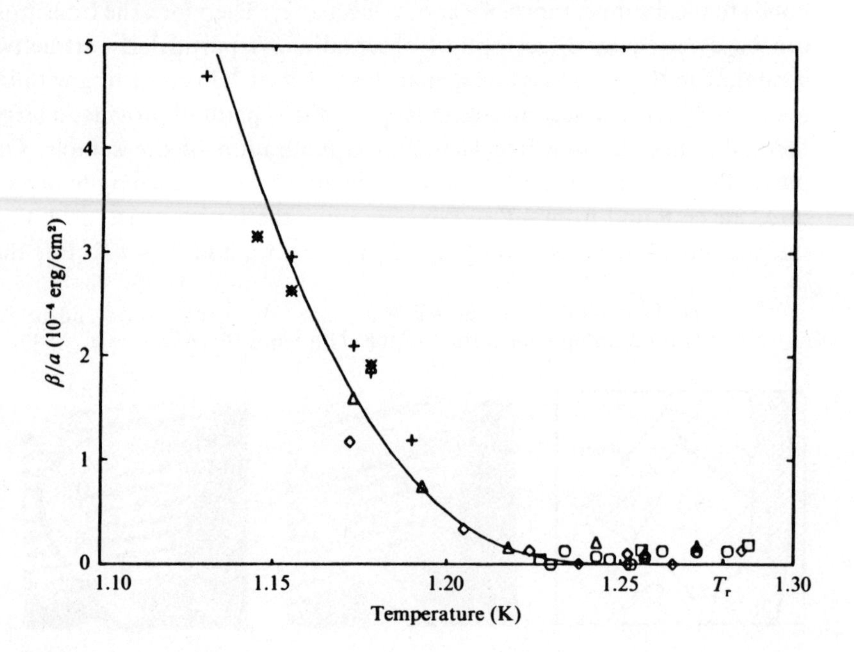

General references

Surface tension and surface stress

Herring, C. (1953). The Use of Classical Macroscopic Concepts in Surface Energy Problems. In *Structure and Properties of Solid Surfaces* (eds. Gomer & Smith), pp. 5–72. Chicago: University Press.

The anisotropy of γ

Landau, L.D. (1965). The Equilibrium Form of Crystals. In *Collected Papers*, pp. 540–5. Oxford: Pergamon.

The roughening transition

Weeks, J.D. (1980). The Roughening Transition. In *Ordering in Strongly Fluctuating Condensed Matter Systems* (ed. Riste), pp. 293–317. New York: Plenum.

2

CHEMICAL ANALYSIS

Introduction

The physics at a solid surface is determined by the identity, concentration and geometrical arrangement of the chemical species present at the surface in question. Unfortunately, surface experiments are conducted in the presence of a very large number of bulk atoms. The number of surface atoms is typically of the order of $N_A^{2/3}$, i.e., $10^{15}/\text{cm}^2$ as compared to $N_A = 10^{23}/\text{cm}^3$ in the bulk. Hence, a small surface-derived signal rides atop a large bulk background signal. Standard methods of chemical analysis simply do not have sufficient sensitivity to provide useful information.

The need for accurate 'titration' methods is particularly acute for surface problems because of the ease with which contamination can occur. The contaminants come from both the ambient atmosphere and from impurities that diffuse to the surface from within the bulk. For example, consider a solid in equilibrium with a gas of molecules of mass m. Elementary kinetic theory provides an estimate of the surface impact rate for a gas at fixed pressure (P) and temperature (T):

$$\text{rate} = \frac{P}{(2\pi mkT)^{1/2}}. \tag{2.1}$$

For nitrogen at 300 K and a pressure of 10^{-8} Torr the surface impact rate is $5 \times 10^{12}/(\text{cm}^2\,\text{s}^1)$. If every molecule that strikes the surface sticks, a 'clean' surface would be covered with a monolayer of nitrogen in three minutes. Experiments on clean surfaces require ultra-high vacuum (UHV) conditions, 10^{-10} Torr or better.

Electron spectroscopy

Essentially all practical surface elemental analysis employs electron spectroscopy in one form or another. The reason for this derives from two experimental facts. First, electrons with kinetic energies in the range 15–1000 eV have a very short mean free path in matter (< 10 Å). Second, the binding energy of a core electron is a sensitive function of atomic identity. Therefore, measurements of the kinetic energy of electrons ejected from a solid after photon or electron bombardment can provide surface-specific elemental information.

The surface sensitivity of electrons is illustrated best with a plot of inelastic mean free path versus electron kinetic energy (Fig. 2.1). The data points scatter around a 'universal curve' that has a broad minimum near 50 eV. This universality is easy to understand. Recall that the dominant electron energy loss mechanism in solids is excitation of valence band electrons. We merely need note that the electron density in the valence band is nearly a constant for most materials – about 0.25 electron/Å^3. Consequently, a Golden Rule calculation of the inelastic mean free path of electrons in a solid modelled as a free electron gas (of this density) gives a good account of the data (dashed curve in Fig. 2.1). We conclude that electrons with kinetic energies in the appropriate range that escape from a solid *without subsequent energy loss* must originate from the surface region.

Perhaps the most common electron-based elemental analysis technique is known as Auger electron spectroscopy (AES). One directs a high energy (> 1 keV) electron beam at a sample and collects the spectrum of backscattered electrons, $N(E)$. $N(E)$ exhibits an elastic peak (electrons

Fig. 2.1. Universal curve of electron mean free path: experiment (Rhodin & Gadzuk, 1979; Somorjai, 1981); theory (Penn, 1976).

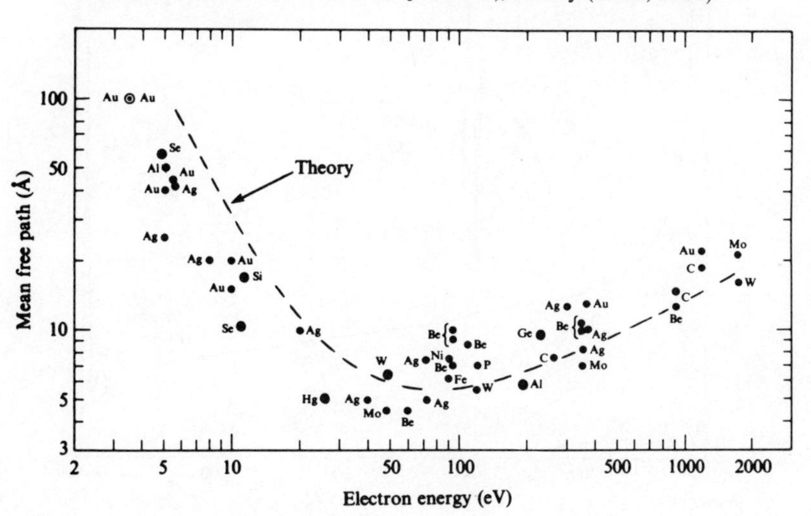

that pass undisturbed through the solid) and a long, seemingly featureless tail of electrons that have lost energy to the solid (Fig. 2.2). Two types of electrons contribute to this tail. Primary electrons exit the sample after losing energy in a single well-defined inelastic event. Other, so-called secondary electrons lose energy through multiple inelastic collisions. The experimental signal from the latter is truly structureless. The former show up as tiny wiggles in $N(E)$ that reveal their origin in the *derivative* signal, $dN(E)/dE$ (inset of Fig. 2.2).

The precise energy position of the sharp structure in the derivative spectrum of Fig. 2.2 is the elemental signature of the surface. To see this, suppose that an electron in the incident beam collides with an atom in the solid and ionizes a 1s electron that was bound with an energy E_{1s}. If E_{1s} is less than about 2000 eV, the hole in the 1s shell is filled preferentially by a radiationless Auger transition, e.g., a 2s electron drops into the hole and the transition energy ejects a *second*, Auger electron, from the 2p level (Fig. 2.3). Energy conservation demands that the kinetic energy of the outgoing electron be

$$E_{KIN} = E_{1s} - E_{2s} - E_{2p}, \tag{2.2}$$

where E_{2s} and E_{2p} are the binding energies of the 2s and 2p atomic levels, respectively, in the presence of the 1s core hole. Notice that the outgoing

Fig. 2.2. $N(E)$ and $dN(E)/dE$ for electrons backscattered from a titanium target after bombardment with 1 keV electrons (Park & den Boer, 1977).

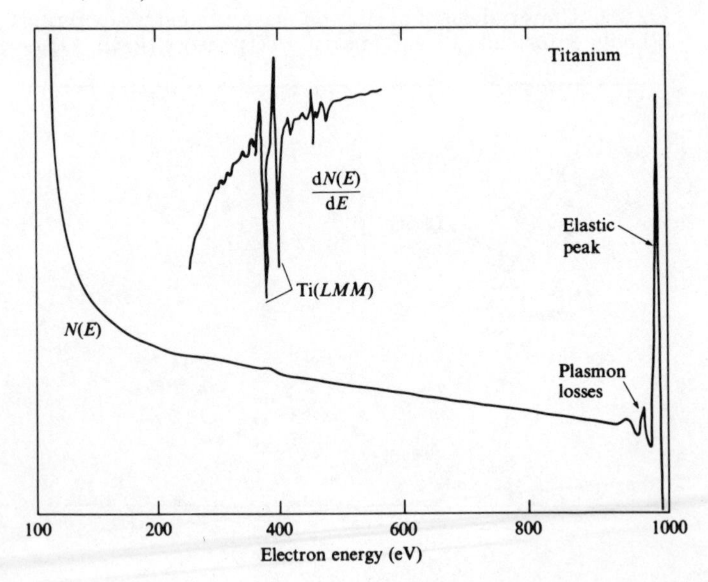

electron's kinetic energy depends only on the properties of the atom. Similar core-hole decay processes occur for all the atoms of the periodic table (except hydrogen and helium) and the characteristic Auger electron energies are well known and tabulated. The key point is that every element

Fig. 2.3. *KLL* Auger decay of a 1s core hole.

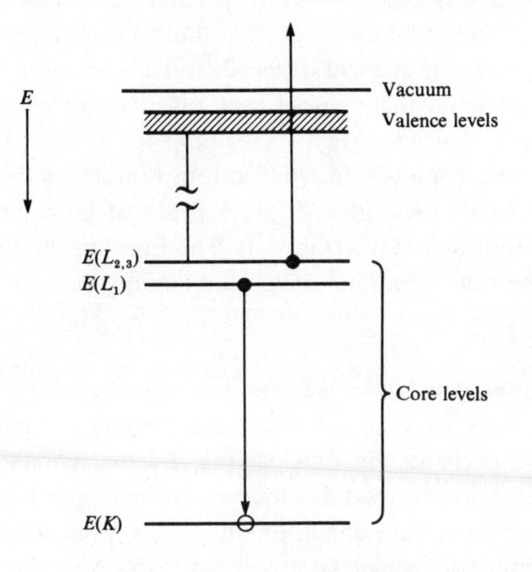

Fig. 2.4. XPS electron energy spectrum from a titanium target illuminated with Mg Kα radiation (Wagner *et al.*, 1978).

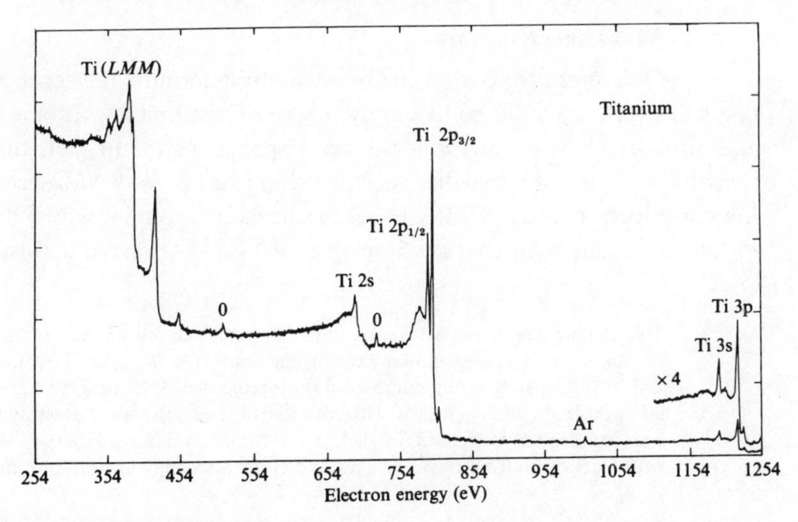

exhibits *some* Auger decay* for which the ejected electron's kinetic energy falls within the critical range for surface sensitivity. Auger spectroscopy is suited perfectly for surface elemental analysis because every surface atom leaves its 'fingerprint' in the kinetic energy spectrum. The principal disadvantage to AES is that the incident electron beam charges up a non-conducting sample.

X-ray photoemission spectroscopy (XPS) is a related surface analysis technique that also takes advantage of the short mean free path of electrons in matter and the elemental specificity of core-hole binding energies. Here, one exploits the photoelectric effect using a source of monochromatic x-rays, typically Mg Kα (1254 eV) or Al Kα (1487 eV) radiation. Again, the spectrum of emitted electrons (known as the *energy distribution curve* or EDC) invariably displays peaks at kinetic energies (E_{KIN}) in the surface sensitive range (Fig. 2.4). The Einstein photoelectric equation connects the peak positions to specific binding energies (E_{B}):

$$E_{\mathrm{KIN}} = \hbar\omega - E_{\mathrm{B}}. \tag{2.3}$$

The atomic species present at the surface are determined by matching the inferred values of E_{B} to a table of elemental core binding energies. Note that additional peaks occur in the EDC of Fig. 2.4 that correspond to electrons ejected by an Auger process that follows a primary photoemission event. XPS is sensitive to surface contaminants at the level of about 1% of a monolayer – similar to Auger spectroscopy. However, there is no sample charging problem and, more importantly, small shifts in the observed core level binding energies can be used to distinguish the same element in different chemical environments (Siegbahn *et al.*, 1967).

Mass spectroscopy

One might think that the best way to determine the composition of a solid surface would be to simply scrape off the first few atomic layers and submit them to conventional mass spectrometry. In fact, the best *sensitivity* to surface impurities is achieved in just this way with secondary ion mass spectroscopy (SIMS). Here, the surface is bombarded by a beam of ions or atoms with energies in excess of 1 keV. Atoms and clusters of

* The atomic levels in an Auger transition are labelled in accordance with conventional X-ray spectroscopic nomenclature, i.e., $K, L, M, \cdots$ for the $n = 1, 2, 3, \ldots$ principal quantum numbers of the atomic shells. Hence, a KLL transition fills a hole in the $n = 1$ shell with an electron from the $n = 2$ shell and ejects a second electron from the $n = 2$ shell. An LMM decay fills a hole in the $n = 2$ shell with an electron from the $n = 3$ shell and ejects a second electron from the $n = 3$ shell, etc.

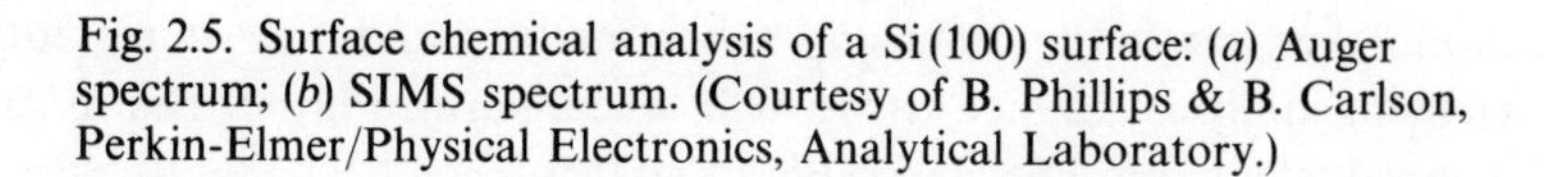

Fig. 2.5. Surface chemical analysis of a Si(100) surface: (*a*) Auger spectrum; (*b*) SIMS spectrum. (Courtesy of B. Phillips & B. Carlson, Perkin-Elmer/Physical Electronics, Analytical Laboratory.)

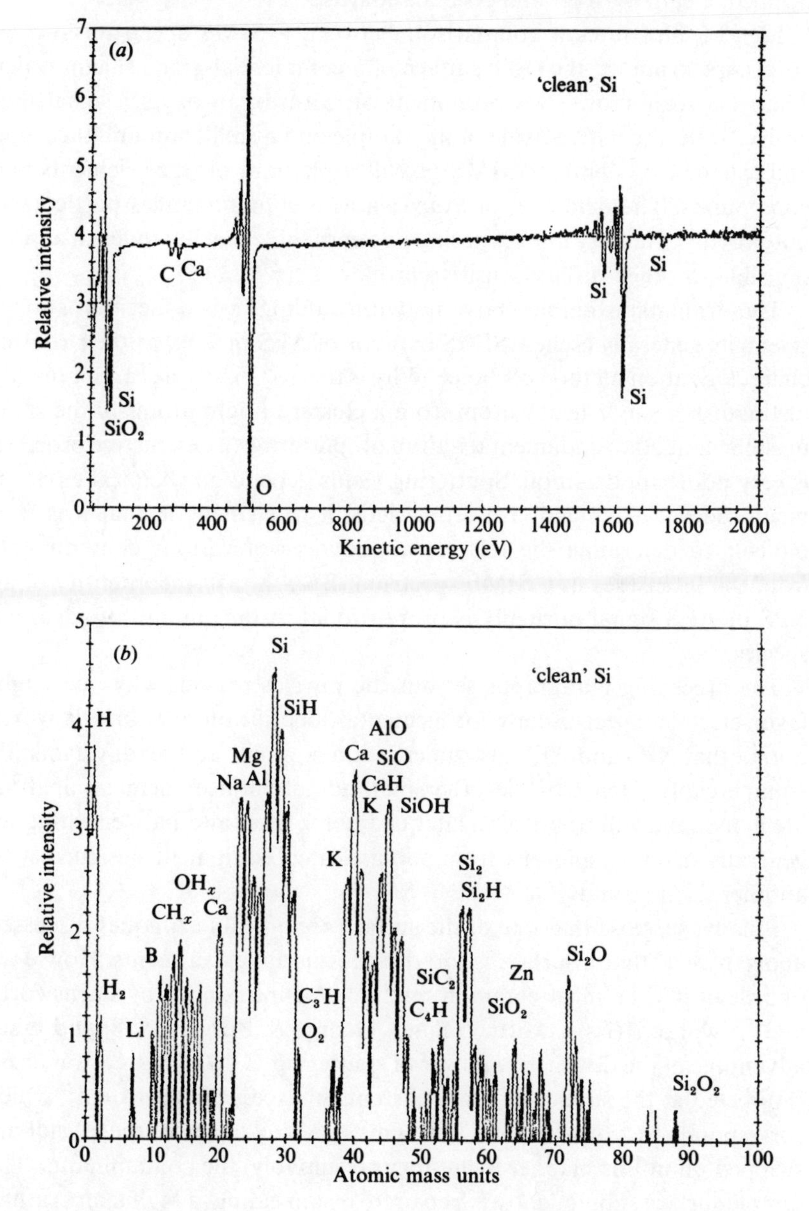

atoms are knocked (sputtered) off the surface and subjected to analysis in a standard mass spectrometer. The limit of detection can be $\sim 10^{-6}$ monolayer, far better than any other method and well suited for trace element analysis. In fact, SIMS will reveal surface impurities in samples deemed 'clean' by AES and XPS standards.

Fig. 2.5 illustrates a comparison between a SIMS spectrum and an Auger spectrum for the (100) surface of a commercial-grade silicon wafer. The AES scan shows two prominent Si features, an oxygen signal that derives from the native oxide of the sample and a small amount of carbon and calcium. By contrast, SIMS reveals a plethora of trace elements and compounds. The detection of *hydrogen* as a contaminant is particularly noteworthy. The very low scattering power of this element renders it almost invisible to other surface sensitive probes.

The dramatic example above notwithstanding, it is a fact that surface scientists generally eschew SIMS in favor of AES or XPS as their routine characterization method of choice. Why is this so? First, one cannot readily distinguish a single heavy atom from a cluster of light atoms of the same mass. Second, the fundamental nature of sputtering (a destructive process!) is very poorly understood. Sputtering yields depend on chemical environment and vary tremendously from element to element so that it is very difficult to determine the relative abundances of various contaminants from the intensities in a SIMS spectrum. By contrast, the intensity of an XPS or AES signal normally is proportional to the amount of adsorbed species.

The preceding paragraphs set out the physics reasons why one might favor electron spectroscopy for elemental identification. It also is worth noting that XPS and AES instrumentation is simple and readily available commercially. Hence, while other methods of surface chemical analysis are in use (and will be noted in later chapters), these alternative techniques generally arise as spin-offs from sophisticated equipment designed with another use in mind.

Finally, suppose that one of the surface analytical techniques discussed above reveals that a surface is too dirty for useful experiments. How does one clean it? The most common method was introduced by Farnsworth and co-workers (Farnsworth, Schlier, George & Burger, 1958) and takes advantage of the destructive power of sputtering. A 400–1000 eV ion beam is directed at the surface at beam currents in excess of $100\,\mu A/cm^2$, which corresponds to a dose of 10^{15} ions/(cm^2 s). Many layers of the crystal are stripped off in this manner including (presumably) the contaminants. The sample surface subsequently is 'repaired' by annealing at high temperature.

An acceptably clean surface generally can be obtained after many sputtering/annealing cycles.

General references

Electron spectroscopy

Carlson, T.A. (1975). *Photoelectron and Auger Spectroscopy*. New York: Plenum.

Mass spectroscopy

Benninghoven, A. (1973). Surface Investigation of Solids by the Statical Method of Secondary Ion Spectroscopy. *Surface Science* **35**, 427–57.

3

CRYSTAL STRUCTURE

Introduction

A complete characterization of a solid surface requires knowledge of not only *what* atoms are present but *where* they are. Just as in the bulk, it is not that the atomic coordinates as such are of much direct interest. Rather, our concerns generally will center on electronic and magnetic properties and it is the geometrical arrangement of the surface atoms that largely determines the near-surface charge and spin density. Put another way, the nature of the surface chemical bond depends in detail on surface bond lengths and bond angles. The corresponding bulk structural issues normally are resolved by x-ray diffraction. Unfortunately, the extremely large penetration depth and mean free path of x-rays severely limits their routine use for surface crystallography. Consequently, much effort has been devoted to the invention and application of alternative experimental approaches to surface-specific structural analysis. Although a number of common techniques will be discussed below, it is a sobering fact that no *single* surface structural tool has emerged that can be used as easily and reliably as x-rays are used for the bulk.

Appeal to theory does not offer much relief. In principle, a solid adopts the crystal structure that minimizes its total energy. We know how to write down an exact expression for this energy; it is a parametric function of the exact position of all the ions in the material. Of course (for an ordered crystal!), translational invariance restricts the number of ion positions that need be independently varied in any computational energy minimization scheme. Even so, reliable first principles prediction of *bulk* crystal structures is possible only for rather simple systems (Cohen, 1985). The problem becomes immensely more difficult for a semi-infinite system because (minimally) translational symmetry is lost in the direction normal to the surface. With very few exceptions, it is impossible to determine surface crystal structures by purely theoretical means. Instead, one relies on simple models and intuition.

The first thing one might guess is that cleavage of a crystal does not perturb the remaining material at all. That is, perhaps the arrangement of atoms is precisely the same as a planar termination of the bulk. As it happens, this so-called 'ideal' surface appears to be the exception rather than the rule. One case where it does seem to occur is at a non-polar (neutral) surface of a cubic insulating compound, e.g., rocksalt. To see why this might be so, recall that the cubic structure of the bulk arises because this particular arrangement of point ions has a lower electrostatic potential energy than other structures. Now consider any two crystal planes of this system that are parallel to the intended cleavage surface. Since both are neutral, there is only a very weak Coulomb interaction between them. Hence, the creation of a surface by removal of half the crystal has almost no effect on the ion positions of the exposed surface plane.

In metals, the ion cores are screened by symmetrical Wigner–Seitz charge clouds formed from the mobile conduction electrons. The residual electrostatic forces are weakly *attractive* and stabilize the familiar close-packed structures of the bulk when hard core Pauli repulsion is included. At a surface, the electrons are free to rearrange their distribution in space to lower their kinetic energy. The resultant *smoothing* of the surface electronic charge density leaves the surface ions out of electrostatic equilibrium with the newly asymmetrical screening distribution. The net force on the ions points primarily into the crystal and a contractive *relaxation* of the surface plane occurs until equilibrium is reestablished (Fig. 3.1). The in-plane structure generally retains the characteristics of an 'ideal' close-packed surface (Fig. 3.2).

Entirely different considerations determine the bulk (and surface) crystal structure of semiconductors. Truly directional chemical bonds between atoms favor the tetrahedral coordination of the zincblende and wurtzite lattices. A highly unstable or metastable state occurs when these bonds are broken by cleavage. The surface (and subsurface) atoms will pay

Fig. 3.1. Electron smoothing at a metal surface (Finnis & Heine, 1974).

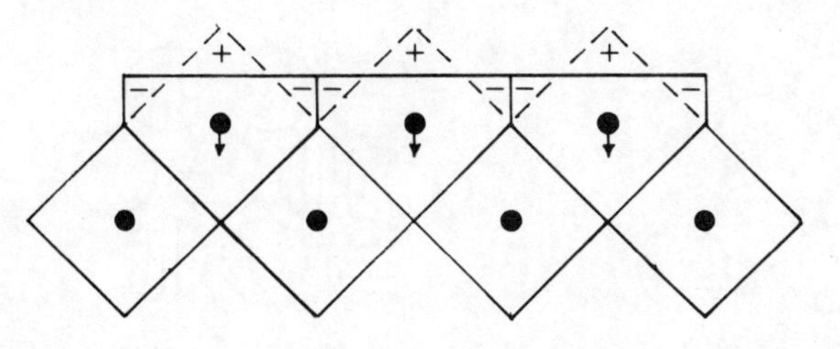

considerable elastic distortive energy in order to reach a structure that facilitates new bond formation. Beyond this, there are very few general predictive principles and the resulting *reconstruction* of the surface commonly yields geometrical structures that are much more complex than the ideal surface termination.

Fig. 3.2. Low-index ideal surfaces of a hard-sphere cubic crystal. Vertical and horizontal markings indicate the second and third atom layers, respectively. Cube face is indicated for (100) to set the scale (Nicholas, 1965).

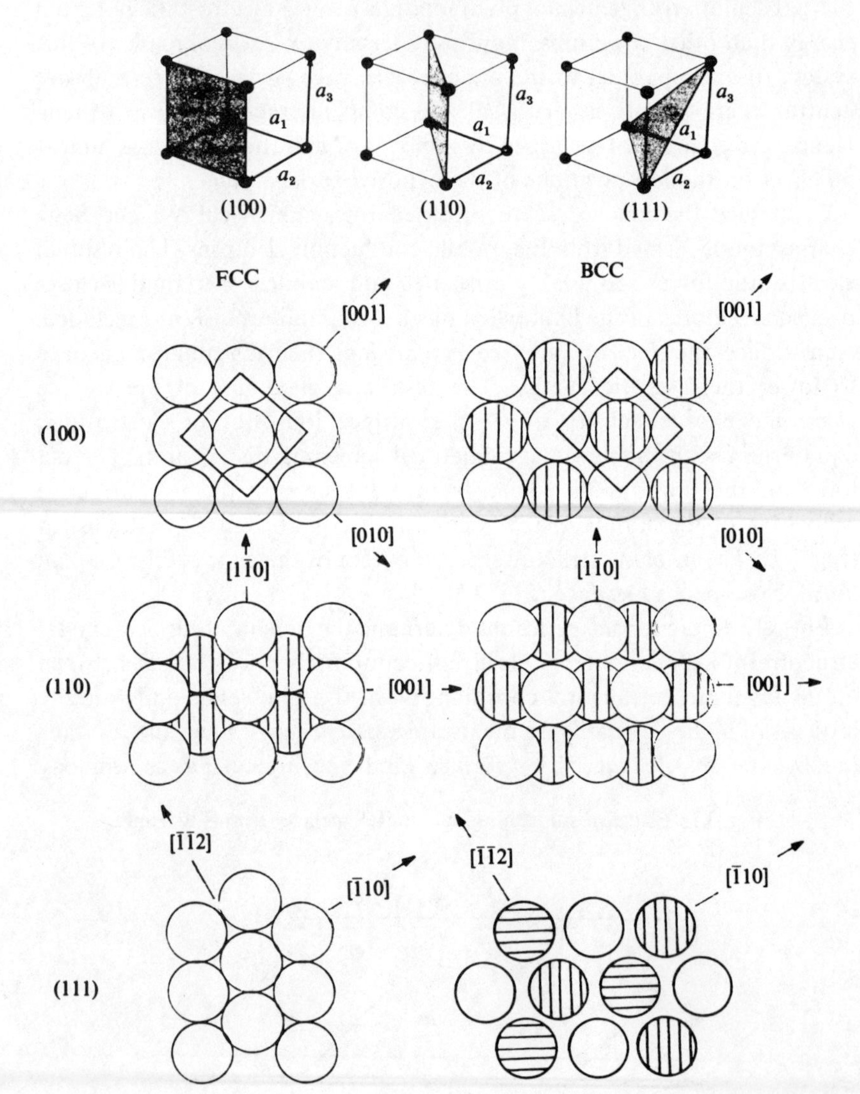

Crystallography and diffraction

The complete translational invariance of a bulk crystal is destroyed by cleavage. At best, one retains periodicity in only two dimensions. For a *strictly* two-dimensional, i.e., planar, periodic structure every lattice point can be reached from the origin by translation vectors, $\mathbf{T} = m\mathbf{a}_s + n\mathbf{b}_s$, where m and n are integers. The primitive vectors, $\mathbf{a}_s$ and $\mathbf{b}_s$, define a unit mesh, or *surface net*. There are five possible nets in two dimensions (Fig. 3.3). The centered rectangular net is simply a special case of the oblique net with non-primitive vectors. It is retained here to conform with longstanding convention.

The specification of an ordered surface structure requires both the unit mesh and the location of the basis atoms. The latter must be consistent with certain symmetry restrictions. In two dimensions, the only operations consistent with the five nets that leave one point unmoved are mirror

Fig. 3.3. The five surface nets (Prutton, 1983).

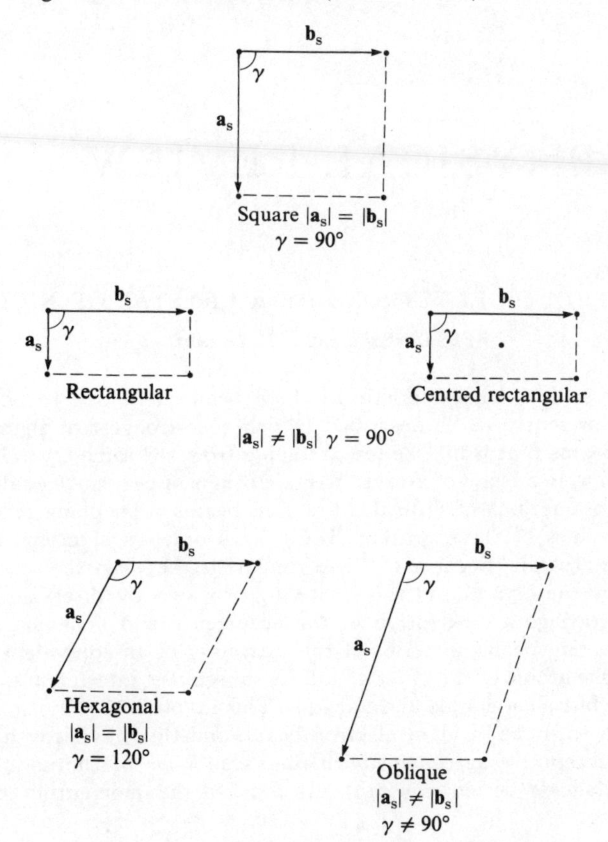

reflections across a line and rotations through an angle $2\pi/p$ where $p = 1, 2, 3, 4$, or 6. The resulting 10 point groups combined with the surface nets yield 13 space groups. The addition of a glide symmetry operation results in a total of 17 two-dimensional space groups.

Ideal or simply relaxed surfaces are identified easily by reference to the bulk plane of termination, e.g., Ni(110) or MgO(100). The periodicity and orientation of the surface net is the same as the underlying bulk lattice; these are called 1×1 structures. However, suppose the primitive translation vectors of the surface differ from those of the ideal surface such that $\mathbf{a}_s = N\mathbf{a}$ and $\mathbf{b}_s = M\mathbf{b}$, as for a typical reconstructed surface. In this case, the common nomenclature is $R(hkl)\ N \times M$, e.g., Au(110) 2×1 or Si(111) 7×7. If the surface net is rotated by an angle ϕ with respect to the bulk net, this angle is appended: $R(hkl)\ N \times M\text{–}\phi$.

Fig. 3.4. Excerpt from Davisson & Germer (1927).

Second Series *December, 1927* *Vol. 30, No. 6*

THE

PHYSICAL REVIEW

DIFFRACTION OF ELECTRONS BY A CRYSTAL OF NICKEL

BY C. DAVISSON AND L. H. GERMER

The most striking characteristic of these beams is a one to one correspondence, presently to be described, which the strongest of them bear to the Laue beams that would be found issuing from the same crystal if the incident beam were a beam of x-rays. Certain others appear to be analogues, not of Laue beams, but of optical diffraction beams from plane reflection gratings—the lines of these gratings being lines or rows of atoms in the surface of the crystal. Because of these similarities between the scattering of electrons by the crystal and the scattering of waves by three- and two-dimensional gratings a description of the occurrence and behavior of the electron diffraction beams in terms of the scattering of an equivalent wave radiation by the atoms of the crystal, and its subsequent interference, is not only possible, but most simple and natural. This involves the association of a wave-length with the incident electron beam, and this wave-length turns out to be in acceptable agreement with the value h/mv of the undulatory mechanics, Planck's action constant divided by the momentum of the electron.

Long experience with diffraction methods in the bulk suggests a search for a similar methodology at the surface. As always, a diffraction experiment designed for crystal structure analysis requires a probe with de Broglie wavelength less than typical interatomic spacings, say, ~ 1 Å. For example, a structure sensitive *electron* must have kinetic energy $E = (h/\lambda)^2/2m \cong 150\,\text{eV}$. However, this energy is very near the minimum of the universal curve (cf. Fig. 2.1)! This fortunate coincidence forms the basis for low energy electron diffraction (LEED) from solid surfaces. Electrons with energies in the range of 20–500 eV that are *elastically* backscattered from a crystal surface will form a Fraunhofer diffraction pattern that is the Fourier transform of the surface atom arrangement. The basic experiment first was performed almost 60 years ago (Fig. 3.4).

Thirty years after his original experiments, Germer returned to the LEED problem and guided the development of the modern LEED display system (Scheibner, Germer & Hartman, 1960). Fig. 3.5 illustrates a typical arrangement. Electrons enter from the left and some fraction backscatter towards a hemispherical grid G_1. A retarding potential difference between G_1 and a second grid G_2 allows only the *elastically* backscattered electrons (about 1% of the total yield) to reach G_2. A fluorescent screen S is held at a large positive potential so that the electrons accelerate and excite the

Fig. 3.5. A display-type LEED system (Clarke, 1985).

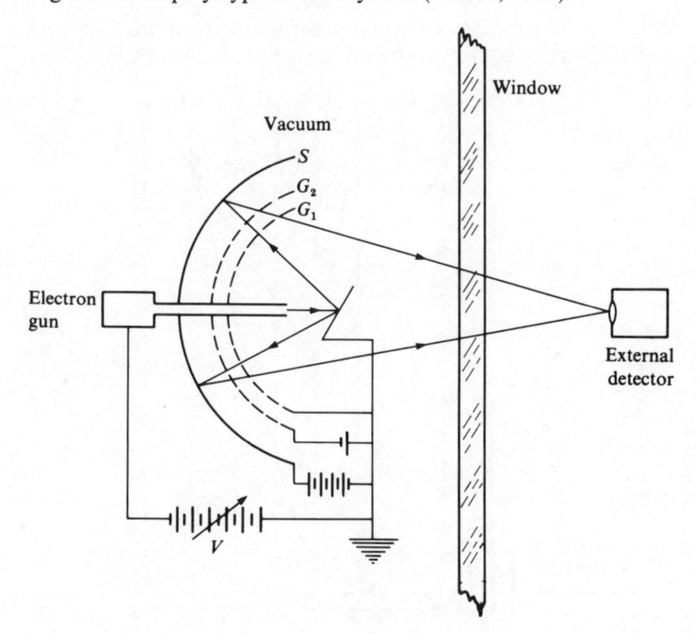

Fig. 3.6. LEED pattern from a Cu(110) crystal surface at 36 eV incident electron energy. (Courtesy of D. Grider, Georgia Institute of Technology.)

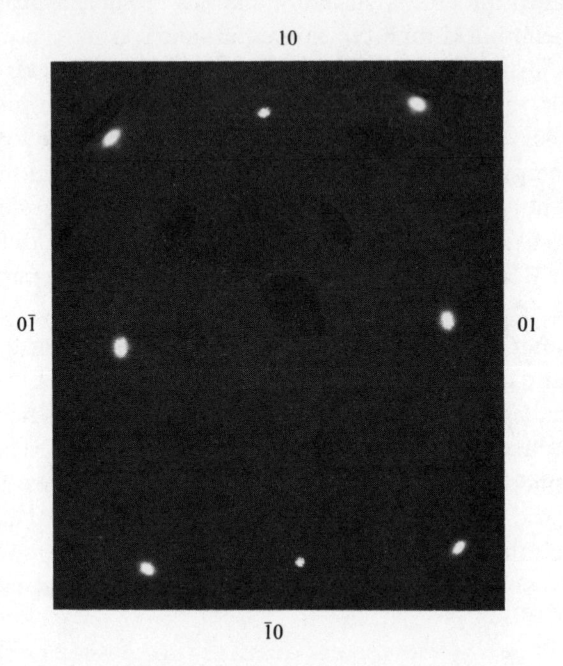

Fig. 3.7. The Ewald construction for an electron incident normal to the surface. Nine backscattered beams are shown (Kahn, 1983).

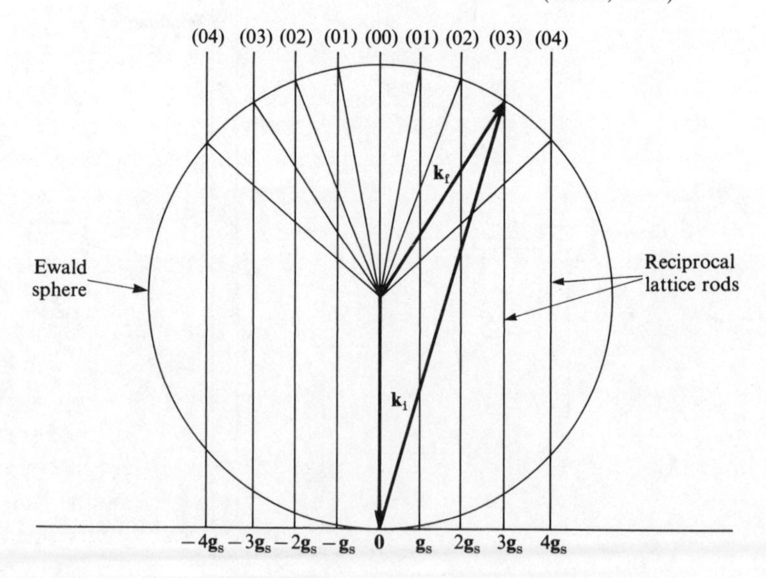

screen phosphors upon impact. A still (or better, video) camera records the image of the diffraction 'spot pattern' (Fig. 3.6).

The LEED pattern is an image of the surface *reciprocal* net when viewed along the surface normal at a great distance from the crystal. To see this, recall that the distance between adjacent points in a reciprocal lattice is inversely proportional to the distance between points in the corresponding direction of the direct lattice. For a purely *planar* lattice mesh the periodic repeat distance is infinite in the z direction. The reciprocal lattice 'points' along the surface normal are therefore infinitely dense – one speaks of a *rod* in reciprocal space. Nonetheless, translational invariance in two dimensions ensures that diffraction occurs if the two-dimensional Laue conditions are satisfied,

$$(\mathbf{k}_i - \mathbf{k}_f)\cdot\mathbf{a}_s = 2\pi m \quad \text{and} \quad (\mathbf{k}_i - \mathbf{k}_f)\cdot\mathbf{b}_s = 2\pi n \tag{3.1}$$

where $\mathbf{k}_i$ and $\mathbf{k}_f$ are the wave vectors of the incident and scattered electron, respectively, and m and n are integers.

The Laue conditions are illustrated best using the familiar Ewald construction (Fig. 3.7). A reciprocal lattice rod passes through every point of the surface reciprocal net, $\mathbf{g}_s = h\mathbf{A}_s + k\mathbf{B}_s$. The magnitude of the electron wave vector sets the radius of the sphere and the diffraction condition is satisfied for every beam that emerges in a direction along which the sphere intersects a reciprocal rod. As in three dimensions, the beams are indexed by the reciprocal lattice vector that produces the diffraction. The beam spots in Fig. 3.6 are labelled by the appropriate $\mathbf{g}_s$.

How does this picture differ from that of ordinary three-dimensional x-ray scattering? If the outgoing electron wave contained contributions from layers deep within the crystal, each reflected wavefront would have to add *in phase* with all the others in order to preserve the coherence of the diffracted beam. This only occurs at certain *discrete* energies. The influence of the crystal structure perpendicular to the surface breaks up the reciprocal rods into a one-dimensional lattice of points, and the usual x-ray Laue conditions are recovered. By contrast, LEED samples only a few lattice planes, and beams are seen at *all* energies as long as the corresponding rod is within the Ewald sphere.

The mere existence of a sharp spot pattern implies the existence of a well-ordered surface and provides direct information about the symmetry of the substrate (Holland & Woodruff, 1973). For this reason, almost every surface science laboratory is equipped with a LEED system. For present purposes, let us simply note that the surface atom arrangement can have *at most* the symmetry indicated by the LEED pattern; the true surface structure could possess a *lower* symmetry. This situation occurs when the surface contains regions (domains) which are oriented with respect to one another by precisely a symmetry operation. For example, two patterns that have three-fold symmetry can be found on a surface rotated 60° with respect to one another. The composite effect (achieved by averaging over the physical size of the electron beam) appears as an apparent six-fold symmetry of the surface.

Additional information can be gleaned by probing the variation in diffracted intensity across the width of a single spot, the so-called *spot profile* (Lagally, 1982). For example, any deviation from perfect two-dimensional periodicity will destroy the delta function character of the reciprocal lattice rods. Broadening and splittings will appear as one

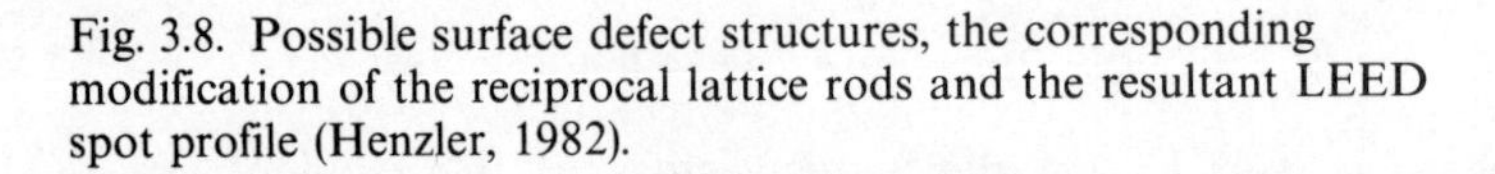
Fig. 3.8. Possible surface defect structures, the corresponding modification of the reciprocal lattice rods and the resultant LEED spot profile (Henzler, 1982).

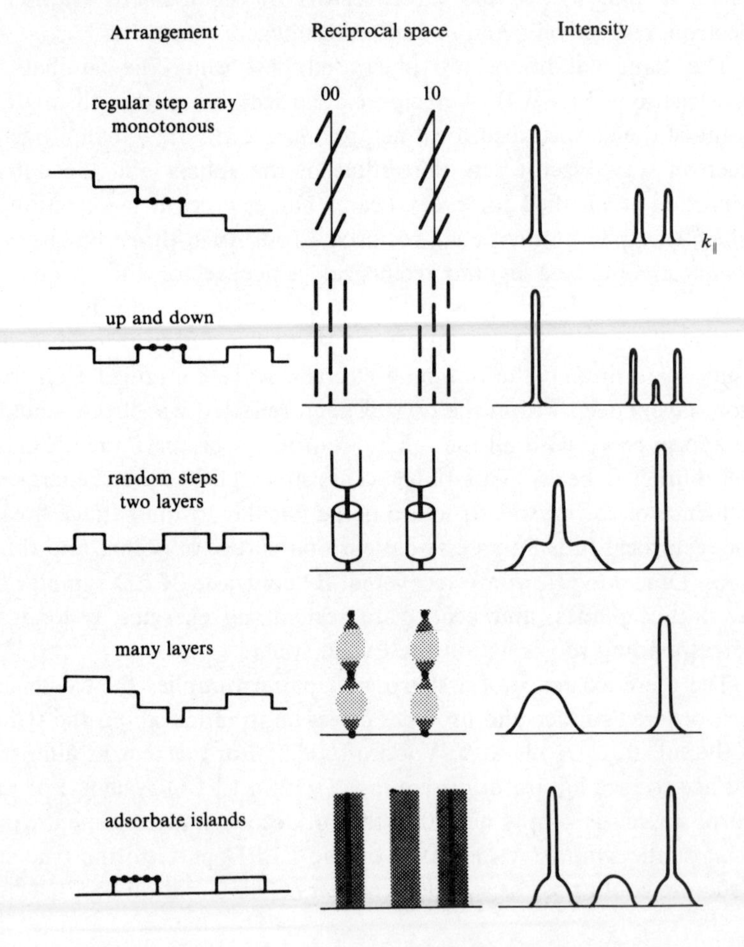

probes $k_{\parallel}$, the wave vector component parallel to the surface. Similarly, small scale variations in the surface topography will break up the rods in the direction normal to the surface. The expected spot pattern is found by superimposing the Ewald sphere on the modified 'rods' (Fig. 3.8). Note that this analysis does not bear on the question of the arrangement of atoms within a surface unit cell.

It is not straightforward to determine surface atom positions from LEED. In x-ray scattering, the intensity of each Laue spot is determined by the product of an atomic scattering factor and a simple geometrical structure factor. The positions of atoms within the unit cell are varied until the geometrical factors predict the correct intensity for each beam. This simplicity occurs because x-rays interact very weakly with matter; each photon is backscattered after a *single* encounter with a lattice ion. Another consequence of this *kinematic* scattering is that spot intensities are independent of both the incident beam energy and the azimuthal angle of incidence. Neither is true in LEED.

Structure determination from LEED is complicated by the fact that every electron undergoes *multiple* elastic scattering within the first few layers of the crystal. Unlike x-rays, the elastic scattering cross section for electrons is very large (~ 1 Å^2) and comparable to the inelastic cross section which makes LEED surface sensitive in the first place. The probability is great that a second (or third, etc.) diffraction event will scatter an electron away from its original diffraction direction.

The energy (voltage) dependence of LEED beam intensities, the so-called $I(V)$ curves, are used in an iterative procedure to determine the geometrical arrangement of surface atoms (Pendry, 1974). First, an arrangement of atoms is postulated that is consistent with the symmetry of the LEED pattern. Second, the intensity of a number of diffracted beams is calculated as a function of incident energy by explicit solution of the Schrödinger equation* for the electron wave function in the first few atomic layers (including the effects of inelastic damping). Third, the resulting $I(V)$ curves are compared to experiment and the process is continued with a refined geometry until satisfactory agreement is obtained. It must be emphasized that this is a highly non-trivial procedure that involves significant computational effort. Even the most experienced practitioners are limited to a very small number of adjustable structural parameters.

As an example, consider a LEED structural analysis designed to study relaxation near a metal surface. Fig. 3.9 shows a comparison between

* The appropriate multiple scattering calculation for this *dynamical* LEED analysis is completely akin to the KKR method of bulk band structure (see, e.g., Ziman, (1972)).

Fig. 3.9. Comparison of LEED theory and experiment for Cu(100) (Davis & Noonan, 1982).

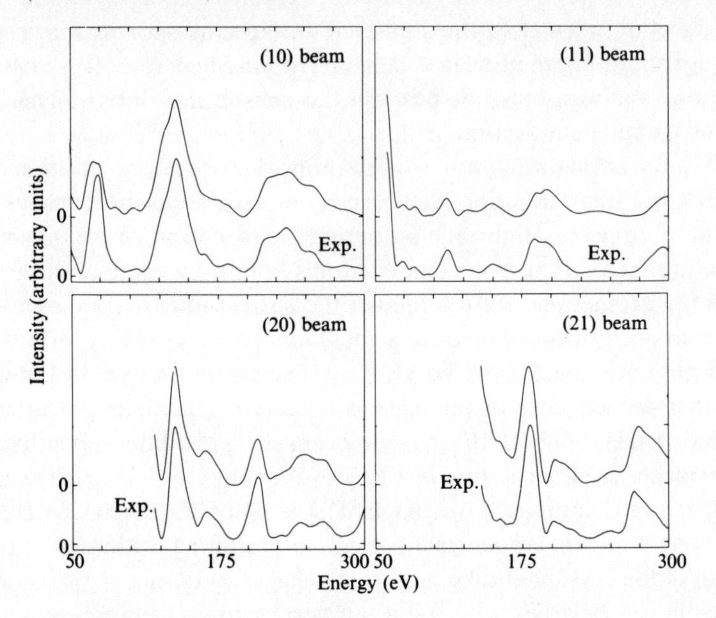

Fig. 3.10. Top layer relaxation for iron versus surface roughness (inverse surface ion density) (Sokolov, Jona & Marcus, 1984).

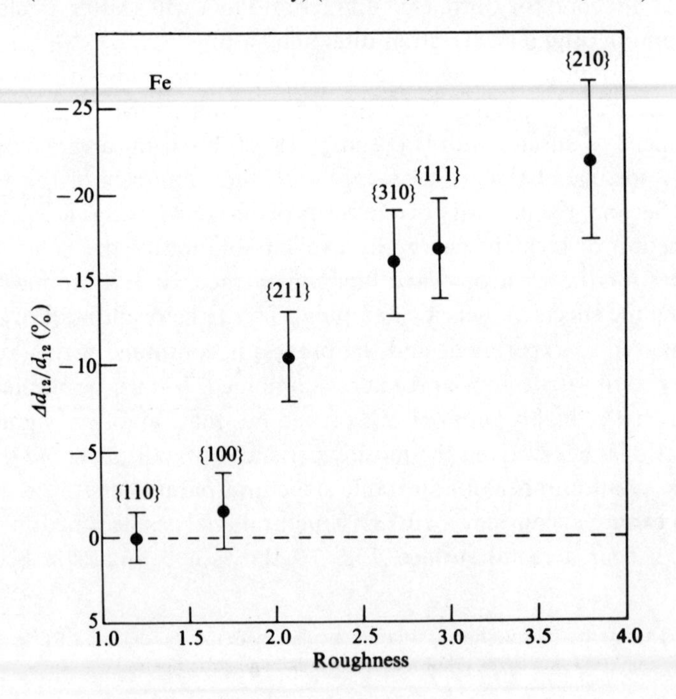

experimental $I(V)$ curves obtained from a Cu(100) single crystal surface and a LEED dynamical calculation. The quality of the agreement for all four beams is evident to the eye. The surface structure that corresponds to this calculation actually exhibits an *oscillatory* relaxation of the interlayer spacings, i.e., the outermost layer spacing is contracted relative to the bulk spacing ($\Delta d_{12}/d_{12} = -1.45\%$) while the spacing between the next deepest pair of planes is *expanded* relative to the bulk ($\Delta d_{23}/d_{23} = +2.25\%$). Careful LEED studies have established the systematics of the outer layer contraction phenomenon. The results are in accord with the charge density smoothing argument given above (Smoluchowski, 1941). The greatest smoothing, and hence the largest surface contraction, occurs for the low density, highly corrugated crystal faces (Fig. 3.10).

LEED cannot be used readily in all situations. As with Auger spectroscopy, the surfaces of insulators are difficult to study because the incident electron beam quickly charges the sample. A more serious possible pitfall of LEED structure analysis can be demonstrated with another relaxation study – this time for the high temperature Si(111) 1 × 1 surface (Fig. 3.11).

Fig. 3.11. Comparison of LEED theory and experiment for Si(111) 1 × 1 (Zehner, Noonan, Davis & White, 1981; Jones & Holland, 1985).

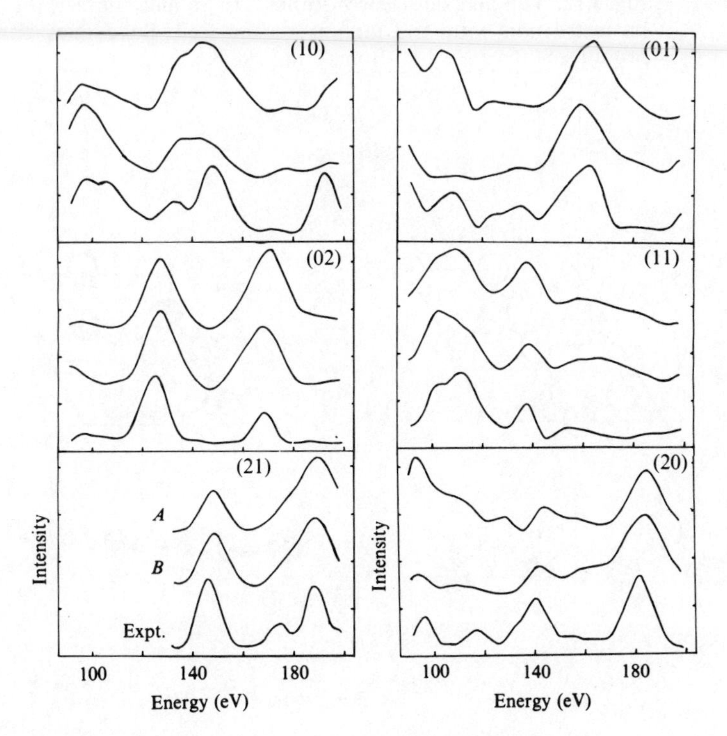

Here, dynamical calculations for two proposed structures are compared to six experimental LEED beam intensity scans. Both models seem to be of comparable quality as judged by eye and also as judged by a least-squares type of criterion. However, the models are *very* different: model A predicts $\Delta d_{12}/d_{12} = -25\%$ and $\Delta d_{23}/d_{23} = +3\%$ whereas model B predicts $\Delta d_{12}/d_{12} = -90\%$ and $\Delta d_{23}/d_{23} = +25\%$! Of course, neither fit is as good as the Cu(100) example but it remains true, in general, that even the very best LEED analyses cannot *unambiguously* determine surface crystal structures.

Let us look again at x-ray scattering. Our initial negative assessment of this conventional technique for use in surface structural analysis was based on the long absorption length of x-rays in matter (~10 microns). However, many years ago, Compton (1923) pointed out that the index of refraction of materials at x-ray wavelengths is very slightly less than unity ($\sim 10^{-6}$) so, by Snell's law, total external reflection of an incoming x-ray beam occurs for glancing incidence angles ($\sim 0.1^{0}$). Consequently, an x-ray photon will be diffracted out of the crystal after penetration of only a few atomic layers. Reciprocal lattice vectors of the surface dominate the scattering and the simplicity of kinematic analysis is recovered although

Fig. 3.12. Top and side view of InSb(111): (*a*) ideal surface; (*b*) 2 × 2 reconstruction with one indium vacancy/cell. Solid lines border the primitive surface meshes (Bohr *et al.*, 1985).

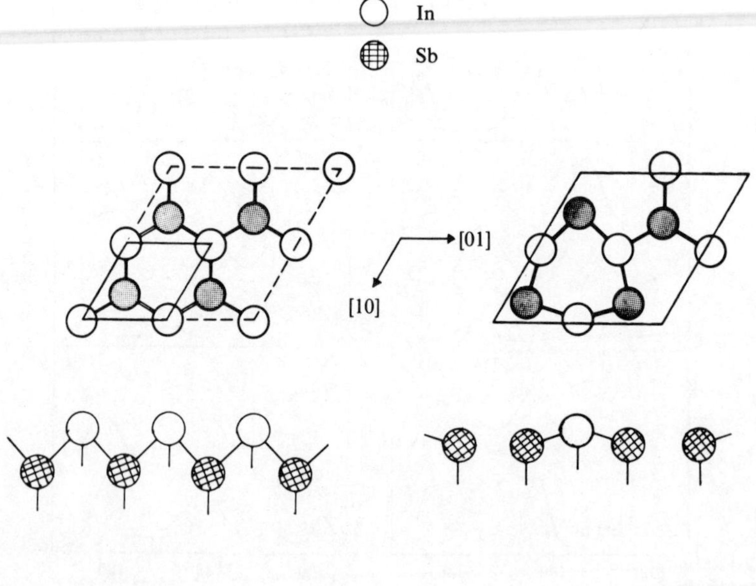

in practice, an intense synchrotron x-ray source ($\sim 10^{11}$ photon/s) is required to produce acceptable counting rates.

As an example, we consider the reconstruction of a *polar* compound semiconductor surface, InSb(111). Parallel to this surface, the crystal is composed of alternating planes of indium and antimony ions. The ideal surface terminated with an indium layer (Fig. 3.12(*a*)) is positively charged and unstable against a reconstruction to a charge neutral surface. A conventional analysis of grazing incidence x-ray data leads to a 2×2 structure where one surface indium atom/unit cell is *missing* (presumably ejected during the cleavage process) and the spacing between the surface indium and subsurface antimony layers strongly contracts (Fig. 3.12(*b*)). This rather extreme reconstruction results in a neutral surface bilayer.

Ion scattering

A completely different approach to surface structural analysis is based on classical Rutherford scattering. Imagine a light ion (H^+, He^+, etc.) beam directed at a solid surface. The crystal presents a target to the ions in the form of columns or 'strings' of atoms that lie parallel to low index directions. Coulomb scattering from the end of such a string at the first atomic layer depends on the impact parameter. The distribution of scattered ions will form a characteristic *shadow cone* behind the surface atom (Fig. 3.13). Atoms within the shadow cone do not contribute to the backscattered signal. If the effects of screening the Coulomb interaction

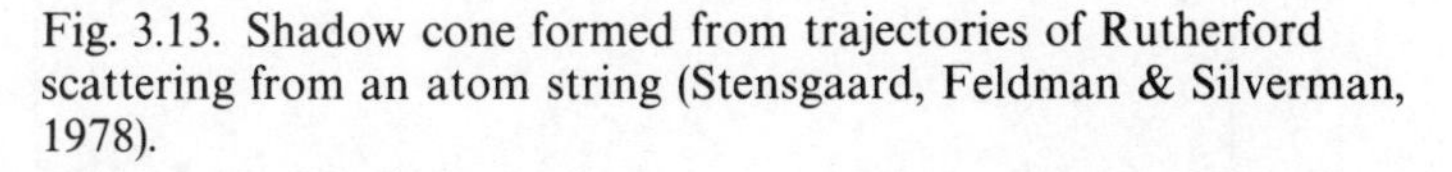
Fig. 3.13. Shadow cone formed from trajectories of Rutherford scattering from an atom string (Stensgaard, Feldman & Silverman, 1978).

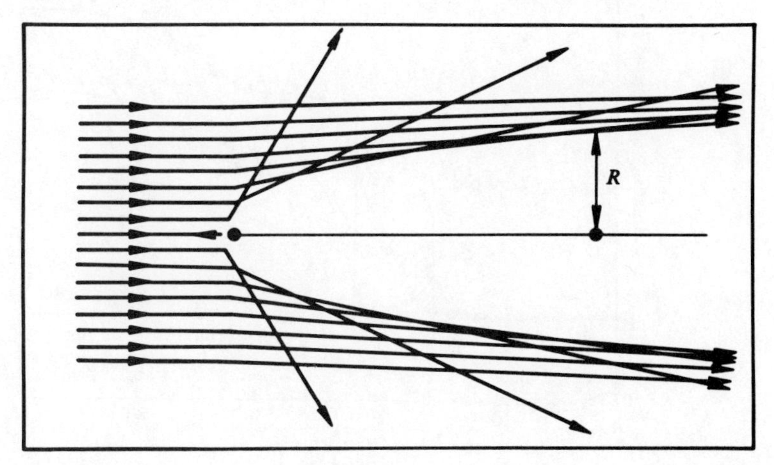

are ignored, the radius of the shadow cone is

$$R = 2\sqrt{Z_1 Z_2 e^2 d/E_0} \tag{3.2}$$

where Z_1 and Z_2 are the atomic number of the projectile and target atoms, respectively, E_0 is the primary beam energy and d is the distance along the string.

A directly backscattered ion suffers a simple binary elastic collision with a surface atom. Conservation of kinetic energy and momentum determine the final ion energy:

$$E = E_0\left(\frac{M_1 - M_2}{M_1 + M_2}\right)^2. \tag{3.3}$$

Equation (3.3) says that the backscattered ion (M_1) suffers an energy shift that depends rather sensitively on the mass of the surface atoms (M_2). Accordingly, analysis of the ion scattering energy spectrum can be counted as another method of surface elemental analysis. (Beam currents are kept

Fig. 3.14. LEIS from TiC(100). Vertical lines denote calculated angles at which Ti shadowing occurs for nearest neighbor C sites and nearest neighbor Ti sites. Inset: scattering geometry at the C site critical angle. A single carbon vacancy is shown (Aono *et al.*, 1983).

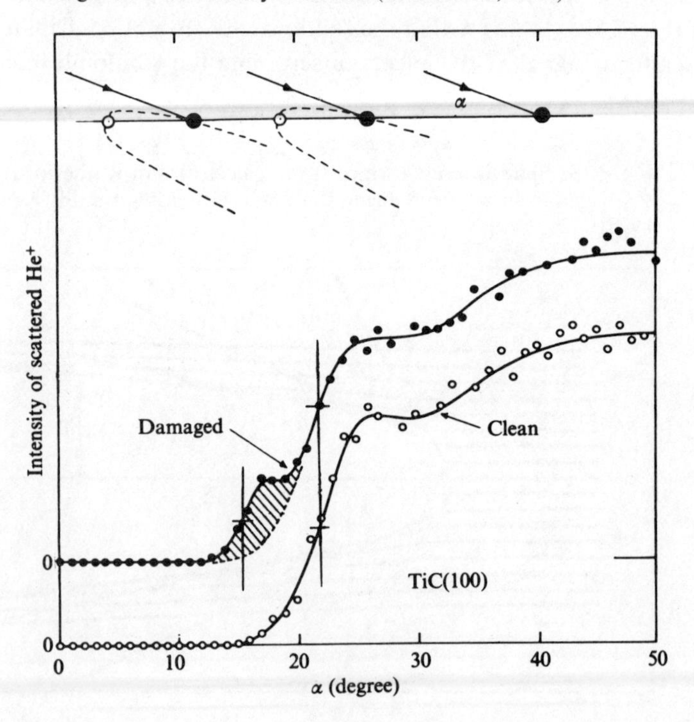

well below the threshold for sputtering damage.) For structural studies, note that both the shadow cone radius and the Rutherford cross section, $d\sigma_R/d\Omega \sim E_0^{-2}$, depend strongly on the primary beam energy. Consequently, discussion of ion scattering divides more or less naturally into three energy regimes: low energy ion scattering (LEIS) (1–20 keV), medium energy ion scattering (MEIS) (20–200 keV) and high energy ion scattering (HEIS) (200 keV–2 MeV).

LEIS is well suited for laboratory surface studies. The large cross section ($\sim 1\,\text{Å}^2$) and shadow cone radius ($\sim 1\,\text{Å}$) guarantee that most ions never get past the surface layer. Those that do are quickly neutralized by electron capture and will not contribute to the experimental signal if only charged particles are collected at the detector (Brongersma & Buck, 1978).

The power and simplicity of the shadow cone concept is illustrated in Fig. 3.14 for the case of 1 keV He^+ scattered from a TiC(100) single crystal surface. The azimuthal angle of incidence is set so that the

Fig. 3.15. HEIS for 2 MeV He^+ from W(100). Spectra are shown for incidence along the ⟨100⟩ channelling direction and a 'random' direction (Feldman, 1980).

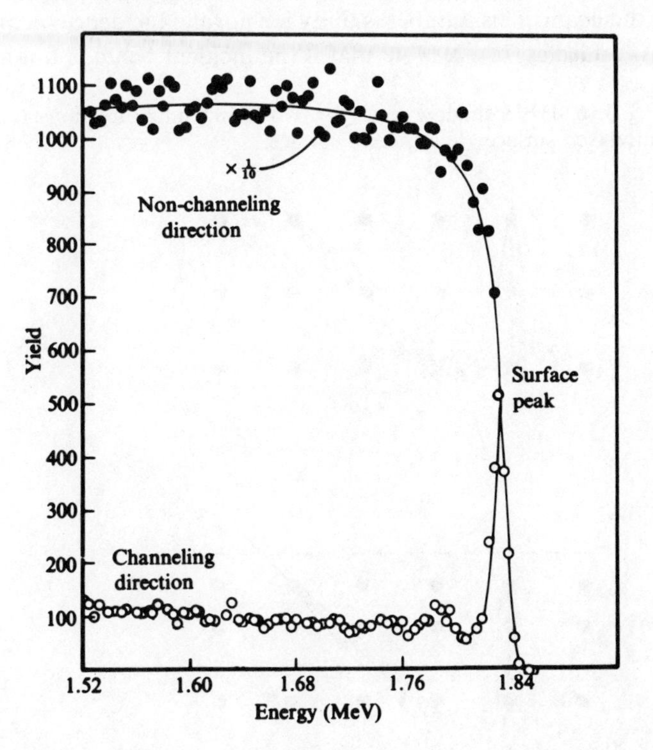

scattering plane contains a surface row where titanium and carbon atoms alternate and, using (3.3), only backscattering from titanium atoms is recorded. At large polar angles of incidence, α, significant scattering is seen. However, the signal rapidly diminishes below a certain critical angle. In this regime, titanium atom backscattering is occluded by a neighboring carbon atom's shadow cone. Since the shadow cone radius is fixed by (3.2) one can precisely calculate this critical angle (here, 22.1°). If the surface intentionally is damaged by preferential sputtering of carbon atoms one observes that residual backscattering occurs below the critical angle from 'unshadowed' atoms.

In the HEIS regime, only a few ions are directly backscattered from the surface. The majority of the ions incident along a low-index atom string penetrate deep into the bulk and undergo a series of correlated collisions with neighboring strings. The behavior is known as *channelling*. Channelled ions collide with loosely bound electrons and lose energy according to the stopping power of the solid. Eventually they are backscattered out of the crystal but with a smaller energy than the ions elastically scattered from the surface. Therefore, an ion energy analysis at fixed collection angle reveals a 'surface peak' (Fig. 3.15). The calibrated area under the surface peak is proportional to the number of atoms/string visible to the beam. This number is unity for normal incidence on an ideal FCC(100) surface at $T = 0$. Note that if the incident beam is not aligned

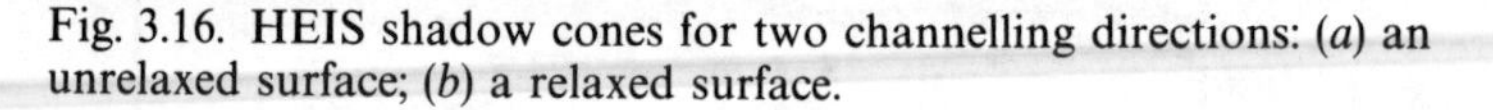

Fig. 3.16. HEIS shadow cones for two channelling directions: (*a*) an unrelaxed surface; (*b*) a relaxed surface.

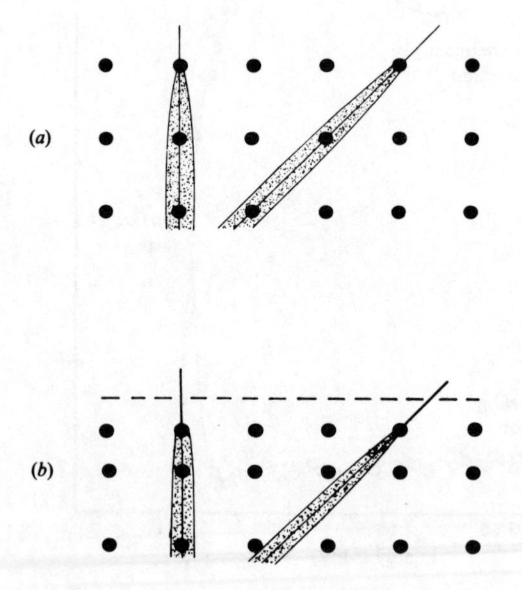

with a channelling direction, ions are backscattered with equal probability from all depths.

High energy ion scattering is particularly sensitive to interplanar relaxation. For an ideal surface, complete shadowing of subsurface atoms occurs for ions incident along channelling directions. However, there will be *incomplete* shadowing along non-normal directions if the outermost atomic plane is displaced (Fig. 3.16). The ion yield will not be symmetric if the crystal is rocked back and forth about this string (Fig. 3.17). The greater-than-unity value of the surface peak in Fig. 3.17 reflects thermal vibrations of the surface atoms. The moving atoms expose deeper lying atoms to the beam which otherwise are shadowed at $T = 0$.

Structural determinations using HEIS also require an iterative procedure. After a structure is proposed, a calculation is performed for comparison with the data. However, because the screened Coulomb interaction potential is so well known, the expected ion yield is calculated easily by simulation of the classical scattering process using the Monte Carlo* method. The solid curve in Fig. 3.17 corresponds to a multilayer oscillatory relaxation model of the Ag(110) surface.

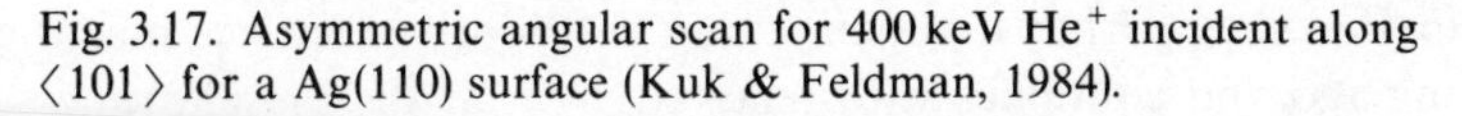

Fig. 3.17. Asymmetric angular scan for 400 keV He^+ incident along ⟨101⟩ for a Ag(110) surface (Kuk & Feldman, 1984).

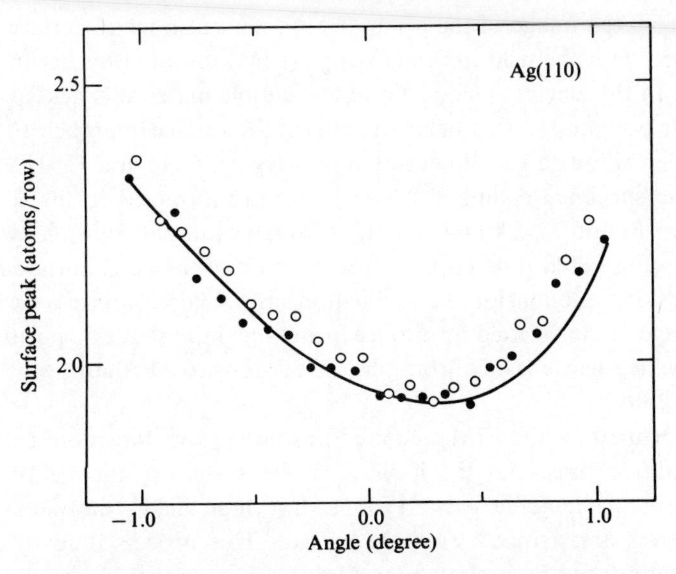

* The Monte Carlo method is a numerical technique for performing averages over specified probability distributions. Here, classical ion trajectories are averaged over a Gaussian distribution of thermal vibrations of the target (Hammersley & Handscomb, 1964).

Our discussion of LEED, glancing incidence x-ray scattering and ion scattering illustrates that surface structure determination is non-trivial. Generally speaking, a structural model can only be accepted definitively if it is consistent with *all* the available high-quality data from a number of different experimental methods. Consequently, the total number of completely 'solved' surface structures is very small. A particularly notorious case is that of the Si(111) 7×7 reconstruction. The basic structure of this surface has only recently become clear – after nearly 30 years of intense effort. Interestingly, the major breakthroughs came from an unexpected direction: microscopy.

Microscopy

The purpose of all microscopy is to produce a faithful image of the experimental specimen. Such images have considerable aesthetic appeal in condensed matter physics: what could be more satisfying than to actually 'see' the arrangement of atoms in a solid? In the present context, any direct representation of the surface topography both builds intuition and imposes severe constraints on model structures proposed to account for data collected by more indirect methods, e.g., LEED, ion scattering, etc. It is then perhaps not surprising that work in surface imaging began long ago and continues with renewed vigor at the present time.

A projected image of the atom arrangement at a metal surface can be obtained with the field ion microscope (FIM) invented by Erwin Muller (1951). In this device, a sharp tip of the sample material is held at a large positive potential so that field strengths at the surface approach 10^9 V/cm. One then admits a gas of neutral atoms, typically He or a He/H mixture, into the specimen chamber. These atoms are attracted to the solid (see Chapter 8) and lose kinetic energy through multiple collisions with the surface (Fig. 3.18(*a*)). Eventually, they remain in the neighborhood of the surface long enough for the ambient electric field to ionize an electron. An image of the facetted tip surface forms when the resulting positive ions rapidly accelerate away from the metal towards a fluorescent screen (Fig. 3.18(*b*)).

Unfortunately, the FIM is limited to study of the transition metals and their alloys since the tip itself must be stable at the fields needed to ionize the imaging gas. At sufficiently high fields, the metal atoms themselves are stripped from the surface. This process (known as field evaporation) can be exploited for alloys to gain chemical specificity if the FIM is coupled to a mass spectrometer. Nevertheless, for general purpose microscopy we must turn to other techniques.

The application of electron microscopy to surface imaging follows recent

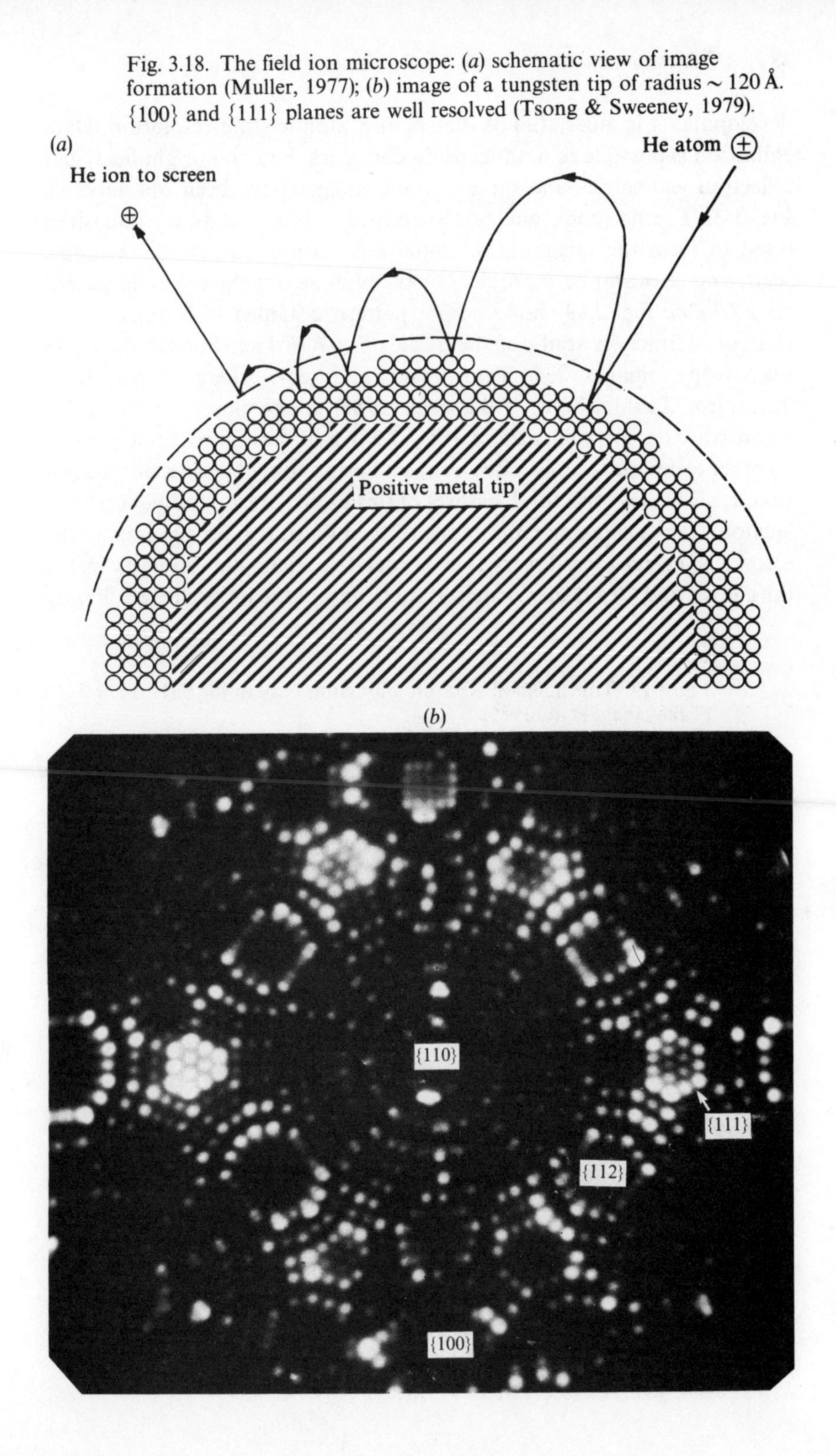

Fig. 3.18. The field ion microscope: (*a*) schematic view of image formation (Muller, 1977); (*b*) image of a tungsten tip of radius ~ 120 Å. $\{100\}$ and $\{111\}$ planes are well resolved (Tsong & Sweeney, 1979).

developments in bulk studies that permit atomic-scale resolution. Data collection is possible in both transmission (dark field and bright field) and reflection geometries and quite striking images have been obtained (cf. Fig. 1.3). Furthermore, one can analyze the diffracted beam intensities (used to form the image) in a simple kinematical framework. Multiple scattering ceases to be a problem at the high electron energies employed (>100 keV). Fig. 3.19 shows a spot pattern obtained in a transmission electron diffraction study of the Si(111) 7×7 surface. Similar data and microscopy images led Takayanagi and co-workers (Takayanagi, Tanishiro, Takahashi & Takahashi, 1985) to propose a remarkable reconstruction for this surface (Fig. 3.20) that has since been verified by other techniques, e.g., surface x-ray scattering, ion scattering, photoemission, etc. The key structural features of this model are: (*a*) twelve top layer 'adatoms', (*b*) a stacking fault in one of the two triangular subunits of the second layer, (*c*) nine dimers that border the triangular subunits in the third layer and (*d*) a deep vacancy at each apex of the unit cell. The driving

Fig. 3.19. Transmission electron diffraction pattern for Si(111) 7×7 (Takayanagi *et al.*, 1985).

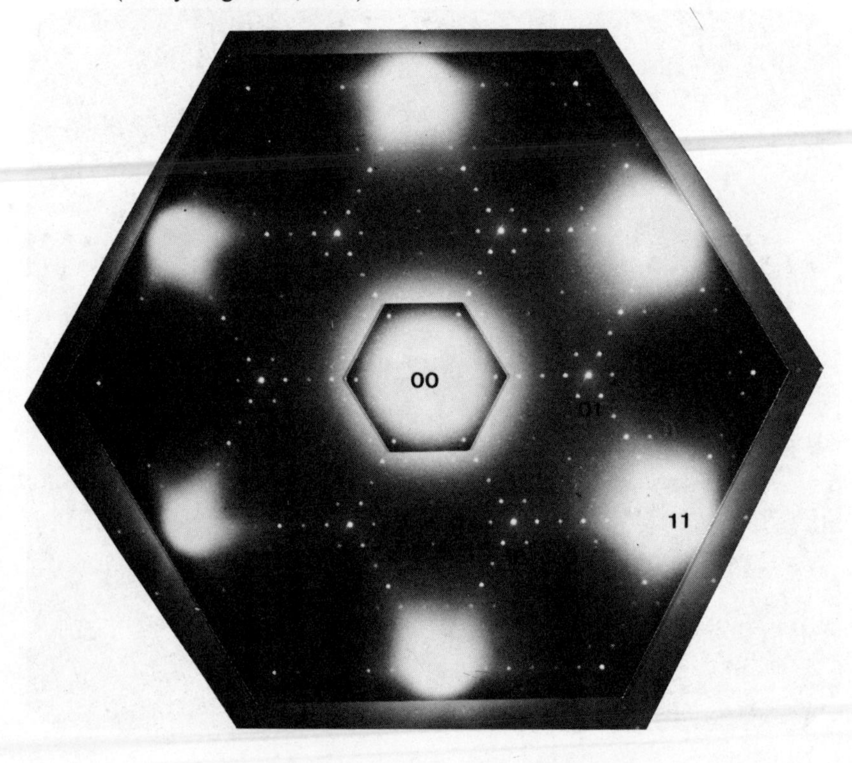

force for this reconstruction is unknown in detail; a plausible argument will be given in Chapter 4.

Perhaps the most exciting new development in surface microscopy comes from an entirely different direction. The scanning tunnelling microscope (STM) is a device that provides direct, real space images of

Fig. 3.20. Surface structure of Si(111) 7 × 7: (*a*) first three layers of atoms shown in top view. The surface unit cell is outlined (Robinson *et al.*, 1986); (*b*) schematic view that indicates the prominent depressions in the surface (round and oval holes), the dimers (double lines) and the stacking fault (shaded region) (McRae, 1984).

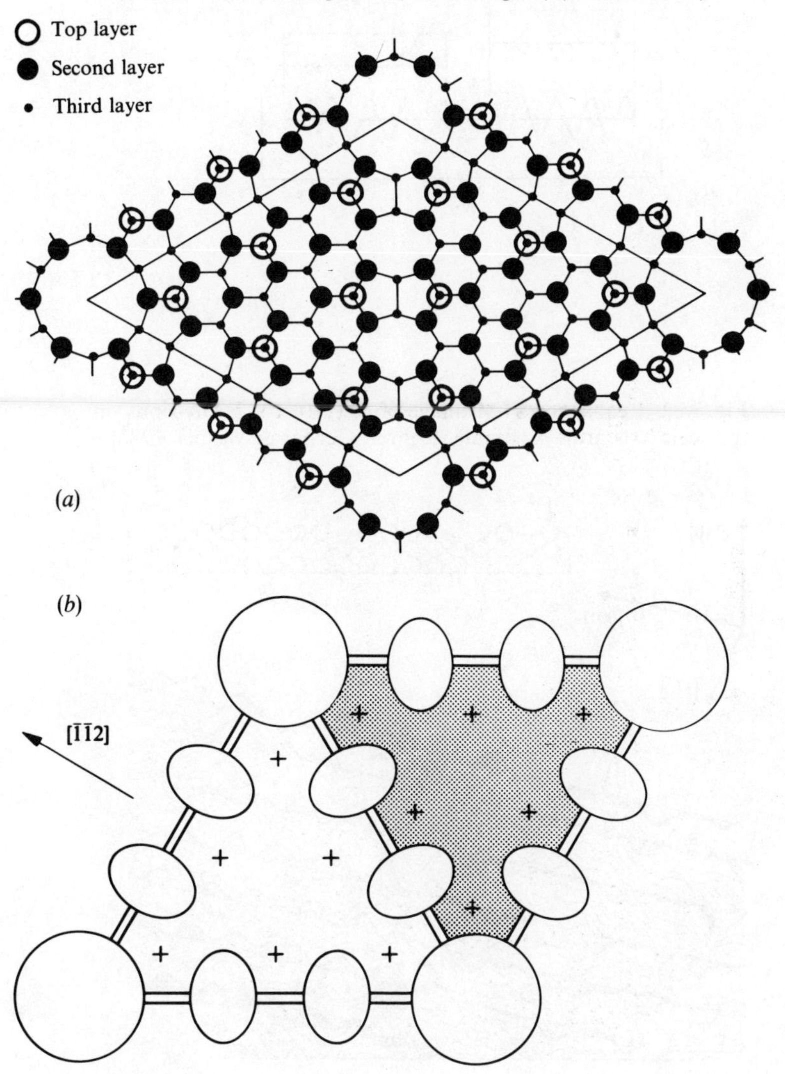

Fig. 3.21. Potential wells and Fermi level wave functions for vacuum tunnelling: (*a*) macroscopic well separation; (*b*) microscopic well separation with an applied bias voltage.

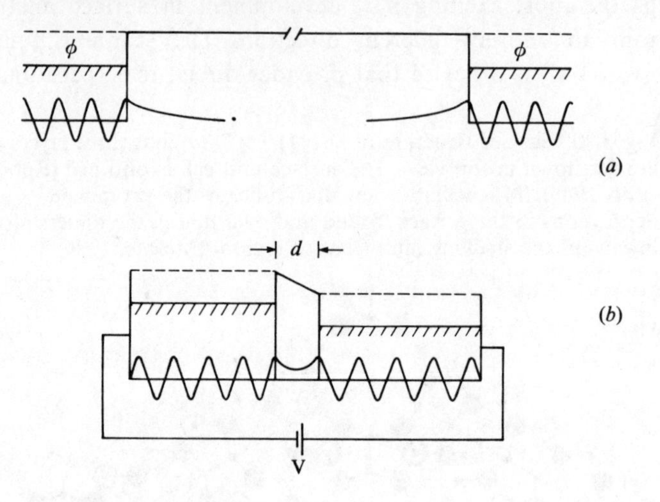

Fig. 3.22. Real-space STM image of Au(110) 1 × 2. Divisions on the scale axis are 5 Å (Binnig, Rohrer, Gerber & Weibel, 1983).

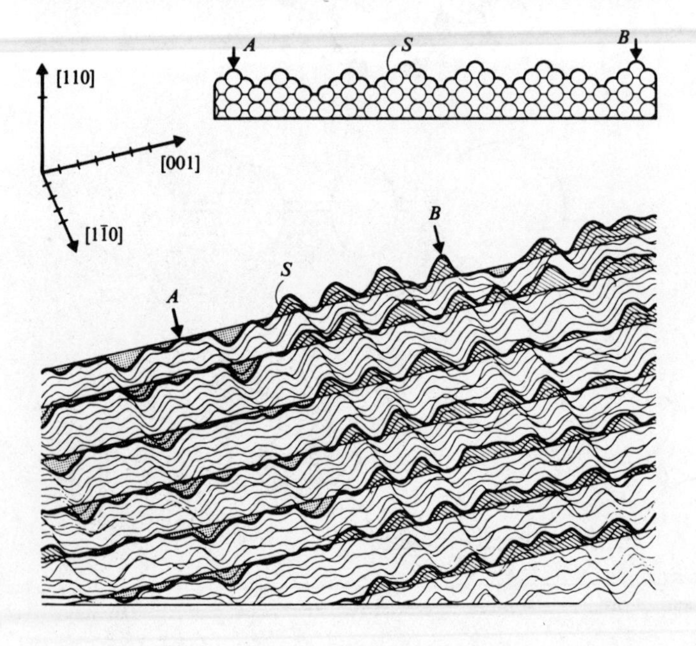

surface topography on the atomic scale. It is non-destructive, does not require periodicity of the surface or even ultra-high vacuum conditions and can provide chemical and electronic structure information as well. The principles of operation are extremely simple and already are apparent from a particle-in-a-box model of the interaction of two nearby solids. Fig. 3.21(*a*) illustrates two finite wells separated by a macroscopic distance. Quantum mechanics tells us that the Fermi level electron wave functions 'leak' out of the confining potential with a characteristic exponential inverse decay length of $\kappa = h^{-1}(2m\phi)^{1/2}$, where m is the electron mass and ϕ is the work function of the solid.

Now reduce the separation between the wells to microscopic dimensions and establish a potential difference, V, between the two (Fig. 3.21(*b*)). The overlap of the wave functions now permits quantum mechanical tunnelling and a current can be driven through the vacuum gap. The magnitude of the tunnelling current is a measure of the wave function overlap and is proportional to $\exp(-2\kappa d)$, where d is the vacuum gap width. In the real microscope (Binnig, Rohrer, Gerber & Weibel, 1982), the probe is a metal

Fig. 3.23. STM image of Si(111) 7 × 7 near an atomic step. Note that the 'deep holes' occur right at the step (Becker, Golovchenko, McRae & Swartzentruber, 1985).

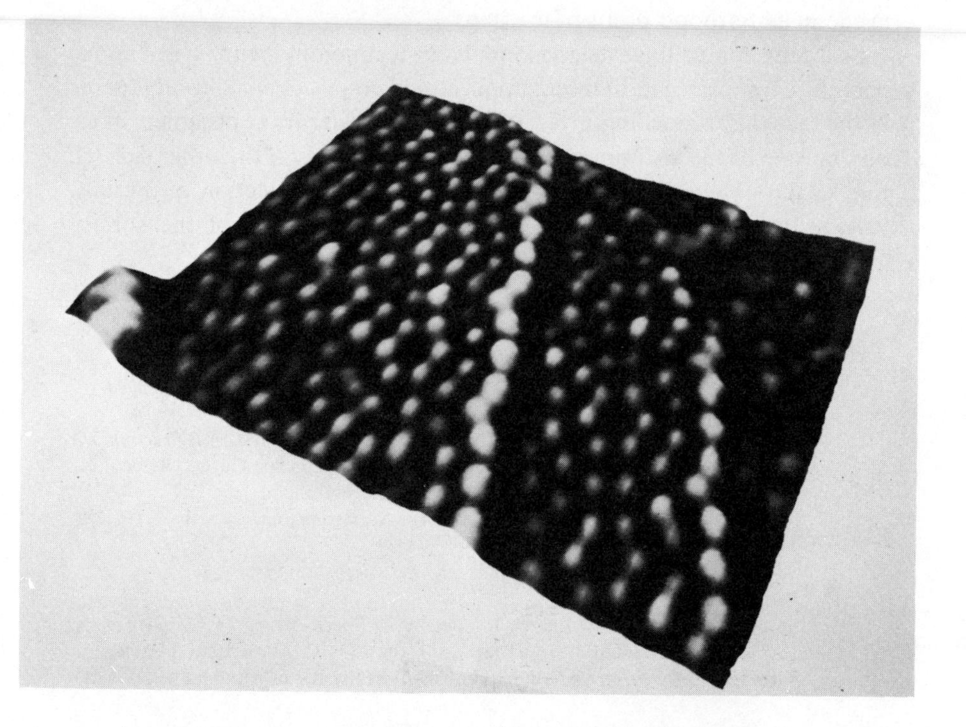

tip (not unlike a FIM tip) that is stabilized above a surface with $d \sim 5$ Å. The tip is scanned across the surface at a fixed bias voltage and a piezoelectric feedback mechanism regulates the vertical motion of the tip so that the tunnelling current (~ 1 nA) is kept *constant*. In this way, the tip traces contours of constant wave function overlap, i.e., the surface topography.

An early application of STM established a 'missing row' model for the surface of Au(110) (Fig. 3.22). This is one of only a handful of metals that exhibit a true reconstruction rather than a simple relaxation. The atomic-scale resolution of the STM image in both the vertical and lateral directions shows a rather disordered surface. Nonetheless, a hard-sphere model of the topography (inset) not only shows the missing rows along [110] but demonstrates that monolayer steps (*S*) expose (111) facets (cf. Fig. 3.2). This result is entirely consistent with the discussion of Chapter 1 if we presume that surface stresses drive the system to expose close-packed, low surface tension (111) planes (cf. Fig. 1.8).

The STM was instrumental in sorting out the complex Si(111) 7×7 structure discussed above. The deep holes and isolated top layer atoms show up very well in the topographs. Particularly striking results emerge if one combines the raw microscope data with computer-assisted image processing (Fig. 3.23). In this example, the Si(111) reconstruction is shown in the neighborhood of a surface step.

Scanning tunnelling microscopy is a technology with tremendous potential. For example, in recent applications practitioners take advantage of the fact that tunnelling involves electron transfer from occupied states on one side of the vacuum gap to unoccupied states on the other side (cf. Fig. 3.21(*b*)). Hence, by varying the magnitude and direction of the bias voltage, one performs scanning tunnelling *spectroscopy* of the surface electronic structure (see Chapter 4).

General references

Crystallography and diffraction

Wood, E. (1964). Vocabulary of Surface Crystallography. *Journal of Applied Physics* **35**, 1306–11.

Estrup, P.J. (1970). Low Energy Electron Diffraction. In *Modern Diffraction and Imaging Techniques in Materials Science* (eds. Amelinckx, Gevers, Remaut & Van Landuyt), pp. 377–406. Amsterdam: North-Holland.

Brennan, S. (1985). Two-Dimensional X-ray Scattering. *Surface Science* **152/153**, 1–9.

Ion scattering

Buck, T.M. (1975). Low Energy Ion Scattering Spectrometry. In *Methods of Surface Analysis* (ed. Czanderna), vol. 1, pp. 75–102. Amsterdam: Elsevier.

Gibson, W.M. (1984). Determination by Ion Scattering of Atomic Positions at

Surfaces and Interfaces. In *Chemistry and Physics of Solids* V (eds. Vanselow & Howe), pp. 427–53. Berlin: Springer-Verlag.

Microscopy

Tsong, T.T. (1980). Quantitative Investigations of Atomic Processes on Metal Surfaces at Atomic Resolution. *Progress in Surface Science* **10**, 165–248.
Binnig, G. & Rohrer H. (1986). Scanning Tunnelling Microscopy. *IBM Journal of Research and Development* **30**, 355–69.
Cowley, J.M. (1986). Electron Microscopy of Surface Structure. *Progress in Surface Science* **21**, 209–50.

4

ELECTRONIC STRUCTURE

Introduction

In this chapter we investigate the electronic properties of clean solid surfaces. Certainly this is a prerequisite to any fundamental understanding of the *electrical* behavior of surfaces and interfaces. However, it also is essential to a coherent view of other surface phenomena, viz., oxidation, heterogeneous catalysis, crystal growth, brittle fracture, etc. There is no question that applications such as these provide most of the impetus behind surface science research. Nevertheless, we restrict ourselves here to only the most basic physics questions. What is the charge density in the neighborhood of the vacuum interface? Are the electron states near the surface different from those in the bulk? How do chemical bonding states in the first few atomic planes rearrange themselves after cleavage? What is the electrostatic potential felt by surface atoms?

The principal experimental probe of these issues is photoelectron spectroscopy and we will have much to say about this technique below. It turns out that the relevant measurements are relatively easy to perform but that the interpretation of the data is not entirely straightforward. It is helpful to have some idea of what to expect. Therefore, we defer our account of the experimental situation and proceed with some rather general theoretical considerations.

The methods of surface electronic structure are the same as those used to analyze the corresponding bulk problem. There are two common approaches. On the one hand, relatively simple constructs such as the nearly-free electron model and the tight-binding model are invaluable to identify gross features and to establish trends. These computations generally are quick and easy. On the other hand, detailed calculations that amount to a precise solution of an appropriate one-electron Schrödinger equation generally resolve ambiguities present in the simpler schemes and pin down the operative physics. Unfortunately, these computations can be very tedious and computer-intensive. Evidently, a trade-off is required.

The same considerations apply to the surface problem. We will make considerable use of the simpler methods to exhibit the most characteristic features of surface electronic structure. However, an over-reliance on these models is dangerous. The number of times that a conclusion based on a detailed calculation has contradicted a conclusion based on a simpler calculation is unnervingly large. In retrospect, the reason for this is simple: many issues of surface electronic structure (and surface physics in general) are decided by a competition between a number of very small energies (often ~ meV's). In general, the model calculations simply lack the precision needed to characterize this competition correctly. Accordingly, we will introduce the language necessary to appreciate both the parameterized model calculations and the more sophisticated calculations.

Let us suppose that the program of the previous chapter has been completed. We know the positions of all the atoms in the *semi-infinite* crystal. Label these positions by the set of vectors **R**. Then, ignoring ion motion, the Hamiltonian that describes the surface electronic structure is:

$$\mathscr{H} = \sum_{i=1}^{N} \frac{p_i^2}{2m} - \sum_{\mathbf{R}} \sum_{i=1}^{N} \frac{Ze^2}{|\mathbf{r}_i - \mathbf{R}|} + \frac{1}{2} \sum_{i,j}^{N} \frac{e^2}{|\mathbf{r}_i - \mathbf{r}_j|}. \tag{4.1}$$

N is the total number of electrons and we recognize the three terms as the kinetic energy of the electrons, the ion–electron attraction and the electron–electron repulsion. As in the bulk, the presence of the final term makes solution of (4.1) intractable in this form. To make progress, one can proceed by means of the Hartree–Fock approximation (Seitz, 1940). This familiar self-consistent field approach leads to a set of coupled integro-differential equations for the eigenstates and energy eigenvalues. Systematic corrections to the Hartree–Fock picture are studied by means of a well-defined perturbation theory. This is the method of choice for much of quantum chemistry but is quite awkward for use in extended systems like solid surfaces.

For present purposes, we rephrase the exact solution to the electronic structure problem in terms of the *density functional* method (Schluter & Sham, 1982; Lundqvist & March, 1983). Therein, it is proved that the ground state energy of the many-body problem, (4.1), can be written as a unique *functional* of the ground state charge density, $n(\mathbf{r})$, viz.,

$$E[n(\mathbf{r})] = T[n(\mathbf{r})] - \sum_{\mathbf{R}} Ze \int \mathrm{d}\mathbf{r} \frac{n(\mathbf{r})}{|\mathbf{R} - \mathbf{r}|} + \frac{1}{2} \iint \mathrm{d}\mathbf{r}\, \mathrm{d}\mathbf{r}' \frac{n(\mathbf{r})n(\mathbf{r}')}{|\mathbf{r} - \mathbf{r}'|} + E_{\mathrm{xc}}[n(\mathbf{r})]. \tag{4.2}$$

In this expression, $T[n(\mathbf{r})]$ is the kinetic energy of a *non-interacting* inhomogeneous electron gas in its ground state with density distribution $n(\mathbf{r})$. The second term is again the ion–electron interaction and the third term is the average electrostatic potential energy of the electrons. All of the many-body quantum mechanics of the problem is lumped into the so-called *exchange-correlation* term, $E_{xc}[n(\mathbf{r})]$.

The great advantage of this formulation is that the density that minimizes (4.2) is found by solution of a set of coupled *ordinary* differential equations:

$$-\tfrac{1}{2}\nabla^2\psi_i(\mathbf{r}) + v_{\text{eff}}(\mathbf{r})\psi_i(\mathbf{r}) = \varepsilon_i\psi_i(\mathbf{r}), \tag{4.3}$$

$$v_{\text{eff}}(\mathbf{r}) = -Ze^2\sum_{\mathbf{R}}\frac{1}{|\mathbf{r}-\mathbf{R}|} + \int d\mathbf{r}'\frac{n(\mathbf{r}')}{|\mathbf{r}-\mathbf{r}'|} + v_{xc}(\mathbf{r}), \tag{4.4}$$

where $n(\mathbf{r}) = \sum|\psi_i|^2$. This result is exact. Of course, the electron–electron interactions have been hidden in the exchange-correlation potential, $v_{xc}[n(\mathbf{r})] = \delta E_{xc}[n(\mathbf{r})]/\delta n(\mathbf{r})$, and practical implementation of this method requires a good approximation for this quantity. The parameters ε_i and ψ_i that enter the Schrödinger-like equation (4.3) formally have no physical meaning. Nevertheless, they frequently are interpreted successfully as one-particle excitation energies and eigenfunctions, respectively (Koelling, 1981).

In practice, it is common to adopt the simple, yet remarkably successful *local density approximation* (LDA) to $v_{xc}[n(\mathbf{r})]$. In this approximation, the exchange-correlation energy density of each infinitesimal region of the *inhomogeneous* electron distribution, $n(\mathbf{r})$, is taken to be precisely equal to the exchange-correlation energy density of a *homogeneous* electron gas

Fig. 4.1. Schematic representation of the local density approximation. $v_{xc}(\mathbf{x}_1) = v_{xc}[n(\mathbf{x}_1)]$ and $v_{xc}(\mathbf{x}_2) = v_{xc}[n(\mathbf{x}_2)]$.

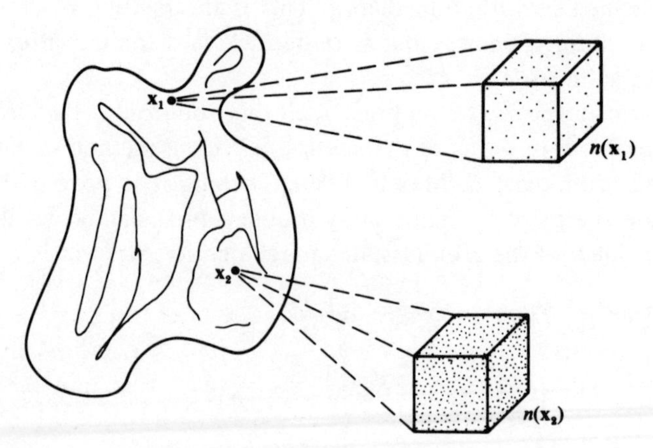

with the same density as the corresponding infinitesimal volume element (Fig. 4.1). The LDA is easy to apply because $v_{xc}[n]$ is known very precisely for the homogeneous electron gas at all densities of physical interest (Ceperley & Alder, 1980). More importantly, experience shows that a solution of the LDA equations (4.3) and (4.4) with $v_{xc}[n(\mathbf{r})] = v_{xc}^{LDA}[n(\mathbf{r})]$ for a semi-infinite system captures most of the essential physics of the surface electronic structure problem. In what follows, we shall specially note those cases where it does not do so.

The jellium model

The LDA equations are difficult to solve for the semi-infinite lattice problem. However, for the case of simple metals, the conduction electrons scatter only very weakly from screened ion core pseudopotentials. The *jellium model* represents an approximation to this situation. The discrete ion cores are replaced by a uniform, positive background charge with density equal to the spatial average of the ion charge distribution. The electrostatic potential created by this charge distribution replaces the ion–electron potential in (4.4). For the analogous surface problem, the semi-infinite ion lattice is smeared out similarly into a uniform positive charge that fills half of space:

$$n_+(\mathbf{r}) = \begin{cases} \bar{n} & z \leqslant 0, \\ 0 & z > 0. \end{cases} \tag{4.5}$$

Here, z is the direction normal to the surface. The positive background charge density, $\bar{n}$, often is expressed in terms of an inverse sphere volume, $(4\pi/3)r_s^3 = 1/\bar{n}$. Typical values of r_s range from about two to five.

The ground state electron density profile for the semi-infinite jellium model is translationally invariant in the x–y plane of the surface. However, the density variation perpendicular to the surface, $n(z)$, reveals two features that are quite characteristic of all surface problems (Fig. 4.2). First, electrons 'spill out' into the vacuum region ($z > 0$) and thereby create an electrostatic dipole layer at the surface. There is no sharp edge to the electron distribution. We can, however, locate an effective surface at:

$$d_{\parallel} = \frac{1}{\bar{n}} \int_{-\infty}^{+\infty} \mathrm{d}z\, z \frac{\mathrm{d}n(z)}{\mathrm{d}z}. \tag{4.6}$$

Second, $n(z)$ oscillates as it approaches an asymptotic value that exactly compensates the uniform (bulk) background charge. The wavelength of these *Friedel oscillations* is π/k_F, where $k_F = (3\pi^2\bar{n})^{1/3}$. They arise because the electrons (with standing wave vectors between zero and k_F) try to screen out the positive background charge distribution which includes a step at $z = 0$. The oscillations are a kind of Gibbs phenomenon since the Fourier decomposition of a sharp step includes contributions from wave vectors of arbitrarily large magnitude (Arfken, 1970).

The formation of a surface dipole layer means that the electrostatic potential far into the vacuum is greater than the mean electrostatic potential deep in the crystal, i.e.,

$$D = v(\infty) - v(-\infty). \tag{4.7}$$

This potential step serves, in part, to keep the electrons within the crystal

Fig. 4.2. Electron density profile at a jellium surface for two choices of the background density, r_s (Lang & Kohn, 1970).

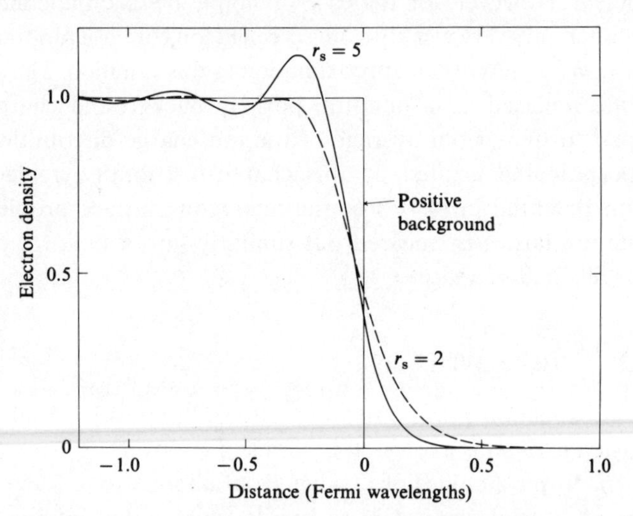

Fig. 4.3. Electrostatic potential, $v(z)$, and total effective one-electron potential, $v_{eff}(z)$, near a jellium surface (Lang & Kohn, 1970).

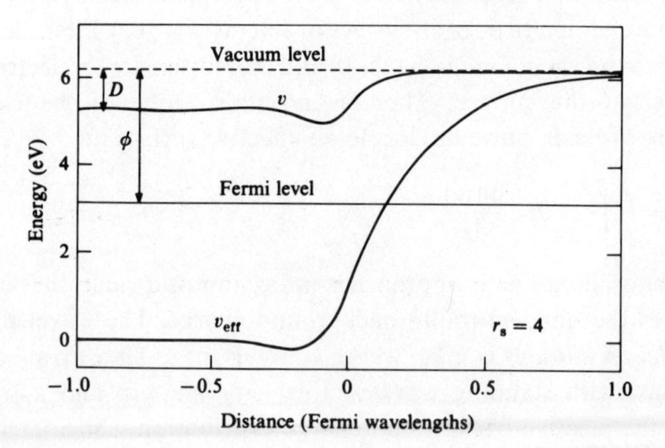

(Fig. 4.3). The remainder of the surface barrier comes from short range Coulomb interactions. The potential energy of each electron is lowered because neighboring electrons tend to stay away. This is a bulk effect which comes entirely from exchange and correlation.

The work function of a crystal surface is, by definition, the minimum energy required to remove an electron from deep within the bulk to a point a macroscopic distance outside the surface. More precisely,

$$\begin{aligned}\phi &= v(\infty) + E_{N-1} - E_N \\ &= v(\infty) - \mu \\ &= D - E_F.\end{aligned} \tag{4.8}$$

In this calculation, both the electrostatic potential and the chemical potential, μ, have been referenced to the mean electrostatic potential deep in the bulk. This choice of a zero is not unique, but has the advantage that the surface dipole enters in a natural way. With this convention, the variation over the periodic table of both the surface (D) and bulk (E_F) contributions to the work function can be substantial (~ 1 Ryd). Nonetheless, the measured work functions* of all the elements cluster around $\phi = 3.5 \pm 1.5$ eV. The substantial *cancellation* between the two terms in (4.8) arises because, crudely speaking, both measure properties of the atom.

Fig. 4.4. Electrostatic potential near a jellium step. The smoothed electron 'surface', $d_{\|}(x)$, is indicated by the heavy solid curve (Thompson & Huntington, 1982).

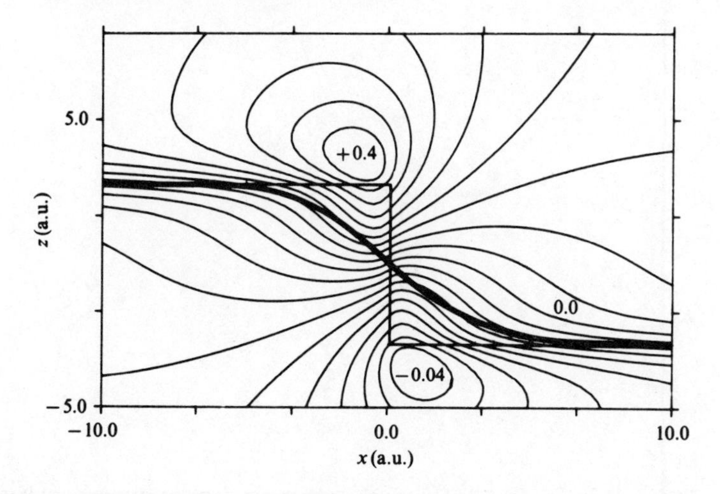

* See, e.g., Holzl & Schulte (1979) for a critical discussion of different techniques for measurement of the work function.

The ionization potential of an atom (E_F) is set largely by the diffuseness of the wave function tails (D).

Surface-specific contributions to the surface dipole also affect the work function of clean surfaces. For example, reconstruction and charge density smoothing will alter D from any free atom value. An extreme example of the latter occurs near a monoatomic step. Let us modify the semi-infinite jellium model so that the positive background has a single step somewhere along the surface. The electrons spill out as before but, in addition, they smooth out the sharp step along the surface. The result is an electrostatic dipole oriented oppositely to the spill-out dipole (Fig. 4.4). The *net* dipole moment is reduced relative to the flat surface value. Work function measurements of regularly stepped metal surfaces confirm this simple picture (Fig. 4.5).

One feature of the effective potential of Fig. 4.3 is qualitatively incorrect. For large positive values of z, $v_{\text{eff}}(z)$ approaches the vacuum level exponentially rapidly whereas, asymptotically, one should recover the power law behavior of the classical *image* potential. This is a failure of the local density approximation. If the exact $E_{xc}[n(\mathbf{r})]$ were known, the corresponding exchange-correlation potential would be

$$\lim_{z\to\infty} v_{xc}(z) = -\frac{e}{4|z-d_\perp|} \tag{4.9}$$

Fig. 4.5. Work function change for stepped metal surfaces (Besocke, Krahl-Urban & Wagner, 1977).

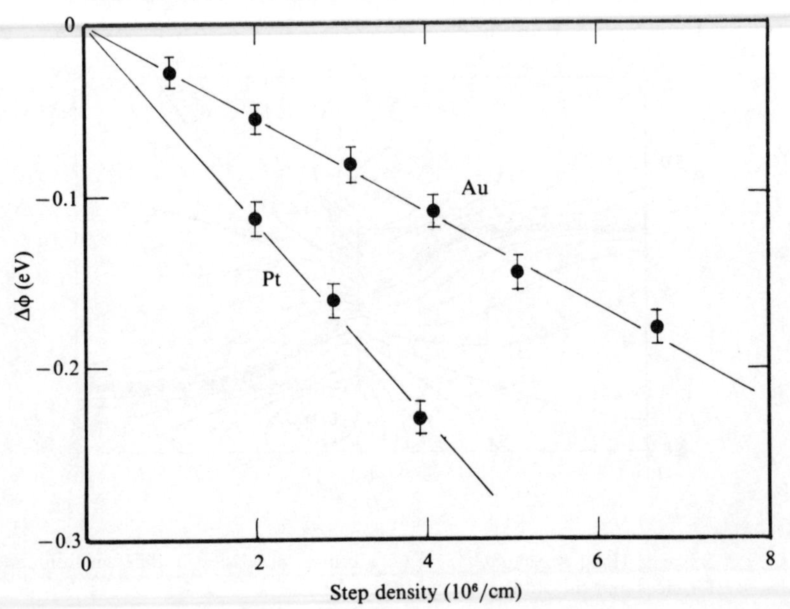

where $d_\perp$ is the *centroid* of the charge distortion $\delta n(z)$ induced by the presence of an electron outside the metal:

$$d_\perp = \int_{-\infty}^{+\infty} \mathrm{d}z\, z\, \delta n(z) \Big/ \int_{-\infty}^{+\infty} \mathrm{d}z\, \delta n(z). \tag{4.10}$$

Both $d_\perp$ and $d_\parallel$ (4.6) are ~ 1 Å although they are not quite identical.*

The semi-infinite jellium model also can be used to estimate the surface tension of simple metals. Fig. 4.6 shows that the surface tension of low electron density metals is well described by the model while high density metals are predicted to spontaneously cleave! The difficulty stems from the complete neglect of the ionic lattice. Using first-order perturbation theory one can take account of the linear response of the electron gas to the weak pseudopotentials of the crystalline lattice. The resulting energy shift brings the jellium theory into much better accord with experiment (Fig. 4.6).

Fig. 4.6. Comparison of surface tension data with the results of the jellium model (Lang & Kohn, 1970).

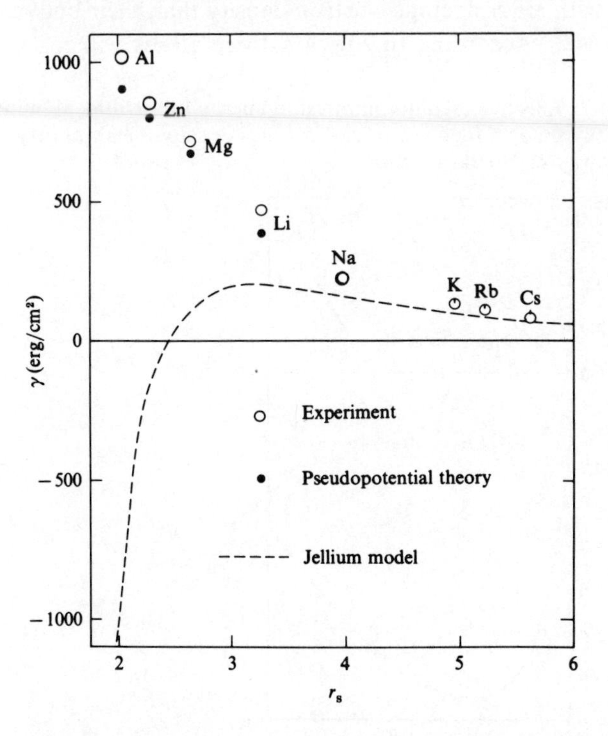

* The distinction between the two and the motivation for the notation used here will emerge in Chapter 6.

The pseudopotential correction to the jellium model discussed above analyzes the effect of the screened ion cores on the semi-infinite electron gas within perturbation theory. An alternative analysis of the same problem examines the converse situation: the effect of the semi-infinite electron gas on the screened ion cores. To be specific, this *effective medium* approximation considers the change in energy that occurs when a free atom is immersed into an otherwise uniform electron gas (Norskov & Lang, 1980; Stott & Zaremba, 1980). In the bulk, the standard jellium model smears out *all* the ionic and electronic charge. The effective medium approach smears out all the charge except that associated with a single atom. The latter is treated exactly (within LDA). The immersion energy is calculated as a function of the density of the effective medium (Fig. 4.7). The results are consistent with our intuition. The closed shell helium atom repels external electrons whereas both aluminum and oxygen favor interaction ('bonding') with electrons over a range of densities.

The effective medium energy curves provide a simple explanation for the oscillatory relaxation phenomenon discussed in Chapter 3. The atoms at the ideal surface of a simple metal, e.g., Al, find themselves embedded in a medium with *lower* average electron density than their bulk counterparts (cf. Fig. 4.2). According to Fig. 4.7, these atoms will relax inward

Fig. 4.7. Effective medium immersion energy for helium, aluminum and oxygen as a function of electron gas density (Chakraborty, Holloway & Norskov, 1985).

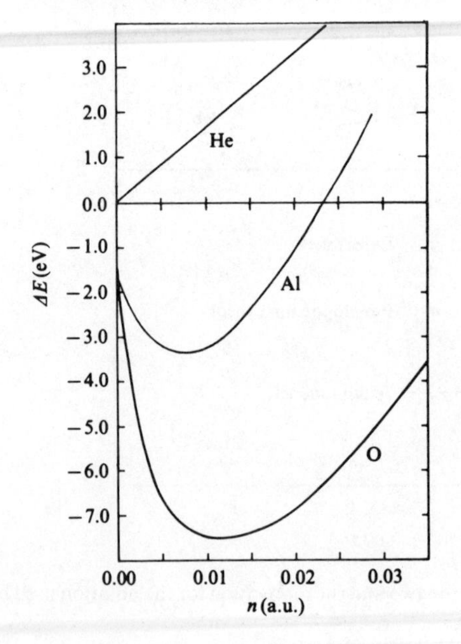

towards a higher density position that brings them nearer to the immersion curve minimum. However, this contraction of the outer layer brings additional charge density into the neighborhood of the second layer of atoms. These atoms now find themselves on the *high* density side of the effective medium minimum. To lower the average electron density around them, the second layer exerts a force which pushes away the third layer of atoms, i.e., the second pair of atomic planes expand relative to the bulk spacing. By the same argument, the next deepest pair of planes contract and a damped oscillatory relaxation proceeds into the bulk.

The final step in the LDA hierarchy beyond the effective medium approach would embed the entire semi-infinite lattice into an electron gas. This is equivalent to a complete solution of (4.3) and (4.4) – a formidable task. However, the basic elements that emerge from such detailed calculations are present in a class of models rather different from any jellium model. These are the band structure models.

One-dimensional band theory

The jellium description of a metal surface can be described as a one-dimensional model that neglects the details of the electron–ion interaction and emphasizes the nature of the smooth surface barrier. The one-dimensional band structure approach to surface electronic structure emphasizes the lattice aspects of the problem and simplifies the form of the surface barrier. The basic theme of the band structure models is the influence of a boundary condition for the Schrödinger equation that reflects the presence of a free surface. In both the nearly-free electron model and the tight-binding model this new boundary condition leads to the existence of *surface states.*

The one-dimensional nearly-free electron model (appropriate to a *metal* surface) neglects the electron–electron interaction and self-consistency effects present in the LDA Schrödinger-like equation (4.3) and (4.4). The effective potential includes only the ion cores and a crude surface barrier (Fig. 4.8, solid curve):

$$\left[-\frac{d^2}{dz^2}+V(z)\right]\psi(z)=E\psi(z). \tag{4.11}$$

The effect of the screened ion cores is modelled with a weak periodic pseudopotential,

$$V(z)=-V_0+2V_g\cos gz, \tag{4.12}$$

where $g=2\pi/a$ is the shortest reciprocal lattice vector of the chain.

The solution to this problem in the bulk is well known (Kittel, 1966).

For present purposes, a two-plane-wave trial function is sufficient:

$$\psi_k(z) = \alpha e^{ikz} + \beta e^{i(k-g)z}. \tag{4.13}$$

Substituting (4.13) into (4.11) leads to the secular equation,

$$\begin{bmatrix} k^2 - V_0 - E & V_g \\ V_g & (k-g)^2 - V_0 - E \end{bmatrix} \begin{bmatrix} \alpha \\ \beta \end{bmatrix} = 0, \tag{4.14}$$

which is readily solved for the wave functions and their energy eigenvalues:

$$\begin{aligned} E &= -V_0 + (\tfrac{1}{2}g)^2 + \kappa^2 \pm (g^2\kappa^2 + V_g^2)^{1/2} \\ \psi_k &= e^{i\kappa z} \cos(\tfrac{1}{2}gz + \delta). \end{aligned} \tag{4.15}$$

In this expression, $e^{i2\delta} = (E - k^2)/V_g$, and the wave vector has been written

Fig. 4.8. One-dimensional semi-infinite lattice model potential (solid curve) and an associated surface state (dashed curve).

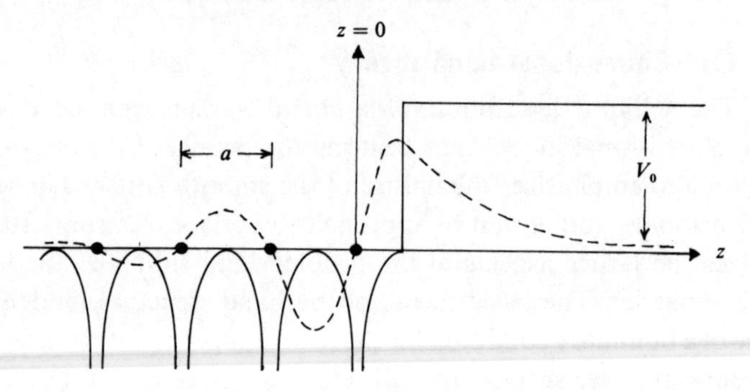

Fig. 4.9. $E(\kappa^2)$ for the one-dimensional semi-infinite nearly-free electron model.

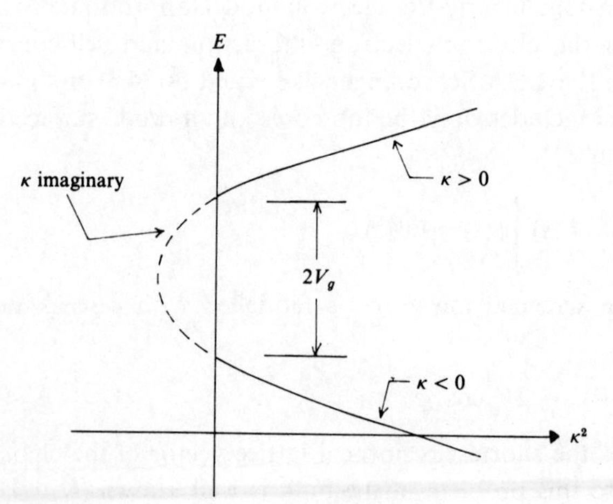

in terms of its deviation from the Brillouin zone boundary: $k = g/2 + \kappa$. The character of the eigenfunctions depends on the sign of V_g. If $V_g < 0$, the lowest energy solution is even with respect to reflection about $z = 0$. If $V_g > 0$, the lowest energy solution is odd with respect to reflection.

A plot of the function $E(\kappa^2)$ reveals the features of this model most relevant to the surface (Fig. 4.9). The familiar energy gap appears at the Brillouin zone boundary, i.e., $\kappa = 0$. However, $E(\kappa^2)$ actually is a continuous function of κ^2 if one permits negative values of the argument. In other words, perfectly valid solutions of the Schrödinger equation exist for *imaginary* values of κ when $0 < |\kappa| < |V_g|/g$. In the bulk, these solutions are discarded because they become exponentially large as $|z| \to \infty$ and cannot satisfy the usual periodic boundary conditions. However, for the semi-infinite problem, the solution that grows for positive z is acceptable since it will be matched (at $z = a/2$) onto a function that describes the decay of the wave function in the vacuum:

$$\begin{aligned} \psi(z) &= e^{\kappa z} \cos\left(\tfrac{1}{2}gz + \delta\right) \qquad && z < a/2 \\ \psi(z) &= e^{-qz} && z > a/2 \end{aligned} \tag{4.16}$$

where $q^2 = V_0 - E$.

If the logarithmic derivative of $\psi(z)$ can be made continuous at $z = a/2$, an electronic state exists that is localized at the surface of the lattice chain. The energy of this *surface state* lies in the bulk energy gap. To see if the necessary wave function match can occur, simply graph the trial solution ($z < 0$) for a sequence of energies within the gap (Fig. 4.10). The different

Fig. 4.10. Wave function matching at $z = a/2$: (*a*) $V_g < 0$; (*b*) $V_g > 0$. The sequence 1, 2, 3, indicates increasing energy starting from the bottom of the gap (Forstmann, 1970).

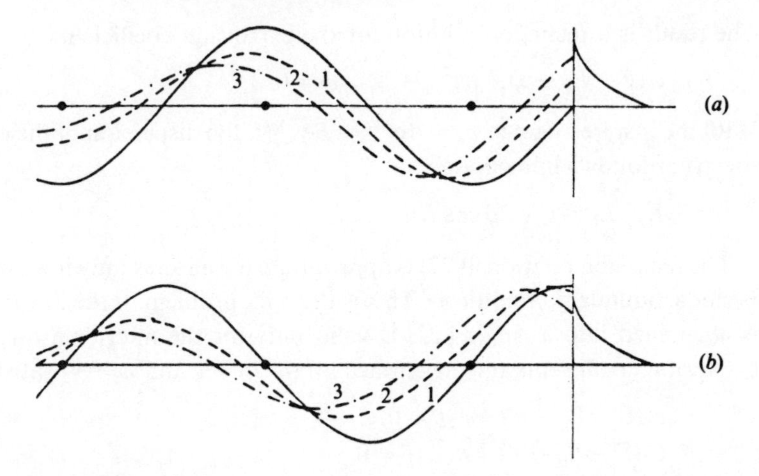

curves reflect a smooth variation of the phase shift, δ. With our choice of a matching plane, the curvature of ψ_k can match the decaying vacuum solution only for $V_g > 0$. The dashed curve in Fig. 4.8 illustrates the resulting surface state wave function. This solution often is called a Shockley (1939) state.

A one-dimensional model most appropriate to a *semiconductor* surface should focus on wave functions constructed from atomic-like orbitals. This is the basis of the tight-binding model (Ashcroft & Mermin, 1976). The lattice potential is constructed from a superposition of N free atom potentials, $V_a(\mathbf{r})$, arranged on a chain with lattice constant a:

$$V_{\mathrm{L}}(\mathbf{r}) = \sum_{n=1}^{N} V_a(\mathbf{r} - n\mathbf{a}), \tag{4.17}$$

where

$$[-\nabla^2 + V_a(\mathbf{r}) - E_a]\phi(\mathbf{r}) = 0. \tag{4.18}$$

The non-self-consistent Schrödinger equation for the bands is

$$\{-\nabla^2 + V_a(\mathbf{r}) + [V_{\mathrm{L}}(\mathbf{r}) - V_a(\mathbf{r})]\}\psi(\mathbf{r}) = E\psi(\mathbf{r}). \tag{4.19}$$

The simplest trial function *ansatz* is a superposition of s-like Wannier orbitals – one on each site:

$$\psi(\mathbf{r}) = \sum_{n=1}^{N} c_n \phi(\mathbf{r} - n\mathbf{a}). \tag{4.20}$$

When (4.20) is substituted into (4.19), a large number of Hamiltonian matrix elements are generated between orbitals centred on different sites. As usual, we retain only the on-site matrix element α and the nearest neighbor hopping matrix element β:

$$\langle l | V_{\mathrm{L}} - V_a | m \rangle = -\alpha\delta_{l,m} - \beta\delta_{l,m\pm 1}. \tag{4.21}$$

The result is a recursion relation for the expansion coefficients,

$$c_n(E - E_0 + \alpha) + (c_{n-1} + c_{n+1})\beta = 0. \tag{4.22}$$

With the inspired choice, $c_n = A\mathrm{e}^{inka} + B\mathrm{e}^{-inka}$, the dispersion of the energy spectrum follows immediately:

$$E = E_0 - \alpha + 2\beta \cos Ka. \tag{4.23}$$

The recursion relation (4.22) is appropriate for all sites only if we impose periodic boundary conditions. This is the bulk problem. If the linear chain is *not* joined into a ring, (4.22) is valid only for the interior atoms. The expansion coefficients on the surface atoms ($n = 1$ and $n = N$) satisfy

$$\begin{aligned} c_1(E - E_0 + \alpha') + c_2\beta &= 0, \\ c_N(E - E_0 + \alpha') + c_{N-1}\beta &= 0, \end{aligned} \tag{4.24}$$

where we have allowed for the possibility that the diagonal Hamiltonian matrix element of the surface atoms, $\langle 1|V_L - V_a|1\rangle = \langle N|V_L - V_a|N\rangle = \alpha'$, might differ from its value in the bulk. The allowed values of the wave vector, k, are found by substituting (4.23) into (4.24) and eliminating A and B in the expression for c_n. The result is a transcendental equation with N roots. Most of these roots correspond to solutions that have equal wave function amplitude on every atom of the chain. However, if $|\alpha' - \alpha| > |\beta|$, two of the roots are complex. For each, the corresponding eigenfunction has appreciable amplitude only on a surface atom (Goodwin, 1939). The energy of these states split off either above or below the bulk continuum (4.23). Notice that these so-called Tamm (1932) surface states occur only if there is a strong enough perturbation ($\alpha' \neq \alpha$) of the potential right at the surface – precisely what one might expect at a semiconductor surface with broken bonds.

Three-dimensional theory

The tight-binding method is particularly well suited to an extension to three dimensions. For example, the density of states, $\rho(E)$, is a familiar concept from bulk condensed matter physics. For surface studies, it is useful to be able to resolve this quantity into contributions from each atomic layer parallel to the surface. Better still, we define a *local density of states* (LDOS) at each point in space that is weighted by the probability density of each of the system eigenfunctions:

$$\rho(\mathbf{r}, E) = \sum_{\alpha} |\psi_\alpha(\mathbf{r})|^2 \delta(E - E_\alpha). \tag{4.25}$$

Alternatively, we can project each $\psi_\alpha(\mathbf{r})$ onto one particular orbital, $\phi_i(\mathbf{r})$, localized at a specific site i:

$$\rho_i(E) = \sum_{\alpha} |\langle i|\alpha\rangle|^2 \delta(E - E_\alpha). \tag{4.26}$$

For present application, the projected LDOS is characterized best by its second moment:

$$\begin{aligned}\mu_i &= \int \mathrm{d}E E^2 \rho_i(E) \\ &= \sum_{\alpha} \langle i|\alpha\rangle E_\alpha^2 \langle \alpha|i\rangle \\ &= \langle i|\mathscr{H}^2|i\rangle \\ &= \sum_j \langle i|\mathscr{H}|j\rangle \langle j|\mathscr{H}|i\rangle. \end{aligned} \tag{4.27}$$

As before, suppose that the Hamiltonian matrix elements are non-zero only for nearest neighbor sites. In that case, the final expression of (4.27) can be interpreted as a sum over all paths that jump to a near neighbor site (j) from the origin (i) and then jump back (Fig. 4.11). Therefore, if $\langle i|\mathscr{H}|j\rangle = \beta$, the second moment of $\rho_i(E)$ is proportional to the coordination number (Z) of the site:

$$\mu_i = Z\beta^2. \tag{4.28}$$

Since, by definition, surface sites are less well coordinated than bulk sites, we expect the LDOS at the surface to narrow compared to the bulk.

Fig. 4.11. Pictorial representation of (4.27) for a bulk site and a surface site.

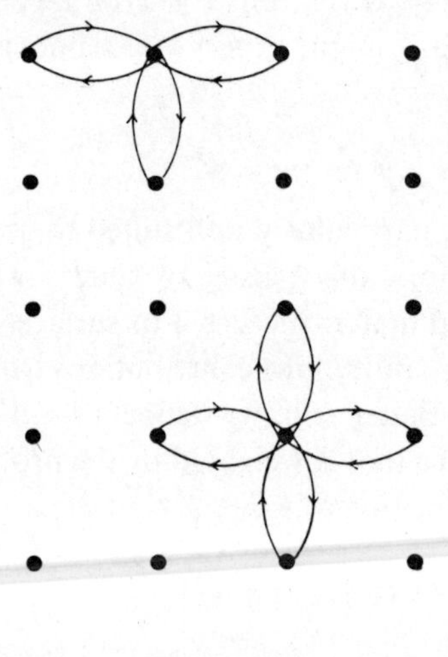

Fig. 4.12. The layer-resolved LDOS for the three uppermost surface planes of a tight-binding solid compared to the bulk density of states (Haydock & Kelly, 1973).

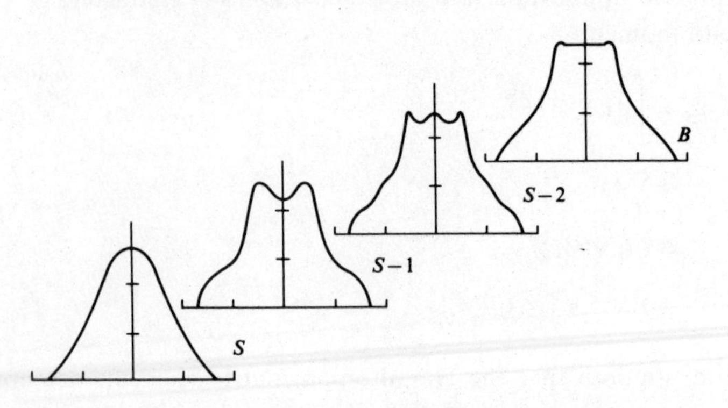

Fig. 4.12 shows the results of a calculation of the layer-resolved local density of states for a three-dimensional tight-binding solid. The narrowing of the surface LDOS is a real (albeit subtle) effect. There is another way to see that the density of electronic states at a surface cannot be identical to that of the bulk. In an infinite crystal, the Bloch functions always are taken to be running waves. By contrast, the eigenfunctions of a finite crystal (with surfaces!) must be standing waves. It is not surprising that the rearrangement of charge needed to accomplish this change will be greatest right at the surface. Fig. 4.12 also demonstrates that the LDOS 'heals' to its bulk value no more than two or three atomic planes from the vacuum interface. The rapid recovery of bulk properties as one proceeds into the crystal is a very general result. The tight-binding model admits surface states if a large enough perturbation occurs at the surface. For our s-band model there is only one state/atom so that surface states must rob spectral weight from the perturbed bulk states. The total normalized LDOS is shared between the two on a layer-by-layer basis (Fig. 4.13). This compensation behavior is very general and illustrates the manner by which a free surface remains *charge neutral.*

As a final example, we return to the question of oscillatory relaxation. The effective medium theory discussed earlier is appropriate for nearly-free electron metals. Now we can deal with the more localized d-states that occur in transition metals. Crudely speaking, the immersion energy of a transition metal has two contributions: $\Delta E = E_{\text{rep}} + E_{\text{band}}$ (Spanjaard & Desjonqueres, 1984). The first piece arises from the pairwise Pauli repulsion between electron clouds on neighboring atoms. Hence, for any single atom, $E_{\text{rep}} \propto Z$, the coordination number. The cohesive energy/atom arises from adding up the contributions from the occupied bonding (and anti-bonding) states that constitute the energy bands:

Fig. 4.13. Layer-resolved LDOS for a tight-binding model including a surface perturbation (Kalkstein & Soven, 1971).

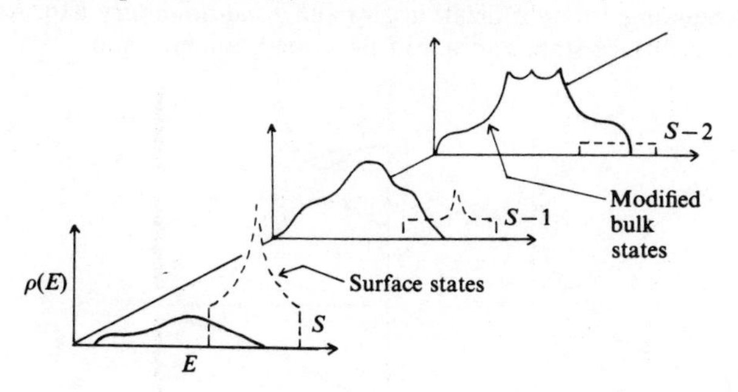

$$E_{\text{band}} = \int^{E_F} \mathrm{d}E\, E\rho_i(E). \tag{4.29}$$

This is the *first* moment of the occupied LDOS so (4.28) suggests that, roughly,

$$E_{\text{band}} \propto -\mu_i^{1/2} = -\sqrt{Z}\beta. \tag{4.30}$$

Consequently, the immersion energy

$$\Delta E \propto AZ - B\sqrt{Z}, \tag{4.31}$$

where A and B are constants. The essential point is that a plot of (4.31) looks just like Fig. 4.7 with the local coordination number playing the role of the effective medium density. The minimum of the curve will occur at Z_{bulk} and the reduced coordination at the surface initiates the same oscillatory chain reaction discussed earlier (Tomanek & Bennemann, 1985a).

To this point, our simple models have demonstrated the electronic origin of a common structural phenomenon at metal surfaces and led us to expect localized states and altered densities of states near all surfaces. Only one additional concept is needed to make contact with spectroscopic measurements: the surface projected band structure. We use the nearly-free electron model to illustrate the main idea. First recall the one-dimensional case. The 'surface' consists of two points, the wave vector **k** is directed along the chain, and the surface states appear in the energy gap at the zone boundary. Fig. 4.14 depicts a more general one-dimensional (multi-) band structure where the notation, $k_\perp$, is used to underscore the fact that this direction is perpendicular to the surface. Hybridization produces an 'avoided crossing' of the energy bands so that a new energy gap appears in the spectrum. The two gaps appear clearly at the extreme right side of

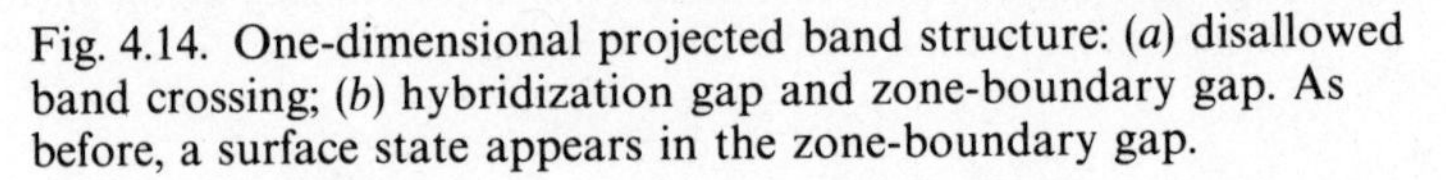

Fig. 4.14. One-dimensional projected band structure: (*a*) disallowed band crossing; (*b*) hybridization gap and zone-boundary gap. As before, a surface state appears in the zone-boundary gap.

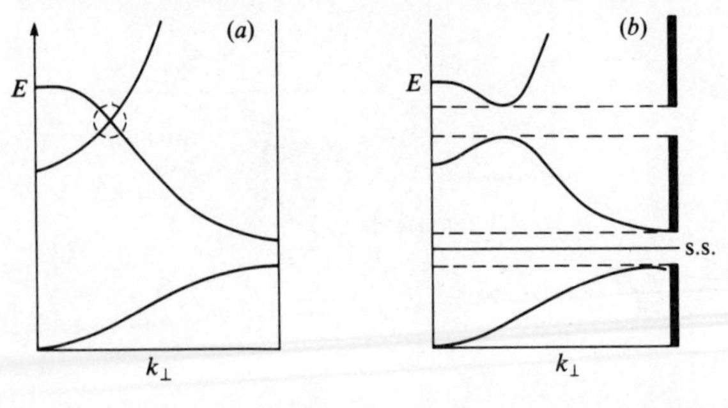

this figure in the *projection* of the bulk state continuum against the surface 'plane'.

In three dimensions, the cleaved crystal can be viewed as a system with a unit cell that is macroscopically large in the direction normal to the surface. Therefore, as in Chapter 3, it is meaningful to speak of a *surface Brillouin zone* (SBZ) which is characterized by two-dimensional wave vectors, $\mathbf{k}_{\parallel}$. The $\mathbf{k}_{\parallel}$ are good quantum numbers if the surface crystal structure is translationally invariant. For each value $\mathbf{k}_{\parallel}$, a $k_{\perp}$ rod extends back into the bulk three-dimensional Brillouin zone. The energy bands along this (and every) rod can be projected onto the SBZ just as in the one-dimensional case (Fig. 4.15).

Fig. 4.15 depicts a hypothetical metal because every energy can be identified with at least one bulk state *somewhere* in three-dimensional **k**-space. By contrast, the projection of the bulk states of a semiconductor show a gap completely across the SBZ. The surface state that appeared in the one-dimensional model now persists for a range of $\mathbf{k}_{\parallel}$ in the SBZ – one speaks of a band of surface states. In addition, this state mixes with a degenerate, propagating, bulk state at the $\mathbf{k}_{\parallel}$ point where its dispersion enters the projected continuum. The resulting hybrid bulk state has an abnormally large amplitude on the surface atoms (compared to normal standing wave bulk states) and is called a *surface resonance*. Surface states also can occur in the hybridization gap. Of course, a quantitative

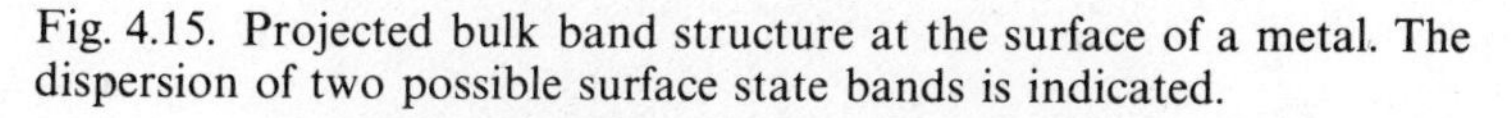

Fig. 4.15. Projected bulk band structure at the surface of a metal. The dispersion of two possible surface state bands is indicated.

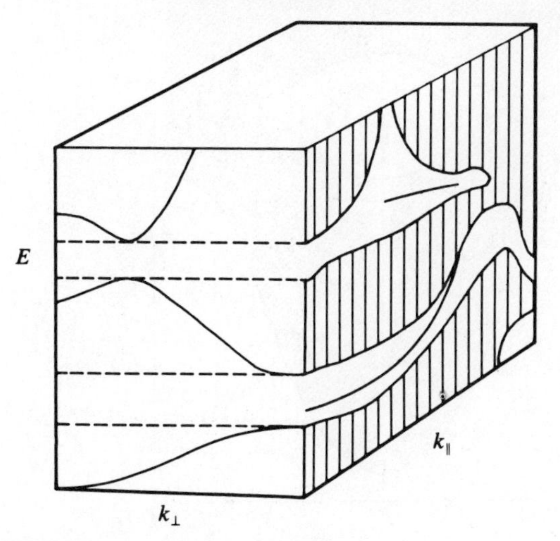

Fig. 4.16. Edge view of a five-layer slab for surface electronic structure calculations. Slab has infinite extent in the x–y plane.

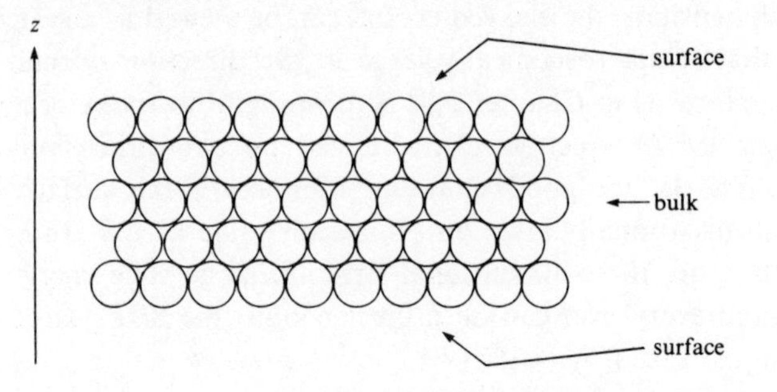

Fig. 4.17. Surface states (dashed curves) and bulk projected bands at a Cu(111) surface according to a six-layer surface band structure calculation (Euceda, Bylander & Kleinman, 1983).

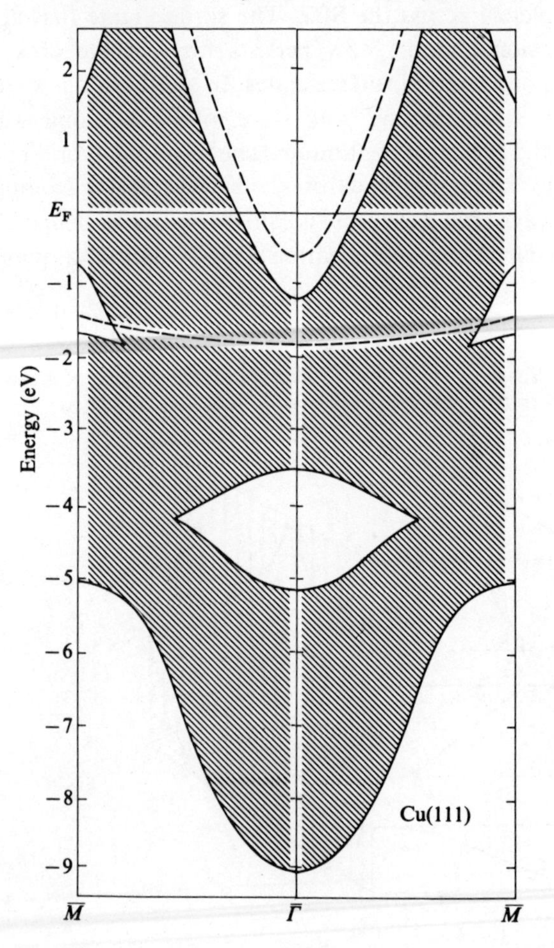

discussion of the bulk projections and surface states requires a nomenclature for labelling points in the SBZ. A standard convention has been catalogued (Plummer & Eberhardt, 1982).

The intuition gleaned from the simple models developed above is crucial for any overall appreciation of the electronic structure physics at surfaces. Nevertheless, high-quality angle-resolved photoemission data (see below) demand quantitative comparison with the most sophisticated theoretical results available. In some cases, particularly for semiconductors, carefully parameterized tight-binding calculations are sufficient. However, in general, the most reliable results come from complete solutions of the LDA equations (4.3) and (4.4).

The first fully self-consistent electronic structure calculation for a semi-infinite lattice was performed for Na(100) using a wave function matching technique not unlike our simple NFE discussion (Appelbaum & Hamann, 1972). However, the LDOS in this calculation (and others) showed the same rapid healing toward bulk-like behavior as observed in the simple models (cf. Fig. 4.12). Consequently, almost all LDA surface band structure calculations now employ a 'slab' geometry (Fig. 4.16). Typically only five to ten atomic layers are sufficient for the central layer to display a bulk LDOS. As an example, consider the Cu(111) surface. The cross-hatched region of Fig. 4.17 denotes the projection of the bulk energy bands of copper onto the Cu(111) SBZ. The free-electron nature of the s–p band is evident from the parabolic shape of the projected band edges at the top and bottom of the figure. The 3d bands of Cu lie between 2 eV and 5 eV below the Fermi level. A six-layer LDA surface calculation predicts two bands of states localized at this surface. The uppermost surface state is a simple Shockley state derived from the s–p band just as in the nearly-free electron model. The lower surface state is a Tamm state split off just above the bulk 3d continuum – similar to our tight-binding model results.* The quality of this calculation is best judged by direct appeal to experiment.

Photoelectron spectroscopy

The kinetic energy distribution of electrons photoemitted from a solid is the primary experimental window on the electronic structure of its surface. We have encountered this type of distribution once before (cf. Fig. 2.4). In that case, the sharp peaks in the spectrum were associated with the binding energy of specific atomic core levels. Here, we slightly

* Most surface states do not fall so easily into the simple Shockley–Tamm classification scheme.

modify the Einstein relation that connects the binding energy E_B, the photon energy $h\nu$, and the outgoing electron kinetic energy E_{KIN}:

$$E_{KIN} = h\nu - E_B - \phi \tag{4.32}$$

Fig. 4.18. Illustration of the relationship between the occupied electronic density of states (*a*) and the photoemitted electron kinetic energy distribution (*b*) (Feuerbacher, Fitton and Willis, 1978).

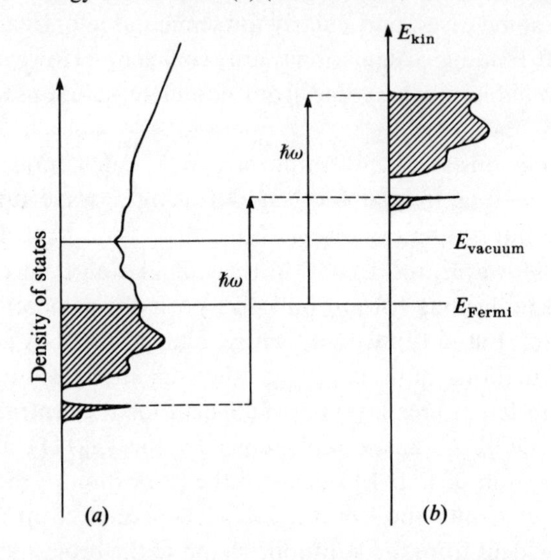

Fig. 4.19. Plot of the second moment (cf. (4.27)) of the experimental XPS density of states versus electron emission angle for a polycrystalline Cu sample. Surface sensitivity increases as the grazing exit angle decreases (Mehta & Fadley, 1979).

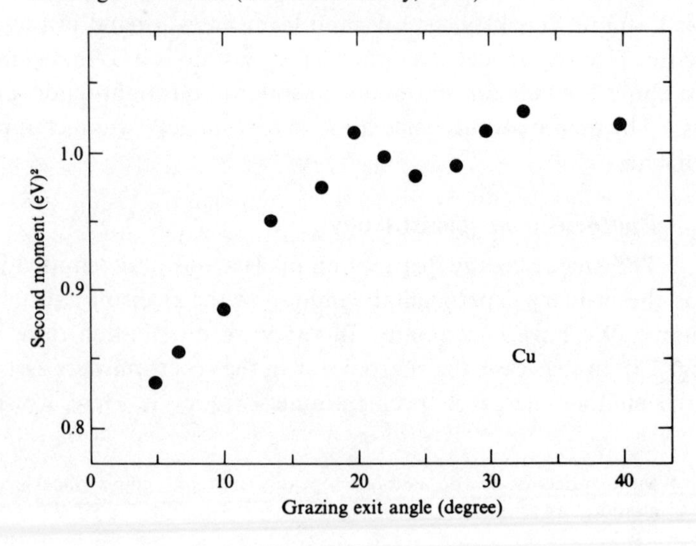

to account for the work function of the material. For present purposes, note that (4.32) applies to electrons of *any* binding energy – not merely those bound in deep core states. X-ray photons eject electrons from the valence band ($E_B < 10\,\text{eV}$) as well. This is important because (if the intrinsic ejection probability is the same for all electron states) the intensity distribution of photoemitted electrons is an image of the occupied electronic density of states (Fig. 4.18). Unfortunately, at typical XPS photon energies, valence band electrons have a very low photoelectric cross section and the kinetic energy of electrons ejected from the valence band is rather large. This means that the spectra are more characteristic of the bulk density of states than the surface LDOS. To get around this problem, one can enhance the sensitivity of XPS to the surface valence band by the use of grazing angle incidence radiation (refraction limits the penetration depth) and grazing angle emission electron collection (reduces the effective mean free path normal to the surface). The narrowing of the LDOS at the surface has been verified using exactly this trick (Fig. 4.19).

Ultraviolet photoelectron spectroscopy (UPS) is ideally suited for study of the surface valence band. This technique is conceptually identical to XPS except that the incident photons are in the range of 20–150 eV. However, there are three significant advantages for surface studies. First, the universal curve of mean free path guarantees that UPS photoelectrons originate from the surface region. Second, the valence band photo-cross section is large at UPS excitation energies. Third, the energy resolution is excellent because typical laboratory line sources (He I (21.2 eV) and He II (40.8 eV) resonance lamps) have natural linewidths three orders of magnitude smaller than laboratory x-ray sources.*

A detailed view of the individual states that constitute the surface LDOS is possible if one does not collect all the photoemitted electrons at once. The technique of *angle-resolved photoemission* maps out the dispersion, i.e., the wave vector dependence of the energy, of individual electron bands. To see how this works, write the kinetic energy that enters (4.32) as $E_{\text{KIN}} = \hbar^2(k_\parallel^2 + k_\perp^2)/2m$, where $\mathbf{k}_\parallel$ and $\mathbf{k}_\perp$ denote the components of the escaping photoelectron's momentum *in the vacuum* parallel and perpendicular to the surface plane, respectively. If θ is the angle between the surface normal and the electron energy analyzer, $k_\parallel = (2mE_{\text{KIN}}/\hbar^2)^{1/2} \sin\theta$.

* It is worth noting that synchrotron radiation from electron storage rings now provides continuously tunable light of great intensity and small bandwidth in both the UV and x-ray portions of the electromagnetic spectrum (Winnick & Doniach, 1980). Unfortunately, storage rings are large machines operated as regional (or national) facilities. Beam time at a synchrotron is limited and expensive compared to laboratory sources.

Translational invariance in the plane of the surface guarantees that

$$\mathbf{k}_{\parallel}\,(\text{outside}) = \mathbf{k}_{\parallel}\,(\text{inside}) + \mathbf{g}_{s}, \tag{4.33}$$

where $\mathbf{g}_s$ is a surface reciprocal latticevector. By contrast, the perpendicular component of the photoelectron momentum bears no particular relationship to $\mathbf{k}_\perp$ (inside) of the initial band state. Minimally, the potential step at the surface retards the photoelectron and decreases the 'component' of kinetic energy perpendicular to the surface. The ejected electron could originate from any value of $\mathbf{k}_\perp$ along the reciprocal lattice rod perpendicular to the SBZ at $\mathbf{k}_\parallel$. Hence, according to (4.32), if one energy analyzes photoelectrons as a function of their *angle* of emission, peaks in the associated energy distribution curve reflect initial states of the solid indexed by $\mathbf{k}_\parallel$. The dispersion of an electron state, $E(\mathbf{k}_\parallel)$, shows up as a smooth variation in the energy of a photoemission peak as the detection angle varies (Fig. 4.20).

Fig. 4.20. Photoemission energy distribution curves from Cu(111) at different collection angles. Equation (4.32) has been used to express the electron kinetic energy in terms of the binding energy of the electron state (Kevan, 1983).

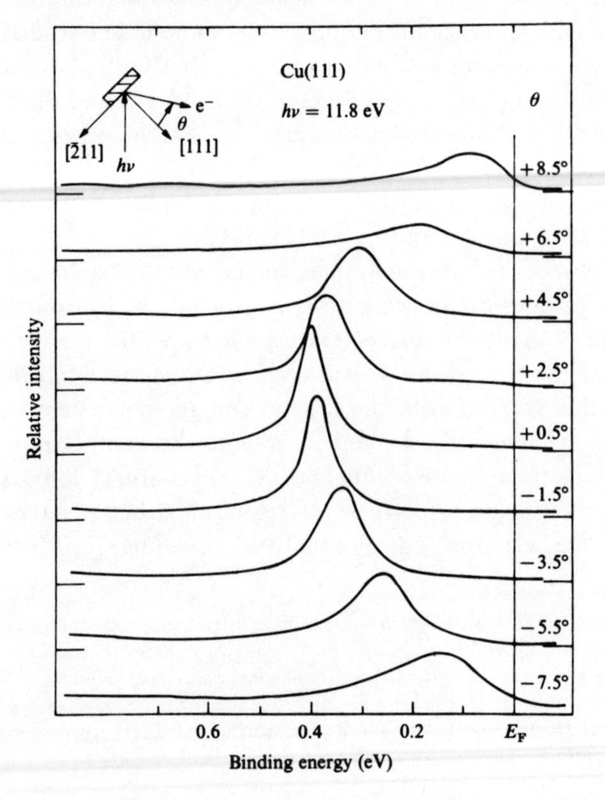

Fig. 4.21. Experimental dispersion of Cu(111) surface states plotted with a projection of the bulk bands: (*a*) Shockley state near the zone center (Kevan, 1983); (*b*) Tamm state near the zone boundary (Heimann, Hermanson, Miosga and Neddermeyer, 1979). Compare with Fig. 4.17.

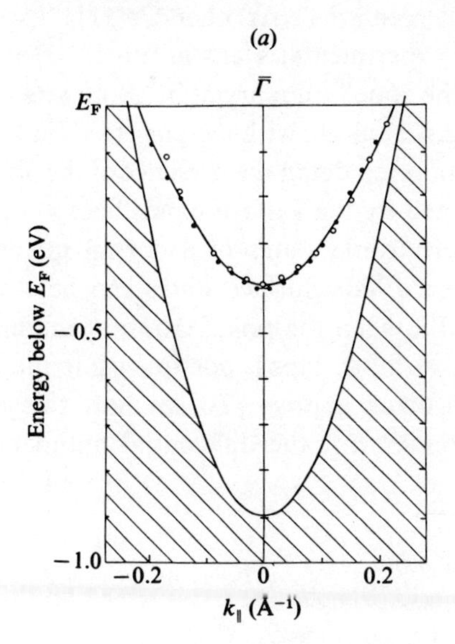

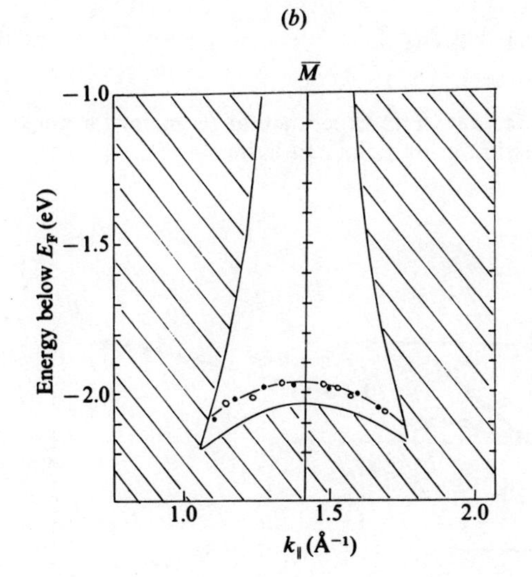

The charge density at a surface has contributions from both standing wave bulk states and legitimate surface states. There are well-defined spectroscopic tests that identify any particular feature in a photoelectron energy distribution curve (EDC) as one or the other (Plummer & Eberhardt, 1982). The data of Fig. 4.20 turn out to derive from a surface state. The small polar angles of emission correspond to probing the SBZ in the neighborhood of the zone center, $\bar{\Gamma}$. Another Cu(111) dispersing surface state appears in similar experimental scans at much larger polar angle that probe the SBZ near the zone boundary at $\bar{M}$. Both sets of data are shown plotted as $E(\mathbf{k}_{\parallel})$ in Fig. 4.21 along with the projected bulk bands of copper. These experiments confirm in detail the presence of the Shockley and Tamm surface states predicted by the surface band theory.

It is fruitful to classify the electronic states of a crystal in terms of their symmetry properties (Heine, 1960). Surface states can be classified similarly with angle-resolved UPS when the polarization of the incident photon beam is an independent variable. This is possible using the highly polarized radiation from synchrotron sources. To see how this works, consider the Golden Rule expression for the differential photoemission cross section:

$$\frac{\mathrm{d}\sigma(\omega)}{\mathrm{d}E_f} = \sum_i |\langle f|\mathscr{H}^{\mathrm{ext}}|i\rangle|^2 \delta(\hbar\omega - E_f + E_i) \tag{4.34}$$

The transition operator has the form:

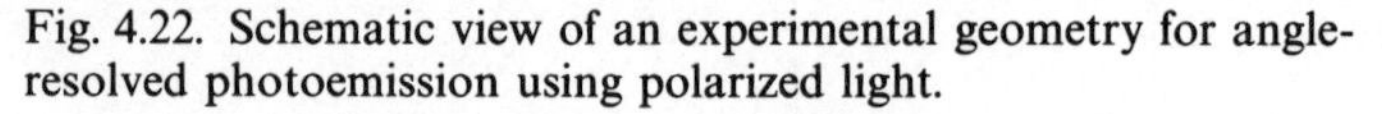

$$\mathscr{H}^{\mathrm{ext}} = \frac{e}{2mc}[\mathbf{A}(\mathbf{r})\cdot\mathbf{p} + \mathbf{p}\cdot\mathbf{A}(\mathbf{r})], \tag{4.35}$$

Fig. 4.22. Schematic view of an experimental geometry for angle-resolved photoemission using polarized light.

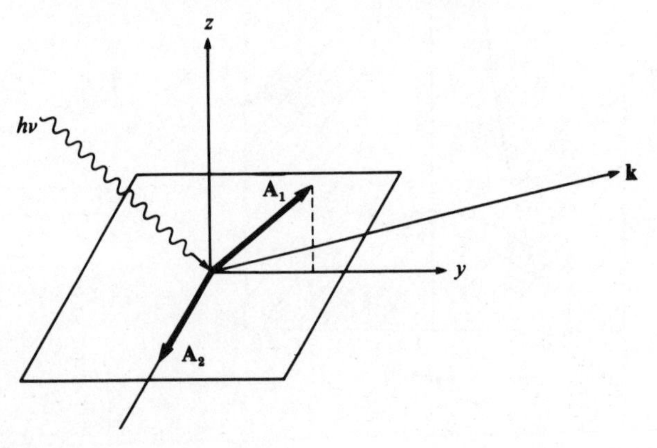

where $\mathbf{A}(\mathbf{r})$ is the external vector potential that characterizes the incident radiation field and $\mathbf{p} = -ih(\hat{x}\partial/\partial x + \hat{y}\partial/\partial y + \hat{z}\partial/\partial z)$ is the momentum operator. $\mathbf{A}$ may be taken as a constant in the ultraviolet part of the electromagnetic spectrum (10–100 eV) since the photon wavelength (> 100 Å) is large compared to atomic dimensions. The matrix element in (4.34) determines both the intensity of the photo-cross section and the symmetry of the states probed (Scheffler, Kambe & Forstmann, 1978).

Fig. 4.22 illustrates a typical experimental geometry and depicts two possible orientations of the photon $\mathbf{A}$ vector for polarized radiation incident in the y–z plane. If this plane is a *mirror* plane of the surface we can classify all the electronic states as either even or odd with respect to reflection in the plane. In particular, if the electron analyzer is situated in this plane, the final spherical plane wave continuum state of the ejected electron must be *even* when $x \to -x$ (otherwise the wave function has a node at the position of the detector and the intensity vanishes). Since the total matrix element, $\mathbf{A}\cdot\langle f|\mathbf{p}|i\rangle$, must have even symmetry, the orientation

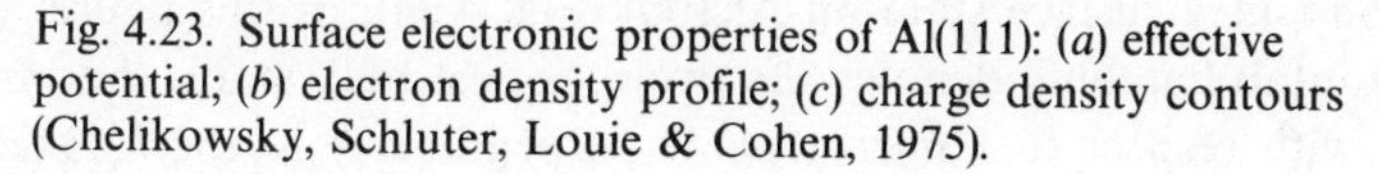
Fig. 4.23. Surface electronic properties of Al(111): (*a*) effective potential; (*b*) electron density profile; (*c*) charge density contours (Chelikowsky, Schluter, Louie & Cohen, 1975).

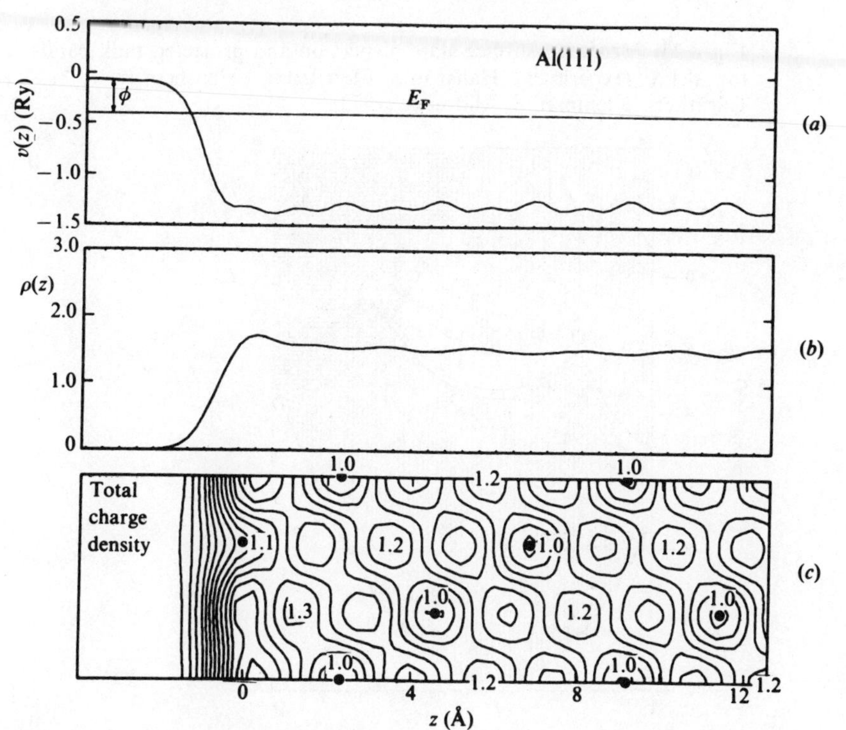

of **A** determines which initial states are excited. For the two cases of Fig. 4.22:

$$\mathbf{A}_1 \parallel \text{mirror plane} \rightarrow \langle f|\partial/\partial y|i\rangle \text{ or } \langle f|\partial/\partial z|i\rangle \text{ must be non-zero} \rightarrow |i\rangle \text{ is even,}$$

$$\mathbf{A}_2 \perp \text{mirror plane} \rightarrow \langle f|\partial/\partial x|i\rangle \text{ must be non-zero} \rightarrow |i\rangle \text{ is odd.}$$

The great value in the ability to pick out initial states of well-defined symmetry will become obvious in a number of examples later in this book.

Metals

Self-consistent surface band structure calculations confirm the picture of simple metal surfaces that emerges from model calculations. The effective potential and electronic charge density obtained for a 12-layer slab appropriate to Al(111) bear a remarkable likeness to the jellium results (Fig. 4.23). Contours of constant charge density clearly demonstrate the charge 'smoothing' effect described earlier. Angle-resolved UPS measurements and LDA calculations are in excellent agreement for the dispersion of a Shockley surface state on Al(100) (Fig. 4.24). In this figure, vertical (horizontal) hatching denotes the projection of bulk states that are even (odd) with respect to a mirror plane perpendicular to the k_y axis along

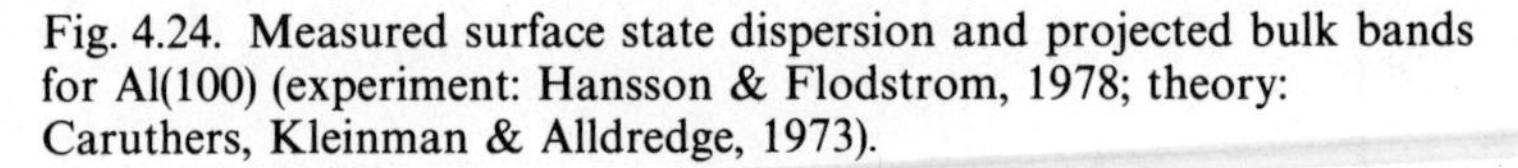

Fig. 4.24. Measured surface state dispersion and projected bulk bands for Al(100) (experiment: Hansson & Flodstrom, 1978; theory: Caruthers, Kleinman & Alldredge, 1973).

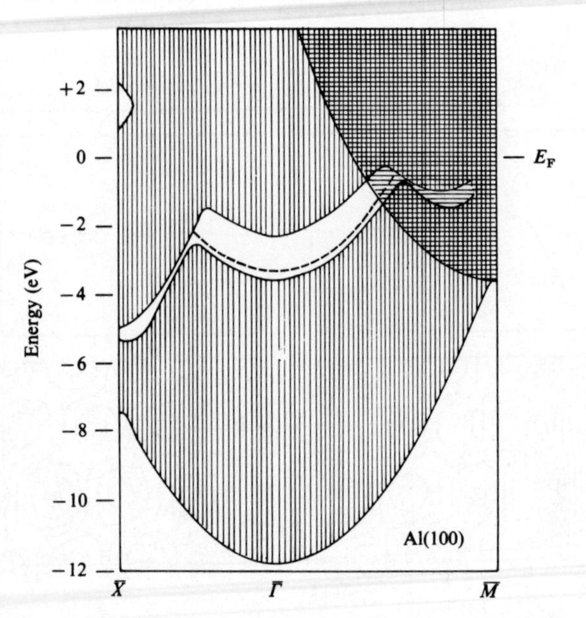

$\bar{\Gamma}\bar{M}$. The surface state (dashed curve) is observed to have even symmetry so it cannot mix with odd symmetry bulk states. The state remains sharp when *odd* symmetry bulk states close the absolute gap along $\bar{\Gamma}\bar{M}$ but transforms to a diffuse surface resonance when even symmetry states close the gap along $\bar{\Gamma}\bar{X}$.

The electronic properties of transition metals are dominated by narrow conduction bands formed from the overlap of fairly localized d-orbitals. Consider an atom at the surface of a corrugated open surface where the coordination number is much less than a bulk site. In that case, a d-level derived surface state retains much of its atomic character and the level remains near the center of the band. This feature shows up in a particularly

Fig. 4.25. Slab calculation of layer-resolved LDOS for W(100). Surface states 'fill in' the bulk LDOS (Posternak, Krakauer, Freeman & Koelling, 1980).

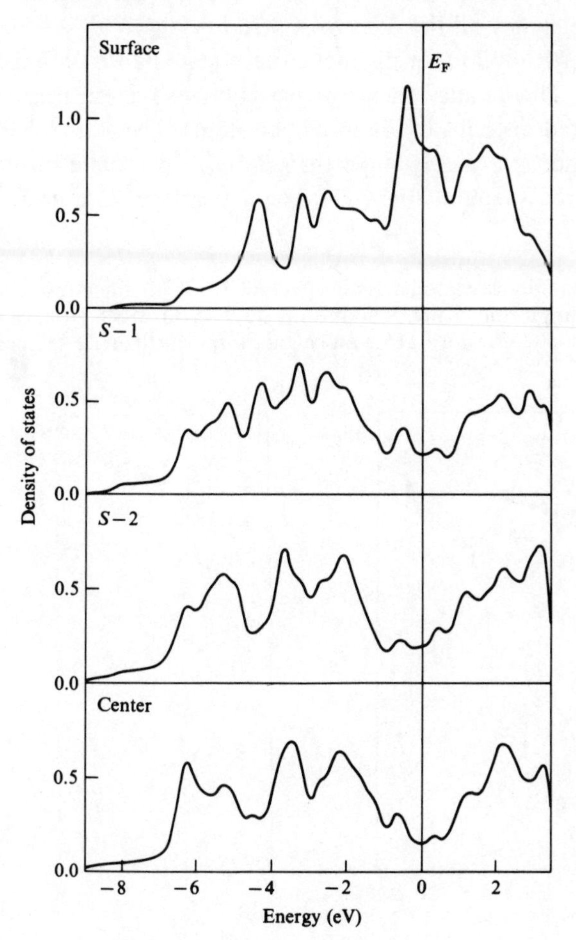

striking way for body-centered cubic metals because the bulk density of states of that structure has a deep minimum right at the band center (Fig. 4.25). The ubiquitous band narrowing at the surface also is evident from this figure.

The physics of W(100) is an excellent example of the richness and complexity of transition metal surface phenomena. For example, the 1×1 structure used in the calculation of Fig. 4.25 does not persist down to low temperature: a reversible phase transition occurs that reconstructs the surface to a ($\sqrt{2} \times \sqrt{2} - R45°$) structure (see Chapter 5). Inquiry into the driving force for this reconstruction has focussed considerable attention on the electronic structure of the room temperature phase. It turns out that the surface Brillouin zone is littered with surface states (Fig. 4.26(*a*)). No simple models reproduce this sort of behavior. Even the most sophisticated, fully relativistic, electronic structure calculations fail to account completely for all the data (Fig. 4.26(*b*)).

The experiments reveal four distinct surface state bands within 5 eV of the Fermi level. The detailed theory suggests how some of these features can be interpreted in terms of the model concepts. The state labelled (*A*) at the SBZ center is a nearly unperturbed $5d_{3z^2-r^2}$ atomic orbital that sticks out into the vacuum (Fig. 4.27). The (*S*) surface state near E_F is a

Fig. 4.26. Surface state bands for W(100) 1×1: (*a*) angle-resolved UPS data (Campuzano, King, Somerton & Inglesfield, 1980; Holmes & Gustafsson, 1981); (*b*) LDA slab calculation (Mattheiss & Hamann, 1984).

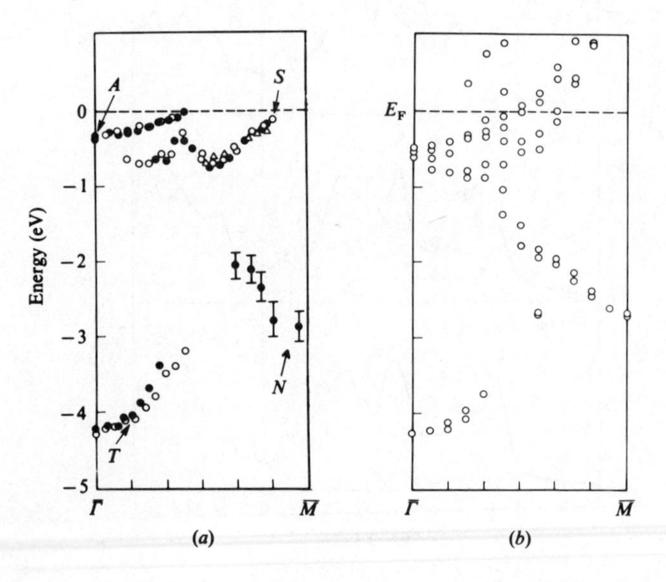

Shockley state whose wave function character derives from the bulk states at the projected band edges (not shown in Fig. 4.26(*b*)). The state labelled (*T*) is akin to a tight-binding Tamm state in the sense that its existence is very sensitive to the potential perturbation at the surface. The (*N*) state near the SBZ boundary defies any simple classification.

A very interesting bit of transition metal surface physics appears in the rare-earth row of the periodic table. The electronic configuration of the corresponding free atoms is [Xe]$4f^n6s^2$. The integer n reflects the fact that the number of electrons in the tightly bound 4f shell increases as the atomic number increases. In the condensed phase, one non-bonding f-electron is transferred to the 5d band (the gain in cohesive energy is greater than the loss of Hund's rule energy), i.e., the elemental solids are trivalent transition

Fig. 4.27. Charge density contours of the (*A*) surface state on W(100) (Posternak, Krakauer, Freeman & Koelling, 1980).

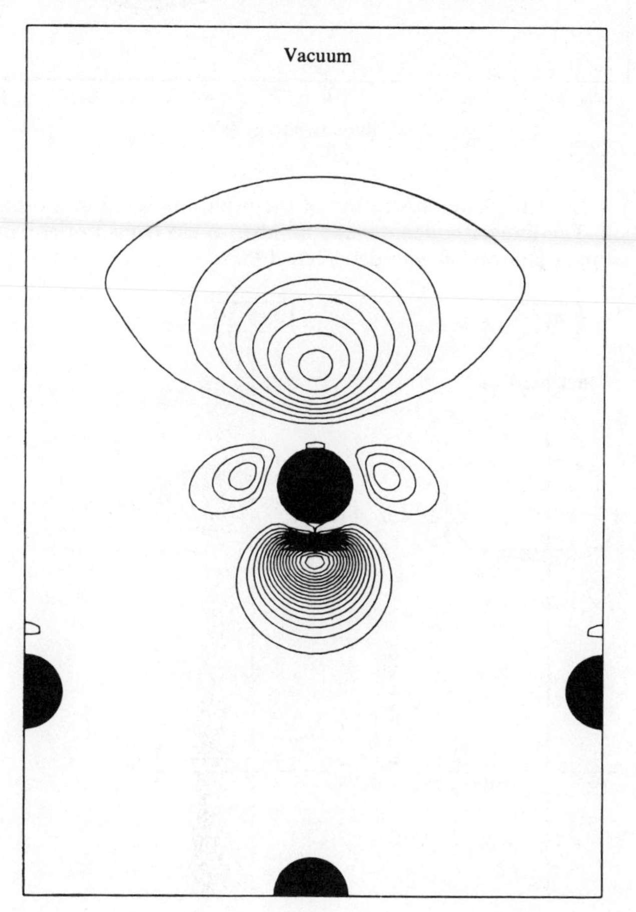

Fig. 4.28. Photoemission EDC's from the f-shell of samarium metal normalized at $E_B = 5\,\text{eV}$ (Gerken *et al.*, 1985). The UPS spectrum (dots) is more surface sensitive than the XPS results (crosses).

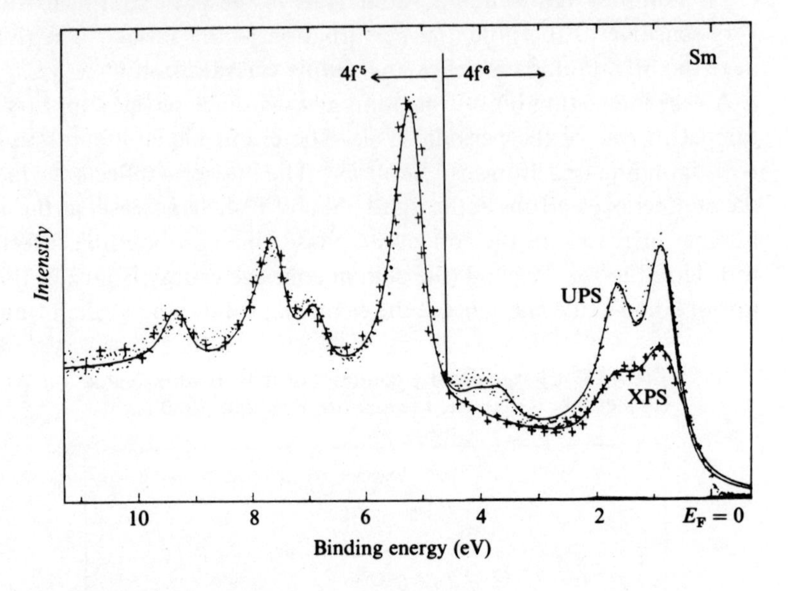

Fig. 4.29. Schematic illustration of the origin of SCLS in a d-band metal. The integer m denotes the number of electrons in the band. (Eastman, Himpsel & van der Veen, 1982).

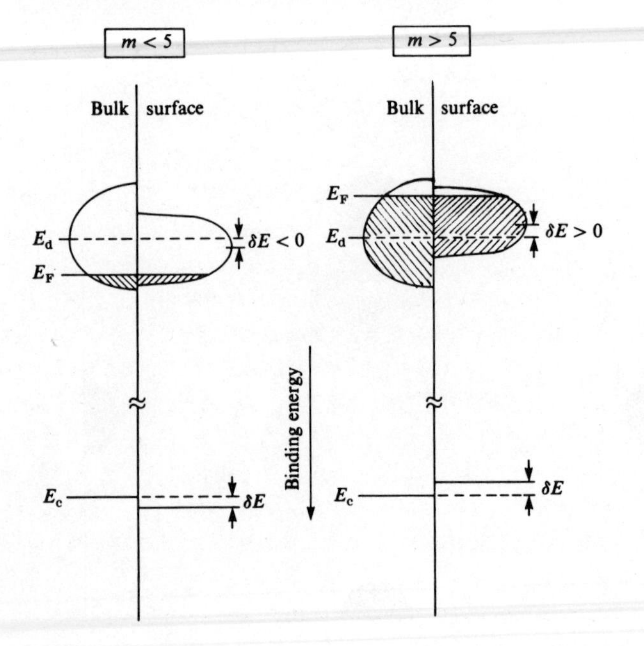

metals.* However, the reduced coordination of the surface atoms renders the gain in cohesive energy much less. If Hund's rule prevails for the last layer of atoms the result is a trivalent bulk metal with a *divalent* surface. Spectroscopic evidence reveals this to be the case for samarium (Fig. 4.28). XPS and UPS both reveal two multiplet–split spectral features in the electron energy distributions – one characteristic of emission from a $4f^5$ trivalent state and one characteristic of a $4f^6$ divalent state. The relatively stronger divalent signal from UPS ($h\nu = 100\,\mathrm{eV}$) indicates a divalent surface. It is interesting to think about the atomic arrangement at this surface. A divalent Sm ion is 12.5% larger in radius than a trivalent Sm ion. How do the big divalent ions sit atop the smaller trivalent ions? The answer is not known.

The d-band narrowing at the surface of all transition metals has an interesting consequence. Charge must flow between the surface atoms and the bulk so that the composite system maintains a common Fermi level. In the simplest picture, the direction of electron transfer is determined by the relative position of the bulk Fermi level and the surface narrowed band center (Fig. 4.29). The change in the amount of charge in the valence orbitals of the surface atoms produces a different electrostatic potential at a surface site than at a bulk site (see also Chapter 12). Consequently, the deep core energy levels of the surface atoms rigidly shift down (up) relative to their bulk values if the bulk band is less (more) than half-filled. Experiments on the transition metals show that the magnitude of this surface core level shift (SCLS) is roughly proportional to the atom's deviation from a half-filled d-shell (Citrin & Wertheim, 1983).

Alloys

The phenomenon of surface core level shifts is not merely a curiosity in the electronic properties of surfaces. The following digression will show that SCLS measurements directly address some fundamental thermodynamic and elemental composition issues at the surface of metallic *alloys*. A bulk binary (AB) alloy is a two-component thermodynamic system for which the surface composition need not be identical to the bulk composition. Four concentration variables ($x_A^{\mathrm{b}}, x_A^{\mathrm{s}}, x_B^{\mathrm{b}}$ and x_B^{s}) characterize the system which has N^{b} total bulk sites and N^{s} total surface sites. By definition, the system is at equilibrium when the free energy is a minimum with respect to small variations in these variables, $\delta F\{x_A^{\mathrm{b}}, x_A^{\mathrm{s}}, x_B^{\mathrm{b}}, x_B^{\mathrm{s}}\} = 0$, subject to the constraint that the total number of A and B atoms is fixed. Explicitly performing the required variations we easily find

* Europium and ytterbium remain divalent in the condensed phase.

that the constrained minimum condition is equivalent to:

$$N^{\mathrm{b}} \frac{\partial F}{\partial x_A^{\mathrm{s}}} = N^{\mathrm{s}} \frac{\partial F}{\partial x_A^{\mathrm{b}}} \quad \text{and} \quad N^{\mathrm{b}} \frac{\partial F}{\partial x_B^{\mathrm{s}}} = N^{\mathrm{s}} \frac{\partial F}{\partial x_B^{\mathrm{b}}}. \tag{4.36}$$

Now consider a completely different calculation. Compute the change in free energy that accompanies the *exchange* of a surface B atom with a bulk A atom:

$$\Delta F = F\left\{\frac{N_A^{\mathrm{s}}+1}{N^{\mathrm{s}}}, \frac{N_A^{\mathrm{b}}-1}{N^{\mathrm{b}}}, \frac{N_B^{\mathrm{s}}-1}{N^{\mathrm{s}}}, \frac{N_B^{\mathrm{b}}+1}{N^{\mathrm{b}}}\right\} - F\{x_A^{\mathrm{s}}, x_A^{\mathrm{b}}, x_B^{\mathrm{s}}, x_B^{\mathrm{b}}\}. \tag{4.37}$$

Expanding the first term to first order yields

$$\Delta F = \left[\frac{\partial F}{\partial x_A^{\mathrm{s}}} \frac{1}{N^{\mathrm{s}}} - \frac{\partial F}{\partial x_A^{\mathrm{b}}} \frac{1}{N^{\mathrm{b}}}\right] + \left[\frac{\partial F}{\partial x_B^{\mathrm{b}}} \frac{1}{N^{\mathrm{b}}} - \frac{\partial F}{\partial x_B^{\mathrm{s}}} \frac{1}{N^{\mathrm{s}}}\right]. \tag{4.38}$$

Therefore, combining (4.36) with (4.38) we find that the free energy of interchange, $\Delta F = \Delta U - T\Delta S$, *vanishes* when the alloy is in thermodynamic equilibrium. To use this result, split the total entropy of interchange into two pieces, an entropy of mixing term and a remainder term (ΔS_0):

$$\Delta S = -k\Delta \sum_{i=1}^{4} N_i \ln x_i + \Delta S_0. \tag{4.39}$$

Substituting (4.39) into the equilibrium condition, $\Delta U = T\Delta S$, immediately leads to a relation that specifies the surface composition of the alloy in terms of the bulk composition:

$$\frac{x_A^{\mathrm{s}}}{x_B^{\mathrm{s}}} = \frac{x_A^{\mathrm{b}}}{x_B^{\mathrm{b}}} \mathrm{e}^{-\Delta U/kT} \mathrm{e}^{+\Delta S_0/kT}. \tag{4.40}$$

The enrichment of one alloy component relative to its bulk concentration is known as *surface segregation.*

What does all this have to do with surface core level shifts? The key point is that the electronic contribution to the *heat of segregation* (ΔU as defined above) can be related directly to the SCLS for an alloy composed of adjacent elements in the periodic table. To see this, consider the response of a metal with atomic number Z to a photoemission event that creates a core hole. Before any Auger decay occurs, conduction band charge rapidly flows into the affected Wigner–Seitz cell to screen the positive charge of the hole. The combination of unit positive charge deep within the core and unit negative charge in a low lying band orbital creates an effective $Z + 1$ 'impurity' atom in the Z-electron host solid (Rosengren & Johansson, 1981). Now, suppose a photoemission experiment is used to measure the energy position of both surface and bulk core levels in the Z-electron metal. Using the impurity atom approximation, the difference in energy between the two *final* states, i.e., the SCLS, is precisely the interchange energy, ΔU, of a binary alloy composed of Z and $Z+1$ atoms (Fig. 4.30)!

The trends in SCLS described earlier and the XPS impurity atom argument suggest that elements near the beginning (end) of a transition metal row will segregate to the surface of alloys formed with elements immediately to their right (left). In the thermodynamic language of Chapter 1, this situation would arise if the *surface tension* of the segregating species were lower than its alloy partner. The experimental surface tension data shown in Fig. 1.4 clearly illustrate this trend across a given transition metal row. Furthermore, a simple generalization suggests a semi-empirical rule to determine the segregating species in an alloy formed from transition metals that are not adjacent in the periodic table: the constituent with lower elemental surface tension enriches the alloy surface (Miedema, 1978). This rule will be valid when other contributions to ΔU, such as elastic strains due to atomic radius mismatch, do not dominate the physics.

As an example, the argument above predicts that gold will segregate to the surface of a NiAu alloy. A graphic illustration of this phenomenon is evident in the ion kinetic energy distributions from a low energy ion scattering study of the (100) surface of a Ni–1.0% Au single crystal (Fig. 4.31). The experiment is performed for two different azimuthal angles of incidence. In the first, ion scattering from the first layer completely shadows the second layer. When the crystal is rotated for the second

Fig. 4.30. The XPS surface core level shift approach to the heat of segregation of a binary alloy (Egelhoff, 1983).

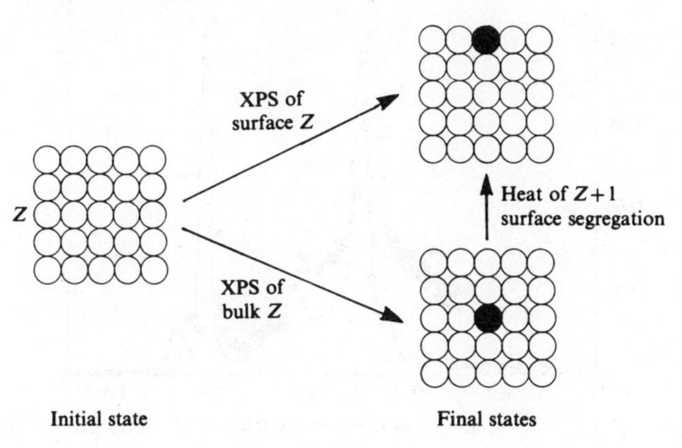

measurement, both the first and second atomic layers are visible to the beam. The LEIS data clearly indicate that more than 90% of the surface atoms are gold whereas the second layer gold concentration differs little from the bulk value.

The *electronic* structure of a binary alloy surface is complicated by the fact that a wide range of stoichiometric order is possible. A completely segregated material will have a single component at the surface whereas

Fig. 4.31. Kinetic energy distributions for 5 keV neon ions scattered from Ni–1%Au(100): (*a*) only top layer visible to beam; (*b*) first and second layers visible to beam. The solid curves are computer simulations of the expected yield (Buck, Stensgaard, Wheatley & Marchut, 1980).

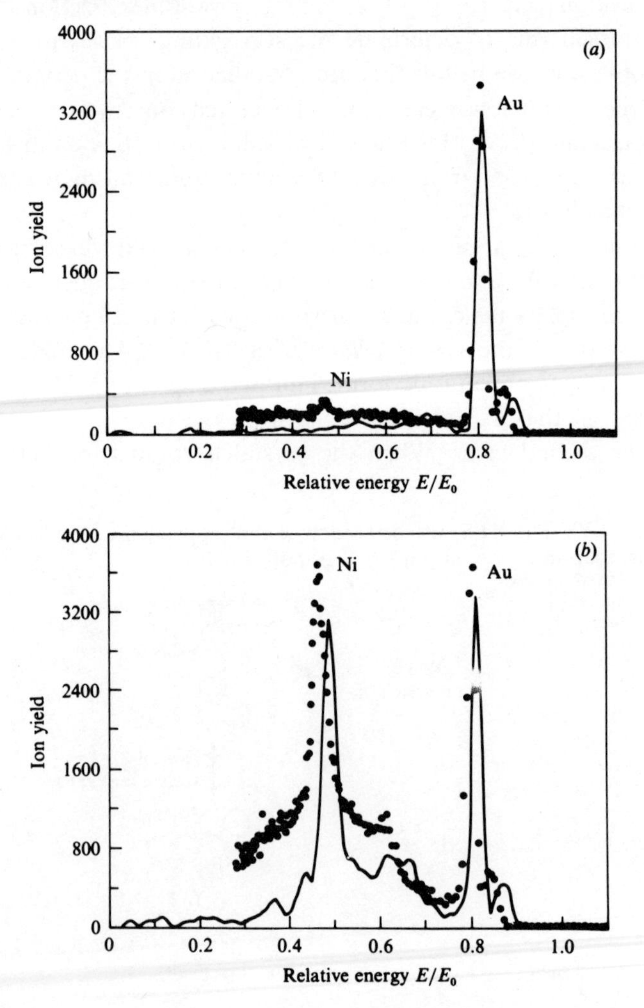

a solid solution in the top layer exhibits a random mixture of A and B atoms distributed on a fixed two-dimensional lattice. The proper treatment of short range clustering and long range ordering in an alloy is a formidable problem – even in the bulk (Faulkner, 1982). At the surface, model calculations reveal some general features for a semi-infinite, substitutional, random alloy. Suppose that each of the alloy constituents possesses a single atomic level. If the two atomic levels, E_A and E_B, are close together, the surface LDOS displays a simple narrowing similar to the case of single component metals. However, for sufficiently large separation of the pristine

Fig. 4.32. Surface (dashed curve) and bulk (solid curve) LDOS for a random substitutional alloy $A_{0.5}B_{0.5}$ for two choices of the separation between constituent atomic levels (Desjonqueres & Cyrot-Lackmann, 1977).

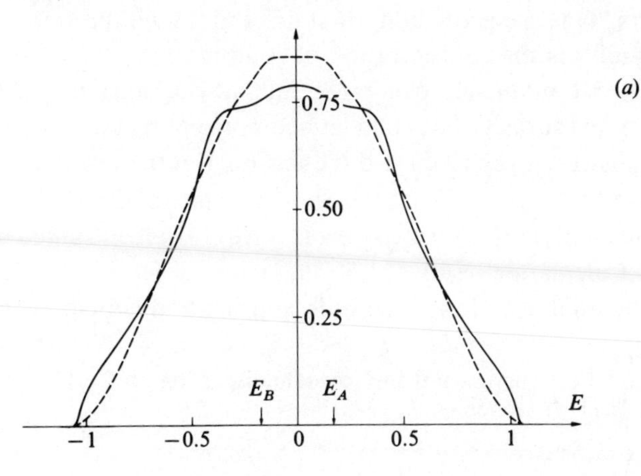

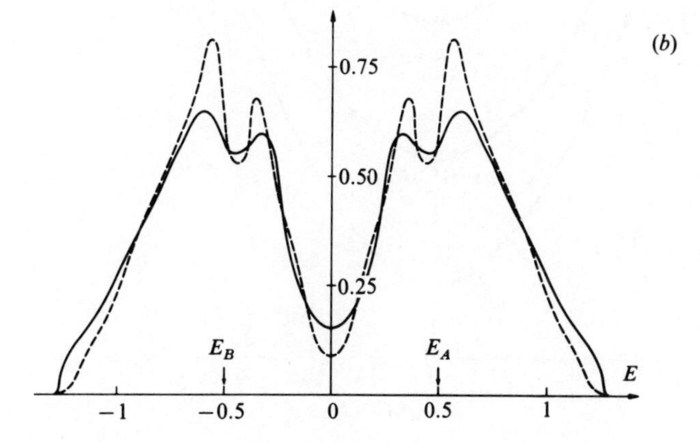

atomic levels, a so-called 'split-band' occurs and the surface LDOS shows considerable structure (Fig. 4.32).

Random alloys contain many small cluster units. For example, an A atom may be totally surrounded by B atoms or vice versa. In isolation, these cage-like structural entities produce sharp 'molecular' levels in the energy spectrum. In the solid, the sharp levels are smeared out by fluctuations in the random potential but some remnant structure remains. This structure is more pronounced at a surface because the reduced coordination of the surface atoms *increases* the probability that such caged atoms will occur. For example, the coordination of a surface atom is reduced by one (relative to the bulk) for a simple cubic lattice so that

$$\begin{aligned} P_{\text{bulk}}[A(6B)] &= x(1-x)^6 \cong 0.8\% \\ P_{\text{surf}}[A(5B)] &= x(1-x)^5 \cong 1.6\%, \end{aligned} \tag{4.41}$$

where $P[A(nB)]$ is the probability that an A atom will be surrounded by nB atoms and x is the concentration of A atoms.

Does the random one-electron potential that characterizes a disordered alloy destroy the surface states of a one-component metal? It appears that no general answer can be given to this question. For the case of Cu–10% Al, angle-resolved UPS reveals that the Shockley state of Cu(111) persists for two different stable surface structures of CuAl(111), albeit shifted in energy (Fig. 4.33). Calculations show that the main effect of the aluminum impurity potential is to lower the bulk s–p band of copper. The edge of

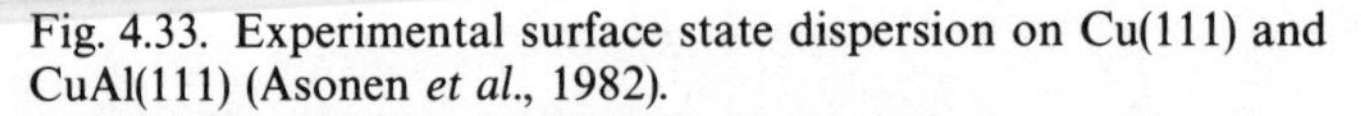

Fig. 4.33. Experimental surface state dispersion on Cu(111) and CuAl(111) (Asonen *et al.*, 1982).

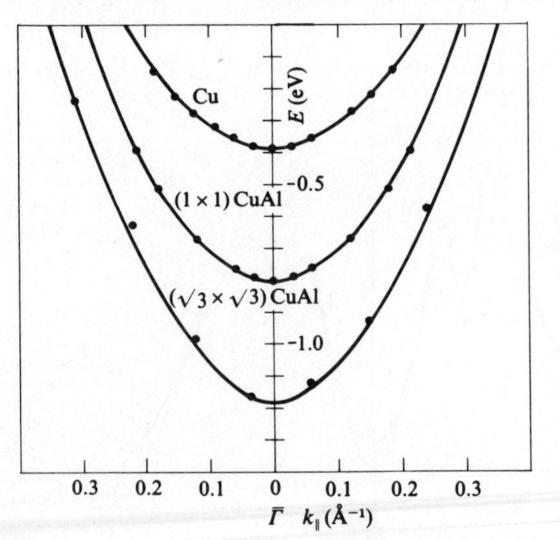

the projected bulk continuum (cf. Fig. 4.17) drops accordingly and the surface state (which is synthesized from these states) tracks this effect. The dispersion of the CuAl 1×1 surface state is very well reproduced by a slab calculation that uses an *ab initio* tight-binding scheme based on the local density approximation (Bullett, 1982).

Semiconductors

The general principles that determine the electronic structure of semiconductor surfaces derive from the familiar notions of local chemical bonding (Cotton & Wilkinson, 1962). In the bulk, significant covalent bond strength arises if *hybrid bond orbitals* are formed from linear combinations of low-lying atomic s- and p-orbitals. The highly directional sp^3 hybrid bonds that result determine the diamond and zincblende crystal structures of most common semiconductors. Overlapping hybrid orbitals on neighboring tetrahedrally coordinated sites produce bonding and anti-bonding levels which ultimately broaden into the semiconductor valence and conduction band, respectively (Fig. 4.34).

An infinite two-dimensional plane that slices through the bulk along a periodic array of tight-binding hybrid orbitals forms an *ideal* semiconductor surface. We have already indicated (Chapter 3) that these surfaces normally reconstruct to more complicated atomic arrangements. Nevertheless, a study of the electronic properties of such ideal surfaces is not unwarranted because it will provide a clue to the origin of the driving force for reconstruction. For example, the three low-index faces of an unreconstructed diamond lattice reveal a striking diversity of geometric and electronic structure (Fig. 4.35).

Fig. 4.34. Successive steps in the formation of the band structure of a semiconductor (Harrison, 1980).

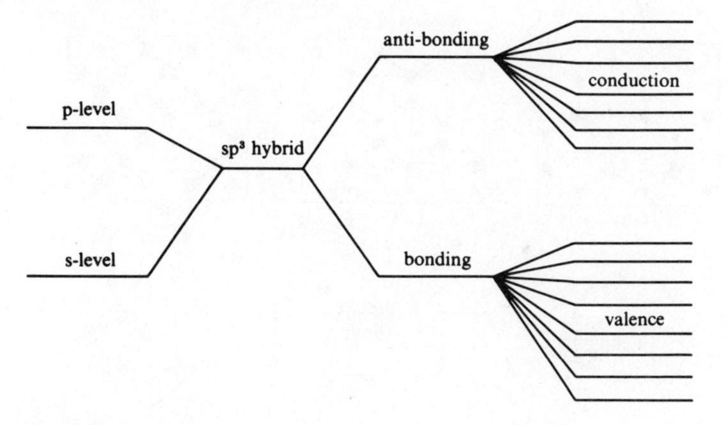

The ideal surfaces of the diamond lattice each expose hybrid orbitals that 'dangle' into the vacuum. Every such orbital is half-occupied if we imagine that the two bonding electrons/orbital of the bulk are shared between the two half-crystals formed by cleavage. The (111) surface contains one dangling hybrid per surface unit cell. The two atoms/cell of the (110) surface each dangle a single hybrid orbital whereas the two dangling hybrids in the (100) surface cell attach to a single atom. It is apparent from inspection of Fig. (4.35(*a*)) that the areal *density* of dangling hybrids is lowest for the (111) surface and highest for (100). According to the arguments of Chapter 1 we expect the surface tension of the (111) face to be lowest. Indeed, this is the natural cleavage plane of both silicon and germanium.

Fig. 4.35. Crystallography of a homopolar semiconductor: (*a*) edge view that illustrates the ideal termination of three low-index faces (Harrison, 1980); (*b*) top view – decreasing atom size indicates increasing distance from the surface. Dashes outline the surface unit mesh; (*c*) corresponding ideal surface Brillouin zone with conventional labelling (Ivanov, Mazur & Pollmann, 1980).

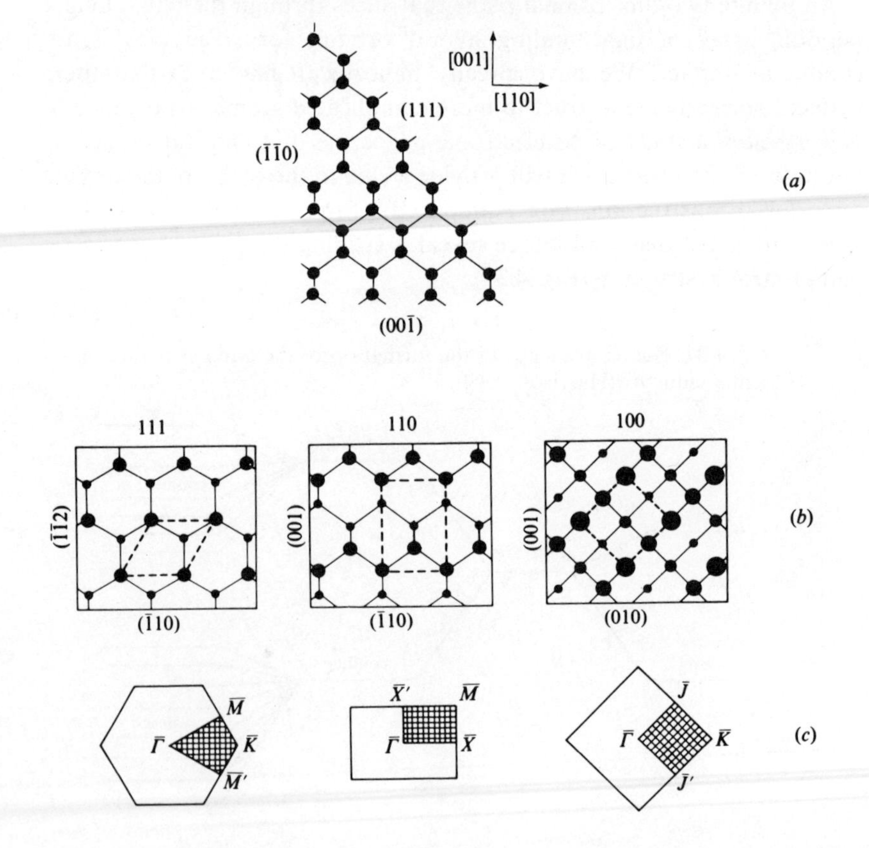

When the surface bonds are broken, hybrid bonding and anti- bonding states revert to the original single atom hybrid energy position. Since this surface localized level lies in the fundamental gap (cf. Fig. 4.34) it qualifies as a *bona fide* surface state. Empirical tight-binding calculations for the ideal surfaces of silicon verify the intuition that interactions among these

Fig. 4.36. Projected bulk band structures (cross-hatched), surface states (solid curves) and resonances (dashed curves) for three ideal silicon surfaces (Ivanov, Mazur & Pollmann, 1980; Casula, Ossicini & Selloni, 1979).

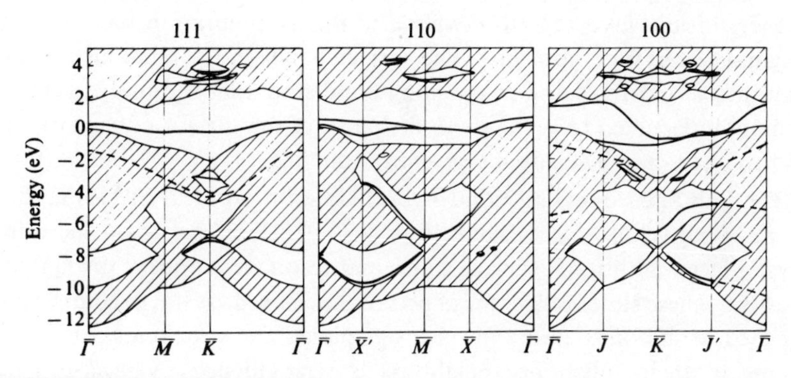

Fig. 4.37. Charge density contours of individual surface states on ideal Si(111) viewed along [110]: (*a*) dangling bond state in the fundamental gap at the SBZ center, Γ; (*b*) back-bond surface resonance at Γ̄; (*c*) low-lying back-bond surface state at *K*. Solid lines connect the ideal atom positions. The vacuum is at the top in all panels (Schluter, Chelikowsky, Louie & Cohen, 1975).

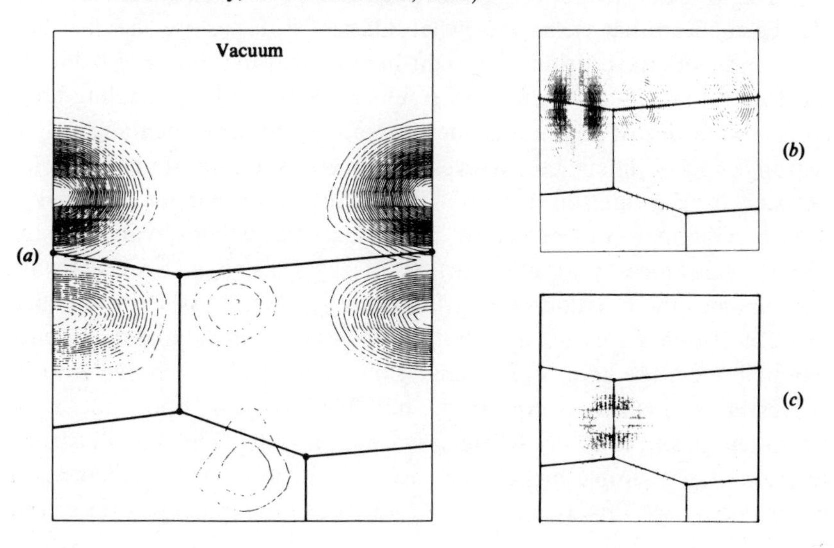

states on neighboring surface sites yield a dispersive energy band (Fig. 4.36). In fact, two bands appear in the fundamental gap for (110) and (100) since these surfaces have two hybrids/surface cell.

As noted above, every dangling surface hybrid has only one electron to contribute to the surface state occupancy. On the ideal (110) surface, each atom contributes its electron to the lowest of the two surface bands. Since this only half fills the band, the surface is *metallic*. The upper band is empty. For (100), each surface atom has two electrons to contribute to the surface bands. However, with *two* tetrahedral neighbors missing, there is no advantage to the original sp^3 hybridization of the bulk orbitals. It is energetically favorable to *dehybridize* this configuration back into its original s- and p-components. The surface bands in the fundamental gap split into a (mostly) s and p_z dangling bond state that is doubly occupied and an unoccupied *bridge* state in the plane of the surface of primarily p_x and p_y character. The surface is semiconducting.

The half-filled dangling bond hybrid state on an ideal Si(111) surface points directly into the vacuum (Fig. 4.37(*a*)). However, as on the other crystal faces, additional surface states and resonances appear throughout the SBZ. These states reflect the efforts of the cleaved crystal to compensate for the loss of bonds across the cleavage plane. The attachment of surface atoms to their subsurface neighbors is strengthened by *back-bonding* surface states. For example, a back-bond surface resonance is localized between the first and second atomic layers whereas a true surface state back-bonds the second and third atomic layers (Fig. 4.37(*b*) and (*c*)). Deeper lying surface states further strengthen this compensatory bonding.

The charge density contours illustrated in Fig. 4.37 come from an LDA slab calculation that yields somewhat different surface state energies than the bands obtained from the tight-binding calculation (Fig. 4.36). In particular, the *self-consistent* LDA results place the Si(111) dangling bond surface state almost precisely at the midpoint of the fundamental gap. The reason for this difference involves a delicate, yet important, interplay between surface states and the self-consistent surface barrier potential.

In our discussion of metals we noted that charge conservation at the surface guarantees that a distortion of the bulk band states always accompanies the creation of a surface state (cf. Fig. 4.13). Here, a deficit of 1/2 electronic state each from both the conduction band and the valence band balances the gain of one band gap surface state. If charge is to be conserved, one electron/atom in the half-filled dangling bond state must be 'stolen' from the newly standing wave bulk states. Fig. 4.38 illustrates the results of a simple model calculation that shows that this balancing occurs via a *layer-by-layer* cancellation between the midgap surface state wave function charge density and the bulk valence band deficit charge density.

Suppose that the surface state did not lie at midgap. From the nearly-free electron model we know that a shift in energy of the surface state involves a phase shift of the corresponding wave function. Since the bulk states are unaffected by this shift, Fig. 4.38 shows that a phase-shifted surface state wave function creates an electrostatic dipole. Depending upon the sign of the dipole, the surface potential barrier is either raised or lowered. However, the surface state energy itself then must change in order to assure wave function matching across the vacuum interface. A self-consistent positive feedback situation ensues which drives the surface barrier and surface state energy to the midgap condition of Fig. 4.38. The entire surface region is metallic since the Fermi level of an intrinsic semiconductor also lies at midgap.

A particularly interesting situation arises if one dopes a semiconductor

Fig. 4.38. Layer-resolved charge density of a midgap surface state and the corresponding band deficit charge density (Appelbaum & Hamann, 1974).

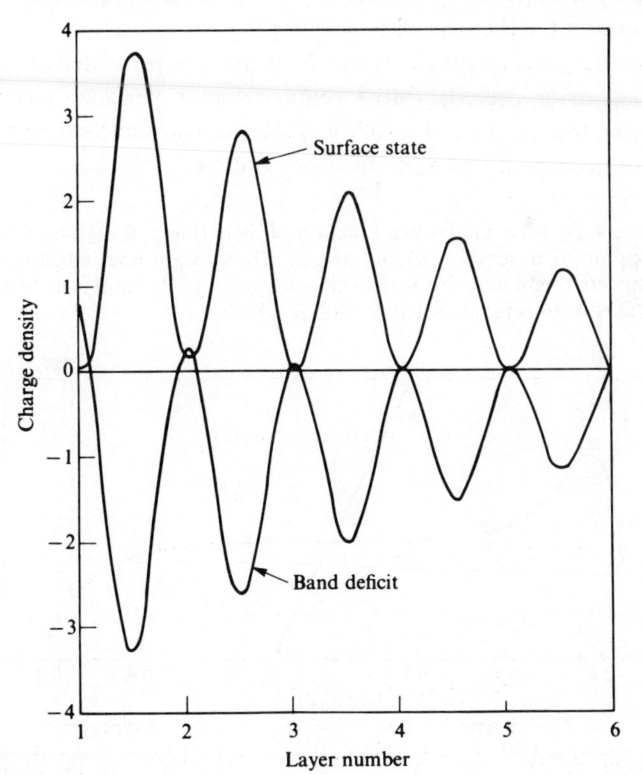

with impurities. In the bulk, it is possible to move the Fermi level from just above the valence band maximum to just below the conduction band minimum (Many, Goldstein & Grover, 1965). The measured work function tracks the variation of E_F. The change that occurs in this situation when surface states are present first was addressed by Bardeen (1947).

Consider the case of n-type doping so the Fermi level is raised above midgap to the position of the bulk impurity level. Electrons from impurity atoms in the near surface region of the crystal flow into the mid-gap surface state to lower their energy. A positively charged *space charge layer* is left behind that extends hundreds of Ångströms into the bulk because the bulk doping density of states/layer parallel to the surface ($\sim 10^8$–10^{12}/cm^2) is much smaller than the intrinsic surface density of dangling bonds ($\sim 10^{15}$/cm^2). This charge flow creates an electrostatic dipole layer that retards the motion of bulk electrons towards the surface. Further, the magnitude of this dipole is such that the electrostatic energy cost during traversal almost exactly cancels the chemical potential reduction due to n-type doping. Therefore, the energy to extract a bulk electron, the work function, is *nearly independent* of the position of the bulk Fermi level – a phenomenon known as *Fermi level pinning* (Fig. 4.39). An analogous situation occurs for the case of p-type doping.

Unfortunately, the relatively straightforward view presented above must be regarded as a prelude. Most semiconductor surfaces reconstruct. Furthermore, the nature and stability of the reconstructed surfaces is very sensitive to preparation conditions. For example,

Fig. 4.39. Measured work function of Si(111) as a function of bulk doping by acceptors (A) and donors (D). Straight line indicates the expected behavior (from the shift of E_F due to doping) if no surface states were present (Allen & Gobeli, 1962).

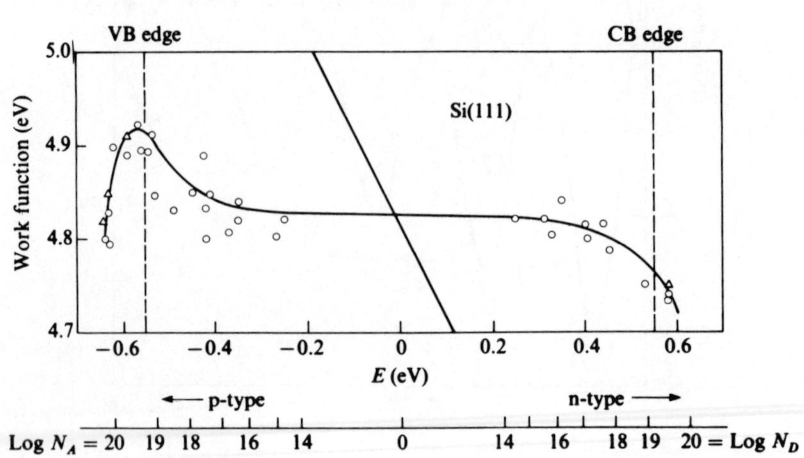

> The cleaved surface of Si(111) exhibits a 2×1 metastable reconstruction which, upon annealing at 380 °C, transforms irreversibly into a 7×7 structure. The 7×7 structure can also be obtained from a chemically polished and sputtered surface annealed at about 1100 °C. Upon annealing the cleaved surface at 900 °C, an unstable 1×1 structure appears which can be quenched or impurity-stabilized to room temperature. Laser annealing of the (111) surface also produces a 1×1 structure. (Kahn, 1983.)

It is fair to state that there is very little understanding of this extraordinarily complicated behavior (Monch, 1979). In the present context, we mostly limit ourselves to a general view of how simple chemical and electronic considerations can destabilize some structures and stabilize others.

Consider again the low-index faces of a diamond lattice at $T = 0$. We saw that the (111) and (110) ideal structures contribute one electron/surface atom to a half-filled dangling bond surface state. However, an energetically equivalent arrangement would place *two* electrons on some sites and leave others empty. According to Jahn & Teller (1937), this type of electronically degenerate situation is unstable with respect to lattice distortions that lower the symmetry and remove the degeneracy (Sturge, 1967). Quite generally, the ensuing distortions tend to *lower* the energy of occupied surface states and *raise* the energy of unoccupied surface states. This action clearly lowers the energy of the entire system and, in some cases, sweeps the surface states completely out of the fundamental gap.

Orbital rehybridization is an important mechanism in covalent semiconductor reconstruction. In the simplest scheme, the surface 'buckles' as atoms alternately rise above and sink below the surface plane. This motion dehybridizes the sp^3 hybrid back toward its s- and p-components. The outward atom fills its deep s-state with two electrons and pyramidally bonds to its neighbors with p-orbitals. The dangling p-state on the inward atom is empty but the atom bonds to its three in-plane nearest neighbors via sp^2 hybrids. The resulting structure no longer is degenerate electronically; it has 1×1 symmetry for the (110) surface and 2×1 symmetry for the (111) surface. Recall that the (100) surface was capable of dehybridizing the sp^3 hybrid to a non-degenerate configuration even in the ideal structure.

We should be chastened to learn that *none* of the structure predictions of the previous paragraph are observed for silicon or germanium surfaces. For example, Si(110) reconstructs to (unknown) 4×5 and 5×1 structures. Si(100) subtly bends pairs of adjacent dangling bond surface atoms toward one another into asymmetric dimers. This motion lowers the energy of the occupied dangling bond surface band still further. There does exist a

2 × 1 structure of Si(111), but this reconstruction is apparently more remarkable than the simple ion motions suggested above.

The Si(111) 2 × 1 surface provides an excellent example of the interplay between structure and electronic properties and experiment and theory in surface physics. The 'buckled' 2 × 1 model must be rejected because its predicted surface state dispersion and energy gap are incompatible with photoemission and optical experiments. Any alternative model both must agree with spectroscopic data and make good chemical sense. Consider the ideal (111) termination in more detail (Fig. 4.40(*a*)). The very weak dispersion of the mid-gap surface state occurs because the dangling bonds interact at *second* neighbor distances. However, if the bond topology of the surface is changed by a shear distortion of the top two layers of atoms, the dangling bond orbitals reside on *nearest* neighbor atoms (Fig. 4.40(*b*)). The zig-zag chain of adjacent p_z orbitals then can π-bond as in organic materials. The energy of this structure is quite low and the occupied and unoccupied surface states are simply the bonding and anti-bonding π states. Predictions based on this π-bonded chain model of Si(111) 2 × 1 agree very well with many experiments, e.g., angle-resolved photoemission (Fig. 4.41).

What about the Si(111) 7 × 7 structure? This very complex reconstruction solves the problem of high energy dangling bonds even more efficiently than the 2 × 1 structure – it simply gets rid of them. There is one dangling

Fig. 4.40. Top and side view of Si(111): (*a*) ideal surface; (*b*) π-bonded chain model. Dashes outline the surface unit cells and shaded circles identify the surface atoms that 'dangle' $3p_z$ orbitals into the vacuum (Pandey, 1981).

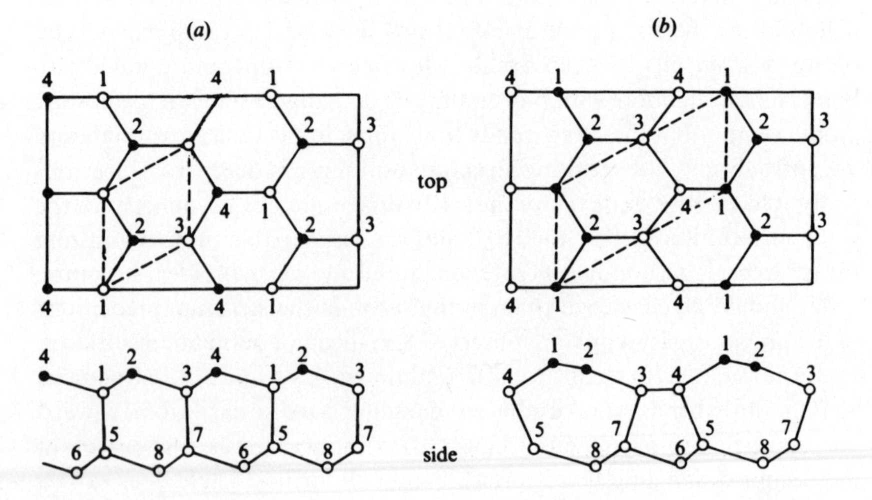

bond on each of the 49 top-layer atoms in an ideally terminated 7×7 superlattice (cf. Fig. 4.35). The purpose of a stacking fault in one triangular subunit of the new unit cell is obvious from Fig. 3.20. Seven atoms of the ideal surface simply vanish along the row where the faulted subunit is matched to the unfaulted subunit. Hence, there are only 42 atoms (and dangling bonds) in what is now the *second* layer of the reconstructed surface. Moreover, one can trade three dangling bonds for one by permitting single silicon atoms (ejected from the original top layer or migrating from elsewhere) to 'adsorb' and bond tetrahedrally to three second-layer atoms each. The 12 top-layer adatoms exhibited by the 7×7 silicon surface reduces the total number of dangling bonds to 18.* The energy gained by halving the number of dangling bonds more than offsets the energy cost required to form a stacking fault.

A number of electronic states have been identified at the surface of this reconstructed silicon surface by photoemission spectroscopy. However, a recent development in scanning tunnelling microscopy now permits one to obtain energy-resolved real-space images of these states. The idea is

Fig. 4.41. Experimental angle-resolved UPS surface state dispersion compared to LDA slab results for the projected bulk bands (cross-hatched), surface states (solid curves) and surface resonances (dashed curves) of π-bonded Si(111) (Northrup & Cohen, 1982).

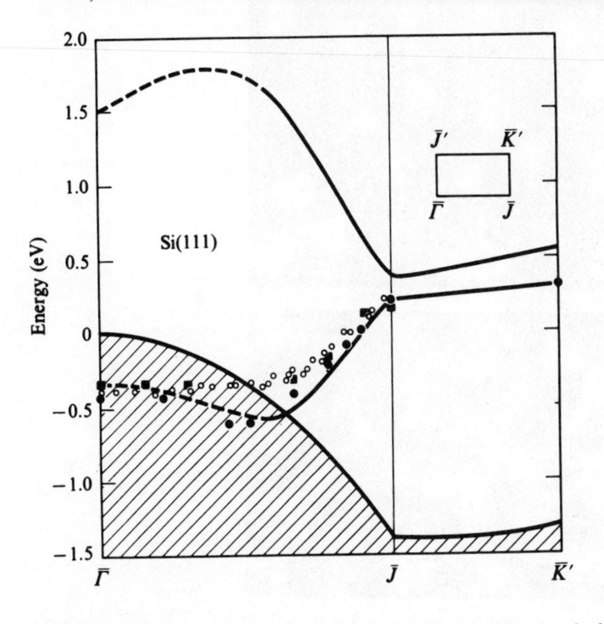

* There actually is one more dangling bond/cell on the atom at the bottom of the deep hole at the unit cell apex.

Fig. 4.42. Scanning tunnelling microscopic images of the topography (top panel) and three electronic surface states of a Si(111) 7 × 7 surface. See text for discussion (Hamers, Tromp & Demuth, 1986).

simple. As we have seen (see Chapter 3), STM topographs are obtained by varying the applied voltage in order to maintain a constant current flow between the microscope tip and the sample. The current couples principally to bulk states that terminate at the surface. However, if the bias voltage happens to match the energy of an occupied (or unoccupied for reverse bias) electronic state localized at the surface we should expect an increase in current whose lateral variations reflect the spatial distribution of the corresponding wave function.

The Si(111) 7×7 images exhibited in Fig. 4.42 were obtained in this scanning tunnelling spectroscopy mode. The top panel shows a conventional topograph similar to that of Fig. 3.23 that provides a top view of the basic unit cell. The succeeding panels display spatial variations in the tunnel current (brighter areas denote greater current flow) for different bias voltages. The second frame corresponds to a surface state 0.35 eV below E_F. This state is localized on the 12 adatoms but exhibits a distinct asymmetry between the faulted and unfaulted portions of the unit cell. The third image shows a state 0.8 eV below the Fermi level that arises from dangling bonds of the six second-layer atoms (see Fig. 3.20(*a*)) that are not directly bonded to adatoms. Note also the dangling bonds at the bottom of the deep 'corner' holes. Finally, a deep (-1.8 eV) 'back-bond' state is imaged in the fourth panel that probably corresponds to $3p_x$ and $3p_y$ orbitals of the adatoms bonded to $3p_z$ orbitals of the atoms directly below them.

The bonding in compound semiconductors is intermediate between covalent and ionic. Mostly covalent materials such as GaAs and InSb crystallize into the zincblende structure where cations and anions are arranged alternately on a diamond lattice. More ionic compounds like CdTe and ZnO form the wurtzite structure – an alternative tetrahedrally bonded lattice. These compound structures possess both *polar* and *non-polar* surfaces. For example, the (110) zincblende surface contains an equal number of cations and anions, i.e., it is electrically neutral. By contrast, the polar (111) and (100) surfaces expose a complete plane of either cations or anions depending on cleavage. These surfaces are nominally charged. In addition, the presence of two distinct chemical species and attendant charge transfer between them guarantees that the local screened potential at the cation site, v_C, differs from its counterpart at the anion site, v_A. In the bond orbital language, each species separately forms sp^3 hybrids which then overlap to form the bulk band structure.

GaAs will serve as a prototype of a weakly ionic semiconductor. In an average sense, we expect the surface electronic structure of GaAs to resemble that of Ge. For the ideal surfaces, this does not differ too much from the results for Si (Fig. 4.36). The precise position of the surface states for the polar surfaces depends upon which species terminates the crystal. For the non-polar (110) surface, the principal effect of the ionic part of the potential, $v_C - v_A$, is to completely *split* apart the two sp^3 hybrid orbital dangling bond surface states in the fundamental gap. The lower band is filled and a semiconducting surface results. We expect that the empty Ga-localized surface state will result in Fermi level pinning for doped GaAs. However, *no* Fermi level pinning is observed for this surface! Again, a reconstruction occurs.

The GaAs (110) surface is not unstable towards a Jahn–Teller distortion. However, sp^3 hybridization is not chemically favorable for a Group III element like gallium. Instead, a trigonal sp^2 bonding configuration would be more appropriate. Consequently, the orbital rehybridization reconstruction scheme advanced earlier for silicon would be favored in this case. LEED studies indicate that this is precisely what occurs. The predicted outward motion of the arsenic atoms and the inward motion of the gallium atoms occur by a rotation of the surface chain of atoms (Fig. 4.43). In this way, the structure avoids the large energy cost associated with the stretching or compression of a covalent bond.

Fig. 4.44 compares angle-resolved UPS data for GaAs (110) with the results of a self-consistent LDA slab calculation for the relaxed 1 × 1 geometry. An almost completely dehybridized arsenic 4s-like surface state

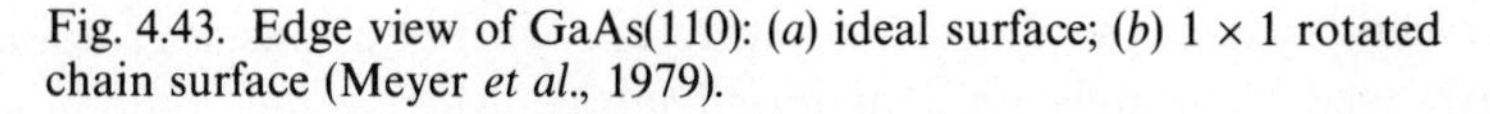

Fig. 4.43. Edge view of GaAs(110): (*a*) ideal surface; (*b*) 1 × 1 rotated chain surface (Meyer *et al.*, 1979).

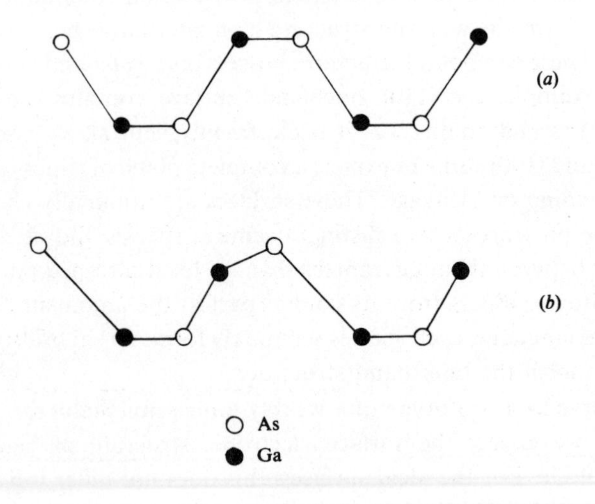

Fig. 4.44. Projected bands and angle-resolved UPS data for GaAs(110). Solid, dashed and dotted curves denote theoretical surface states, surface resonances and experimental points, respectively. The anion states A_1 and A_3 (not discussed in the text) are localized on the second layer of arsenic atoms (Chelikowsky & Cohen, 1979).

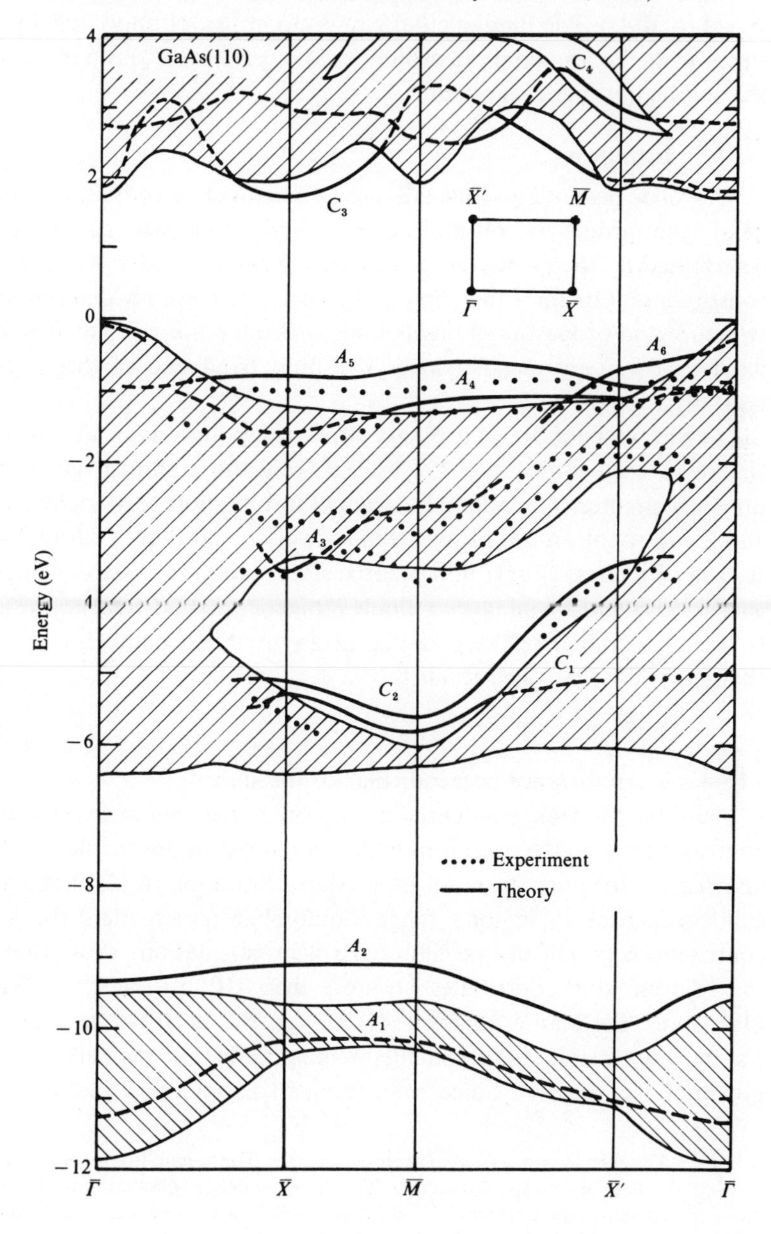

(A_2) appears deep in the valence band gap.* Surface bonding occurs through the Ga-localized sp^2 hybrids near -6 eV (C_1 and C_2). The remaining sp^2 hybrid is anti-bonding (C_3) while p-like back-bonds to the raised surface arsenic atoms appear just below the valence band maximum (A_4, A_5 and A_6). The fundamental gap is swept free of unoccupied surface states since the $3p_z$ dangling orbital band of gallium (C_4) is pushed above the conduction band minimum.

Insulators

The insulating solids fall into two distinct categories. On the one hand, the properties of molecular crystals and rare gas solids are determined by the characteristics of their weakly van der Waals bonded constituents. On the other hand, classical electrostatics determines the structure and properties of highly ionic rocksalt materials like most alkali halides and many metal oxides. The bulk band gap of these systems typically exceeds 4 eV.

The simplest model of a binary ionic compound associates a formal charge to each lattice site. Let us elaborate the argument given in the previous chapter. We follow the original formulation of Madelung (1918) and focus attention on a single thin slab of crystal that is parallel to the surface of interest. For a non-polar surface, every bulk slab of thickness a (~ 1 Å) is neutral and has no dipole moment perpendicular to the plane. It is easy to calculate the potential at any *external* point due to all the charges in this slab. The result is

$$V(z) \sim e^{-2\pi z/a}, \tag{4.42}$$

where z is the distance perpendicular to the slab.

The total Madelung potential at *any* site is the sum of two pieces: the contribution from the other ions in the slab that contains the site of interest and the contribution from all other slabs. Equation (4.42) shows that a subtle cancellation of long range Coulomb forces renders the second contribution practically negligible. Explicit calculations show that even the adjacent slab contributes no more than 10% to the total (Watson, Davenport, Perlman & Sham, 1981). In fact, the electrostatic potential at a surface site is only slightly smaller in magnitude than the bulk Madelung potential (see below). Since ionic bonds are compressible, any small

* The energetics work out in the following way. The atomic 4s level of arsenic lies at -17 eV with respect to vacuum. The *electron affinity* of GaAs is about 4 eV (i.e., the bottom of the conduction band lies 4 eV below the vacuum level) and the band gap is about 2 eV. Hence, the atomic level lies about 11 eV below the valence band maximum.

rebalancing of Coulomb forces at the surface requires only minor relaxations. For this reason, LEED generally finds an ideal crystal termination for non-polar ionic surfaces.

The situation changes at a polar surface. Fig. 4.45(*a*) and (*b*) illustrates a polar termination of the wurtzite and zincblende lattices. Each microscopic slab parallel to the surface now carries a net charge. It is reasonable to model the electrostatics of these structures by a stack of parallel plate capacitors (Fig. 4.45(*c*)). The electric field within the stack is a constant:

$$E = 4\pi\sigma \frac{a}{d+a}. \tag{4.43}$$

Obviously, this field completely destabilizes the crystal since the cation and anion planes are forced in opposite directions. However, the field can be cancelled* if the surface slabs are partly neutralized by an areal charge density, $\sigma' = \sigma(a/d + a)$.

How does nature arrange matters so that the surface layer of a polar crystal has a different charge than the corresponding bulk layers? The simplest artifice is a reconstruction. According to Fig. 4.45, a stable (0001) wurtzite surface or (111) zincblende surface requires a compensating charge of magnitude $\sigma' = \sigma/4$. One approach is simply to remove one fourth of the surface atoms. Indeed, this is precisely what we saw for the indium-terminated surface of InSb(111) (cf. Fig. 3.12). This solution is particularly elegant since the remaining indium atoms relax inward to satisfy their

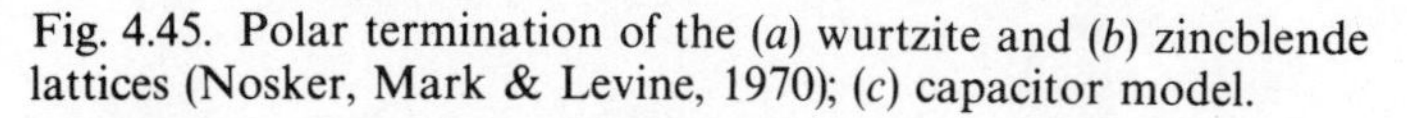

Fig. 4.45. Polar termination of the (*a*) wurtzite and (*b*) zincblende lattices (Nosker, Mark & Levine, 1970); (*c*) capacitor model.

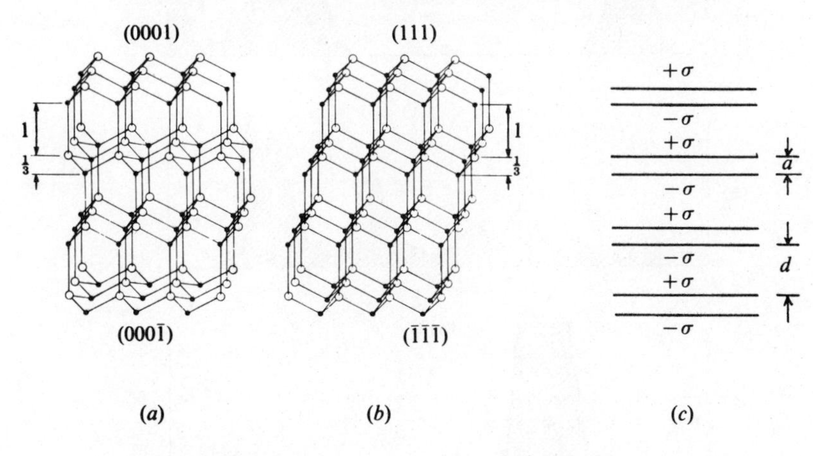

* The cancellation occurs to within a constant of the order of unity – the bulk Madelung constant.

covalency requirements by rehybridizing to an sp^2 bonding configuration. For more ionic species, a polar surface spontaneously *facets* to expose non-polar faces of low surface energy.

At an unreconstructed polar surface, charge neutralization can occur by a change in the local electronic structure that tends to deplete charge at one face and accumulate charge at the opposing face. Self-consistent LDA calculations for a four-layer model suggest that this mechanism may be operative for MgO(111) (Fig. 4.46). The LDOS of the two interior layers are bulk-like whereas both the cation (111) and anion $(\bar{1}\bar{1}\bar{1})$ terminated layers become *metallic*. Unfortunately, this prediction has not been tested to date – bulk charging of the sample severely distorts photoelectron spectra.

The probable electronic structure of a *non-polar* ionic surface can be inferred without sophisticated calculations. Consider the energy levels of a cation and anion as a function of their relative separation (Fig. 4.47). Begin at infinite separation. It costs energy I to ionize a typical cation and we gain back an energy A when this charge is transferred to the anion. The two atoms are neutral at large separations because $I-A$ generally is positive. However, as a crystal forms, anions surround cations and vice versa. The electrostatic potential is positive at the negative anion site and

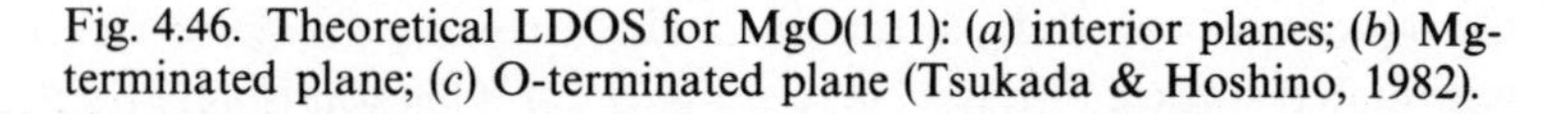

Fig. 4.46. Theoretical LDOS for MgO(111): (*a*) interior planes; (*b*) Mg-terminated plane; (*c*) O-terminated plane (Tsukada & Hoshino, 1982).

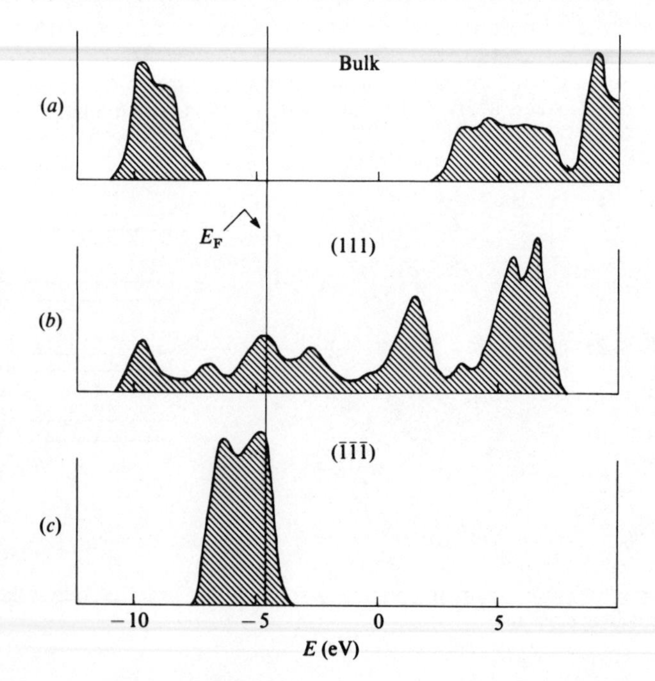

negative at the positive cation site. At sufficiently small separation, the electrostatic energy shifts reverse the ordering of I and A. At this point it is energetically favorable to form an ionic solid. The bulk band gap is determined by the magnitude of the Madelung potential at a typical lattice point.

As we have seen, the more nearly atomic surface sites experience a slightly reduced Madelung potential. The energy levels on these atoms follow the dashed path of Fig. 4.47 and appear as surface states split off into the fundamental gap from the bulk band edges. These are Tamm states since they derive from a perturbation of the bulk potential in the surface region. It is important to realize that there is a one-to-one correspondence between these states and the cation and anion localized dangling bond states at the surface of covalent compound semiconductors. The dangling bond states smoothly evolve into Tamm states as the ionicity of the compound increases.

Perhaps the most salient point to raise concerning ionic insulator surfaces is that they are rarely perfect. Defects and vacancies are very common and it is difficult to prepare a stoichiometric surface. General experience suggests that localized electronic states will be associated with such defects. Unfortunately, unambiguous experimental evidence for their existence is difficult to obtain. Indeed, it is difficult to probe the surface electronic structure of insulators by any means. As we have said, conventional photoelectron spectroscopy is of limited value due to the ubiquitous charging problem. Nonetheless, some information has been obtained by study of the distribution of energy *lost* by an electron backscattered by a single crystal insulator (Fig. 4.48). This energy loss spectrum is similar to that of Fig. 2.2 except that here we focus on relatively small energies compared to Auger processes. Notice that a feature grows

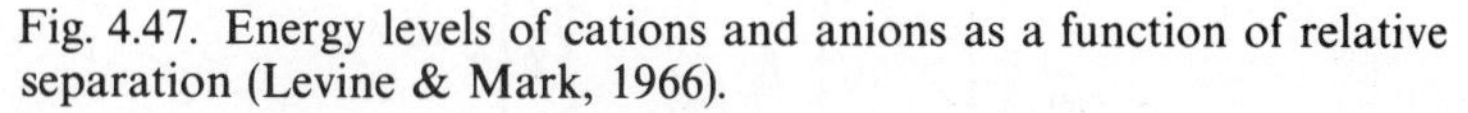
Fig. 4.47. Energy levels of cations and anions as a function of relative separation (Levine & Mark, 1966).

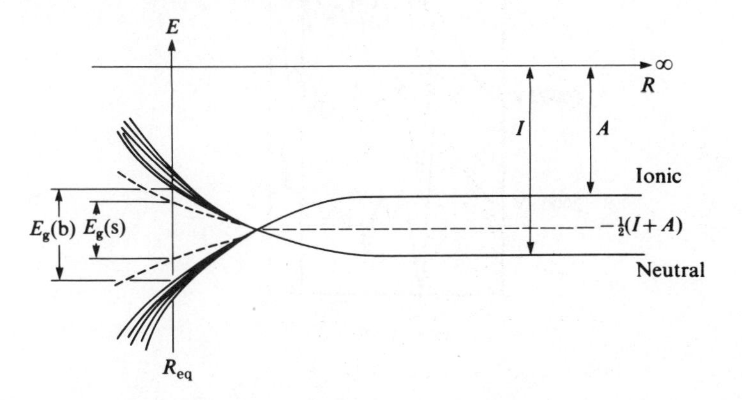

into the derivative spectrum of MgO(100) as the primary electron energy E_p approaches the minimum of the universal curve of mean free path. Since the bulk band gap of MgO is 7.8 eV, this structure probably corresponds to excitations into a Tamm state split off below the conduction band.

Finally, let us return to the surface properties of van der Waals bonded insulators, specifically the rare gas solids. For present purposes, the most important electronic property of these materials is illustrated by the effective medium curve for helium in Fig. 4.7. The positive immersion energy reflects the fact that a rare gas atom *repels* an external electron whenever their wave functions overlap. However, at greater distances, an electron is *attracted* to a solid (of dielectric constant ε) by the long range, classical image force,

$$V(z) = -\frac{\varepsilon - 1}{\varepsilon + 1}\frac{e^2}{4z}. \tag{4.44}$$

Fig. 4.48. Electron energy loss spectra from non-polar MgO(100) as a function of incident electron energy. The vertical line indicates the position of a surface sensitive feature (Henrich, Dresselhaus & Zeiger, 1980).

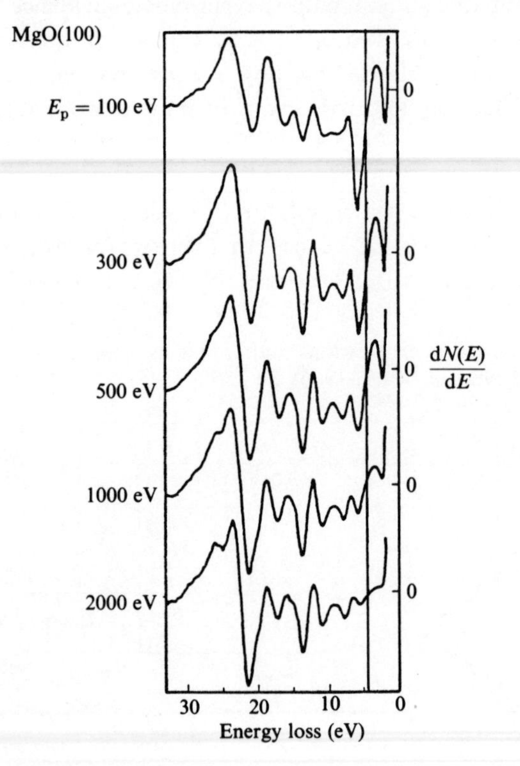

Therefore, a well is formed by a one-dimensional Coulomb potential (normal to the surface) terminated by a steep repulsive barrier near the crystal edge. This well can support bound states (the so-called 'image' surface states) that normally are unoccupied.* However, *excess* electrons that one manages to trap into these states move as free electrons along the plane of the surface. In fact, the surface of a rare-gas crystal charged in this manner is a physical realization of a *two-dimensional electron gas*. In the next chapter, we will explore the phase diagram of this system – a subject of considerable theoretical and experimental interest.

General references

The jellium model

Lang, N.D. (1973). The Density Functional Formalism and the Electronic Structure of Metal Surfaces. In *Solid State Physics* (eds. Seitz, Turnbull & Ehrenreich), vol. 28, pp. 225–300. New York: Academic.

One-dimensional band theory

Garcia–Moliner, F. & Flores, F. (1979). *Introduction to the Theory of Solid Surfaces*, Chapter 4. Cambridge: Cambridge University Press.

Three-dimensional theory

Forstmann, F. (1978). Electron States at Clean Surfaces. In *Photoemission and the Electronic Properties of Surfaces* (eds. Feuerbacher, Fitton & Willis), pp. 193–226. Chichester: Wiley.

Photoelectron spectroscopy

Plummer, E.W. & Eberhardt, W. (1982). Angle-Resolved Photoemission as a Tool for the Study of Surfaces. In *Advances in Chemical Physics* (eds. Prigogine & Rice), vol. 49, pp. 533–656. New York: Wiley.

Metals

Inglesfield, J.E. (1982). Surface Electronic Structure. *Reports on Progress in Physics* **45**, 223–84.

Alloys

Wille, L.T. & Durham, P.J. (1985). Surface States on Random Substitutional Alloys. *Surface Science* **164**, 19–42.

Semiconductors

Lieske, N.P. (1984). The Electronic Structure of Semiconductor Surfaces. *Journal of Physics and Chemistry of Solids* **45**, 821–70.

Insulators

Henrich, V.E. (1985). The Surfaces of Metal Oxides. *Reports on Progress in Physics* **48**, 1481–541.

* Image states occur on metal surfaces as well. Their energy levels lie between the Fermi level and the vacuum level (see, e.g., Hulbert *et al.*, 1985).

5

PHASE TRANSITIONS

Introduction

The study of phase transitions plays a central role in modern condensed matter physics. Changes of phase often are very dramatic events and certainly one wants a good understanding of such transformations. However, the stature of this field derives mostly from the recognition that the fundamental concepts, language and methodology developed to attack the phase transition problem have far-reaching utility in other areas of physics. In this chapter, we take advantage of the successes in this branch of statistical physics as a part of a two-pronged program. On the one hand, we apply the phenomenological methods of the modern theory (which focus on notions such as symmetry and order) to highlight those aspects of surface phase transitions that do not depend on the details of the system. On the other hand, a few specific examples are examined in more depth to illustrate that an appreciation of these details can significantly deepen our understanding of surface processes. We begin with a brief review.

Phase transitions occur because all systems in thermodynamic equilibrium seek to minimize their free energy, $F = U - TS$. One phase will supplant another at a given temperature because different states (e.g., liquid/vapor, magnetic/non-magnetic, cubic/tetragonal) partition their free energy between the internal energy $U(T)$ and the entropy $S(T)$ in different ways. It is useful to characterize competing phases in terms of a so-called *order parameter*. Consider two such phases for simplicity. By construction, the order parameter has a non-zero value in one phase (usually the low temperature/low symmetry state) and vanishes in the other (high temperature/high symmetry) phase. For the familiar cases of a liquid–vapor phase transition and a ferromagnet, appropriate order parameters are the density difference between the liquid and the vapor and the homogeneous magnetization, respectively. For a structural phase transition, the order parameter might be the amplitude of a specific phonon mode.

The behavior of the order parameter near the transition temperature, T_c, distinguishes two rather different transformation scenarios. A discontinuous change in the order parameter occurs at a *first-order* transition. In this case, two independent free energy curves simply cross one another. The system abruptly changes from one distinct equilibrium phase to a second distinct equilibrium phase. First-order transitions exhibit the familiar phenomena of phase coexistence and nucleation and growth. By contrast, two competing phases become indistinguishable at T_c for a *continuous* phase transition. Here, the order parameter rises smoothly from zero as the temperature is lowered although there are large fluctuations in its value around the mean. One typically finds that the order parameter at a continuous transition varies as $(T-T_c)^\beta$ for T very near T_c. Moreover, we now understand that the numerical value of the *critical exponent* β (and a few other related exponents) only depends on a few physical properties, e.g., the symmetry of the system, the dimensionality of the order parameter (scalar, vector, etc.) and the dimensionality of space. This property is called *universality* and suggests that interesting things can happen at a surface – the effective dimensionality is two rather than three.

Reconstruction

Solid surfaces undergo a wide variety of reconstructive phase transitions as a function of temperature.* Unfortunately, there are very few cases where one can confidently display a structural phase diagram as one would do for the analogous bulk problem. The problem is twofold. First, it is difficult to perform the necessary surface crystallography to establish the true structures (see Chapter 3). Second, many surface phases are actually metastable, i.e., the surface is not in true thermodynamic equilibrium. There is a simple reason for this. The cleavage process only liberates a fixed amount of (ruptured bond) energy and this may not be enough to move the surface atoms around to the configuration of lowest free energy. It is easy to get 'hung-up' in a metastable state. Significant thermal annealing may be needed to find the true equilibrium state. This explains why one often finds discussions of a surface phase diagram that more nearly resemble a processing history (cf. the discussion of Si(111) in the previous chapter).

The (100) surface of iridium (a 5d transition metal) undergoes a

* See Chapter 11 for a discussion of adsorbate-induced reconstructions.

reconstructive $1 \times 1 \rightarrow 1 \times 5$ transformation at temperatures in excess of 800 K. This appears to be a good example of a first-order metastable-to-stable phase transition. The metastable 1×1 structure exhibits a typical metal surface structure: an ideal lattice termination with top layer (at least) relaxation. The ground state 1×5 structure actually is best described as a close-packed atomic arrangement sitting atop an ideal face-centered cubic (100) substrate (Fig. 5.1). It is quite plausible that the energy barrier between these two arrangements arises as rigid atomic rows shift over subsurface atoms as the transition proceeds. This scenario is based on measurements from a state-of-the-art LEED video study that permits 20 ms real time resolution of the reconstruction (Heinz, Schmidt, Hammer & Muller, 1985). The time and temperature dependence of the growth of

Fig. 5.1. Atomic geometry of the Ir(100) $1 \times 1 \rightarrow 1 \times 5$ phase transition; (*a*) the ideal 1×1 surface; (*b*) possible 'intermediate' structure; (*c*) the reconstructed 1×5 quasi-hexagonal structure (Heinz, Schmidt, Hammer & Muller, 1985).

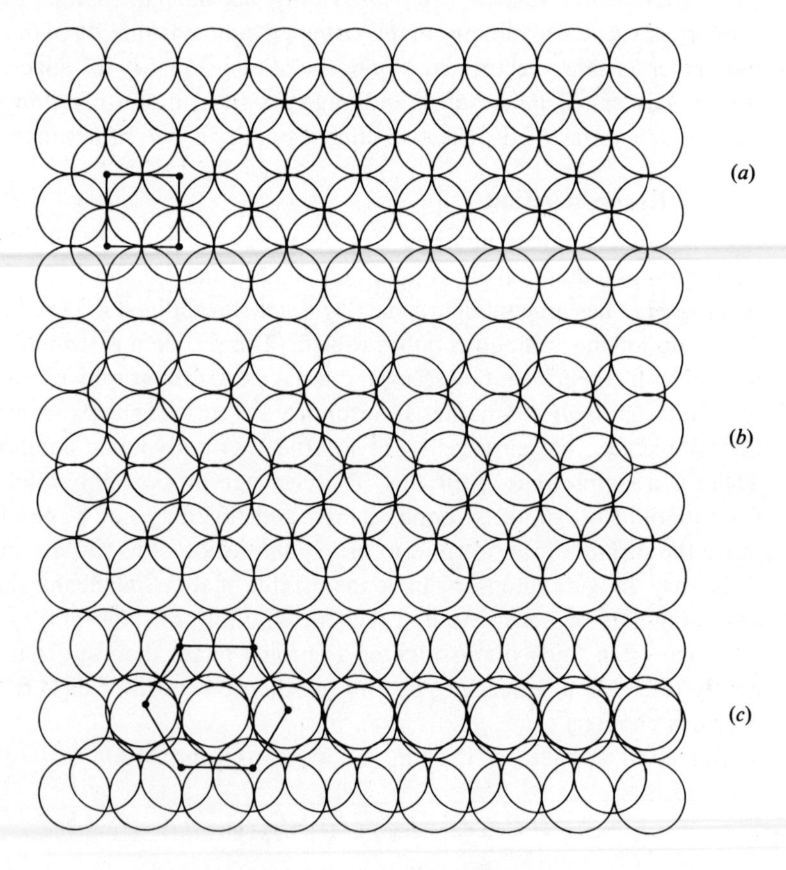

1×5 superlattice reflection intensity suggests a transition activation energy of about 0.9 eV/atom. The driving force for this reconstruction is the reduced surface energy of a close-packed metal surface compared to a more open surface (cf. Fig. 1.8). This is a big effect (in absolute magnitude) for iridium since it has one of the highest surface tensions among the elements (cf. Fig. 1.4). The countervailing energy cost is associated with the misfit between the square substrate and the hexagonal overlayer. The trade-off between these competing effects is central to the phenomenon of *epitaxy* – a subject dealt with at length in Chapter 16.

Unlike the Ir(100) example, one usually does not have detailed atomic geometry information for typical cases of reconstructive phase transitions. Diagnostic LEED patterns provide only the *symmetry* of the high and low temperature phases. However, according to Landau & Lifshitz (1969), this information alone is nearly sufficient to determine whether a transition is first order or continuous. A summary of the argument goes as follows. Let $\rho(\mathbf{r})$ denote the surface atom density corresponding to the crystal structure of the high symmetry phase. This function is invariant under the symmetry operations of the corresponding surface space group – call it G_0. After the phase transition, the reconstructed crystal surface is described by a new density function $\rho'(\mathbf{r}) = \rho(\mathbf{r}) + \delta\rho(\mathbf{r})$ which is invariant under the symmetry operations of a new space group G. The Landau–Lifshitz rule states that the transition *can* be continuous only if G is a subgroup of G_0 and the function $\delta\rho(\mathbf{r})$ transforms according to a single irreducible representation of G_0. The transition will be first order if this condition is not met.

Consider the example of a surface with a hexagonal Bravais surface net. Si(111) falls into this category. A straightforward symmetry analysis of the sort sketched above demonstrates that the $1 \times 1 \rightarrow 7 \times 7$ reconstruction cannot proceed via a continuous phase transformation. Therefore, this transition ought to exhibit the characteristic features of a first-order transition. Fig. 5.2 illustrates a striking confirmation of this prediction. These electron microscopy images (obtained in reflection) of a stepped Si(111) surface clearly show the nucleation and growth of regions of 7×7 reconstruction as the sample is cooled below the transition temperature. The complete transformation occurs over a temperature range of 20–30 K (below T_c). This 'sluggish' behavior is not at all uncommon for first-order solid-state phase transitions where strain plays an important role (Khachaturyan, 1983).

Let us return to the $1 \times 1 \rightarrow (\sqrt{2} \times \sqrt{2}\text{-}R45^\circ)$ reconstructive phase transition of W(100) mentioned in Chapter 4. This is a very subtle transformation compared to the previous examples. The low temperature structure consists of tungsten atoms displaced from their ideal positions by a small amount to form zig-zag chains (Fig. 5.3). A Landau–Lifshitz symmetry analysis indicates that this transition can be continuous and this appears to be the case experimentally (Wendelken & Wang, 1985).

Fig. 5.2. Reflection electron micrographs of the $1 \times 1 \rightarrow 7 \times 7$ transition (upon cooling) for a stepped Si(111) surface: (*a*) initial 1×1 structure; (*b*)–(*e*) (dark) regions of 7×7 nucleate at the top of monoatomic steps and expand across the terraces. Full arrows in (*c*) and (*e*) indicate the direction of growth; (*f*) completed 7×7 structure (Osakabe, Tanishiro, Yagi & Honjo, 1981).

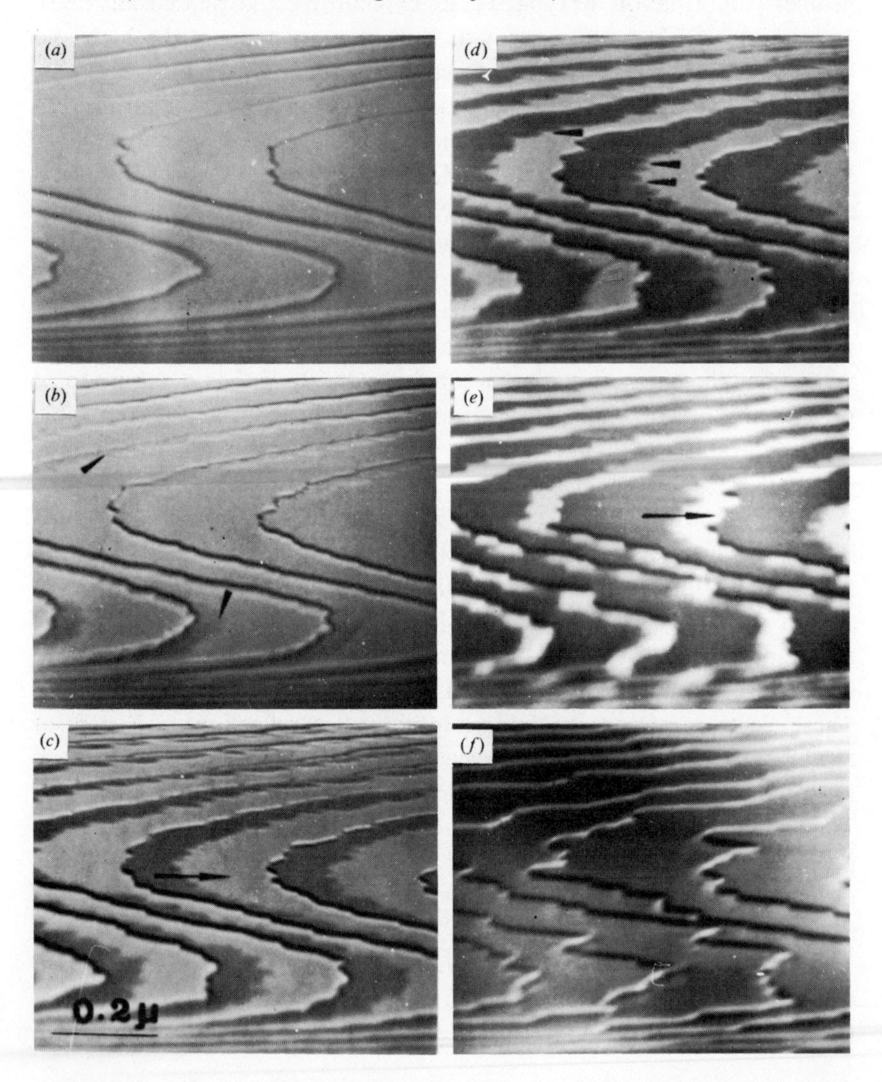

The atomic displacements that transform the ideal surface into the reconstructed geometry correspond to a longitudinal phonon mode with wave vector $\mathbf{q} = \pi/a(1,1)$, where a is the lattice constant. An appealing picture of the phase transition presumes that the frequency of this phonon mode becomes 'soft', i.e., approaches zero as the temperature drops. The crystal spontaneously distorts when the mode frequency reaches zero at T_c. Anharmonic effects keep the amplitude of the distortion finite. However, it is also possible that the surface tungsten atoms are displaced above T_c as well – but in random directions. At sufficiently low temperature, surface interactions favor an ordering of the displacements patterns into the observed structure. At present, available experiments do not unambiguously favor one mechanism over the other and it is unclear which one actually prevails (Inglesfield, 1985).

In any event, it *is* clear that the reconstructed surface has a lower internal energy than the ideal surface. Accurate LDA slab calculations show that the zig-zag reconstruction splits the peak in the local density of states at E_F characteristic of the ideal surface (cf. Fig. 4.25) so that the Fermi level finally resides in a minimum between two local maxima in the surface LDOS. This redistribution of the electronic state density reduces the total energy because the energies of some occupied states are lowered while the energies of some unoccupied states are raised (Fu, Freeman, Wimmer & Weinert, 1985). Notice that this behavior is reminiscent of the driving force for reconstruction that sweeps surface states out of the gap for some semiconductor surfaces.

Fig. 5.3. Surface structure of W(100): (*a*) high temperature 1×1 phase (open circles); (*b*) low temperature $\sqrt{2} \times \sqrt{2}$–$R45°$ phase (closed circles).

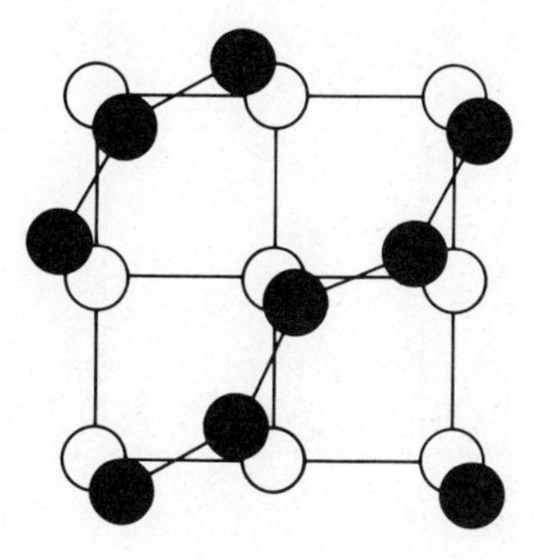

The $2 \times 1 \rightarrow 1 \times 1$ reconstructive phase transition that occurs for a Au(110) surface provides an elegant example of the concept of universality. Experiments show that the high temperature 1×1 structure reversibly transforms to the 'missing row' structure depicted in Fig. 3.22 at about 650 K. However, even without knowing the detailed structure, the symmetry of the (110) surface alone demands that a continuous phase transition (if it occurs) must exhibit the critical properties of the two-dimensional Ising model (Bak, 1979). None of the messy details of the surface (or even of gold itself) enter this argument. We predict that the temperature dependence of the order parameter (the intensity of LEED superlattice reflections unique to the 2×1 phase) near T_c will exhibit a value of the critical exponent β in accord with the famous exact result of Onsager (Ma, 1985), i.e., $\beta = 1/8$. The experimental result is $\beta = 0.13 \pm 0.02$ (Fig. 5.4).

Melting

At sufficiently high temperature, reconstructive structural transformations give way to a different class of equilibrium surface phase transitions: roughening or melting. The roughening transition was discussed in Chapter 1. Recall that this transition is characterized by a temperature at which the free energy of a monoatomic step vanishes. Spontaneous creation of such steps generates an instability of a crystalline

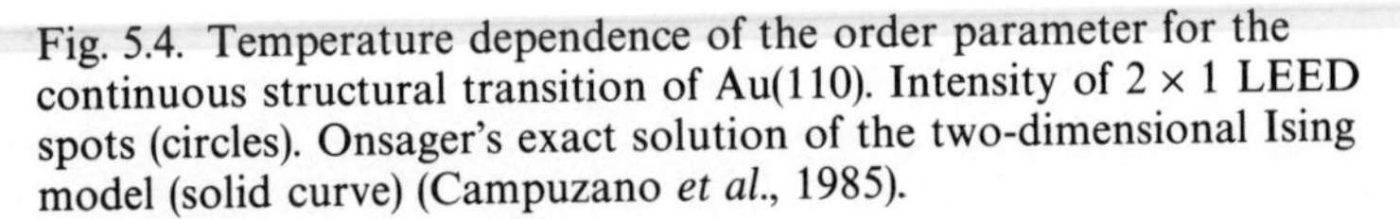

Fig. 5.4. Temperature dependence of the order parameter for the continuous structural transition of Au(110). Intensity of 2×1 LEED spots (circles). Onsager's exact solution of the two-dimensional Ising model (solid curve) (Campuzano *et al.*, 1985).

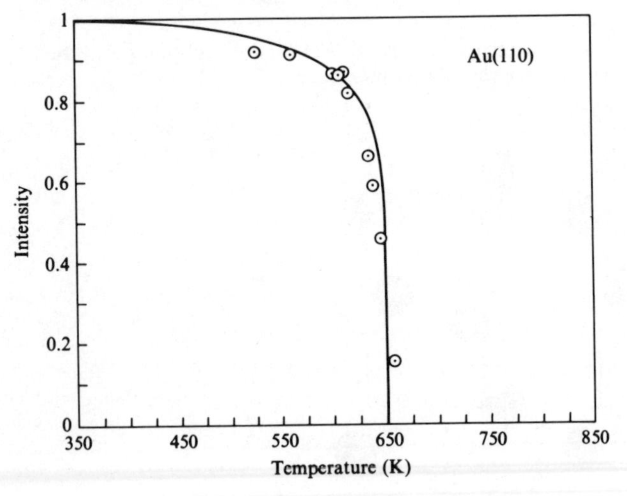

facet against long wavelength fluctuations in the local height of the surface. By contrast, one associates melting with a loss of order through short wavelength fluctuations in atomic displacements. In this section we take up the melting transition in more detail.

Melting is perhaps the most familiar example of a first-order phase transition. At the bulk melting point, T_m, discontinuities occur in the thermodynamic variables as the symmetry abruptly changes from that of a crystalline space group to the rotationally invariant state of a liquid. The simplest estimate of T_m comes from the Lindemann criterion: the crystal melts when the atomic mean square displacement due to thermal vibrations, $\langle u^2 \rangle$, is a significant fraction ($\sim 25\%$) of the lattice constant. It is instructive to carry out this calculation explicitly.

The total displacement of a single atom in a crystal may be written as a superposition of individual contributions from each of the independent phonon modes. Thus, $\langle u^2 \rangle = \sum |u_q|^2 = \sum (n_q + \frac{1}{2})\hbar/NM\omega_q$ for a monoatomic harmonic crystal of atomic volume $\Omega = V/N$. In the high temperature limit, the Bose–Einstein occupation factor $n_q = kT/\hbar\omega_q$, so

$$\langle u^2 \rangle = \frac{kT}{NM} \sum_q \frac{1}{\omega_q^2}. \tag{5.1}$$

In the Debye model (Ziman, 1972), $\omega_q = cq$ up to a cutoff energy $k\theta_D = \hbar c q_D$ and $\Omega q_D^3 = 6\pi^2$. Hence,

$$\langle u^2 \rangle = \frac{kT\Omega}{M(2\pi)^3} \int \frac{d^3q}{c^2 q^2} = \frac{3\hbar^2 T}{Mk\theta_D^2}. \tag{5.2}$$

Diffraction experiments directly measure the mean square atomic displacement. This is so because thermal vibrations attenuate diffracted beam intensities by the so-called Debye–Waller factor $\exp(-|\Delta\mathbf{k}|^2 \langle u^2 \rangle/4)$, where $\Delta\mathbf{k}$ is the scattered beam momentum transfer. Therefore, a comparison of x-ray and LEED data for the same system reveals the relative amplitude of thermal vibrations at the surface as compared to the bulk. This information usually is presented in the form of a 'surface' Debye temperature (Fig. 5.5). Typically, experiments show that the thermal excursion of surface atoms perpendicular to the surface is 50–100% greater than a bulk atom at the same temperature. The lower value follows immediately if we assume that a surface atom experiences exactly half the restoring force of a bulk atom.

A naive application of the Lindemann criterion would suggest that surface atoms disorder ('melt') at significantly lower temperature than the bulk melting temperature. If this is the case, a microscopic chain reaction could ensue as follows. A disordered layer at the surface exerts a perpendicular restoring force on a *second* layer atom that is intermediate between that of an ordered surface layer and the vacuum. Accordingly, the second layer melts at a slightly higher temperature than the surface but still lower than the bulk value. A similar argument applies to the third layer, and so on. Each layer melts abruptly when its local Lindemann condition is met. The melt front propagates into the crystal with increasing temperature until the process is completed at T_m.

A graphic confirmation of this idea is possible with the theoretical technique of molecular dynamics simulations (Abraham, 1984). In this method, one numerically integrates the classical equations of motion for a set of N particles interacting via a specified law of force. The system is kept in thermal equilibrium at temperature T by renormalizing the particle velocities at each time step so that $\langle KE \rangle = (3/2)kT$. Fig. 5.6 illustrates real-space particle trajectories for a molecular dynamics simulation using a realistic interaction potential appropriate to silicon. At moderate temperatures, all the atoms in planes parallel to a free Si(100) 1 × 1 surface execute small harmonic discursions around their equilibrium positions. However, at an elevated temperature still *below* the bulk melting point one clearly sees a disordered surface layer. It is interesting to note that this layer is not completely disordered – the underlying ordered layers induce some residual short range order in the melted region. The melt front is observed to move into the crystal as the temperature is raised as indicated above.

Experimental evidence for the 'surface initiated' model of melting sketched above comes from ion scattering spectroscopy. Fig. 5.7 displays ion kinetic energy distributions at four different temperatures for 97.5 keV

Fig. 5.5. Surface Debye temperature for Pb(110) derived from LEED measurements at different primary beam energies. The data approach the bulk value as the sampling depth of the electron increases (Farrell & Somorjai, 1971).

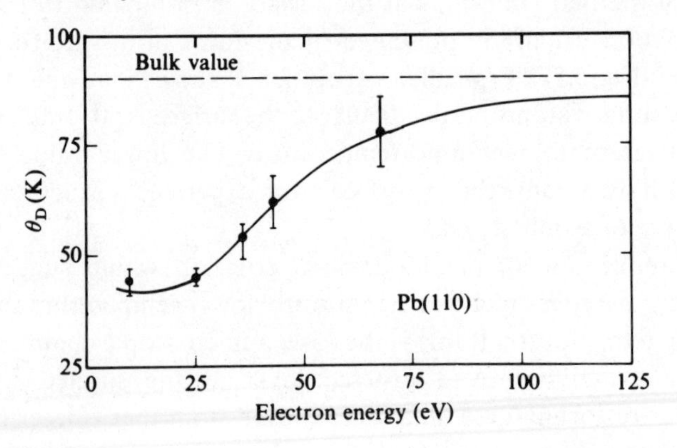

protons incident along the [101] channelling axis of a Pb(110) surface.* The highest temperature spectrum (*d*) was recorded just above the bulk melting temperature (600.7 K) of lead. It has exactly the appearance we found in Chapter 3 for ions incident along a random, non-channelling

Fig. 5.6. Particle trajectories from a molecular dynamics simulation of Si(100): (*a*) side view at $T = 1003$ K; (*b*) side view at $T = 1683$ K. (Courtesy of U. Landman and D. Luedtke, Georgia Institute of Technology).

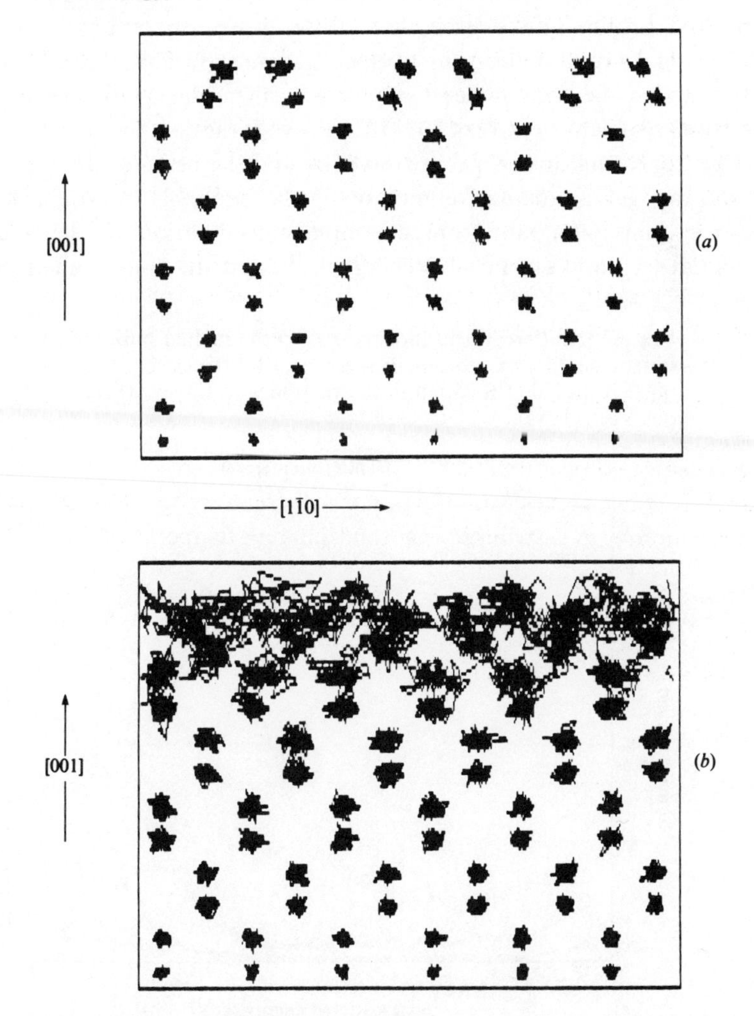

* This domain of ion kinetic energy was designated medium energy ion scattering in Chapter 3. The special features associated with it (to be discussed in Chapter 10) are not crucial to the analysis here.

direction (cf. Fig. 3.15). This reflects the fact that the totally disordered bulk sample exposes atoms from all depths to the ion beam. The lowest temperature spectrum (*a*) fits well to a surface peak calculated from a Monte Carlo simulation with an rms vibration amplitude given by (5.2). However, at all temperatures above 560 K, the spectra can be fit only by assuming that the incident beam traverses a disordered liquid layer whose thickness increases with temperature. The liquid layer thickness diverges as $T \to T_m$ from below.

Melting also has been observed at the surface of helium. However, rather than the helium lattice itself, the phase transition occurs in a two-dimensional layer of electrons trapped in image surface states localized 100 Å above the helium (see Chapter 4)*. Typical experiments achieve electron areal densities (N) of 10^5–$10^9/\text{cm}^2$ with the compensating positive charge appearing on a capacitor plate behind the helium slab.

We can get a qualitative picture of the probable two-dimensional electron phase diagram from a simple consideration of the average potential $\langle V \rangle$ and kinetic $\langle K \rangle$ energy of the system. As a rough estimate,

Fig. 5.7. Scattered yield (normalized to the melted bulk value) vs. ion kinetic energy for protons incident on Pb(110): (*a*) 295 K; (*b*) 561 K; (*c*) 600.5 K; (*d*) 600.8 K. Solid lines are guides to the eye (Frenken & van der Veen, 1985).

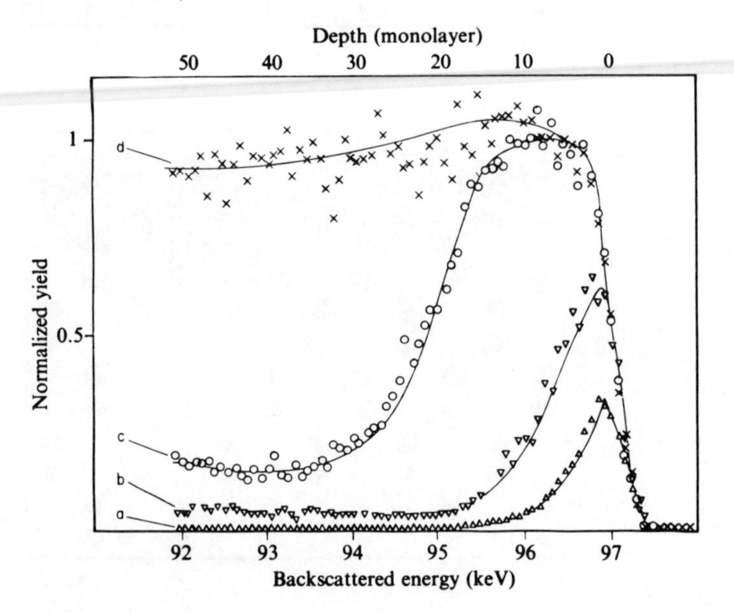

* The experiments actually trap electrons on a *liquid* ^{4}He substrate. The distinction is not relevant to the discussion at hand.

$\langle V \rangle \cong e^2(\pi N)^{1/2}$ since the electrons interact via the Coulomb potential. At $T=0$ (quantum limit), $\langle K \rangle \cong E_F = \pi h^2 N/m$. The kinetic energy dominates at high density and we expect an electron liquid (or gas). However, at sufficiently low density, the potential energy dominates and the liquid should condense into a solid. This is the Wigner (1934) crystal. Theoretical estimates suggest that the areal densities quoted above are well within the solid phase (Ceperley, 1978).

The Wigner crystal melts when the average kinetic energy of the lattice exceeds some fraction of the average potential energy. The conventional dimensionless parameter is called $\Gamma = \langle V \rangle / \langle K \rangle$. $\langle V \rangle$ is defined exactly as above but now, in the high temperature classical limit, the average kinetic energy is taken as that of an ideal two-dimensional gas, $\langle K \rangle = kT$. Consequently, classical melting should occur along a phase boundary in (N, T) space such that

$$\frac{e^2(\pi N)^{1/2}}{kT} = \Gamma_m. \tag{5.3}$$

Grimes & Adams (1979) discovered that a two-dimensional Wigner solid of electrons suspended above ^{4}He melts at $\Gamma_m = 137 \pm 15$ from a triangular lattice. The solid phase was detected by driving the electron layer up and down against the helium surface with a radio frequency electric field. This motion excites capillary waves on the liquid surface. The radio frequency absorption is resonantly enhanced when the inverse wavelength of the capillary waves at the driving frequency matches the reciprocal lattice vectors of the solid. The phase transition occurs around 0.5 K for the range of densities investigated.

Is melting in two dimensions the same as melting in three dimensions? The answer is *no* because a two-dimensional solid is fundamentally different from a three-dimensional solid. Suppose we try to use the Lindemann criterion. Our earlier calculation must be modified slightly since the integral over wave vectors in (5.2) now is restricted to two dimensions, i.e., $d^3q \rightarrow d^2q$. However, this small change causes the integral to diverge logarithmically at the lower limit! The long wavelength phonons destroy positional order in a two-dimensional 'solid'. Of course, for any finite sized sample, $\langle u^2 \rangle$ is not truly infinite. Nevertheless, a two-dimensional solid is characterized by long range *orientational* order rather than long range translational order (Mermin, 1968). That is, the relative orientation of the crystalline axes is maintained at large distances although strict periodicity of the lattice sites along the axes is not.

Melting in two dimensions proceeds by thermal excitation of topological defects of the lattice. Consider the triangular lattice appropriate to the

Wigner solid. An elementary dislocation of this lattice is depicted in Fig. 5.8(*a*). The Burger's vector of the dislocation is the amount by which a path around the dislocation fails to close. The energy of such a dislocation is calculable from elasticity theory (Friedel, 1964),

$$U = \frac{\mu(\lambda + \mu)}{\lambda + 2\mu} \frac{a_0^2}{4\pi} \ln \frac{A}{A_0}, \tag{5.4}$$

where μ and λ are the Lamé constants of the material, a_0 is the lattice constant, and $A_0 \sim a_0^2$. The long range strain field of the dislocation leads

Fig. 5.8. Defects in a two-dimensional triangular lattice: (*a*) an isolated dislocation; (*b*) a bound dislocation pair (Nelson & Halperin, 1979).

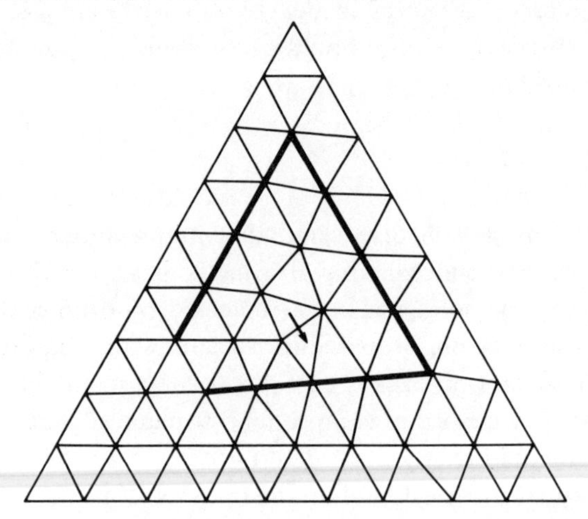

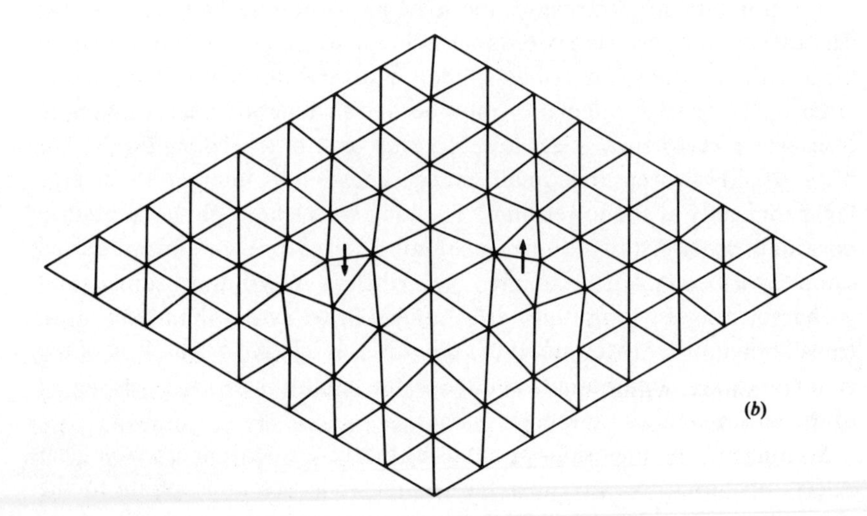

to a logarithmic dependence of the energy on the area (A) of the solid. Normally, we would expect that this strain energy is too large for thermal excitation of such defects. However, the entropy of a dislocation *also* depends logarithmically on the area since there are approximately A/A_0 possible positions for the defect, i.e., $S = k \ln(A/A_0)$. Consequently, the crystal 'melts' due to the spontaneous creation of free dislocations above a melting temperature defined by $U - T_m S = 0$ or

$$kT_m = \frac{a_0^2}{4\pi} \frac{\mu(\mu + \lambda)}{\lambda + 2\mu}. \tag{5.5}$$

Below T_m, the dislocations are bound in pairs with opposite Burger's vectors (Fig. 5.8(*b*)). The energy of such a pair is rather modest since it depends only on the logarithm of the distance *between* the two dislocations. Melting occurs via thermal unbinding of the pairs. This is the same Kosterlitz–Thouless (1973) mechanism invoked in Chapter 1 in connection with the roughening transition.

Let us apply the Kosterlitz–Thouless theory of two-dimensional melting to the Wigner electron solid. That is, we want a theoretical estimate of Γ_m. To do this we need values of the Lamé constants in order to insert (5.5) into (5.3). A longitudinal elastic wave in an electron lattice is equivalent to a plasmon excitation. In three dimensions, the Wigner lattice is incompressible since $\omega_p(q)$ never goes to zero. A similar argument is possible in two dimensions, so that λ (which measures compressibility) may be taken as infinite. By contrast, any solid supports shear modes. Indeed, the velocity of long wavelength transverse acoustic phonons is directly related to the shear coefficient: $c^2 = \mu/mN$. This number is difficult to calculate at finite temperature due to phonon anharmonicity and the presence of bound dislocation pairs. The best estimate of $\mu(T)$ comes from molecular dynamics simulations of the phonon modes of a Wigner crystal (Morf, 1979). It turns out that $\mu \propto N^{1/2}$ so that Γ_m is independent of electron density. The final theoretical estimate of $\Gamma_m = 128$ is in quite reasonable agreement with experiment.

Magnetism

The Ising model invoked earlier in our account of a structural phase transition on Au(110) is most commonly encountered in discussions of magnetism. It is natural to ask whether this model reappears in its more familiar context within the realm of surface physics. Until recently, many workers would have answered in the negative. Early experiments were interpreted to demonstrate that there is no magnetism at a clean solid surface. One spoke of magnetically 'dead' surface layers. Recent progress in both experiment and theory clearly shows that this is not the case. It is possible to observe and study both surface magnetism and surface magnetic phase transitions.

Magnetism is a quantum mechanical phenomenon that depends in a subtle manner on the electron–electron interaction in solids. The crucial quantities that enter are the *intra-atomic* Coulomb energy, U, which is responsible for Hund's rules in atoms, and the *inter-atomic* exchange energy, J, which appears in the Heitler–London model of the hydrogen molecule. The detailed physics of ferromagnetism in even the most familiar magnets, iron, cobalt and nickel, is not completely understood. The altered environment of surface atoms gives one every reason to believe that magnetism at a crystal surface may be rather different from bulk magnetism.

Fig. 5.9. Exchange split density of states for the Stoner model (Ziman, 1972).

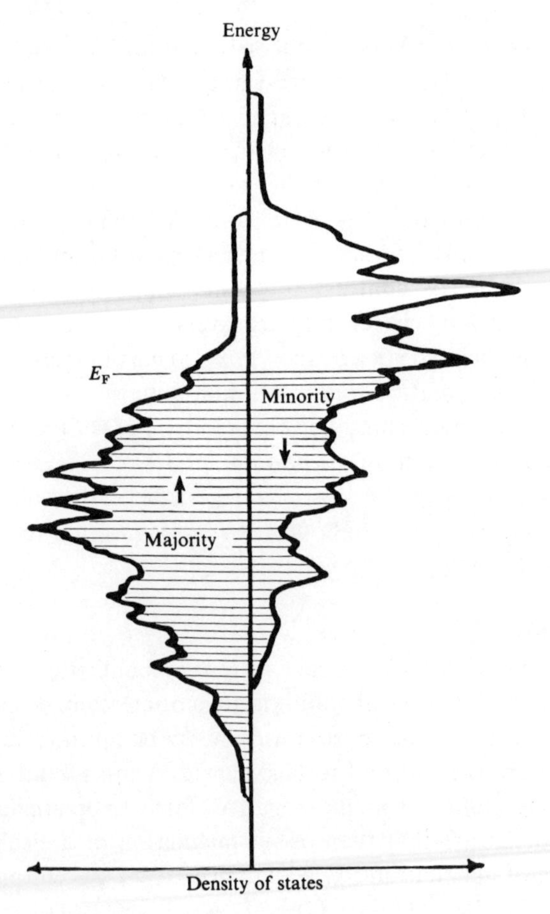

We first briefly review the situation in the bulk. The self-consistent LDA electronic structure calculations presented in this book tacitly assume that independent electrons occupy Bloch band states. In this situation, magnetism emerges from the familiar Stoner model, because it takes account of the interplay between itinerant electrons and an atomic-based direct interaction between electrons of opposite spin (Ziman, 1972):

$$\mathscr{H} = \sum_k \varepsilon(k) c_k^\dagger c_k + U \sum_k n_{k\uparrow} n_{k\downarrow}. \tag{5.6}$$

In this second-quantized expression, $\varepsilon(k)$ represents the dispersion of a single energy band and U is the Coulomb interaction noted above. Both U and a small external magnetic field (H) split the band eigenstates into up and down spin components,

$$\left.\begin{aligned} \varepsilon_{k\uparrow} &= \varepsilon(k) - \mu_0 H + U n_{k\downarrow}, \\ \varepsilon_{k\downarrow} &= \varepsilon(k) + \mu_0 H + U n_{k\uparrow}, \end{aligned}\right\} \tag{5.7}$$

where μ_0 is the magnetic moment of the electron (Fig. 5.9). If the up and down spin populations are not too different we may solve (5.7) for the spin-occupancies and compute the macroscopic magnetic moment (M),

$$\begin{aligned} M &= \mu_0 (n_\uparrow - n_\downarrow) \\ &= U\rho(E_F) M + \mu_0^2 \rho(E_F) H, \end{aligned} \tag{5.8}$$

where, as usual, $\rho(E)$ is the electronic density of states. The magnetic susceptibility, $\chi = \partial M / \partial H$, becomes

$$\chi = \frac{\mu_0^2 \rho(E_F)}{1 - U\rho(E_F)}. \tag{5.9}$$

If the *Stoner criterion* ($U\rho(E_F) > 1$) is satisfied, the assumption that $\langle n_\uparrow \rangle \cong \langle n_\downarrow \rangle$ breaks down. This is the signal that one spin population dominates and that the paramagnetic state is unstable towards ferromagnetism at zero temperature and external field.

How does this argument carry over to the surface? In Chapter 4 we saw that the LDOS at a transition metal surface can differ substantially from its value in the bulk (cf. Fig. 4.25). Therefore, the presence of the density of states at the Fermi level, $\rho(E_F)$, in the Stoner criterion immediately suggests possible new behaviour at such surfaces. We can investigate surface ferromagnetism in this case by combining values of $\rho(E_F)$ at the surface of the 3d transition metals (calculated with a tight-binding model) with values of the intra-atomic exchange parameter (fitted, to reproduce the *bulk* magnetic moment). The resulting 'surface Stoner criterion' predicts that iron, cobalt and nickel will retain their magnetic moments at the surface (Fig. 5.10). In addition, this simple theory also suggests that paramagnetic vanadium and anti-ferromagnetic chromium will order ferromagnetically at their respective surfaces. Manganese is a marginal case in this analysis.

A more quantitative theoretical picture of surface magnetism requires a more sophisticated approach to surface electronic structure than the tight-binding method used above. Slab calculations that employ the local density approximation can be used if one somehow includes the spin dependence of the electron–electron interaction. Unfortunately, the precise way to do this is not known. Practitioners typically modify the exchange-correlation potential that enters (4.4) so that the exact spin dependence of the exchange energy (which *is* known) is reproduced. In this way, electrons of different spin move in different effective potentials. Systematic calculations with this spin-dependent $v_{xc}(\mathbf{r})$ for the 3d and 4d transition metals correctly predict that only Fe, Co and Ni satisfy the bulk Stoner criterion (Williams & von Barth, 1983). In accordance with the *surface* Stoner prediction, the corresponding calculation for a seven-layer slab of iron finds that the ferromagnetic state has lower energy than the paramagnetic state.

Fig. 5.11 shows contours of constant *spin* density near the (100) surface of this iron slab. The majority spin electron density (solid curves) clearly resembles an atomic 3d orbital that spills into the vacuum. This suggests that the ratio of the intra-atomic Coulomb energy to the bandwidth (W) may be quite large, i.e., $U \gg W$. In that case, a Wannier orbital basis that emphasizes local moment formation may be more appropriate than the Bloch basis assumed in (5.6). The calculated magnetic moment at a surface site is about $3\mu_B$ compared to $2.25\mu_B$ in the bulk. This is not too surprising. In the simplest view, surface atoms interpolate between the properties of

Fig. 5.10. Test of the Stoner criterion for 3d transition metal surfaces (Allan, 1981).

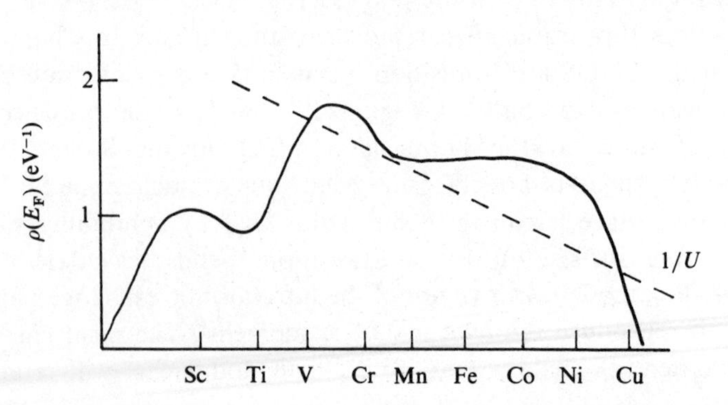

bulk atoms and free atoms. The magnetic moment of a free iron atom is $4\mu_B$.

Experimental evidence for the existence of surface magnetism comes primarily from electron spectroscopy. For example, angle-resolved UPS spectra exhibit two surface sensitive features near the Fermi energy of Ni(100). The symmetry selection rules discussed in Chapter 4 show that these states are respectively odd and even with respect to the (100) and (110) mirror planes. Projections of the bulk energy bands onto the surface Brillouin zone show that each state exists in regions of the zone where there is a gap in only *one* spin band of the requisite symmetry (Fig. 5.12). Consequently, we infer that the nickel (100) surface supports two magnetic surface states – one of majority spin (parallel to the direction of bulk spontaneous magnetization) and one of minority spin (anti-parallel to the bulk moment).

A more direct measurement of local magnetization exploits the intrinsic

Fig. 5.11. Edge view of spin density contours for Fe(100). Solid (dashed) lines indicate majority (minority spin) (Ohnishi, Freeman & Weinert, 1983).

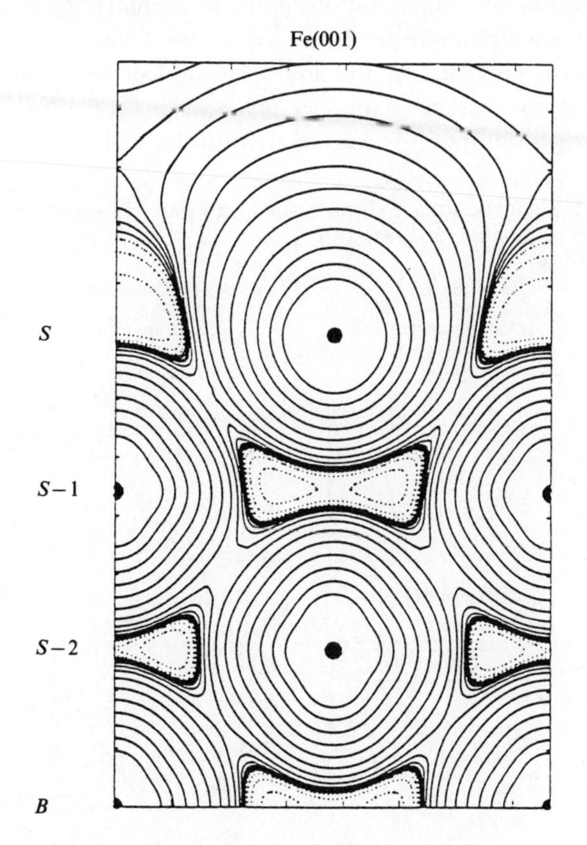

spin of the electron, i.e., we imagine an energy, angle- and spin-resolved UPS experiment. The spin can be measured in a number of different ways (Siegmann, Meier, Erbudak & Landolt, 1984). For example, a conventional Mott detector (familiar from nuclear physics) can be used if the photo-ejected electrons are post-accelerated to a kinetic energy of about 100 keV. Spin-resolved data for Fe(100) obtained in this way are shown in Fig. 5.13. This angle-resolved experiment measured only those electrons photoemitted normal to the crystal surface. Hence, the spin up (majority) and spin down (minority) energy distribution curves reflect electronic initial states at the center of the SBZ ($k_{\parallel} = 0$). The magnitude of the experimentally determined exchange splitting U is in good agreement with the band structure results. Similar spectra result if one samples other regions of the SBZ and it is obvious to the eye that the number of majority spin electrons (below E_F) is not equal to the number of minority spin electrons. This imbalance is responsible for the net magnetic moment of the Fe(100) surface.

From the point of view of phase transitions, the demonstration that a surface can support a net magnetization should immediately be followed by a study of its temperature dependence. To do this, focus attention on any one initial state energy in Fig. 5.13. Label the photoemission intensity from the majority and minority spins at this energy $I_+(T)$ and $I_-(T)$, respectively. It turns out that it is reasonable to suppose that the surface

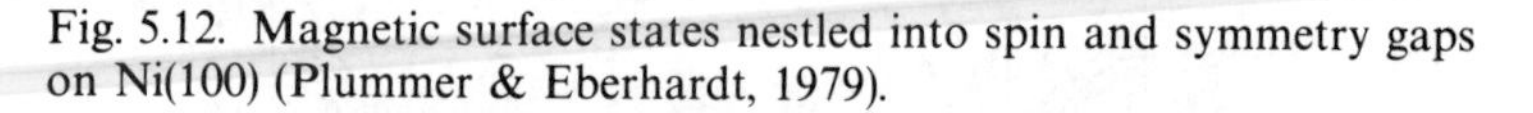

Fig. 5.12. Magnetic surface states nestled into spin and symmetry gaps on Ni(100) (Plummer & Eberhardt, 1979).

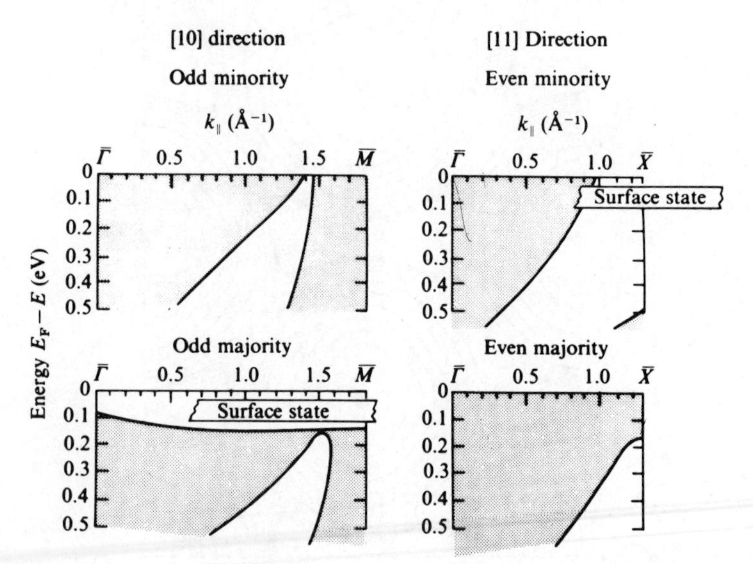

magnetization is proportional to the normalized net spin polarization, $P(T) = (I_+ - I_-)/(I_+ + I_-)$. Temperature dependent spin-resolved UPS data analyzed in this way lead to a surprising result: the surface magnetization falls to zero nearly linearly as T approaches T_c (Fig. 5.14). This is in distinct contrast to the temperature dependence commonly observed for the *bulk* saturation magnetization (Kittel, 1966). How can we understand this?

The simplest picture of magnetic order in a solid posits that electron–electron interactions generate an effective *exchange interaction* that favors parallel (ferromagnetism) or anti-parallel (anti-ferromagnetism) alignment of local moments. As we have seen, both theory and experiment support the idea that local moments can exist at the clean surface of a crystal. A crude estimate of the relevant interaction energy between nearest neighbors is $J = U^2\rho(E_F)/2S$, where S is the spin (Mathon, 1983). For iron, this estimate gives $J \cong 0.15\,\text{eV}$. The exchange interaction between sites **R**

Fig. 5.13. Spin- and angle-resolved UPS spectra for Fe(100) at low temperature ($T/T_c = 0.3$). The energy distribution curves for the majority spins (open symbols) and minority spins (closed symbols) are separated by the exchange splitting U (Kisker, Schroder, Gudat & Campagna, 1985).

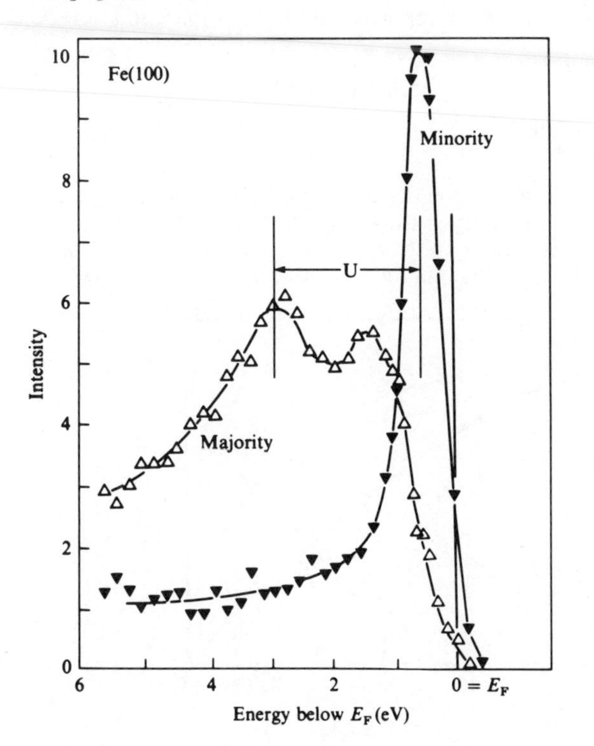

and $\mathbf{R}'$ at distances greater than nearest neighbors either falls off exponentially or decreases slowly and oscillates in sign. The two cases correspond to large and moderate values of U/W, respectively. In either case, we can write an *effective* Hamiltonian that describes pairwise interaction between local moments, $S(\mathbf{R})$:

$$\mathscr{H} = -\sum_{\mathbf{R},\mathbf{R}'} J_{\mathbf{RR}'}\mathbf{S}(\mathbf{R})\cdot\mathbf{S}(\mathbf{R}') - \mathbf{H}\cdot\sum_{\mathbf{R}}\mathbf{S}(\mathbf{R}). \tag{5.10}$$

This is the famous Heisenberg model with the addition of a term to represent the effect of an external magnetic field.

We describe the thermodynamics of magnetism at a free surface by the Heisenberg model with the lattice sites, $\mathbf{R}$, restricted to the $z > 0$ half-space. Our goal is to find the magnetization at an arbitrary site $\mathbf{R}$ – defined as the average value of the fluctuating local moment $\mathbf{S}(\mathbf{R})$ with respect to the Hamiltonian (5.10), i.e., $M(\mathbf{R}) = \langle \mathbf{S}(\mathbf{R}) \rangle$. There is no need to perform this calculation exactly. The essential differences between surface and bulk magnetism already appear at the level of the *mean field approximation* (Stanley, 1971). Herein, the influence on a given spin, $\mathbf{S}(\mathbf{R})$, due to the fluctuating spins on other sites is replaced by the effect of the average moment at each neighbor site. That is, one of the $\mathbf{S}(\mathbf{R})$ factors in the interaction term of (5.10) is replaced by its average value, viz.,

$$\mathscr{H} = -\left[H + \sum_{\mathbf{R}'} J_{\mathbf{RR}'}M(\mathbf{R}')\right]\sum_{\mathbf{R}} S(\mathbf{R}). \tag{5.11}$$

Fig. 5.14. Temperature dependence of the spin polarization of photoelectrons emitted from Fe(100). The solid line through the data (symbols) is a guide to the eye. The solid curve is the result expected for the bulk (Kisker, Schroder, Gudat & Campagna, 1985).

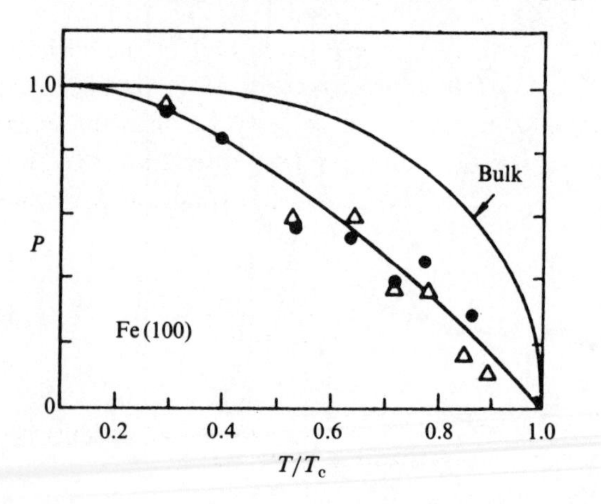

For simplicity, $\mathbf{S}(\mathbf{R})$ has been replaced by the scalar quantity $S(\mathbf{R})$. The calculation of the magnetization now is an elementary problem in statistical mechanics. The result is a transcendental equation for the magnetization:

$$M(\mathbf{R}) = \tanh\left[\frac{H + \sum_{\mathbf{R}'} J_{\mathbf{RR}'} M(\mathbf{R}')}{kT}\right]. \tag{5.12}$$

Consider the case of a simple cubic lattice where each spin interacts only with its nearest neighbors (at a distance a_0) with a strength J. If only small values of H are permitted, the magnetization will be nearly uniform. Accordingly, we treat $\mathbf{R}$ as a continuous variable and expand the argument of (5.12) as

$$\sum_{\mathbf{R}'} J_{\mathbf{RR}'} M(\mathbf{R}') = J\{6M(\mathbf{R}) + a_0^2 \nabla^2 M(\mathbf{R}) + \cdots\}. \tag{5.13}$$

The first derivative terms vanish because of the symmetry of the bulk lattice. We are interested in the behavior of the magnetization as it approaches the Curie point from below. Hence, $M(\mathbf{R})$ is small in absolute magnitude and the hyperbolic tangent in (5.12) can be expanded to low order:

$$AM(\mathbf{R}) + BM^3(\mathbf{R}) - C\nabla^2 M(\mathbf{R}) = H. \tag{5.14}$$

In this expression, $A = kT - 6J = k(T - T_c)$ and B and C are constants as $T \to T_c$.* In zero field, (5.14) predicts that the uniform ($\nabla^2 M = 0$) magnetization of the bulk decays to zero according to

$$M = \sqrt{\frac{-A}{B}} \propto (T_c - T)^{1/2}. \tag{5.15}$$

This is the behavior of the curve labelled 'bulk' in Fig. 5.14.

Let us see how the results change if $\mathbf{R}$ is chosen at the surface of a semi-infinite lattice. First, the remaining symmetry of the problem suggests that $M(\mathbf{R}) = M(z)$. Second, it is natural to suppose that the exchange coupling constant between spins within the first atomic layer, $J_\parallel$, may be different from the bulk value. In this case, the expansion performed in (5.13) must be replaced by

$$\sum_{\mathbf{R}'} J_{\mathbf{RR}'} M(\mathbf{R}') = 4J_\parallel M_s + JM_s + a_0 J \left.\frac{\mathrm{d}M}{\mathrm{d}z}\right|_{z=0} + \cdots \tag{5.16}$$

* Technically, we have reduced the Heisenberg model on a lattice (5.10) to an Ising model with a continuous spatial degree of freedom (5.14).

where M_s is the magnetization of a surface site. The first derivative term is present in this case because the surface atom is not situated symmetrically with respect to its nearest neighbors. Now comes the essential point: the fundamental formula (5.14) still can be used at the surface if we provide an additional constraint that effectively transforms (5.16) into (5.13). With $J_\parallel = J(1 + \Delta)$, it is easy to see that this *boundary condition* is

$$M_s = \lambda \left. \frac{dM}{dz} \right|_{z=0}. \tag{5.17}$$

The *extrapolation length*, $\lambda = a_0/(1 - 4\Delta)$, is a convenient measure of the difference between the surface and bulk magnetic exchange interaction.

The temperature dependence of the surface magnetization now follows directly. Multiply (5.14) by dM/dz and integrate over z from deep within the bulk (where $M^2 = -A/B$ from (5.15)) to the surface ($M = M_s$). The boundary condition (5.17) enters after an integration by parts. The final result (after setting $H = 0$) is a quartic equation for the surface magnetization,

$$M_s^4 + \frac{2}{B}\left(A - \frac{C}{\lambda^2}\right)M_s^2 + \frac{A^2}{B^2} = 0, \tag{5.18}$$

which has a simple solution as $T \to T_c$, i.e., when M_s is small:

$$M_s = \frac{-A\lambda}{\sqrt{2BC}} \propto T_c - T. \tag{5.19}$$

The surface transition is again continuous, but the magnetization vanishes with a different power law from the bulk. It is in fact just the linear dependence found in experiment (Fig. 5.14).

A continuous phase transition is characterized by fluctuations in the order parameter – the magnetization in the present case. Below (above) T_c, finite regions of the disordered (ordered) phase fluctuate in and out of the dominant ordered (disordered) phase volume. The size of these regions is measured by the so-called *correlation length*, $\xi(T)$, which diverges at the transition point. However, the preceding analysis showed that a second length scale is relevant to the surface problem: the extrapolation length λ that describes the curvature of the magnetization profile $M(z)$ near the surface. A more careful study of the mean field theory shows that both the precise form of $M(z)$ and the nature of the surface magnetic phase diagram depend critically on the *interplay* between these two length scales (Kumar, 1974; Lubensky & Rubin, 1975). The details are beyond the scope of the present discussion, so in the following we merely sketch the main results.

Suppose the exchange interaction between two surface spins is less than the corresponding interaction between two bulk spins. In that case, $\Delta < 0$ so that $\lambda > 0$ and according to (5.17) the magnetization profile must look like Fig. 5.15(*a*). M_s is less than M_{bulk} and the scale length for the recovery of the magnetization to its bulk value is set by the correlation length, i.e., $M(z) = M(z/\xi)$. Completely different behavior is found if the exchange interaction between surface spins is sufficiently large so that $\lambda < 0$. In this case, the surface orders magnetically at a surface transition temperature, $T_{cs} > T_c$. This is a striking prediction. Spontaneous bulk magnetization develops only when the temperature is lowered below T_c (Fig. 5.15(*b*) and (*c*)). Even then, the surface boundary condition guarantees that the magnitude of M_s always exceeds the bulk saturation magnetization.

So far we have considered only the surface properties of systems that undergo a continuous phase transition in the bulk. A mean field analysis also is possible for the semi-infinite analog of (5.14) appropriate to a system for which the bulk undergoes a *first*-order transition (Lipowsky & Speth, 1983). In this case, if $J_{\parallel}$ is little different from J, a first-order transition

Fig. 5.15. Magnetization profiles for the semi-infinite nearest neighbor Ising model: (*a*) $\lambda > 0$, $T < T_c$; (*b*) $\lambda < 0$, $T_c < T < T_{cs}$; (*c*) $\lambda < 0$, $T < T_c$ (Kumar, 1974).

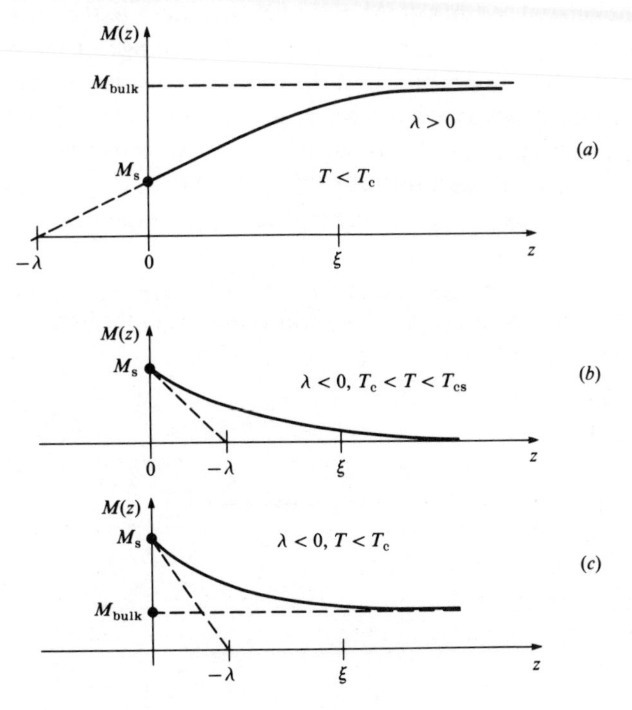

occurs at the surface at the bulk transition temperature, T_c. However, if $|\Delta| \gg 1$, the surface phase transition becomes *continuous*. This latter case corresponds to a significant weakening of the surface exchange constant relative to the bulk. As $T \to T_c$ from below, the theory predicts that the surface region disorders over a distance l from the surface while the bulk remains perfectly ordered (Fig. 5.16). Moreover, the disordered layer width diverges in a characteristic fashion:

$$l = l_0 \ln\left(\frac{T_0}{T_c - T}\right), \tag{5.20}$$

so that the entire sample is disordered for $T > T_c$. T_0 and l_0 are constants that depend on the details of the system. This scenario should sound familiar; the analysis of ion scattering data used earlier to support a model of the surface melting of Pb(110) (cf. Fig. 5.7) required just such a disordered surface layer in the relevant ion trajectory Monte Carlo simulations. In fact, (5.20) was found to provide an excellent fit to the temperature dependence of the requisite layer widths. It seems reasonable to conclude that the presumed first-order melting of (at least) a lead crystal can proceed instead via a continuous transition initiated at one of its crystalline surfaces.

Critical phenomena

The general features of the mean field solutions to the semi-infinite Ising model are likely to be correct over a wide range of temperature. However, for continuous transitions, the specific power law prediction for the temperature dependence of the magnetization (and other thermodynamic quantities) is usually *not* correct in the so-called *critical region* very near the transition temperature ($|(T - T_c)/T_c| \ll 1$). For example, (5.15) and (5.19) predict that the order parameter critical exponent β takes the

Fig. 5.16. Spatial variation of the order parameter for $T < T_m$ and $J_\parallel \ll J$ for a system with a first-order transition in the bulk (Lipowsky & Speth, 1983).

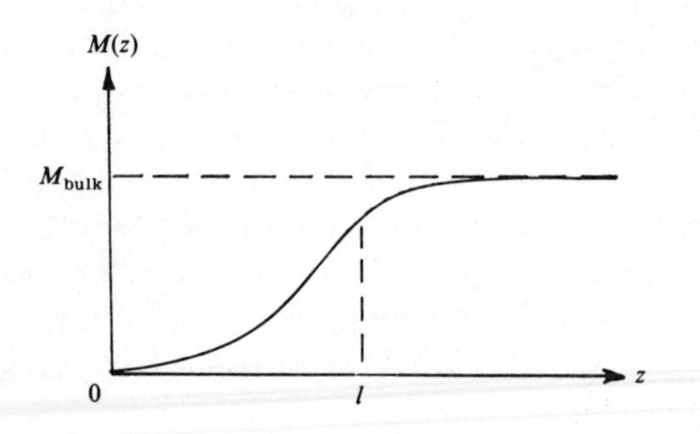

value 1/2 and 1 for the bulk and surface case, respectively. By contrast, it is well known that the correct bulk exponent is actually quite close to 1/3 for systems that exhibit the symmetry of the Ising model. This discrepancy is related directly to the neglect of fluctuations in the mean field theory. *A priori*, it is not obvious whether such fluctuations will play an equally important role at the surface. It is best to appeal directly to experiment.

Unsurprisingly, it is difficult to perform accurate measurements that combine the temperature control needed to explore the critical region with the ultra-high vacuum requirements of clean surface work. To date, the best results come from spin-polarized LEED measurements. In these experiments, elastic scattering intensities are measured for an incident electron beam polarized parallel (I_+) and antiparallel (I_-) to the bulk magnetization axis.* These intensities are different because of the exchange interaction between the incoming electron and the electrons of the target. As in the spin-resolved photoemission case, the scattering polarization asymmetry, $A = (I_+ - I_-)/(I_+ + I_-)$, is proportional to the magnetization for temperatures sufficiently close to T_c (Feder & Pleyer, 1982). For the 3d transition metals, the surface appears to order at the bulk value of T_c. According to our earlier analysis, this means that $\lambda > 0$, i.e., the surface exchange constant $J_\parallel$ is less than the bulk value J. In particular, for Ni(100), a surface magnetization exponent of $\beta_s = 0.825 \pm 0.03$ emerges from a power law fit to the LEED asymmetry data (Fig. 5.17). The corresponding theoretical prediction for the semi-infinite Heisenberg model appropriate to nickel (including the effect of fluctuations) is $\beta_s = 0.878$.

The Heisenberg model also describes the interactions between the large 4f moments of the rare-earth metals. Experiments are feasible as well. For example, the magnetization at the surface of polycrystalline gadolinium is shown in Fig. 5.18. This measurement was performed by combining surface sensitive *ion scattering* with a spin polarization probe. In detail, a 10 keV deuteron beam is scattered from the surface at extreme grazing incidence (0.2°). Some of the incident deuterons are *neutralized* by capture of a surface electron that is polarized either parallel or anti-parallel to the magnetization axis of the sample. The hyperfine interaction communicates the spin polarization information to the deuterium nucleus and the polarization of the latter is interrogated by standard methods of nuclear physics.

The electron capture results for gadolinium show that $M_s \propto T_c - T$ for temperatures far below the critical point. This is consistent with the mean

* The production of a spin-polarized beam of electrons turns out to be a problem in surface physics! See Chapter 9.

Fig. 5.17. Log–log plot of spin-polarized LEED exchange-scattering asymmetry versus reduced temperature for Ni(100) (Alvarado, Campagna & Hopster, 1982).

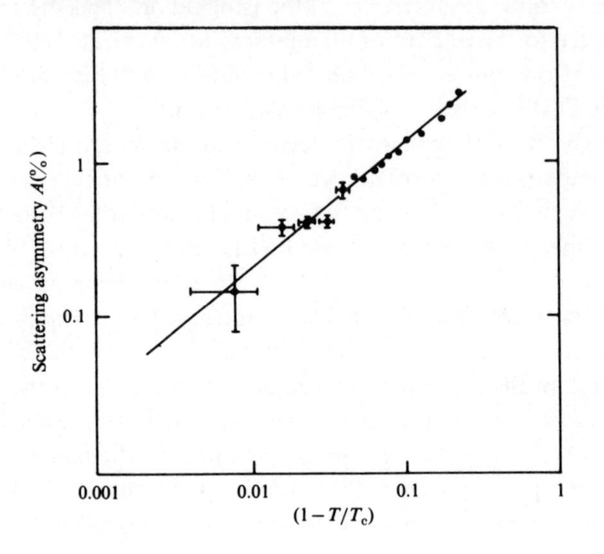

Fig. 5.18. Surface magnetization of Gd(0001) derived from the relative nuclear polarization of deuteron electron capture (Rau, 1982). Inset: scattering asymmetry from spin-polarized LEED in the critical region (Weller *et al.*, 1985).

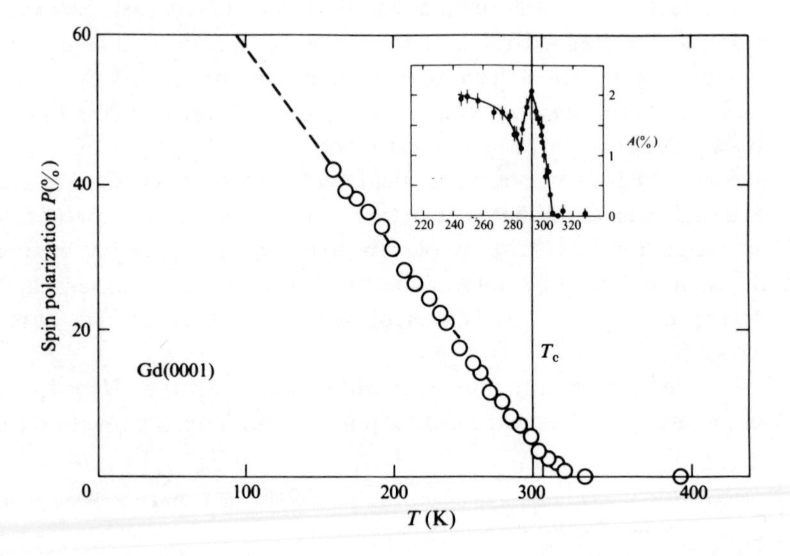

field theory. However, the magnetization appears to persist *above* the bulk critical temperature. This inference is confirmed by high resolution spin-polarized LEED measurements. The asymmetry data show that T_{cs} is more than 20 K greater than T_c (inset to Fig. 5.18). Fig. 5.15(*b*) then implies that this surface must exhibit an appreciable enhancement of the bulk interatomic exchange in the surface layer ($\lambda < 0$).

Phenomenological theories of phase transitions cannot tell us why the surface critical properties of nickel and gadolinium are so different. Any quantitative estimate of either $J_{\parallel}$ or T_{cs} requires a detailed microscopic analysis although it is natural to suppose that these quantities will differ from their bulk values if the surface is relaxed or reconstructed. Even if this is not the case, we know that long wavelength spin wave excitations reduce the net magnetization above $T = 0$ (see Chapter 6) and ultimately destroy long range order at the Curie point. Consequently, a surface susceptibility that contains the surface spin wave spectrum must be used to determine microscopic parameters such as $J_{\parallel}$ and T_{cs} for use in a realistic model calculation. Progress in this direction has been slow.

General references

Reconstruction

Willis, R.F. (1985). Surface Reconstructive Phase Transitions. In *Dynamical Phenomena at Surfaces, Interfaces and Superlattices* (eds. Nizzoli, Rieder & Willis), pp. 126–47.

Melting

Brinkman, W.F., Fisher, D.S. & Moncton, D.E. (1982). Melting of Two-Dimensional Solids. *Science* **217**, 693–700.

Magnetism

Allan, G. (1981). Itinerant Electron Surface Magnetism. *Surface Science Reports* **1**, 121–55.

Siegmann, H.C., Meier, F., Erbudak, M. & Landolt, M. (1984). Spin-Polarized Electrons in Solid State Physics. In *Advances in Electronics and Electron Physics* (ed. Hawkes), vol. 62, pp. 1–99. Orlando: Academic.

Critical phenomena

Binder, K. (1983). Critical Behavior at Surfaces. In *Phase Transitions and Critical Behavior* (eds. Domb & Lebowitz), vol. 8, pp. 1–144. London: Academic.

6

ELEMENTARY EXCITATIONS

Introduction

The preceding chapters have focused entirely on the equilibrium free energy state of an isolated clean crystal surface. Unfortunately, many of the most interesting conceptual (and commercial) issues in surface physics intimately involve the *interaction* of a solid surface with foreign matter. If the interaction is strong, it is necessary to treat the surface and the foreign material as a single combined system. This is the subject of Part 2 of this book. However, if the interaction is weak, the surface merely responds to the external perturbation while retaining its individual identity. In fact, any real experiment designed to probe the properties of even an isolated surface invariably perturbs the system and invokes a characteristic response. This response is determined by the low-lying excited states of the system.

For example, consider an experiment designed to determine the binding energy and dispersion of an electronic surface state. In practice, one uses photoemission spectroscopy to measure the kinetic energy and propagation vector of an electron ejected from the sample into the vacuum. However, what one actually measures is the energy and relative momentum of an excited electronic state (with a finite lifetime) that consists of 10^{23} electrons in the presence of the surface-localized hole left behind by the photoelectric event. *A priori*, this could be a horribly complex state of the interacting many-body system. As it happens, low energy excitations such as this one turn out to have a very simple character (Anderson, 1963). Our naive intuition remains useful because photoemission (and any similar probe) creates weakly interacting 'quasiparticle' states that are recognizably derived from the simplest one-particle band structure eigenstates. More generally, it will be useful to think of the interaction of a surface with external perturbations (absorbates, external fields, electron beams, etc.) in terms of the creation and destruction of quasiparticles and other sorts of elementary excitations. The latter come in a number of varieties.

In many cases the dominant response of a many-particle system occurs through excitations of its normal modes, i.e., its collective excitations. As a first approximation it is conventional to study separately the charge, ion and spin variables of the problem. In this way, the familiar plasmon, phonon and magnon excitations result, respectively. These modes are well studied in the bulk. Our analysis proceeds as in Chapter 4. There, we showed that the surface places a boundary condition on the Schrödinger equation that permits the existence of localized single-particle electronic excitations, i.e., the surface states. In this chapter, surface-localized collective excitations will emerge from similar considerations applied to the classical field equations of electromagnetism and elasticity and the equation of motion for spins interacting via a Heisenberg Hamiltonian. We begin with the electronic degree of freedom.

Excitons and plasmons

As noted above, the spectroscopy of surface energy band quasiparticle states requires a photon or electron to knock an electron out of the crystal. Consider now a gentler process whereby an electron merely is excited from an occupied state to a previously unoccupied (bound) state above the Fermi level. The screening properties of a metal ensure that such an electron and hole have little mutual interaction. However, in a semiconductor or an insulator, the potential energy of this two-body system is screened only by the static dielectric constant of the material,

$$V_{\text{Coul}}(\mathbf{r}-\mathbf{r}') = -\frac{e^2}{\varepsilon|\mathbf{r}-\mathbf{r}'|}. \tag{6.1}$$

The Coulomb attraction between an excited electron–hole pair in a dielectric medium leads to a hydrogen atom-like bound state – the exciton. Let us denote the *intrasite* interaction in (6.1) by the symbol α. In addition, there generally is an *intersite* Coulomb interaction that permits correlated hopping of the electron and hole onto an adjacent site. We denote the nearest neighbor hopping integral by the symbol β. Just as for the single-particle excitations, this hopping broadens the energy of the exciton into a band (Knox, 1983).

At the free surface of a crystal, it is reasonable to suppose that the value of the intrasite Coulomb interaction will differ from its bulk value, call it α'. The notation introduced here should make it clear that the mathematical formalism of surface excitons completely parallels the tight-binding model of single-particle excitations developed in detail in Chapter 4; simply change the meaning of the symbols as indicated above. In the earlier case, a localized surface state split off from the bulk states if $|\alpha - \alpha'| > |\beta|$. Precisely the same thing occurs here. In particular, if the inequality is satisfied, a surface exciton splits off below (above) the bulk exciton band if $\alpha > (<) \alpha'$.

Fig. 6.1(*a*) illustrates excitons associated with solid argon observed by optical absorption. The prominent features of the upper spectrum correspond to transitions from the ground state into (spin-orbit split) hydrogenic exciton states of the bulk labeled by principal quantum numbers $n = 1$

Fig. 6.1. Excitons at the surface of insulators: (*a*) argon (Saile *et al.*, 1970); (*b*) anthracene (Philpott & Turlet, 1976). Note the change in the energy scale.

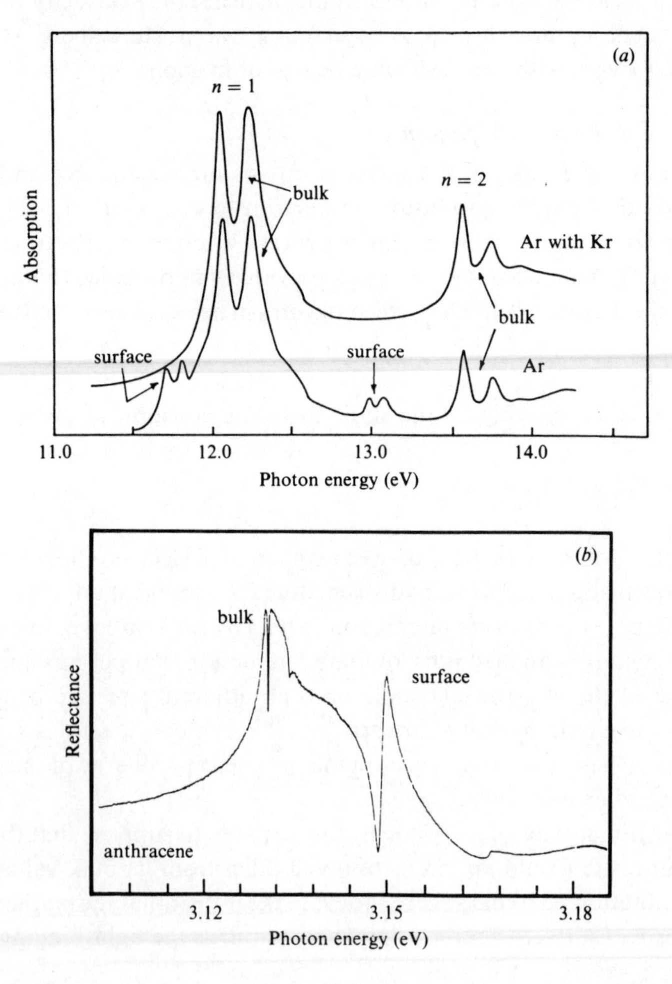

and $n = 2$. The lower spectrum clearly shows absorption features split off below each of the bulk peaks. The latter are surface excitons since they only appear when the surface of the bulk sample is free of contaminants – a krypton overlayer for the case of the upper spectrum. A similar 'crud' test reveals the presence of excitons at the surface of condensed anthracene, a molecular crystal (Fig. 6.1(*b*)). In this case, the optical reflectance data show a surface exciton split off *above* the bulk exciton band.

It is interesting to pursue the analogy to tight-binding electronic states a bit further. We saw earlier that bulk running wave states must be converted into standing wave states in the presence of free surfaces. The same occurs in this case as well. But, there is a new, more profound effect that is particularly relevant to those bulk excitons that propagate directly towards the surface. An exciton in a dielectric medium ($\varepsilon > 1$) that approaches the vacuum ($\varepsilon = 1$) feels a long range repulsive force from its own image (for any orientation of the exciton dipole). Of course, the image force actually derives from the polarization of atoms in the near surface region. Nevertheless, this repulsion establishes an exciton 'dead zone' where the density of excitons is depleted relative to the bulk. This layer can extend hundreds of Ångströms and is observed by shifts and broadenings in reflectivity spectra like Fig. 6.1(*b*).

The surface exciton alters the charge density at a particular site of a semi-infinite crystal by creation of a single electron–hole pair. In a metal, a coherent superposition of electron–hole pairs can be formed that represents a wave-like disturbance of the charge density localized at the surface. This is the *surface plasmon*. Like its bulk counterpart, the surface plasmon is a longitudinal mode. Let us characterize the charge compressions and rarefactions along the surface associated with this mode by a wave vector $\mathbf{q}_{\parallel}$. Then, Laplace's equation demands that the accompanying electrostatic potential decay exponentially away from the surface according to:

$$\phi(\mathbf{r}) = \phi_0 \, e^{i\mathbf{q}_{\parallel}\cdot\mathbf{r}_{\parallel}} e^{-q_{\parallel}|z|}. \tag{6.2}$$

Notice that the characteristic length scale for this decay is set by the magnitude of the surface wave vector. It is easy to demonstrate that when one superposes equal amplitude contributions like (6.2) from all two-dimensional vectors $\mathbf{q}_{\parallel}$ the result is a simple dipole potential. This is unsurprising since the half-crystal is charge neutral.

The tangential component of the electric field associated with (6.2) is continuous. The normal component of $\mathbf{E}(\mathbf{r})$ is discontinuous,

$$\left.\begin{aligned} E_z(z = 0^+) &= \phi_0 q_{\parallel} e^{i\mathbf{q}_{\parallel}\cdot\mathbf{r}_{\parallel}}, \\ E_z(z = 0^-) &= -\phi_0 q_{\parallel} e^{i\mathbf{q}_{\parallel}\cdot\mathbf{r}_{\parallel}}. \end{aligned}\right\} \tag{6.3}$$

Therefore, in the usual way, Laplace's equation is satisfied everywhere in space except at the surface plane $z=0$ where a charge sheet exists:

$$\rho(\mathbf{r}) = \frac{1}{4\pi}[E_z(z=0^-) - E_z(z=0^+)]. \tag{6.4}$$

Suppose the metal that occupies the positive half-space ($z>0$) is characterized by the dielectric function $\varepsilon(\omega)$. In order to ensure that $\nabla \cdot \mathbf{D} = 0$ we must require

$$\varepsilon(\omega) = -1. \tag{6.5}$$

If we choose the standard form, $\varepsilon(\omega) = 1 - \omega_p^2/\omega^2$, where ω_p is the bulk plasma frequency, (6.5) admits only the *surface plasmon* solution: $\omega_s = \omega_p/\sqrt{2}$. This result will be true for long wavelength oscillations of the surface charge layer.* A schematic real-space view of the surface plasmon and its attendant electric field is shown in Fig. 6.2. As the excitation wavelength begins to approach atomic dimensions we expect the surface plasmon to exhibit dispersion, i.e., $\omega = \omega(q_\parallel)$. Electron energy loss spectroscopy (EELS) shows this phenomenon quite beautifully (Fig. 6.3). In this experiment, high energy (50 keV) electrons lose energy to the plasmons as they traverse thin (~ 100 Å) metal films. The wave vector parallel to the surface is varied by adjusting the angle of incidence – just as in angle-resolved photoemission.

The theory of surface plasmon dispersion is rather complicated. However, a simple, comprehensible result emerges for the jellium model of a

Fig. 6.2. Schematic view of the electric field induced by a surface plasmon (courtesy of H. Ibach).

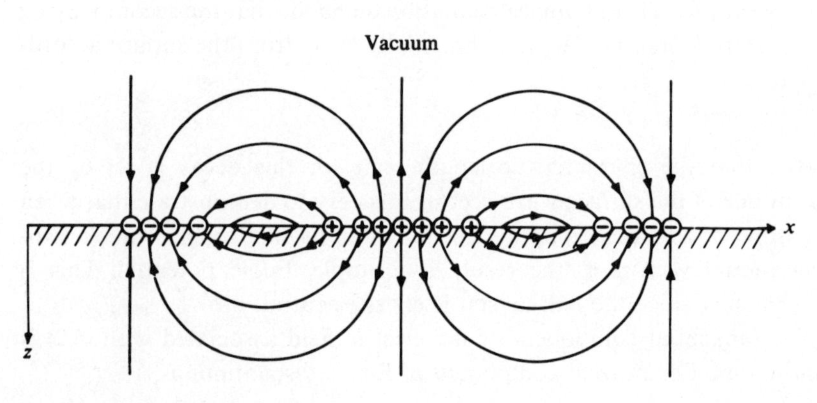

* The wavelength cannot be *too* long. If $q_\parallel^{-1} \gtrsim c/\omega_p$ retardation effects must be included (see Chapter 7).

metal surface (Feibelman, 1982):

$$\omega_s(q_\parallel) = \frac{\omega_p}{\sqrt{2}}\{1 + \tfrac{1}{2}q_\parallel(d_\parallel + d_\perp) + \cdots\}. \tag{6.6}$$

In this expression, $d_\parallel$ and $d_\perp$ are precisely the effective surface 'positions' defined earlier (cf. (4.6) and (4.10)). The former sets the surface location for the tangential component of the electric field. Since this field is continuous, $d_\parallel$ simply marks the centroid of the bare jellium electron spill-out. By contrast, the normal component of the field is discontinuous in a manner that depends on the charge induced at the surface (6.4). Of course, this charge distribution is not a delta function on the microscopic scale. The quantity $\delta n(z)$ that enters (4.10) is the charge profile of the surface plasmon. Note that the group velocity, $d\omega/dq_\parallel$, implied by (6.6) is very small compared to the speed of light. This justifies the electrostatic approximation used here.

We now are in a position to demonstrate that a familiar static property of a metal surface can be represented by (virtual) excitations of its charged normal modes. To see this, consider the interaction of a single electron with the electrostatic field produced outside a metal surface by the plasmon. The interaction energy is simply $e\phi(\mathbf{r})$, where $\mathbf{r}$ points to the position of the external electron. To be quantitative, we must evaluate the constant, ϕ_0 in (6.2). To begin, recall the elementary oscillator model for the *bulk* plasmon (Kittel, 1966). A displacement of the electron gas by a distance u_0 relative to the fixed positive background generates an electric field

Fig. 6.3. EELS results for plasmon dispersion at the surface of polycrystalline films: (*a*) aluminum; (*b*) indium (Krane & Raether, 1976).

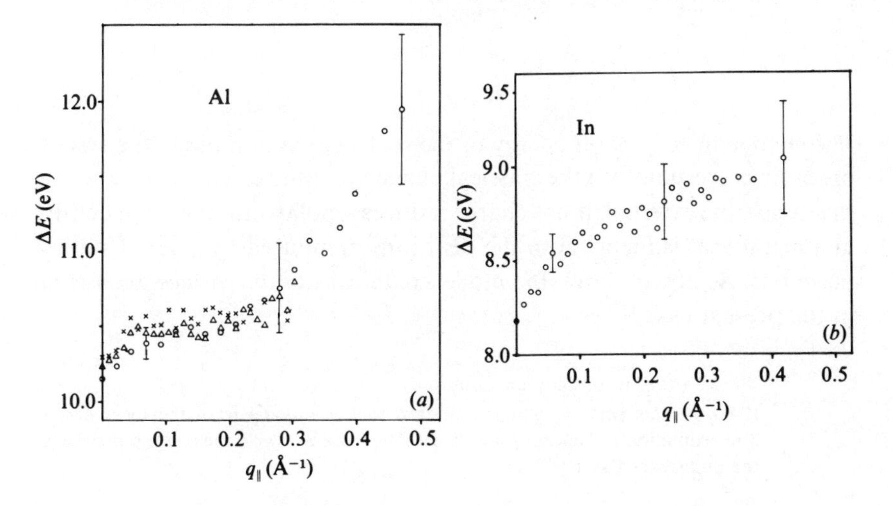

$E = 4\pi Neu_0$, where N is the number density of free electrons. This field acts as a restoring force so that the gas oscillates with a frequency $\omega_p^2 = 4\pi Ne^2/m$. At the surface, the characteristic frequency is $\omega_s^2 = \omega_p^2/2$. Consequently, reversing the argument, the electric field within the surface of the metal is $E = 2\pi Neu_0$. Equating this with the interior field given by (6.3) yields $\phi_0 = 2\pi Neu_0/q_\parallel$.

It is simplest to represent the charge displacement $u_{\boldsymbol{\kappa}}$ associated with a surface plasmon of wave vector $\boldsymbol{\kappa}$ in terms of a harmonic oscillator in second-quantized form.* The result is

$$u_{\boldsymbol{\kappa}} = \sqrt{\frac{\hbar}{2m\omega_s}}\sqrt{\frac{\kappa}{NA}}(a_{\boldsymbol{\kappa}} + a_{-\boldsymbol{\kappa}}^\dagger)\,e^{i\boldsymbol{\kappa}\cdot\mathbf{r}_\parallel}e^{-\kappa z} \tag{6.7}$$

where A is the area of the surface and the extra dimensionless factor κ/NA is required to normalize the oscillator amplitude (Evans & Mills, 1973). The total energy of the interacting system follows by combining (6.7) with (6.2):

$$\mathscr{H} = \sum_{\boldsymbol{\kappa}} \hbar\omega_s(a_{\boldsymbol{\kappa}}^\dagger a_{\boldsymbol{\kappa}} + \tfrac{1}{2}) + \sum_{\boldsymbol{\kappa}} g_\kappa(a_{\boldsymbol{\kappa}} + a_{-\boldsymbol{\kappa}}^\dagger)\,e^{i\boldsymbol{\kappa}\cdot\mathbf{r}_\parallel}e^{-\kappa z} \tag{6.8}$$

where the constants have been lumped into

$$g_\kappa = 2\pi e^2\sqrt{\frac{N\hbar}{2m\omega_s\kappa A}}. \tag{6.9}$$

Equation (6.8) can be simplified since it is a quadratic function of the oscillator operators. In particular, we can 'complete the square' by making the substitution $a_\kappa = b_\kappa - (g_\kappa/\hbar\omega_s)\exp(-i\boldsymbol{\kappa}\cdot\mathbf{r}_\parallel - \kappa z)$:

$$\begin{aligned}\mathscr{H} &= \sum_{\boldsymbol{\kappa}} \hbar\omega_s(b_{\boldsymbol{\kappa}}^\dagger b_{\boldsymbol{\kappa}} + \tfrac{1}{2}) - \sum_{\boldsymbol{\kappa}} \frac{g_\kappa^2}{\hbar\omega_s}e^{-2\kappa z} \\ &= \sum_{\boldsymbol{\kappa}} \hbar\omega_s(b_{\boldsymbol{\kappa}}^\dagger b_{\boldsymbol{\kappa}} + \tfrac{1}{2}) - \frac{e^2}{4z}. \end{aligned} \tag{6.10}$$

The change in zero-point energy of the surface plasmon oscillator system precisely corresponds to the classical image potential energy! We conclude that a nearby, external, point charge induces a polarization charge density in a metal that is identical to the distribution induced by a set of surface plasmons. As noted above, the mode excitations of the surface are virtual in the present case.†

* We set $\kappa \equiv q_\parallel$ to simplify the notation.

† It is just this sort of virtual excitation that is missing from the local density approximation and accounts for its failure to properly yield the image potential (cf. the discussion above (4.9)).

Phonons

In this section we study elementary excitations of a semi-infinite crystal that involve correlated motion of ions whose vibrational amplitude decreases exponentially with increasing depth into the crystal. These surface density waves fall naturally into three categories:

(*a*) long wavelength elastic waves,
(*b*) long wavelength optical waves,
(*c*) short wavelength acoustic and optical waves.

The first discussion of such phenomena goes back to Lord Rayleigh (1885).

We treat long wavelength elastic surface waves with classical continuum elastic theory (Landau & Lifshitz, 1970). Let $\mathbf{u}$ denote the displacement from equilibrium of an element of a semi-infinite, isotropic, elastic medium. In general, this vector can be decomposed into the sum of a transverse displacement ($\nabla \cdot \mathbf{u}_t = 0$) and a longitudinal displacement ($\nabla \times \mathbf{u}_l = 0$). The Cartesian components of $\mathbf{u}_t$ and $\mathbf{u}_l$ satisfy the usual wave equation with their respective velocities of sound:

$$\begin{aligned} \frac{\partial^2 \mathbf{u}_t}{\partial t^2} - c_t^2 \nabla^2 \mathbf{u}_t &= 0, \\ \frac{\partial^2 \mathbf{u}_l}{\partial t^2} - c_l^2 \nabla^2 \mathbf{u}_l &= 0. \end{aligned} \tag{6.11}$$

In the bulk, these orthogonal displacements are uncoupled. Near the free surface we seek solutions to (6.11) of the form

$$\mathbf{u}_i = \mathbf{a}_i \mathrm{e}^{\mathrm{i}(q_\parallel x - \omega t)} \mathrm{e}^{-\kappa z} \tag{6.12}$$

where now $\kappa = (q_\parallel^2 - \omega^2/c^2)^{1/2}$ – a positive quantity. Note that these are *macroscopic* waves in the sense that the displacements persist quite far into the bulk in the limit when $q_\parallel \rightarrow 0$.

As always, the boundary conditions determine the nature of the solutions. In an infinite medium, equilibrium stresses correspond to forces that cancel on either side of an infinitesimal volume element. At the surface, all the components of the forces that cross the cleavage plane must vanish (Fig. 6.4):

$$\mathrm{d}F_i = \sigma_{iz}\, \mathrm{d}A_z = 0. \tag{6.13}$$

We assume that the components of the stress tensor, σ_{ij}, are related to the components of the strain tensor, $\varepsilon_{ij} = (\partial u_i/\partial x_j + \partial u_j/\partial x_i)/2$, by Hooke's law. For an isotropic medium, the elastic constants of proportionality can be written in terms of the Young's modulus, Y, and Poisson's ratio, P.

Thus, applying one of the conditions (6.13):

$$\sigma_{yz} = \frac{Y}{1+P}\delta_{yz} = 0 \Rightarrow \frac{\partial u_y}{\partial z} = 0 \Rightarrow u_y = 0. \tag{6.14}$$

This boundary condition requires that the surface modes have displacements confined to the *sagittal plane*, i.e., the plane that contains the propagation direction and the surface normal. Since, by definition,

$$\left.\begin{aligned} \nabla \cdot \mathbf{u}_{\mathrm{t}} &= \mathrm{i}q_{\parallel}u_{\mathrm{t}x} - \kappa_{\mathrm{t}}u_{\mathrm{t}z} = 0, \\ (\nabla \times \mathbf{u}_{\mathrm{l}})_y &= \kappa_{\mathrm{l}}u_{\mathrm{l}x} + \mathrm{i}q_{\parallel}u_{\mathrm{l}z}, \end{aligned}\right\} \tag{6.15}$$

the components of the total displacement follow by inspection:

$$\begin{aligned} u_x &= \{a\kappa_{\mathrm{t}}\mathrm{e}^{-\kappa_{\mathrm{t}}z} + bq_{\parallel}\mathrm{e}^{-\kappa_{\mathrm{l}}z}\}\,\mathrm{e}^{\mathrm{i}(q_{\parallel}x-\omega t)}, \\ u_z &= \{aq_{\parallel}\mathrm{e}^{-\kappa_{\mathrm{t}}z} + b\kappa_{\mathrm{l}}\mathrm{e}^{-\kappa_{\mathrm{l}}z}\}\,\mathrm{i}\mathrm{e}^{\mathrm{i}(q_{\parallel}x-\omega t)}. \end{aligned} \tag{6.16}$$

Using the remaining two boundary conditions, $\sigma_{xz} = \sigma_{zz} = 0$, it is easy to show that this wave is acoustic, i.e., $\omega = c_{\mathrm{R}}q_{\parallel}$, where c_{R} is determined by the bulk elastic constants. The fact that $\kappa > 0$ guarantees that c_{R} is less than both c_{t} and c_{l}. This low velocity acoustic surface mode is called the Rayleigh wave. A more complete discussion takes account of the fact that real materials generally possess anisotropic elastic constants. In this case, one finds either exactly the monotonically damped behavior of (6.16) or an underdamped *generalized* Rayleigh mode that decays sinusoidally into the bulk. Fig. 6.5 indicates the specific type of Rayleigh wave that occurs on the (100) surface of some common materials. The interior area bounded by the dashed lines is the domain of elastic stability.

Another class of long wavelength surface modes exist in ionic compounds. Here, the familiar transverse optic (TO) and longitudinal optic

Fig. 6.4. Unbalanced forces at the surface of a semi-infinite elastic medium.

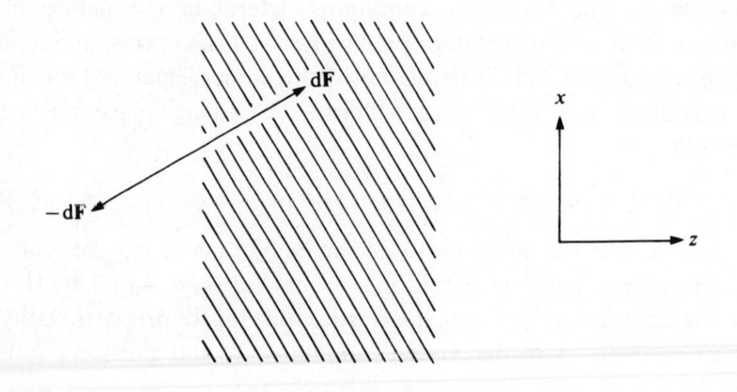

phonons induce an oscillating dipole moment in every unit cell of the crystal. Each cell can be characterized by an *effective charge* that describes the change of the dipole moment with the atomic displacement of the optical mode:

$$e^*_{ij} = \frac{\partial M_i}{\partial u_j}. \tag{6.17}$$

The optical properties of the bulk then are described conveniently in terms of a dielectric function (Kittel, 1966),

$$\varepsilon(\omega) = \varepsilon(\infty) + \frac{\Omega^2}{\omega_{\text{TO}}^2 - \omega^2} \quad \text{with} \quad \Omega^2 = \varepsilon(\infty)\frac{4\pi e^{*2}}{Ma_0^2}. \tag{6.18}$$

We seek an analog to the bulk optical modes which is confined to the surface region. The macroscopic fields produced by this mode should be derivable from an electrostatic potential of the form (6.2). The analysis proceeds as before so the surface mode satisfies $\varepsilon(\omega) = -1$ with (6.18) as the dielectric function. Consequently,

$$\omega_s = \sqrt{\frac{\varepsilon(0) + 1}{\varepsilon(\infty) + 1}}\,\omega_{\text{TO}}. \tag{6.19}$$

Electron energy loss spectroscopy is perfectly suited for direct observation of this long wavelength surface optical phonon. The physical process is inelastic Coulomb scattering from the long range dipole field above the crystal. This scattering is strongly peaked in the forward direction. To see

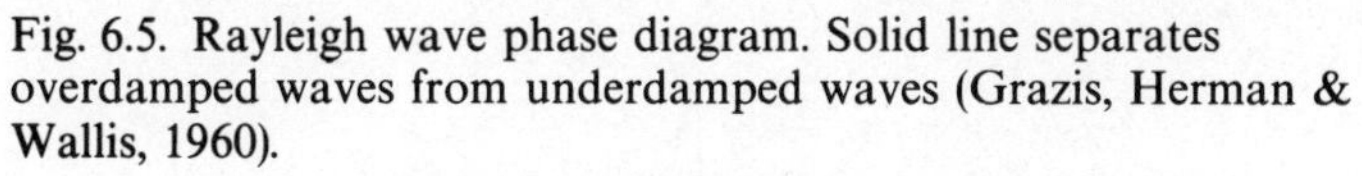
Fig. 6.5. Rayleigh wave phase diagram. Solid line separates overdamped waves from underdamped waves (Grazis, Herman & Wallis, 1960).

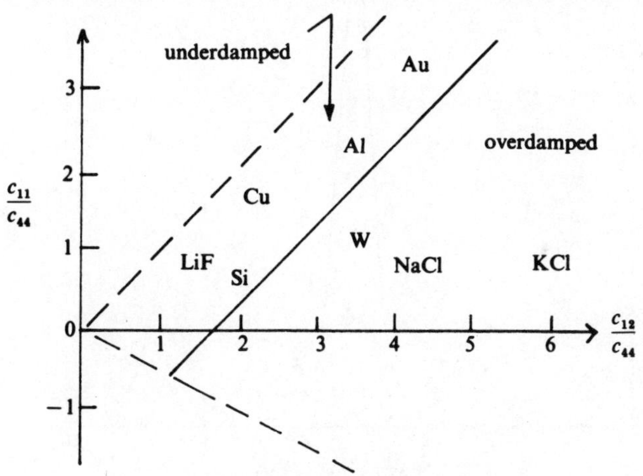

this, consider an incident electron of wave vector **k** and kinetic energy E. The magnitude of the scattered wave vector differs little from $|\mathbf{k}|$ since we consider only long wavelength modes, i.e., $\mathbf{q}_{\|} \to 0$. Accordingly, if the electron indeed scatters through a small angle, θ, we will have $q_{\|} = k\theta$. Since the range of the potential is of the order of $q_{\|}^{-1}$, the time the electron spends within the dipole field is $t \sim 2m/hkq_{\|} = h/E\theta$. Further, a mode of energy $h\omega_s$ will be most strongly excited for an electron transit time of order ω_s^{-1}. Therefore, the strongest scattering occurs into an angle $\theta = h\omega_s/E$. Our initial assumption of small angle scattering is justified for the excitation of phonons ($h\omega \sim 100\,\text{meV}$) by low energy electrons ($E \sim 1$–$100\,\text{eV}$). Since we have learned that low energy electrons diffract from a crystal surface, the strongest EELS signal appears in the specular beam.

An energy loss spectrum for electrons scattered from the (1100) surface of ZnO shows a substantial peak at about 69 meV (Fig. 6.6). The peak position is in excellent agreement with the prediction of (6.19) using the bulk dielectric properties of zinc oxide. The inelastic signal is particularly strong in this case because e^* is large and elementary dipoles down to $q_{\|}a_0$ atomic layers below the surface contribute to the field. The spectrum also shows a small energy *gain* feature at -69 meV. This peak represents the absorption of a pre-existing surface phonon by the electron beam. The gain-to-loss peak ratio can be used to deduce the surface temperature.

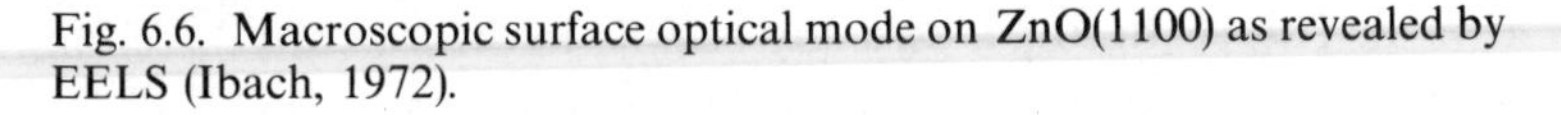

Fig. 6.6. Macroscopic surface optical mode on ZnO(1100) as revealed by EELS (Ibach, 1972).

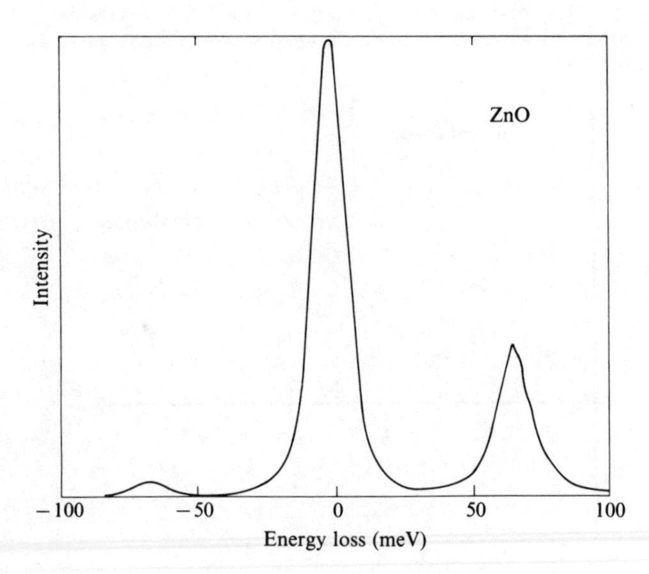

A very similar gain and loss peak appears at 56 meV for Si(111). Here, the very existence of such features in an EELS spectrum is remarkable at first glance because the inversion symmetry of the silicon lattice requires that e^* vanish identically. There are no bulk excitations that produce macroscopic fields throughout the crystal. However, silicon atoms near the surface certainly do *not* sit in inversion symmetric sites. Accordingly, we can expect the effective charge to decay from some finite value at the surface to zero deep in the bulk. It is just these near-surface atoms that produce the scattering field for the EELS experiment.

The observation of macroscopic surface optical modes at the SBZ center by EELS depends crucially on the small momentum transfer associated with Coulomb dipole scattering. Therefore, this mechanism cannot probe excitations at finite wave vector, i.e., the surface phonon dispersion. However, electrons also can scatter directly from the short range atomic potential. In this case, phonons are excited by direct impact, and momentum transfers that span the Brillouin zone are accessible experimentally. The resolution of the data obtained in this way is limited principally by the energy spread of the incident electron beam. With the best present technology, phonon dispersion can be mapped by EELS with about 5 meV resolution.

It is desirable to identify an alternative to electron scattering for the study of surface phonon dispersion for several reasons. First, one always seeks to improve the resolution. Second, the electron impact scattering cross section may not be sufficiently large for all surfaces of interest. Finally, the recurring problem of sample charging persists for insulating surfaces. In Chapter 3, we were drawn to low energy electrons because the relation $E = (h/\lambda)^2/2m$ guarantees that surface sensitive electrons exhibit de Broglie wavelengths that match to typical crystal lattice constants. Notice now that the same wavelength match occurs for neutral helium atoms with kinetic energies in excess of 10 meV. To investigate the surface sensitivity of very low energy helium atoms we return to the effective medium immersion energy curves of Fig. 4.7. We saw there that excess electrons are repelled by a single helium atom. If the argument is reversed, we expect an external He atom to be repelled by the electrons that spill out from a crystal surface. This is also 'impact' scattering.

A thermal beam of helium atoms scatters from the exponentially decaying outer fringes of the surface charge density profile. In fact, conventional diffraction occurs because the in-plane charge distribution naturally reflects the surface reciprocal net. Accordingly, He atom scattering is useful for surface crystallography. For present purposes, we focus on the fact that the energy spread of a thermal He atom beam can be

Fig. 6.7. Yield of He atoms scattered from NaF(100) as a function of relative scattering angle. The elastic Bragg peaks are labelled. Their intensities approach 100 on the scale of the figure (Doak & Toennies, 1982).

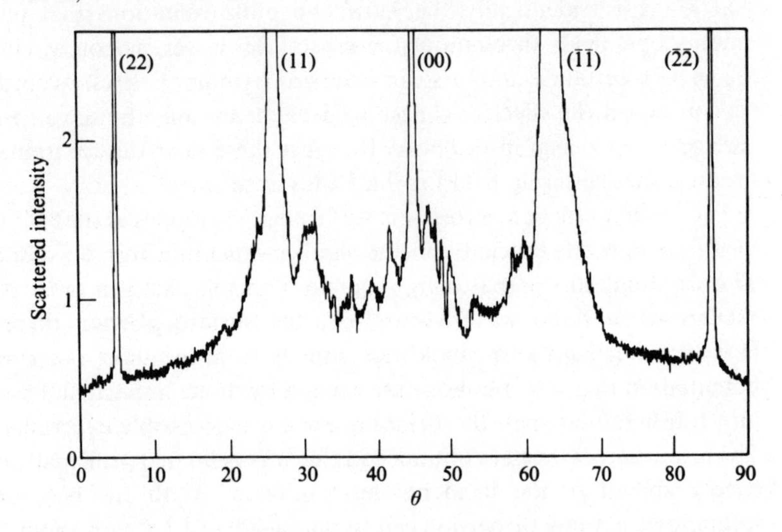

Fig. 6.8. Experimental (filled and open circles) and theoretical (heavy and dashed curves) dispersion of surface phonons on NaF(100). Only modes polarized in the sagittal plane are shown (Brusdeylins *et al.*, 1985).

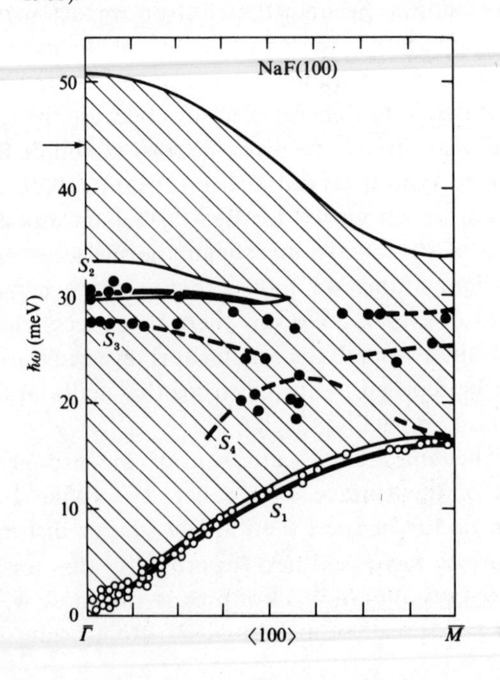

kept as low as 0.25 meV and study *inelastic* scattering events where the magnitude of the parallel momentum transfer is less than that of a surface reciprocal lattice vector. In other words, the surface phonons appear *between* the Bragg peaks (Fig. 6.7).

The measurement of surface phonon dispersion, $\omega(\mathbf{q}_{\parallel})$, by He atom scattering is very similar to bulk phonon mapping with neutron scattering. The momentum transfer is set by fixing the incoming and outgoing beam direction with respect to the crystal normal while the scattered atom's kinetic energy is measured by a time-of-flight analysis. Experimental results for NaF(100) are shown in Fig. 6.8. The data are plotted along with the projection of the bulk phonon modes onto the surface Brillouin zone (hatched region) – precisely the same representation we encountered for surface electronic states in Chapter 4. The lowest frequency branch (S_1) is the Rayleigh mode described above. The frequency of the long wavelength optical phonon (6.19) is indicated by a horizontal arrow. We now see that it is more properly termed a *resonance* because it is embedded in the bulk continuum. This fact, combined with its very long intrinsic amplitude decay length ($\sim q_{\parallel}^{-1}$) guarantees thorough mixing with the bulk modes. A short range impact probe like He atom scattering cannot distinguish it from the bulk phonon background.

NaF(100) apparently supports at least three other surface-localized phonon modes (S_2, S_3 and S_4). These are truly *microscopic* surface states (resonances) in the sense that they originate from short range harmonic forces between ions. To interpret them, we must study the lattice dynamics of a semi-infinite crystal. The appropriate language again resembles the tight-binding calculation of electronic states of Chapter 4.

Begin with the bulk and suppose that the total energy of the system is known as a function of the ionic coordinates. Let $\mathbf{u}(\mathbf{R})$ be the displacement of each ion away from its equilibrium lattice site. In the harmonic approximation, the motion of each ion is governed by Newton's second law:

$$M(\mathbf{R})\ddot{u}(\mathbf{R}) = -\sum_{\mathbf{R}'} \langle \mathbf{R} | \mathbf{D}_0 | \mathbf{R}' \rangle \mathbf{u}(\mathbf{R}') \tag{6.20}$$

where $\langle \mathbf{R} | \mathbf{D}_0 | \mathbf{R}' \rangle$ is a double-indexed *dynamical matrix* of force constants,

$$\langle \mathbf{R} | (\mathbf{D}_0)_{ij} | \mathbf{R}' \rangle = \frac{\partial^2 E_{\text{total}}}{\partial u_i(\mathbf{R}) \partial u_j(\mathbf{R}')}. \tag{6.21}$$

The explicit vector notation in $\langle \mathbf{R} | \mathbf{D}_0 | \mathbf{R}' \rangle$ indexes the Cartesian coordinates of the displacements. In anticipation of the coming surface problem, we use two-dimensional periodicity to simplify the equations of motion. If we set

$$\mathbf{u}(\mathbf{R}) = \frac{1}{\sqrt{N_{\parallel} M(\mathbf{R})}} \mathbf{u}(\mathbf{q}_{\parallel}, z)\, \mathrm{e}^{\mathrm{i}(\mathbf{q}_{\parallel}\cdot\mathbf{R} - \omega t)} \tag{6.22}$$

and define

$$\langle z | \mathbf{D}_0(\mathbf{q}_{\parallel}) | z' \rangle = \sum_{\mathbf{R}'_{\parallel}} \langle \mathbf{R} | \mathbf{D}_0 | \mathbf{R}' \rangle \mathrm{e}^{\mathrm{i}\mathbf{q}_{\parallel}\cdot\mathbf{R}'_{\parallel}}, \tag{6.23}$$

(6.20) can be written,

$$\sum_{z'} \langle z | \mathbf{D}_0(\mathbf{q}_{\parallel}) | z' \rangle \mathbf{u}(\mathbf{q}_{\parallel}, z') = \omega^2(\mathbf{q}_{\parallel}) \mathbf{u}(\mathbf{q}_{\parallel}, z). \tag{6.24}$$

In these formulae, we assume that $\mathbf{R} = z\mathbf{a}_0 + \mathbf{R}_{\parallel}$, where z is an integer and $\mathbf{a}_0$ is the bulk direct lattice vector normal to the surface.

Equation (6.24) is appropriate for calculation of the bulk phonons. The phonon eigenfrequencies follow by requiring that $\det|\omega^2\mathbf{1} - \mathbf{D}_0(\mathbf{q}_{\parallel})| = 0$. Here, $\langle \mathbf{R} | \mathbf{D}_0 | \mathbf{R}' \rangle$ depends only on $\mathbf{R} - \mathbf{R}'$ and the size of the matrix to be inverted is set by the *range* of the harmonic restoring forces. However, suppose we now introduce a perturbation that cuts the infinite crystal into two pieces by severing the restoring forces across a plane parallel to the surface of interest, i.e.,

$$\langle z | \mathbf{V} | z' \rangle = -\langle z | \mathbf{D}_0 | z' \rangle \delta_{z,0} \delta_{z',1}. \tag{6.25}$$

With this artifice, the phonons of the semi-infinite crystal are found from

$$\sum_{z'} \langle z | \mathbf{D} | z' \rangle \mathbf{u}(z') = \sum_{z'} \langle z | \mathbf{D}_0 + \mathbf{V} | z' \rangle \mathbf{u}(z') = \omega^2 \mathbf{u}(z). \tag{6.26}$$

Unfortunately, the additive nature of the perturbation means that (6.26) directs us to invert a matrix that is semi-infinite in rank. There are two common solutions to this problem. In the first approach, one cuts the crystal in *two* places to create a finite sized slab. This is the lattice dynamical analog to our slab approach to the electronic structure problem (cf. Fig. 4.16). Just as in that case, the solution to the secular equation yields standing wave bulk modes, true surface states and surface resonant modes that mix with the bulk but have greatest amplitude on the surface atoms. It is even possible to trace the fate of the macroscopic surface optical mode (Chen, de Wette & Alldredge, 1977).

The alternative approach to the solution of (6.26) is particularly attractive since it retains the true semi-infinite character of the problem. Define the inverse, or *Green function* matrix, $\mathbf{G} = (\omega^2\mathbf{1} - \mathbf{D})^{-1}$ and an analogous matrix $\mathbf{G}_0$ that corresponds to $\mathbf{D}_0$. It is easy to verify that the relationship between the two is simply

$$\mathbf{G} = \mathbf{G}_0 + \mathbf{G}_0 \mathbf{V} \mathbf{G}, \tag{6.27}$$

where matrix multiplication over the planar index is implied. Our problem requires $\det|\mathbf{G}^{-1}| = 0$. However, from (6.27),

$$G^{-1} = G_0^{-1}(1 - G_0 V). \tag{6.28}$$

Therefore, the phonon frequencies of the cleaved crystal are determined from $\det|\mathbf{1} - \mathbf{G}_0\mathbf{V}| = 0$. Notice that this matrix is not larger than the original $\mathbf{G}_0$ matrix of the infinite bulk since the perturbation now appears *multiplicatively*.

The calculated dispersion of surface phonon states and resonances in Fig. 6.8 (solid curves) were found using this method.* It is no accident that comparison to experiment is made only for those modes polarized in the sagittal plane. Helium atom scattering is relatively insensitive to modes that lack appreciable ion motion normal to the surface (or along the scattering direction) because the impinging atoms interact only with the low density tails of the surface electronic charge distribution. Nevertheless, good agreement is found between theory and experiment for NaF(100). This suggests that a reasonable model was used to calculate the force constants (6.21). In fact, the force constants used here come from a semi-empirical scheme that uses bulk data as input: elastic constants, constituent polarizabilities, $\varepsilon(0)$, $\varepsilon(\infty)$ and ω_{TO}. The surface perturbation merely splits the crystal as indicated above. Consequently, the agreement may be unsurprising since, as we have learned (Chapter 4), the surface represents a relatively minor perturbation to a highly ionic crystal.

Vibrational excitations at the surface of a covalent semiconductor must be treated in a different way. There can be little doubt that the nature of microscopic surface modes will strongly depend on the details of any reconstruction and/or unusual electronic properties specific to the surface. A properly microscopic approach would calculate the total energy functional that enters (6.21) as the expectation value of an appropriate *electronic* Hamiltonian. In particular, ionic screening begins to play a role, as does any bond rehybridization that accompanies ionic displacements. Fig. 6.9 illustrates the results of such a calculation for the phonon dispersion at the surface of Si(100) 2×1 reconstructed into the asymmetric dimer geometry discussed in Chapter 4. The long wavelength Rayleigh wave splits off below the acoustic continuum as before. However, a short wavelength mode at the SBZ corner provides an excellent picture of a truly microscopic optical mode (Fig. 6.10). The lattice displacements consist of a simple *rocking* motion of the surface dimer.

* The Green function method can be used for the surface electronic structure problem if the dynamical matrix is replaced by the Hamiltonian. In fact, the results shown in Figs. 4.13 and 4.36 were obtained just this way.

Conduction electron screening plays a particularly important role in determining the phonon spectrum of metals (Devreese, 1983). Therefore, a first-principles calculation of metal surface phonons minimally should account for this phenomenon. Indeed, rather sophisticated slab calculations for simple metals indicate that if the screening is *not* correctly modelled, e.g., by neglect of the exchange–correlation term in the total energy (see (4.2)), the Rayleigh mode frequency can become negative (Calandra, Catellani & Beatrice, 1985)! On the other hand, metals seldom reconstruct and our effective medium theory showed that relaxations tend to return the surface atoms to an environment as nearly isotropic and

Fig. 6.9. Surface phonon dispersion of Si(100) 2 × 1 and the projected bulk modes (Allan & Mele, 1984).

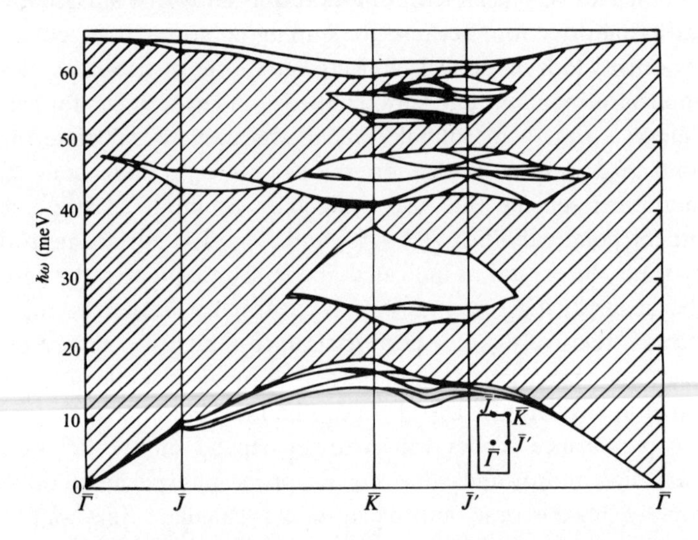

Fig. 6.10. Atomic displacements for the surface rocking mode (R) of Si(100). The arrow lengths are proportional to the relative amplitudes (Allan & Mele, 1984).

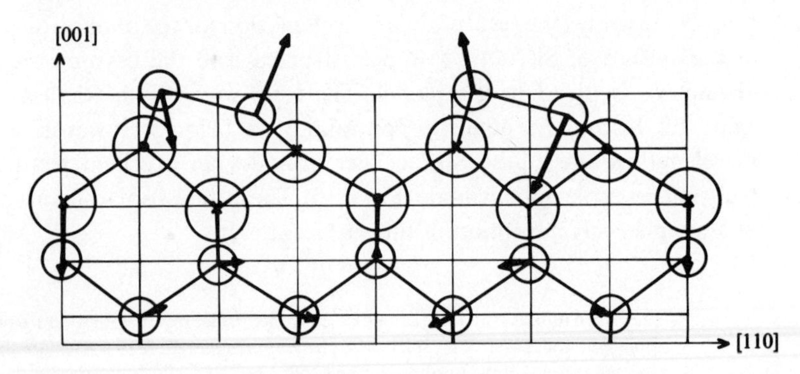

bulk-like as possible. This suggests that the qualitative features of the surface excitation spectrum again may be described adequately by an empirical bulk scheme with a simple cleavage perturbation *à la* (6.25). Atom scattering results for the (111) face of noble metals provide support for this notion (Fig. 6.11). The Rayleigh mode is well described by this theory whereas the energy of a longitudinal acoustic surface resonance (dashed curve) is overestimated, particularly at short wavelengths. It is worth noting that agreement would be considerably improved if one simply reduced the magnitude of the force constant that couples the surface plane to the first subsurface plane from its bulk value.

Magnons

Elementary excitations of the magnetization of a ferromagnet or anti-ferromagnet are known as spin waves, or magnons. Consider a system for which a well-defined spin moment, $\mathbf{S}_i$, is identified with each lattice site. In accordance with the discussion of Chapter 5, we know there is a short range, quantum mechanical, exchange interaction between neighboring spins of the form:

$$-J_{\mathbf{RR}'}\mathbf{S}_{\mathbf{R}}\cdot\mathbf{S}_{\mathbf{R}'}. \tag{6.29}$$

Fig. 6.11. Surface phonon dispersion for Ag(111) measured by He atom scattering compared to a terminated-bulk force constant calculation (Doak, Harten & Toennies, 1983).

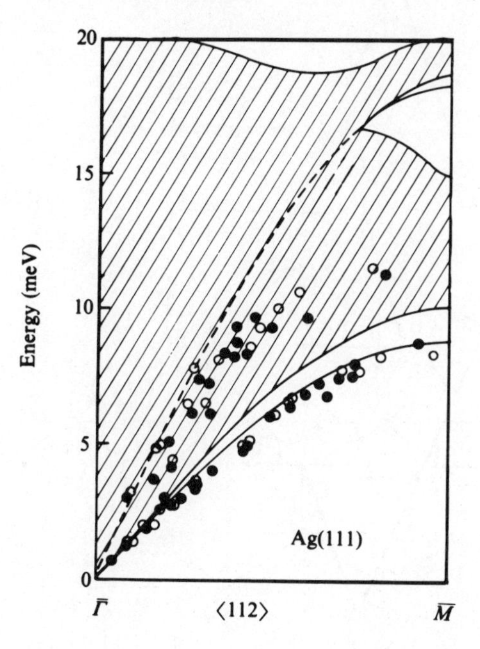

In addition, the moments interact via the classical, long range, dipole–dipole interaction:

$$\frac{\mathbf{S}_{\mathbf{R}}\cdot\mathbf{S}_{\mathbf{R}'}}{|\mathbf{R}-\mathbf{R}'|^3}-\frac{3\mathbf{S}_{\mathbf{R}}\cdot(\mathbf{R}-\mathbf{R}')\mathbf{S}_{\mathbf{R}'}\cdot(\mathbf{R}-\mathbf{R}')}{|\mathbf{R}-\mathbf{R}'|^5}. \tag{6.30}$$

At very long wavelengths the classical dipole energy dominates the problem and the spin wave analysis requires only macroscopic physics – just as for Rayleigh waves in the phonon case. Of course, we are concerned with the situation where the deviation of the magnetization from its average value is confined to the surface.

A long wavelength spin wave can be viewed as the precession of the total magnetization, $\mathbf{M}$, around the direction of an external, static magnetic field, $\mathbf{H}_0$. To fix ideas, take $\mathbf{H}_0$ parallel to the saturation magnetization and aligned with the x-axis (Fig. 6.12). The magnetization moves in response to the applied torque,

$$\frac{d\mathbf{M}}{dt}=\gamma(\mathbf{M}\times\mathbf{H}), \tag{6.31}$$

Fig. 6.12. Geometry for dipolar magnon discussion.

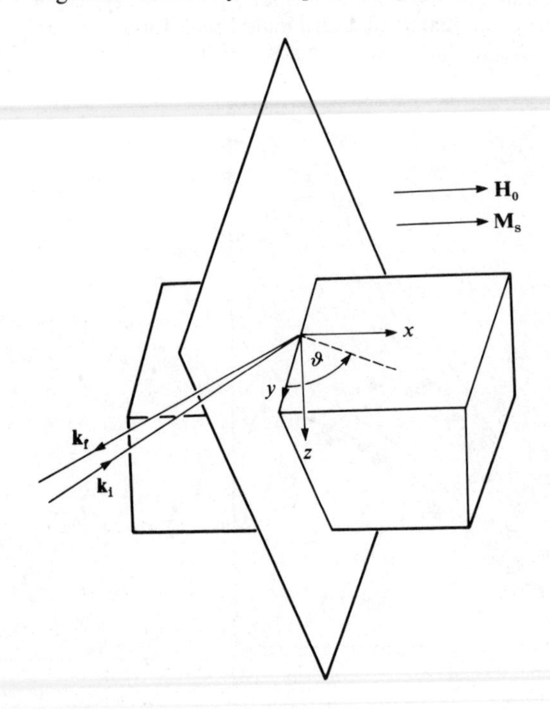

subject only to the condition that $\nabla \cdot \mathbf{B} = \nabla \cdot (\mathbf{H} + 4\pi\mathbf{M}) = 0$. The field $\mathbf{H}$ includes both $\mathbf{H}_0$ and the demagnetizing fields in the y–z plane that rotate $\mathbf{M}$ away from $\mathbf{M}_s$. In the bulk, the precession frequency depends on the angle between the spin wave vector, $\mathbf{q}$, and $\mathbf{M}_s$ and varies continuously from γH_0 ($\mathbf{q} \parallel \mathbf{M}_s$) to $\gamma\sqrt{H_0 B}$ ($\mathbf{q} \perp \mathbf{M}_s$) (Kittel, 1966). The excitation frequencies are in the microwave portion of the electromagnetic spectrum so that the fields may properly be regarded as *magnetostatic*, i.e., $\nabla \times \mathbf{H} = 0$.

We now seek a magnetization wave that is localized at the surface. The total magnetic field manifestly satisfies the magnetostatic condition if we set $\mathbf{H} = \nabla\phi$ and ϕ ought to have a form similar to (6.2) for a localized mode. Although this is indeed the case, the algebra connected with ensuring that the fields satisfy (6.31) and $\nabla \cdot \mathbf{B} = 0$ gets a bit complicated. More importantly, the results one finds are distinctly different from the other elementary excitations we have studied.

The energy of the surface magnetostatic spin wave is dispersionless and always splits off *above* the bulk continuum. This is in sharp contrast to the Rayleigh mode that always splits off *below* the corresponding elastic continuum. In detail, the frequency of this so-called Damon–Eshbach (1961) mode varies from $\gamma\sqrt{H_0 B}$ to $\gamma(H_0 + B)/2$ depending on the propagation direction of the wave in the surface plane. For certain directions, $\mathbf{q}_\parallel$, the frequency of the mode plunges below $\gamma\sqrt{H_0 B}$ and completely mixes with the bulk continuum, i.e., becomes a resonance. This type of behavior is familiar. However, there is one property of this excitation that is completely unfamiliar from previous experience. Heretofore, surface states have always come in pairs: a mode that propagates from left to right across a crystal face is accompanied by a time-reversed mode of equal energy that propagates from right to left. However, in this case, the surface waves are limited to propagation directions defined by $\theta = \pi \pm \cos^{-1}\sqrt{H/B}$ (Fig. 6.12). This means that, for $\mathbf{H}_0 \parallel \mathbf{M}_s \parallel \hat{x}$, a dipolar spin wave can propagate from $+y$ to $-y$ but not from $-y$ to $+y$! The reason for this unusual behavior is *not* that the presence of the magnetic field breaks time-reversal symmetry (bulk magnon modes satisfy the reciprocity relation $\omega(\mathbf{q}) = \omega(-\mathbf{q})$). Rather, it is the fact that the magnetization is an *axial* vector while the surface breaks the reflection symmetry of the bulk (Scott & Mills, 1977).

High resolution Brillouin scattering experiments dramatically illustrate the non-reciprocity of long wavelength dipolar spin waves. In the usual case, incident light of frequency $\omega_i(\mathbf{k}_i)$ is scattered into a final state, $\omega_f(\mathbf{k}_f)$, with both creation (Stokes process) and absorption (anti-Stokes) of excitation quanta, $\omega(\mathbf{q})$. The kinematic relations are $\omega_f = \omega_i \mp \omega$ and $\mathbf{k}_f = \mathbf{k}_i \mp \mathbf{q}$, where the upper sign refers to Stokes scattering. Surface excitations are studied in back-reflection and, of course, only the wave vector components parallel to the surface are conserved. The scattered intensity from magnetic quanta is determined by the strength of the magneto-optic interaction. Fig. 6.13(*a*) shows the experimental spectrum

Fig. 6.13. Brillouin scattering from long wavelength surface magnons on nickel: (*a*) Stokes scattering for $\mathbf{H} \parallel \hat{x}$; (*b*) anti-Stokes scattering for $\mathbf{H}$ anti-parallel to $\hat{x}$ (Sandercock & Wettling, 1979).

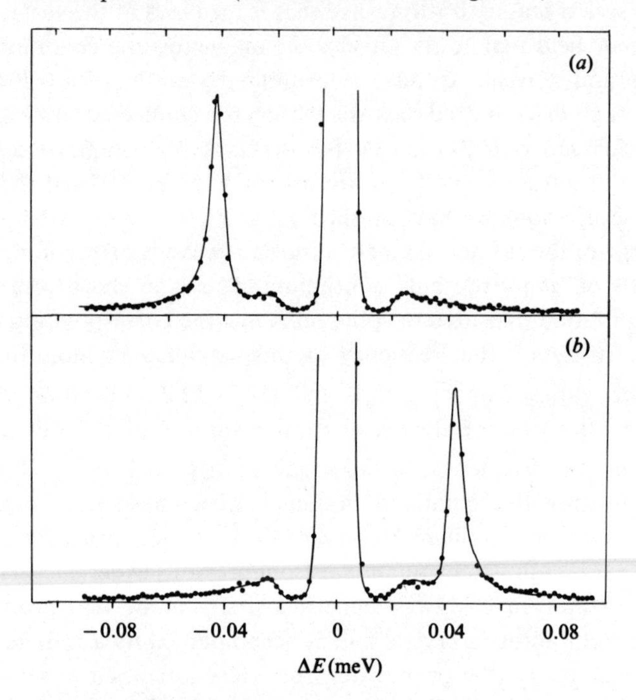

Fig. 6.14. Dispersion of the Damon–Eshbach mode into the bulk continuum.

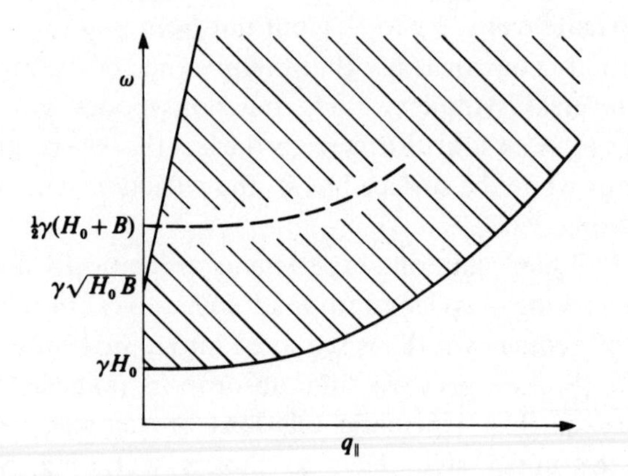

for scattering from a polycrystalline nickel surface at room temperature. The light is incident from the $+y$ half of the scattering plane (Fig. 6.12). Only the Stokes line is observed. When the direction of the magnetic field is reversed, only the magnon absorption process occurs.

We must permit larger and larger deviations from uniformity as the wave vector of the magnetic elementary excitation increases. Equation (5.13) suggests that the first correction to the Damon–Eshbach mode due to the short range exchange interaction is quadratic in $\mathbf{q}$. Hence, the exchange contribution to the energy of a magnon of wave vector $\mathbf{q}$ is approximately $J(qa_0)^2$. The dipolar contribution is $4\pi\gamma M_s$. Therefore, for typical values of J (~ 10 meV) and M_s($\sim 10^3$ Oe), the two energies are comparable for wavelengths of about 1000 Å. Since the energy of the bulk magnetostatic modes increases with both $q_\parallel^2$ and $q_\perp^2$, the surface mode rapidly leaks into the bulk continuum and becomes a resonance (Fig. 6.14).

The estimate above shows that most of the magnetic SBZ is the domain of exchange-dominated spin waves – the lowest energy excitations of, say, (5.10), the Heisenberg model. In the bulk, these magnons are well described by an approximate Hamiltonian (Ziman, 1972),

$$\mathcal{H} = \sum_{\mathbf{q}} J(\mathbf{q}) c_{\mathbf{q}}^{\dagger} c_{\mathbf{q}}, \tag{6.32}$$

where $J(\mathbf{q})$ is the Fourier transform of $J(\mathbf{R})$. The creation and annihilation operators in this expression refer to deviations of the spin moment transverse to the magnetization axis. However, (6.32) is mathematically equivalent to a Hamiltonian for electrons in an energy band if we make the identification $J(\mathbf{q}) \rightarrow \varepsilon(\mathbf{q})$. Better still, if the eigenstates are rewritten in

Fig. 6.15. Magnons at the (100) surface of a nearest neighbor Heisenberg model (Wolfram & DeWames, 1972).

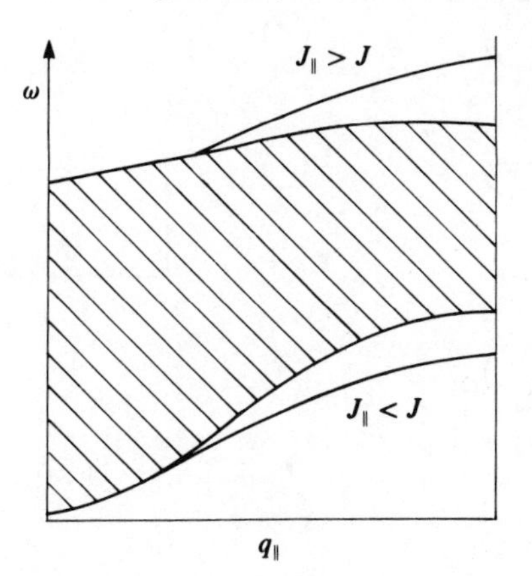

terms of Wannier orbitals we will be back to the familiar tight-binding model. We know what happens in that case. Surface states split off from the bulk band depending on the strength of a perturbation at the surface. In the spin wave case, an acoustic (optical) state splits off below (above) the bulk continuum if the exchange interaction in the surface plane, $J_{\parallel}$, is less (greater) than the bulk value, J (Fig. 6.15).

The nature of these microscopic surface spin waves is easy to understand using a semi-classical picture of magnon excitation. We consider only the ferromagnetic case. In the bulk, the phase of a canted spin advances along the propagation direction (Fig. 6.16(*a*)). For the acoustic surface wave, the phase similarly advances along this direction ($\hat{x}$). However, the magnitude of the canting decreases exponentially (with constant phase) in the direction normal to the surface (Fig. 6.16(*b*)). By contrast, the phase of the Heisenberg *optical* surface magnon advances by π for every layer of penetration into the bulk (Fig. 6.16(*c*)).

Now alter the exchange constant at the surface. The motion of any spin is determined by the analog of (6.31), i.e., the torques produced by neighboring spins:

$$\frac{d\mathbf{S}_a}{dt} = -\sum_i J_{ia}(\mathbf{S}_i \times \mathbf{S}_a). \tag{6.33}$$

Fig. 6.16. Semi-classical representation of spin waves: (*a*) a bulk excitation; (*b*) a surface acoustic wave; (*c*) a surface optical wave.

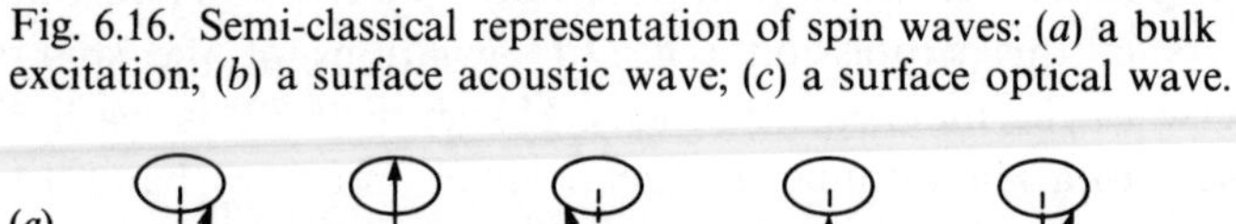

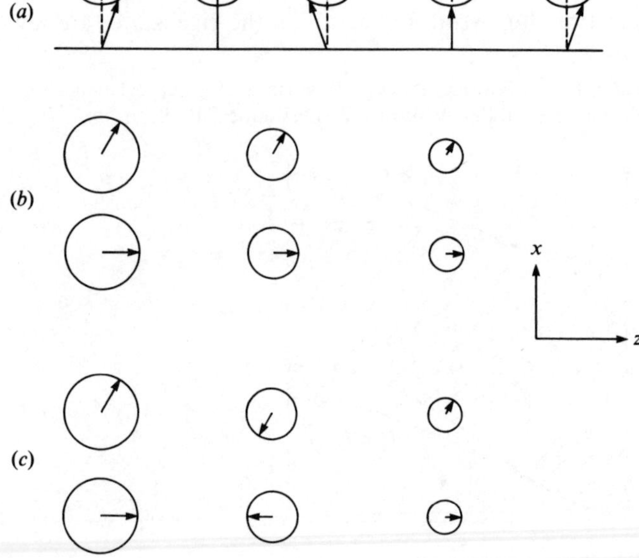

Consider a surface spin, $\mathbf{S}_a$, that is part of an acoustic surface wave on a cubic (100) face. It is coupled to its nearest neighbors in the plane by $J_\parallel$ and to a single neighbor immediately below by J (Fig. 6.17). In addition, the figure shows a 'missing' spin that would be coupled to $\mathbf{S}_a$ in the bulk. The trick is to compute the total torque on $\mathbf{S}_a$ assuming that $\mathbf{S}_{miss}$ is *present* and $J_\parallel = J$ and then find what value of $J_\parallel$ is needed to compensate for the

Fig. 6.17. Contributions to (6.33) for a surface spin of an acoustic spin wave (Sparks, 1970).

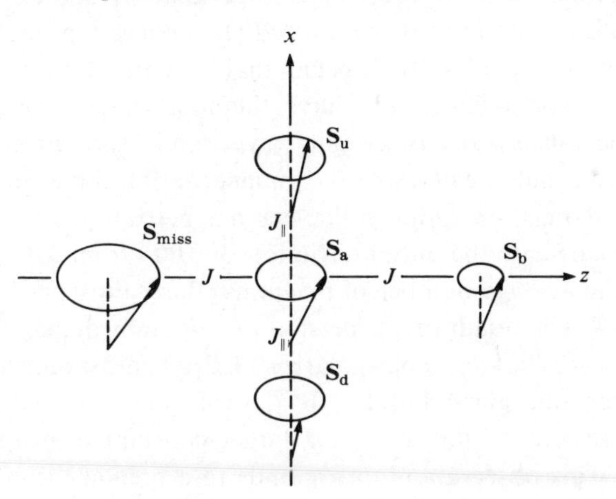

Fig. 6.18. Surface vs. bulk magnetization for $Ni_{40}Fe_{40}B_{20}$ (Pierce, Celotta, Unguris & Siegmann, 1982).

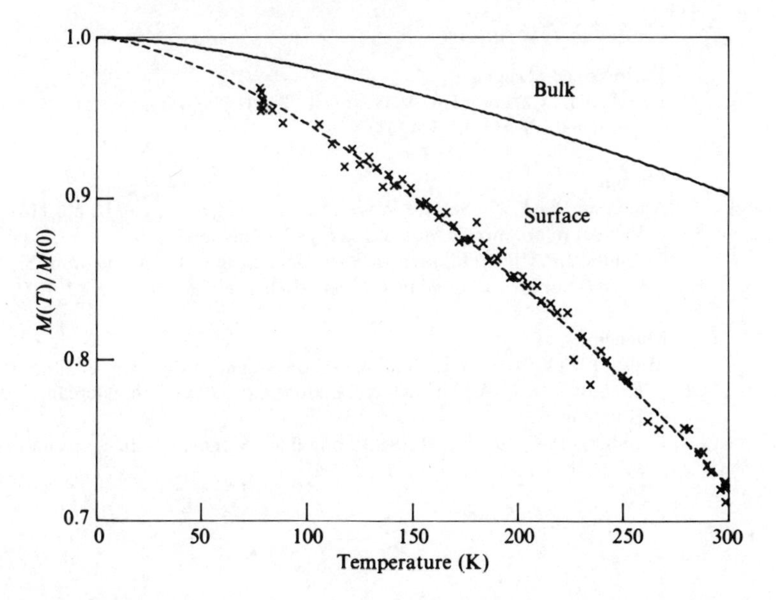

fact that $\mathbf{S}_{miss}$ is not present. The two situations lead to the same torque and, hence, to the same normal mode. Taking account of the properties illustrated in Fig. 6.16 it is easy to see that in the first scenario, $\mathbf{S}_{miss} + \mathbf{S}_b$ produces a net torque along $-\hat{y}$ while $\mathbf{S}_u + \mathbf{S}_d$ produces a net torque along $+\hat{y}$. Therefore, when $\mathbf{S}_{miss}$ is removed, $J_{\parallel}$ must be made *less* than J in order to produce a comparable weakening of the torque along $+\hat{y}$. An exactly equivalent argument shows that an optical surface magnon can exist only if $J_{\parallel} > J$.

To date, there has been no direct experimental observation of microscopic surface spin waves for either localized Heisenberg-type systems or itinerant Stoner-like systems. It appears that a high resolution spin-polarized EELS experiment would have the best chance for success. Nevertheless, the *influence* of surface spin waves can be seen directly from the temperature dependence of the surface magnetization curve. Since each thermally excited magnon reduces the net magnetization of the bulk sample, we can calculate the initial deviation of $M(T)$ from $M(0)$ simply by computing the average number of magnons (Bose–Einstein statistics) at temperature T. The quadratic dispersion of $\omega(\mathbf{q})$ immediately leads to $M(T)/M(0) = 1 - AT^{3/2} + \cdots$. Spin-polarized LEED measurements of a ferromagnetic metallic glass, $Ni_{40}Fe_{40}B_{20}$, verify that this bulk result holds at the surface as well, but with a larger constant of proportionality (Fig. 6.18). This observation corresponds to a greater mean square deviation from equilibrium for surface magnons than for their bulk counterparts – just the same as in the case of phonons (cf. Fig. 5.5).

General references

Excitons and plasmons

Ferrell, T.L., Callcott, T.A. & Warmack, R.J. (1985). Plasmons and Surfaces. *American Scientist* **73**, 344–53.

Phonons

Andersson, S. (1975). Surface Waves. In *Surface Science*, vol. II, pp. 113–44. Vienna: International Atomic Energy Commission.

Toennies, J.P. (1984). Phonon Inelastic Scattering of He Atoms from Single Crystal Surfaces. *Journal of Vacuum Science and Technology* **A2**, 1055–65.

Magnons

Mills, D.L. (1984). Surface Spin Waves on Magnetic Crystals. In *Surface Excitations* (eds. Agranovich & Loudon), pp. 379–439. Amsterdam: North-Holland.

Grunberg, P.A. (1985). Light Scattering from Magnetic Surface Excitations. *Progress in Surface Science* **18**, 1–58.

7

OPTICAL PROPERTIES

Introduction

The interaction of light with the first few atomic layers of a solid is relevant to (at least) two rather different aspects of surface physics. First, the strength of the experimental signal for many of the spectroscopic tools at our disposal such as photoemission, electron scattering, Raman scattering, etc., depends crucially on the intensity of the electric and magnetic fields near the surface. One typically calculates these field amplitudes by use of the classical Fresnel formulae. It is important to inquire whether this approach is sufficient if one is interested in phenomena within an Ångström or two of the crystal surface. These considerations alone suggest that a thorough understanding of the nature of near-surface electromagnetic fields is of both practical and fundamental interest. A rather different motivation to study these fields comes from the realization that very long wavelength elementary excitations of the surface will couple directly to the ambient field. To account for this, we must generalize the results of the previous chapter beyond the static limit to include the finite propagation velocity of light. These excitations, known as *surface polaritons*, are coupled modes of the surface + electromagnetic field system. It is convenient to use the language of optics to discuss surface polaritons, although, in principle, no external driving field is needed to excite them.

Reflection and refraction

The theory of reflection and refraction, embodied in Snell's laws, is one of the most familiar results in classical physics (Halliday & Resnick, 1966). Suppose a plane wave of unit amplitude, $\mathbf{E} = \exp(i\mathbf{q}_{\parallel} \cdot \mathbf{r}_{\parallel} + iq_z z)$, impinges at an angle θ_i on the flat surface of a medium of dielectric constant ε. A reflected wave emerges in the specular direction and a transmitted wave enters the medium at an angle θ_r with a wave vector of magnitude

$q' = \sqrt{\varepsilon} q$ (Fig. 7.1). Snell's law states that $q \sin\theta_i = q' \sin\theta_r$. Maxwell's equations require continuity of both the parallel component of **E** and the perpendicular component of $\mathbf{D} = \varepsilon\mathbf{E}$. For the depicted case where the electric field vector is parallel to the plane of incidence (so-called 'p'-polarization), we are led immediately to one of Fresnel's equations for the reflection coefficient:

$$R_p(\omega) = \frac{\varepsilon(\omega)q_z - q_z'}{\varepsilon(\omega)q_z + q_z'}. \tag{7.1}$$

A similar formula holds when the incident radiation is 's'-polarized, i.e., when **E** is perpendicular to the plane of incidence,

$$R_s(\omega) = \frac{q_z - q_z'}{q_z + q_z'}. \tag{7.2}$$

It is interesting to learn that as early as 1890 Drude worried whether the presence of surface contaminants might not invalidate Fresnel's results. He showed, for example, that light reflected at the Brewster angle from a rocksalt crystal is elliptically (rather than plane) polarized unless the experiment is performed immediately after cleavage (Drude, 1890). An entire field of 'ellipsometry' has arisen around this observation that is particularly useful for the accurate measurement of the thickness of thin coatings. Equations (7.1) and (7.2) are used with a dielectric function that abruptly changes from the properties of the substrate to the properties of the coating as a function of distance. This suggests the following crude model as a first approximation to the optical properties of a solid that takes explicit account of its surface. Identify the 'coating' with the first

Fig. 7.1. Reflection and refraction at a dielectic surface.

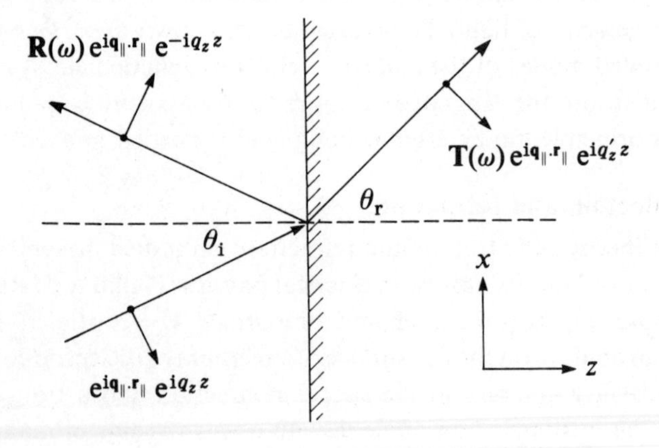

atomic layer and ascribe to it an effective dielectric function that can differ from the rest of the bulk.

Ellipsometry measurements have been performed for clean metals that purport to extract 'surface dielectric constants' for clean metals in just this way (Habraken, Gijzeman & Bootsma, 1980). Unfortunately, it is difficult to bring these results into contact with a microscopic interpretation in an unambiguous way. This does *not* mean that ellipsometry and differential reflectance measurements are useless for detailed study of surface physics (see below). It simply means that the dielectric properties of the surface must be treated more carefully. An alternative approach suggests itself following closer inspection of the conventional classical theory.

The macroscopic fields described by Maxwell's theory have some undesirable features from the point of view of microscopic surface physics. Most striking, of course, is the discontinuity in the normal component of the electric field at the surface required by $\nabla \cdot \mathbf{D} = 0$ and the attendant delta function charge sheet, (6.4). Even if we admit that the field is not truly discontinuous (perhaps merely very rapidly varying on an atomic scale), it is easy to see that the above treatment is internally inconsistent if we ask questions about the nature of the near-surface field. The problem is not with Maxwell's equations; it is with the assumed constitutive relation between **D** and **E**.

The most general linear constitutive relation that can be admitted to classical electrodynamics is of the form

$$\mathbf{D}(\mathbf{r}\,|\,\omega) = \int d\mathbf{r}'\boldsymbol{\varepsilon}(\mathbf{r}, \mathbf{r}'\,|\,\omega)\cdot\mathbf{E}(\mathbf{r}'\,|\,\omega). \tag{7.3}$$

This *non-local* dielectric function allows for the possibility that the field at one point in space may affect the field at other points in space. In order to recover the conventional formula, we assume that the electric field is slowly varying in space, i.e., $\mathbf{E}(\mathbf{r}') \cong \mathbf{E}(\mathbf{r})$. Then the usual frequency-dependent dielectric function emerges as a spatial average:

$$\mathbf{D}(\mathbf{r}\,|\,\omega) = \left[\int d\mathbf{r}'\boldsymbol{\varepsilon}(\mathbf{r}, \mathbf{r}'\,|\,\omega)\right]\cdot\mathbf{E}(\mathbf{r}\,|\,\omega). \tag{7.4}$$

However, this local relation leads back to the discontinuity of E_z at the surface, which contradicts the assumption of a slowly varying field. Therefore, if the fields in the vicinity of the surface are of interest, it is essential to use a non-local constitutive relation.*

* When the incident **E** field is *perpendicular* to the plane of incidence, E_z vanishes identically and Fresnel theory is adequate.

What is the microscopic replacement to the Fresnel field? We first take advantage of translational invariance in the plane to Fourier transform one of the Maxwell equations (cf. (6.23)):

$$i\mathbf{q}_{\parallel}\cdot\mathbf{D}(\mathbf{q}_{\parallel},z|\omega)+\frac{d}{dz}\mathbf{D}_z(q_{\parallel},z|\omega)=0. \tag{7.5}$$

Substituting from (7.3) and taking the long wavelength limit ($\mathbf{q}_{\parallel}\to 0$) we obtain

$$\frac{d}{dz}\int dz'\varepsilon_{zz}(z,z'|\omega)E_z(z')=0. \tag{7.6}$$

Hence,

$$\int dz'\varepsilon_{zz}(z,z'|\omega)E_z(z')=E_z(\text{out}), \tag{7.7}$$

where the constant of integration, $E_z(\text{out})$, is the classical field in the vacuum far from the surface. Equation (7.7) is an integral equation for E_z that yields a perfectly continuous solution for all values of z. Note that the field obtained by solution of (7.7) can be augmented by any solution of

$$\int dz'\varepsilon_{zz}(z,z'|\omega)E_z(z')=0 \tag{7.8}$$

without loss of generality. The reader will recognize (7.8) as a non-local expression of the elementary existence criterion for a *bulk* plasmon (Ziman, 1972). At first glance this seems odd. Translational invariance normally forbids the coexistence of a transverse electromagnetic field (the trans-

Fig. 7.2. Induced charge density at a metal surface (Hanke & Wu, 1977).

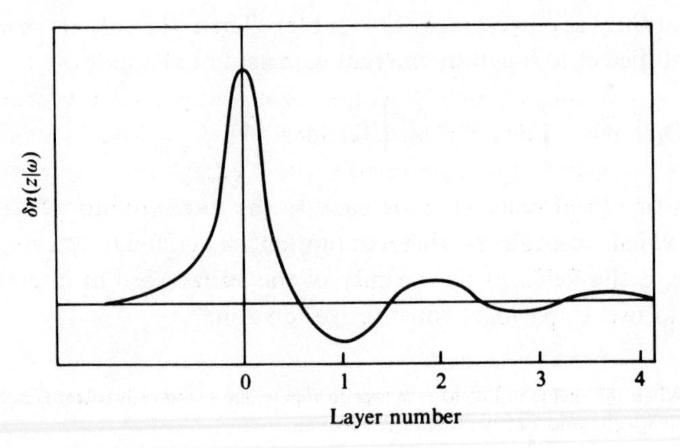

mitted wave) with a longitudinal electromagnetic field (the plasmon wave). But this symmetry is broken at the surface! An incident plane wave can indeed excite bulk plasmons as long as $\omega > \omega_p$.

The preceding formulation makes clear that any explicit calculation of the fields requires a reasonable approximation to the dielectric function, $\varepsilon_{zz}(z, z'|\omega)$. This is a formidable problem. Nonetheless, for metals, most of the essential features appear if we adopt the simple jellium model (see Chapter 4) and treat the dielectric response at the level of mean field theory. The latter can be thought of in the following way. Allow an external field of frequency ω to impinge on the jellium surface. The mobile electrons respond to this perturbation and a charge density distortion $\delta n(z|\omega)$ (relative to the ground state distribution (Fig. 4.2)) appears in the medium (Fig. 7.2). The induced charge consists of two parts. First, a large peak appears in the surface region. This feature corresponds to the delta function charge sheet in the macroscopic description and actually derives from the excitation of electron–hole pairs and (virtual) plasmons. Second, damped oscillations of the induced charge extend deep into the bulk. These are Friedel oscillations which reflect the efforts of the jellium to screen out the external field. Now, according to Poisson's equation, the total induced charge is the source of an induced *field* that again will be concentrated in the surface region. Since the electrons cannot distinguish the external field from a self-generated induced field we must take the *sum* of the two and repeat the entire procedure. Eventually one reaches a *self-consistent* solution where the total input field is equal to the total output field.*

Fig. 7.3 illustrates the fields that emerge from a self-consistent calculation of the sort outlined above. The in-phase part of the field offers no surprises. The results look very much like a smoothed out version of the classical step discontinuity for frequencies below ω_p. Above the plasma frequency, the field of the bulk plasmons dominates. The striking new behavior is connected with a piece of the electric field that oscillates 90° out of phase with the incident wave. Just below the plasma frequency, the out-of-phase electric field develops a sharp peak right at the surface. Most correctly, this peak signals the presence of an efficient energy loss mechanism. The surface saps power from the incident wave by a photoabsorption process that creates copious electron–hole pairs. Rather crudely, the origin of the peak can be understood by use of a simple model dielectric function for

* This scheme is equivalent to the random phase approximation (Ehrenreich & Cohen, 1959).

the semi-infinite jellium system,

$$\varepsilon(z|\omega) = 1 - \frac{\omega_p^2}{\omega^2}\frac{n(z)}{n(\infty)}, \tag{7.9}$$

that interpolates between the vacuum and the metal. Here, $n(z)$ is the jellium ground state charge density profile. The key point is that, for $\omega < \omega_p, \varepsilon(z|\omega)$ must pass through zero somewhere in the surface region. Consequently, the classical screened field diverges at this point. In the more correct quantum mechanical calculation it merely becomes large.

Photoemission measurements provide the clearest demonstration that the fields depicted in Fig. 7.3 are physically meaningful. The microscopic theory predicts fields which exhibit significant intensity variations on a scale comparable to atomic dimensions. This spatial dependence (parti-

Fig. 7.3. In-phase (solid curves) and out-of-phase (dashed curves) contributions to the normal component of the electric field near a jellium surface (Feibelman, 1982). The positive z-axis points into the bulk.

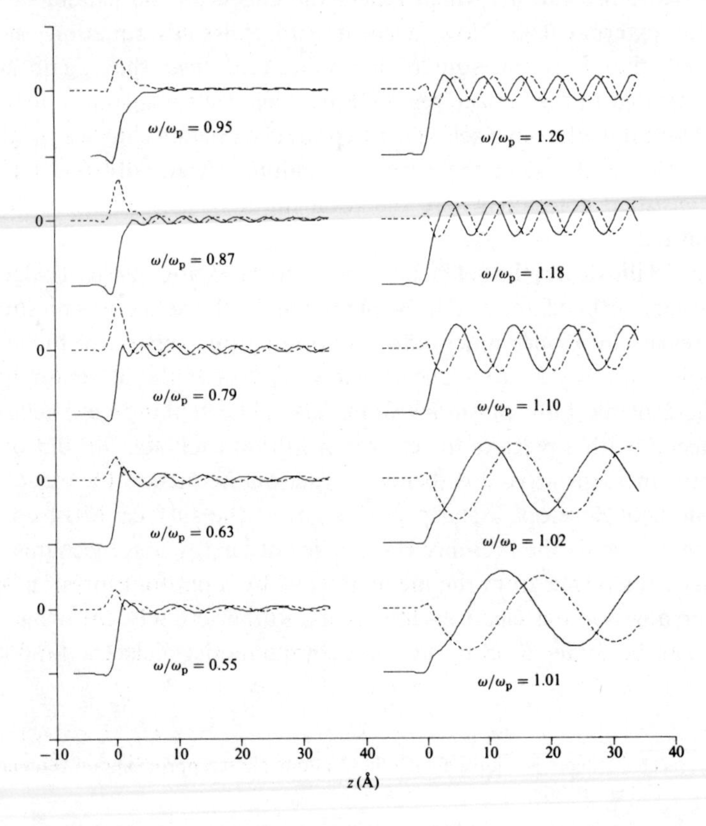

cularly the surface peak) ought to show up in a photoemission process described by (4.34) if the initial state is localized in the surface region. The jellium model is most appropriate to free electron metals. Therefore, Fig. 7.4 compares predicted cross sections with measured cross sections for photoemission from surface electronic states on Al(001) and Be(0001). The agreement is striking – particularly when compared to the results expected from the classical Fresnel fields.

The detailed shape of the surface photoeffect cross section reveals additional interesting physics. We already discussed the origin of the peak. What about the apparent zero in the cross section right at ω_p? In Chapter 6, we saw that the centroid of the induced charge associated with a surface plasmon was at $d_\perp(\omega_s)$, where

$$d_\perp(\omega) = \int_{-\infty}^{+\infty} \mathrm{d}z\, z\, \delta n(z|\omega) \Big/ \int_{-\infty}^{+\infty} \mathrm{d}z\, \delta n(z|\omega). \tag{7.10}$$

In the present case, the frequency dependence of $d_\perp(\omega)$ sets the scale length over which the metal responds to a 'p'-polarized external radiation field. This is shown in Fig. 7.5, measured relative to the jellium positive background (cf. Fig. 4.2). Below the plasma frequency, most of the response is in the vacuum tail of the surface charge distribution. This is easy to understand since (7.9) defines a 'local' plasma frequency, $\omega_p(z) = 4\pi n(z)e^2/m$, and every metal responds best right at its plasma frequency. The entire system responds as a whole at the *bulk* plasma frequency so $d_\perp(\omega)$ appropriately recedes deep into the metal interior. At this point the surface is invisible and the photo-cross section vanishes. Finally, we must recover the Fresnel result at high frequencies. $d_\perp(\omega)$ asymptotically approaches zero (the jellium edge) and the cross section is again finite.*

It now is reasonable to presume that $d_\perp(\omega)$ will enter as a long wavelength/low frequency correction to *all* the standard results of classical refraction theory. As a single example, we merely quote the result for the reflectivity (7.1) (Apell, Ljungbert & Lundqvist, 1984):

$$R_p(\omega) = \frac{\varepsilon(\omega)q_z - q_z' - \mathrm{i}q_\parallel^2[\varepsilon(\omega) - 1]d_\perp(\omega)}{\varepsilon(\omega)q_z + q_z' + \mathrm{i}q_\parallel^2[\varepsilon(\omega) - 1]d_\perp(\omega)}. \tag{7.11}$$

This formula can be used directly in the interpretation of optical reflectance data. However, the explicit surface signal often is swamped by an enormous background signal from the bulk. To suppress the latter one typically

* The phase lag response of the metal surface suggests that $d_\perp(\omega)$ ought to be a complex number at finite frequency. Fig. 7.5 shows the real part only. Im $d_\perp(\omega)$ is related to power absorption (see Chapter 13).

measures the *difference* in reflectivity between the clean surface of interest and the same surface covered by, say, an oxide layer. These so-called differential reflectance data can be compared with dielectric function calculations that treat the surface correctly. That is to say, calculations that employ a reasonable model for $\varepsilon(z, z'|\omega)$. This approach seems to

Fig. 7.4. Comparison of microscopic theory (solid curve) and Fresnel theory (dashed curve) with experiment for surface state photoemission from free electron metals. The magnitude of the Fresnel field has been multiplied by a factor of 10. The incident photon energy is scaled to the plasma frequency and the experimental intensities normalized at the peak (Plummer, 1985).

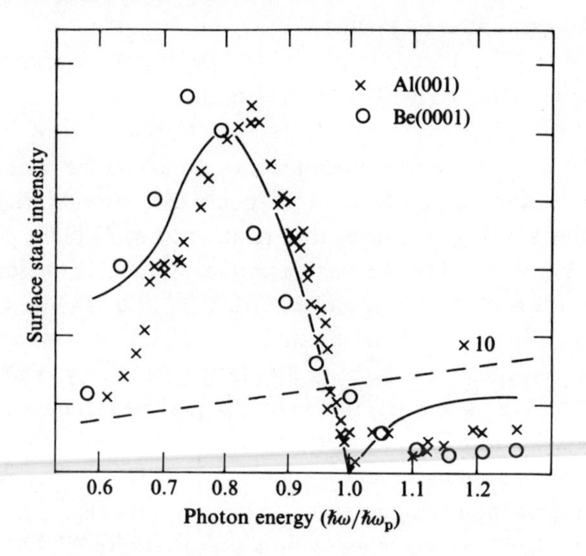

Fig. 7.5. Surface dielectric response scale length normalized to its static value (Apell, Ljungbert & Lundqvist, 1984).

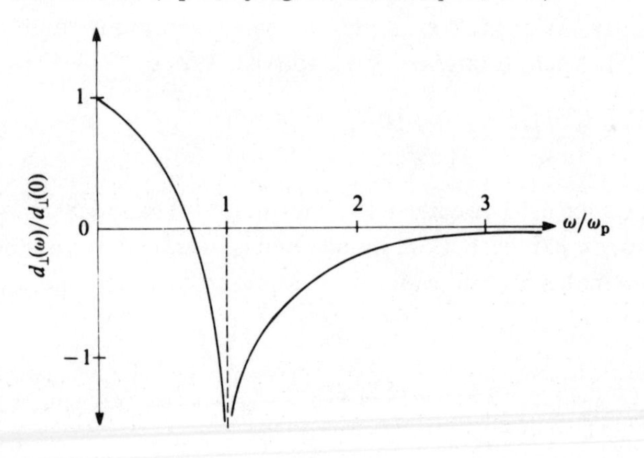

work well for localized systems like semiconductors, for which the tight-binding model is well adapted (Del Sole & Selloni, 1984).

Polaritons

Surface electrodynamics differs from surface electrostatics only in the extreme long wavelength (optical) limit where it is necessary to take account of the finite dispersion of light, $\omega = cq$. This aspect of the problem was ignored previously because, for example, the light curve would appear as a vertical line at $q = 0$ on the scale of the surface plasmon dispersion illustrated in Fig. 6.3. However, on an expanded scale such that $q_{\parallel} \leqslant \omega/c$, we must consider the modes of the radiation field on an equal footing with the modes of the electronic system. Among other things, this analysis will explain why the surface plasmon did not enter our discussion of reflection, refraction and the surface photoeffect.

The electrodynamics of non-magnetic materials is described by the Maxwell equations:

$$\begin{aligned} \nabla\cdot\mathbf{B} = 0, \quad \nabla\times\mathbf{E} + \frac{1}{c}\frac{\partial H}{\partial t} = 0, \\ \nabla\cdot\mathbf{D} = 0, \quad \nabla\times\mathbf{H} - \frac{1}{c}\frac{\partial D}{\partial t} = 0. \end{aligned} \tag{7.12}$$

For harmonic fields, the two curl equations combine to yield

$$\nabla^2\mathbf{E} + \left(\frac{\omega}{c}\right)^2 \varepsilon(\omega)\mathbf{E} + \nabla(\nabla\cdot\mathbf{E}) = 0. \tag{7.13}$$

For the geometry of Fig. 7.1, we seek a solution of (7.13) for which $E_y = 0$ and E_x and E_z do not depend on the y-coordinate. In that case,

$$\begin{aligned} \frac{\partial^2 E_x}{\partial z^2} + \left(\frac{\omega}{c}\right)^2 \varepsilon(\omega)E_x - \frac{\partial^2 E_z}{\partial x \partial z} = 0, \\ \frac{\partial^2 E_z}{\partial x^2} + \left(\frac{\omega}{c}\right)^2 \varepsilon(\omega)E_z - \frac{\partial^2 E_x}{\partial z \partial x} = 0, \end{aligned} \tag{7.14}$$

which, from (7.12), establishes that the accompanying magnetic field is polarized in the transverse direction:

$$H_y = -\mathrm{i}\frac{c}{\omega}\left(\frac{\partial E_x}{\partial z} - \frac{\partial E_z}{\partial x}\right). \tag{7.15}$$

Propagating solutions to (7.14) and (7.15) exist for which the spatial dependence of all the fields is the same: $\exp[\mathrm{i}(q\,x - \omega t)]\,\exp[-\kappa|z|]$,

where

$$\kappa^2 = q_\parallel^2 - \varepsilon(\omega)\frac{\omega^2}{c^2} \tag{7.16}$$

and with specific relations between the amplitudes. In particular, the boundary condition, $\nabla\cdot\mathbf{D} = 0$, leads to

$$-\mathrm{i}\frac{q_\parallel}{\kappa_{\text{out}}}E_x^{\text{out}} = E_z^{\text{out}} = \varepsilon(\omega)E_z^{\text{in}} = \mathrm{i}\frac{q_\parallel}{\kappa_{\text{in}}}\varepsilon(\omega)E_x^{\text{in}}, \tag{7.17}$$

where the notation 'in' ('out') refers to quantities just inside (outside) the dielectric medium. Finally, since the parallel component of the electric field is continuous across the boundary, $\varepsilon(\omega) = -\kappa_{\text{in}}/\kappa_{\text{out}} < 0$, which can be rewritten in the form

$$q_\parallel^2 = \left(\frac{\omega}{c}\right)^2 \frac{\varepsilon(\omega)}{1+\varepsilon(\omega)}, \tag{7.18}$$

subject to the proviso that $1 + \varepsilon(\omega) < 0$.

The implicit dispersion relation (7.18) is valid for any choice of the dielectric function. For the plasmon case, we choose $\varepsilon(\omega) = 1 - \omega_p^2/\omega^2$ and the resulting coupled mode of the radiation field and the surface plasmon excitation is called a *surface plasmon polariton*, or plasmon SP. It was first studied by Sommerfeld (1909) in connection with the propagation of radio waves along the Earth's surface.

Fig. 7.6. Dispersion of a plasmon SP and the projected bulk continuum. See text for discussion (Ritchie, 1973).

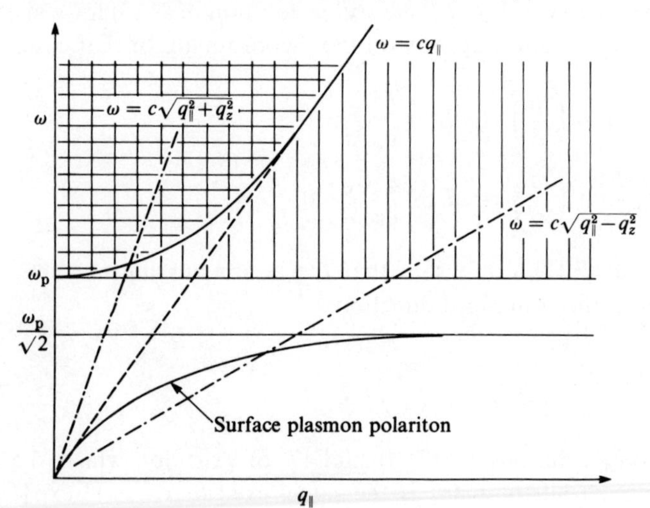

The dispersion relation for the plasmon SP is drawn as a solid curve in Fig. 7.6. Its two branches display a familiar 'avoided crossing' behavior and asymptotically approach the light curve (drawn as a dashed line for the case of $q_z = 0$, e.g., grazing incidence) and the dispersionless (on this scale) surface plasmon curve. Since the plasmon SP dispersion curve *nowhere* intersects the dispersion line for pure radiation ($\omega = cq_{\parallel}$), a grazing incidence external electromagnetic wave does not excite this polariton, energy and momentum cannot be conserved simultaneously.

More generally, Fig. 7.6 illustrates all the excitations possible in a semi-infinite free electron metal. No excitations of any kind exist in the frequency range between ω_s and ω_p. Above ω_p, longitudinal bulk plasmons propagate (vertical lines) whereas bulk transverse electromagnetic waves exist only in the region where $\omega^2 = \omega_p^2 + c^2q^2$ is satisfied (cross hatching). For glancing incidence, only bulk plasma waves can be excited by the external field (Fig. 7.3). If the incident beam is tilted toward the normal, the external radiation dispersion line tilts toward the vertical (upper dashed-dot line) and intersects the continuum of transverse bulk waves. This is normal refraction. The light line also now intersects the upper branch of the plasmon SP dispersion curve. However, no localized surface mode is excited. The energy immediately leaks into the degenerate bulk continuum. There is *no* angle of incidence for which the true plasmon SP (lower branch) can be excited by an external electromagnetic wave. It is not a radiative mode and for this reason the surface plasmon did not figure into the microscopic theory of refraction from a smooth surface.

There are two principal mechanisms that couple external radiation to polaritons. The first of these was recognized by Fano (1941) as the origin of unexpectedly large diffracted beam intensities (Wood's anomalies) for light incident onto plane gratings at certain angles of incidence. Simply put, a grating etched onto a smooth surface with regular spacing, d, introduces a periodicity along the surface with reciprocal lattice vectors $2n\pi/d$. External light can couple to the polariton via an 'umklapp' process that supplies the requisite momentum.* In a metal, the surface plasmon frequency is in the ultraviolet and the requisite grating spacing is rather smaller than can be conveniently etched. Instead, typical experiments focus on an alternative free electron system – a doped *semiconductor*. The number density of free carriers in such a material is much less than in a typical metal so that the characteristic plasma frequency is in the infrared.

* This implies that direct coupling to a *rough* surface should be possible where the parallel momentum boost derives from the Fourier decomposition of the roughness profile (Raether, 1982).

Fig. 7.7 illustrates the measured dispersion curve for a plasmon SP on an InSb surface ruled with a 30 μm grating. The experimental signal of the SP is a dip in the direct reflectivity, which indicates that a portion of the incident intensity has been coupled into the resonant mode.

The most widely used methods to excite surface polaritons use some variation of a technique called attenuated total reflection (ATR). The necessary ingredient can be read off from Fig. 7.6. The lower dashed-dot line shown there is the light curve for a 'slow' photon with an *imaginary* perpendicular (z) component to its wave vector. This situation can be achieved experimentally in a particularly simple and elegant manner (Fig. 7.8). A prism is suspended at a well-defined distance ($\sim$ 1000 Å) above the crystal surface of interest. Light enters the prism so that it is totally internally reflected from the bottom surface. However, in such a case, an evanescent electric field always penetrates into the air gap (Hecht & Zajac, 1974). The field intensity decays exponentially away from the prism surface,

Fig. 7.7. Dispersion of surface plasmon polaritons in InSb (Marschall, Fischer & Queisser, 1971).

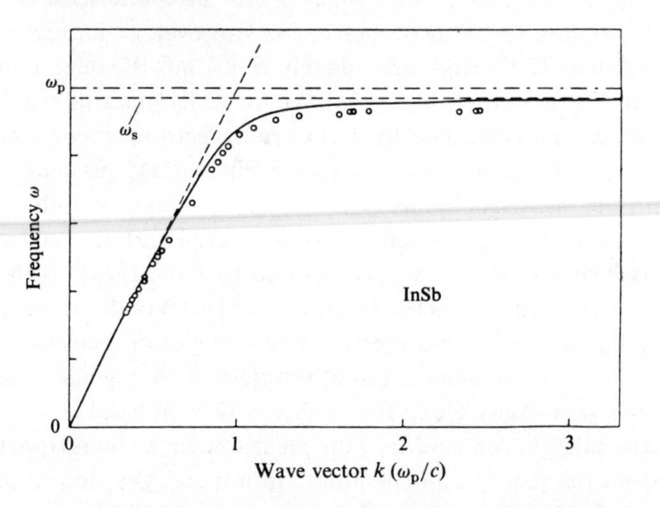

Fig. 7.8. Schematic arrangement of an ATR experiment (Otto, 1968).

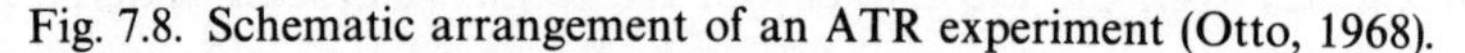

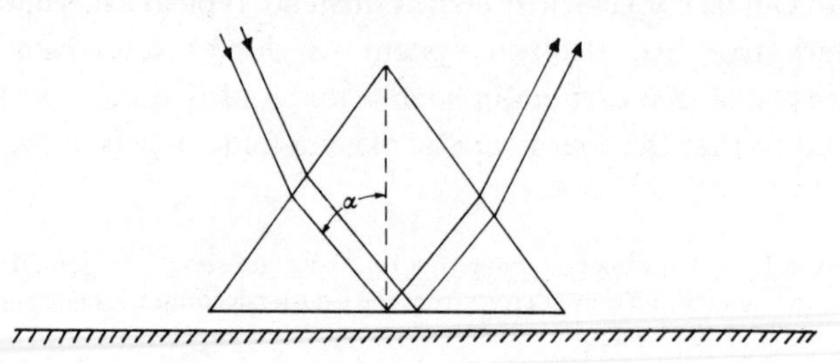

i.e., it possesses an imaginary wave vector in the z-direction. Direct coupling to the surface polariton occurs when the wave strikes the dielectric surface.

Before passing on to other types of polaritons, let us develop a bit of intuition about their nature. The z-component of the plasmon SP electric field has the form shown earlier (Fig. 6.2). According to (7.15), this field is accompanied by a transverse magnetic field (Fig. 7.9). We can compute the flow of energy in this mode by appeal to the time-averaged Poynting vector, $\mathbf{S} = (c/8\pi)\,\mathbf{E} \times \mathbf{H}^*$. The energy flow within the dielectric is readily seen to be in the $-\hat{x}$ direction while the flow in the vacuum is along $+\hat{x}$. However, (7.16) tells us that the fields decay to zero much more rapidly inside the dielectric so that the *net* energy flow is along $+\hat{x}$, the direction of propagation. In fact, an explicit calculation reveals that $|S_{\text{out}}|/|S_{\text{in}}| =$

Fig. 7.9. The magnetic field, H_y, of the plasmon SP.

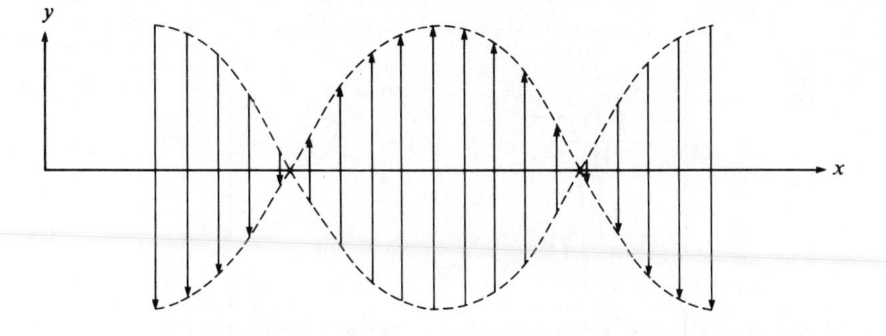

Fig. 7.10. Dispersion of an exciton SP (solid curve) and the projected bulk continuum (shaded region). The steep dashed line denotes the external light curve $\omega = cq_{\parallel}$ (Lagois & Fischer, 1978).

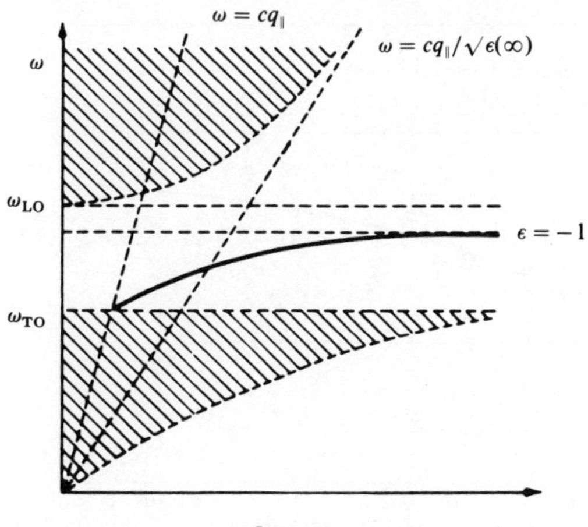

$|\varepsilon(\omega)|^2$ so that, in typical cases, the vast majority of the energy flow occurs in the vacuum (Nkoma, Loudon & Tilley, 1974).

The dispersion properties of a coupled mode of the radiation field with an *exciton* follow immediately from (7.18). All we need is an expression for the exciton dielectric function. In fact, $\varepsilon(\omega)$ in this case has exactly the form of (6.18) if we identify ω_{TO} with the exciton energy and $e^* = e$. We see that $\varepsilon(\omega_{\text{TO}}) = -\infty$ and it is convenient to define $\omega_{\text{LO}}^2 = \omega_{\text{TO}}^2 + \Omega^2/\varepsilon(\infty)$ so that $\varepsilon(\omega_{\text{LO}}) = 0$. Freely propagating *bulk* exciton polaritons exist in the shaded regions indicated on Fig. 7.10. These are radiative modes. The exciton SP must be sought within the 'stop band' where no bulk excitations

Fig. 7.11. ATR detection of a CuBr exciton SP: (*a*) reflectivity spectra for both 's' and 'p' polarization; (*b*) mode dispersion. The solid curve is calculated from (7.18) and the five dashed curves are identical to those shown in Fig. 7.10 (Hirabayashi *et al.*, 1976).

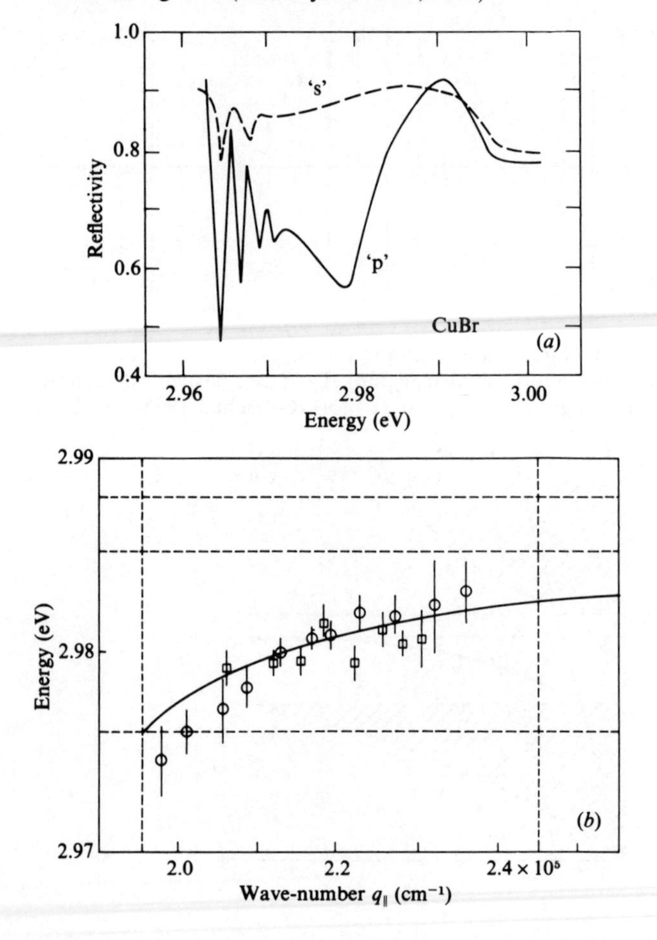

propagate. In fact, the requirement that $1 + \varepsilon(\omega) < 0$ restricts the exciton SP to an even smaller frequency domain. Again, no direct coupling to external radiation is possible.

The ATR method is particularly convenient for experimental study of exciton polaritons. Fig. 7.11(*a*) illustrates the reflectivity measured from a CuBr surface for both 's' and 'p' polarized incident light. Since there is no z-component to the incident **E** field in 's' polarization (real or imaginary), no coupling to the SP is expected. However, as the angle of incidence changes, the dip observed in 'p' polarization traces out the dispersion of the exciton SP depicted in Fig. 7.11(*b*).

The *phonon* surface polariton is a coupled mode of a surface localized periodic lattice distortion and the electromagnetic field. A new feature enters the discussion here because most interesting materials have several atoms in the bulk unit cell. In that case, the dielectric response is determined by more than one bulk mode and (6.18) must be generalized to a *sum* over each of these excitations – each with its own characteristic ionic plasma frequency, Ω_i. We expect several phonon SP branches because an avoided crossing occurs at each point within the bulk stop band where the light curve intersects a macroscopic surface optical phonon branch. Yttrium iron garnet provides a nice example of this phenomenon (Fig. 7.12).

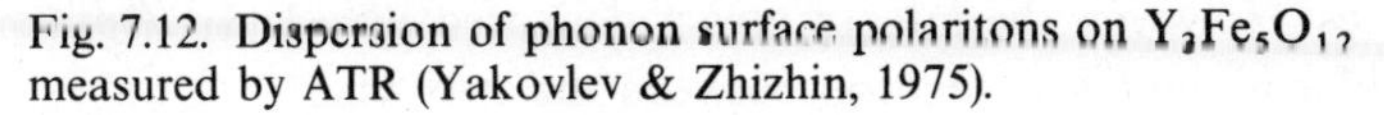

Fig. 7.12. Dispersion of phonon surface polaritons on $Y_3Fe_5O_{12}$ measured by ATR (Yakovlev & Zhizhin, 1975).

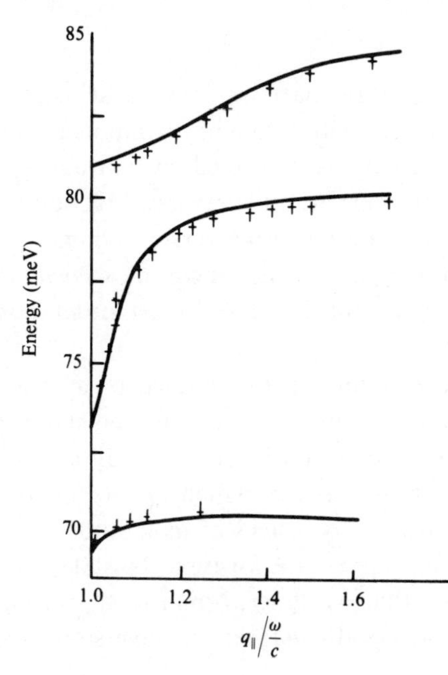

Finally, we come to the *magnon* surface polariton. The analysis of this mode depends on the magnetic permeability that enters the constitutive relation between **B** and **H**, viz., $\mathbf{B} = \mu(\omega)\mathbf{H}$, but otherwise parallels the derivation given above. The resulting dispersion relation is (Camley & Mills, 1982):

$$q_\parallel^2 = \left(\frac{\omega}{c}\right)^2 \frac{\mu(\omega)}{1+\mu(\omega)}. \tag{7.19}$$

At present, there is no experimental evidence for the existence of this variety of polariton. For ferromagnets, the dispersion of the mode would only become apparent well below the magnetostatic Damon–Eshbach frequency, i.e., in the microwave. An ATR experiment would require a sample smooth on the scale of 1 centimeter or larger. The best hope for experiments probably lies with the *anti-ferromagnetic* magnon SP. Here, the corresponding magnetostatic resonance frequency is in the infrared.

Non-linear phenomena

All of the optical effects examined thus far in this chapter presume a linear constitutive relation between the polarization and the electric field. This is entirely appropriate for most common radiation sources: Nernst glowers (infrared), arc lamps (visible and ultraviolet) and even synchrotron sources. However, the intense power delivered by a laser requires one to consider a *non-linear* relation between the two, viz.,

$$\mathbf{j} = \boldsymbol{\sigma}\cdot\mathbf{E} + \boldsymbol{\chi}:\mathbf{EE} + \cdots \tag{7.20}$$

This expansion suggests that a uniform, harmonic electric field of frequency ω quite generally induces a current with frequency components at ω, 2ω, etc. In practice, non-linear activity is restricted by various symmetry considerations. For example, the current must reverse direction when the direction of **E** is reversed in any material with inversion symmetry. In that instance, (7.20) implies that the third-rank tensor of *second harmonic generation* (SHG), $\boldsymbol{\chi}$, vanishes identically. The lowest order non-linear behavior occurs in third order.

Non-linear optical response occurs at the surface of a semi-infinite system even if it is forbidden in the bulk. Both the broken symmetry and the non-uniformity of electric fields near the surface play a role. Let us restrict ourselves to a cubic metal for which $\chi = 0$ in the bulk. Electrons within the sample driven by a field of magnitude E_0 oscillate with an amplitude of about $eE_0/m\omega^2$ (~ 1 Å for typical visible laser power densities). Therefore, electrons that oscillate normal to the surface from within the first few atomic layers do *not* see an inversion symmetric environment and will contribute to a second harmonic current according to (7.20). Moreover, a glance back at Fig. 7.3 shows that the fields in the surface region are highly non-uniform on an atomic scale. Hence, we may generalize (7.20) to include gradient terms in addition to the symmetry-breaking term:

$$\mathbf{j}(2\omega) = \chi_{\text{bulk}}: \mathbf{E}(\omega)\,\nabla\mathbf{E}(\omega) + \chi_{\text{surf}}: \mathbf{E}(\omega)\mathbf{E}(\omega) \tag{7.21}$$

Theory and experiment to date suggest that the first term in (7.21) often dominates surface non-linear optical response. Our previous results then imply an important role for $E_z(\omega)$ – the component of the electric field normal to the vacuum interface.

The simplest experimental test for SHG at a metal surface merely directs an incident beam (ω) toward the crystal. The argument given above predicts that a part of the reflected beam will oscillate at frequency 2ω and will have an intensity that depends on the angle of incidence according to $I \propto |E_z^2|^2 \propto \cos^4 \theta_i$. This expectation is borne out in detail for a clean silver surface (Fig. 7.13). Bear in mind that the efficiency for non-linear conversion of power from frequency ω to frequency 2ω in this case is about 10^{-15}. Clearly, it would be desirable if second harmonic generation were a less energy-intensive venture.

The ATR method provides a simple method to resonantly enhance non linear power generation. We know from the preceding section that a

Fig. 7.13. Second harmonic radiation, in reflection, from a silver surface (Brown, Parks & Sleeper, 1965).

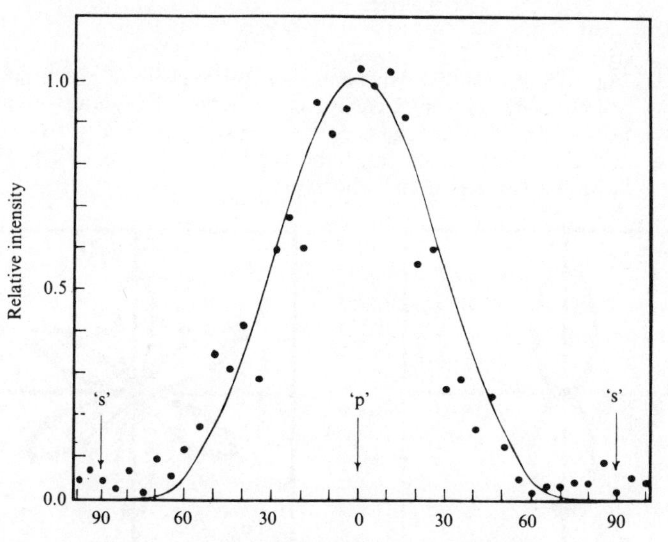

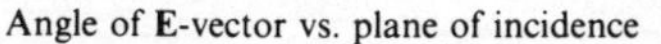

plasmon SP can be 'launched' at a surface with free carriers if the incident beam energy and angle of incidence to a prism-coupled surface match the polariton's dispersion curve. The collective mode steals electric field intensity from the fundamental reflected beam and localizes it at the surface – precisely where the non-linearity of the metal is greatest. Consequently, this particular geometry should resonantly enhance the intensity

Fig. 7.14. Intensity of SHG (relative to direct reflection) during ATR excitation of a plasmon surface polariton (Simon, Mitchell & Watson, 1974).

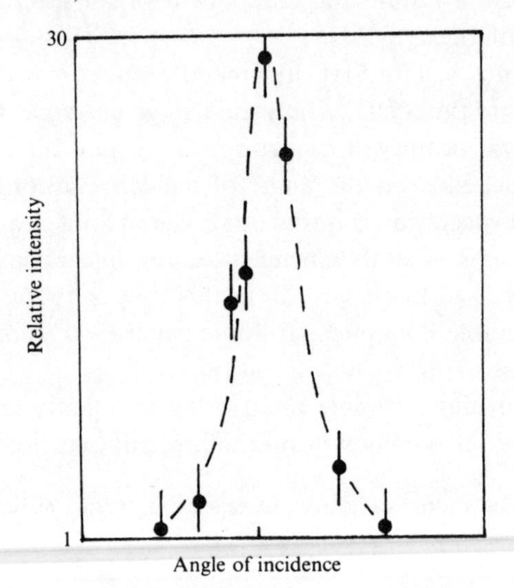

Fig. 7.15. Azimuthal dependence of the intensity of radiation emitted at frequency 2ω from a clean Si(111) surface. The pump laser ($\omega = 1.06\,\mu$m) strikes the crystal at normal incidence: (*a*) as-cleaved surface; (*b*) surface annealed above 400 K. (Courtesy of T.F. Heinz, IBM Watson Research Laboratory.)

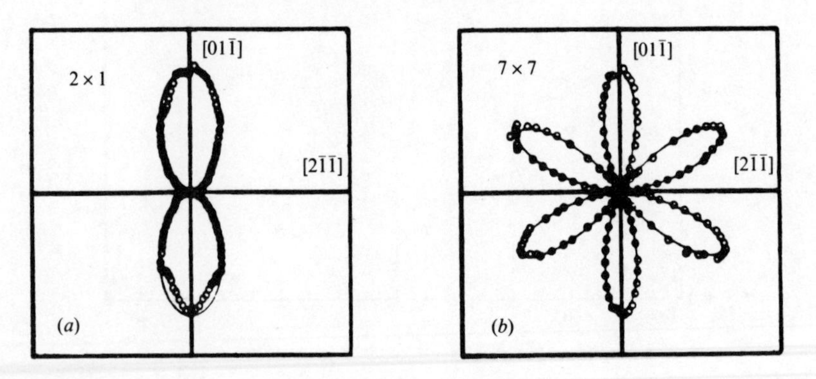

of the outgoing second harmonic radiation. More than an order of magnitude enhancement is observed for silver relative to the direct reflection geometry (Fig. 7.14).

It is appropriate to ask whether our new surface phenomenon – second harmonic generation – is good for anything. That is, can this radiation be used as a probe of independent surface properties? The answer is yes because the tensor properties of χ are very sensitive to the symmetry properties of the electronic structure of the surface (Guyot–Sionnest, Chen & Shen, 1986). For example, the azimuthal dependence of the intensity of emitted SH radiation reflects the presence (or absence) of mirror planes in the surface net. Consider the case of Si(111). As noted in Chapter 4, the as-cleaved surface exhibits a metastable 2×1 reconstruction which transforms to 7×7 upon annealing. Moreover, we have presented specific geometrical models of these surfaces (Figs. 3.20 and 4.40) that possess one-fold and three-fold reflection symmetry, respectively. To test this, Fig. 7.15 shows the azimuthal dependence of SHG from this surface both in its as-cleaved and annealed states. The experiment confirms that the symmetry assumed by the models is correct for both phases.

To close this chapter we revisit the combination of non-linear optical activity with surface polariton excitation in an intriguing (and possibly important) application. This time, the non-linear dielectric medium is the *prism* of an ATR configuration rather than the underlying solid surface. Recall that the parallel component of the light wave at the bottom surface of the prism is $q_{\parallel} = n(\omega/c)\sin\alpha$, where n is the index of refraction of the

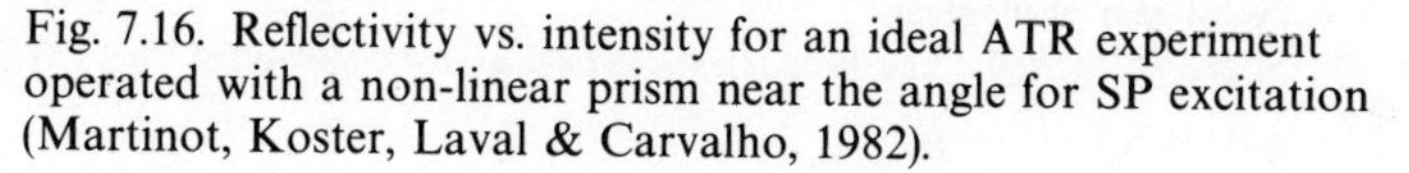

Fig. 7.16. Reflectivity vs. intensity for an ideal ATR experiment operated with a non-linear prism near the angle for SP excitation (Martinot, Koster, Laval & Carvalho, 1982).

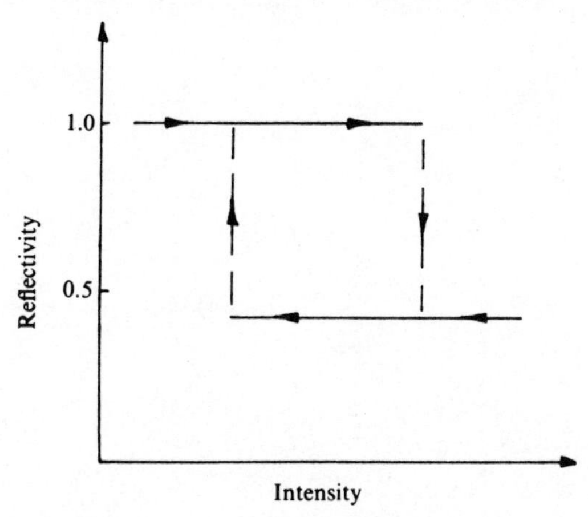

prism (Fig. 7.8). Now suppose that the prism is constructed from a material that exhibits the Kerr effect, i.e., its index of refraction depends on the intensity of the electric field within the medium, $n = n_0 + KE^2$. More precisely, choose a material so that $K < 0$, such as silicon in the infrared.

The reflectivity will be nearly unity if the angle of incidence is set so that $q_\parallel$ is slightly *larger* than the value necessary to couple into the plasmon SP. As the incident beam intensity increases, n and $q_\parallel$ decrease. Abruptly, the reflectivity drops when the surface polariton is excited. The total electric field at the prism surface now is the sum of the external field and the evanescent polariton field. Hence, the wave vector will remain tuned within the resonance width of the plasmon SP even as the incident beam intensity is reduced. Eventually, at a second, lower, critical intensity, the polariton switches off and the reflectivity returns to a value near unity (Fig. 7.16). We observe that the reflectivity achieves two distinct values at the same incident intensity – an example of a phenomenon known as optical bistability (Gibbs, 1985). In principle, this simple configuration can be viewed as a very fast switch.

General references

Reflection and refraction

Feibelman, P.J. (1982). Surface Electromagnetic Fields. *Progress in Surface Science* **12**, 287–408.

Polaritons

Agranovich, V.M. & Mills, D.L. (1982). *Surface Polaritons.* Amsterdam: North-Holland.

Non-linear phenomena

Shen, Y.R. (1984). *The Principles of Non-Linear Optics.* New York: Wiley.

PART 2

ADSORPTION

8

PHYSISORPTION

Fundamentals

The weakest form of adsorption to a solid surface is called physical adsorption, or *physisorption*. It is characterized by the lack of a true chemical bond between adsorbate and substrate. If this is true, some other attractive force must exist that binds a gas phase species to a solid. One possibility that suggests itself is the ubiquitous van der Waals interaction. To see its origin, consider a closed shell atom that sits a distance, z, above a solid surface. We restrict our attention to distances $z \ll c/\omega_p$ (~ 1000 Å) so that the finite propagation velocity of light can be ignored. Even at these distances, a mutual attraction between the atom and the surface exists that arises from the interaction of the polarizable solid with dipolar quantum mechanical fluctuations of the atomic charge distribution. Put another way, the atomic electrons are attracted to their images in the solid.

A one-dimensional harmonic oscillator model of the hydrogen atom is sufficient to capture the essential physics of the van der Waals, or *dispersion*, force between an atom and a solid. Let the oscillator coordinate, r, represent the projection of the electron's orbital motion along the normal to the surface. Consider first the image system appropriate to a perfectly conducting substrate (Fig. 8.1). The total electrostatic energy for this situation is the sum of four terms,

$$U = \frac{1}{2}\left[-\frac{e^2}{2z} - \frac{e^2}{2(z-r)} + \frac{e^2}{2z-r} + \frac{e^2}{2z-r} \right], \tag{8.1}$$

where the factor of one-half takes account of the fact that the electric field within the conductor vanishes. Expanding (8.1) in powers of r/z,

$$U = -\frac{1}{8}\frac{e^2 r^2}{z^3} - \frac{3}{16}\frac{e^2}{z^4}r^3 - \cdots \tag{8.2}$$

Notice that the leading term in (8.2) is proportional to the square of the oscillator displacement coordinate. This means that this term effectively renormalizes (lowers) the frequency of the atomic oscillator by an amount inversely proportional to the cube of the atom–surface separation. That is the basic van der Waals effect. More precisely, the interaction energy is exactly the difference in zero-point energy between a free atom and an atom near a substrate. The argument is reminiscent of the discussion given in Chapter 6 for the conventional image force in terms of virtual excitation of surface plasmons.

One cannot expect the simple oscillator model to predict correctly the absolute magnitude of the dispersion force. Usually one writes

$$V(z) = -\frac{C_V}{z^3}. \tag{8.3}$$

An explicit expression for the constant of proportionality, C_V, must involve some measure of the ability of the adsorbate and the substrate to polarize one another. Qualitatively, the numerator of the z^{-3} term in (8.2) has the form of a product of dipole moments – one for the atom and one for its image. Therefore, if we combine the correct formula for the image dipole appropriate to a dielectric substrate,

$$\mathbf{P}_{\text{image}}^{(\omega)} = \frac{1 - \varepsilon(\omega)}{1 + \varepsilon(\omega)} \mathbf{P}_{\text{atom}}^{(\omega)}, \tag{8.4}$$

with the fact that the atomic polarizability, $\alpha(\omega)$, describes the dielectric response of the atom, it is plausible that

$$C_V = \frac{\hbar}{4\pi} \int_0^\infty \mathrm{d}\omega\, \alpha(\mathrm{i}\omega) \frac{\varepsilon(\mathrm{i}\omega) - 1}{\varepsilon(\mathrm{i}\omega) + 1}. \tag{8.5}$$

Fig. 8.1. Hydrogen atom and its image near a perfect conductor.

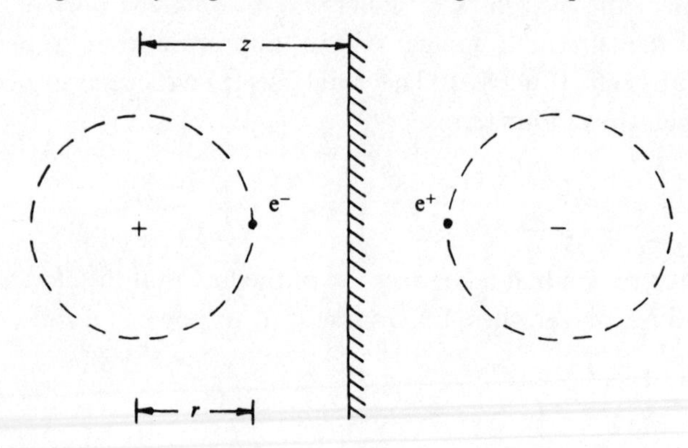

The appearance of Planck's constant reminds us that the van der Waals attraction ultimately derives from zero-point fluctuations while the imaginary frequencies result from a minor mathematical trick (Barash & Ginzburg, 1984).

There is good experimental support for this picture of the long range force between an atom and a solid. Suppose a beam of atoms is directed near the edge of a solid cylinder (Fig. 8.2). The interaction potential can be deduced from a classical analysis of the angle through which the beam deflects. For distances of closest approach from about 200 to 800 Å, the form (8.3) fits the data best for alkali atoms scattered from gold. However, the measured value of the van der Waals constant, $C_V \sim 4\,\text{eV}\,\text{Å}^3$, is at least 60% smaller than the value calculated from (8.5) using measured values for $\alpha(\omega)$ and $\varepsilon(\omega)$. The source of this discrepancy is unknown at present (Mehl & Schaich, 1975).

The second term in (8.2) becomes important if the oscillator atom is brought near to the surface. To first order in perturbation theory, the even parity of the ground state wave function, $\psi_0(\mathbf{r})$, guarantees that this term produces no shift in the atom's total energy. However, the wave function itself changes in first order to ψ_0' and it is easy to verify that $\langle \psi_0' | \mathbf{r} | \psi_0' \rangle \neq 0$ even though $\langle \psi_0 | \mathbf{r} | \psi_0 \rangle = 0$. This is a new effect. The substrate *induces* a permanent dipole moment whose magnitude increases as the atom approaches the surface.

A second consequence of the z^{-4} correction term cannot be captured by our crude harmonic oscillator model. In fact, there *is* a first-order correction to the interaction energy (Zaremba & Kohn, 1976). It is just this correction that defines the reference plane from which the van der Waals interaction should be measured:

$$V(z) = -\frac{C_V}{|z - z_V|^3} = -\frac{C_V}{z^3}\left(1 + \frac{3z_V}{z} + \cdots\right). \tag{8.6}$$

The explicit expression for z_V looks just like (8.5), except that $d_\perp(i\omega)$ enters

Fig. 8.2. Experimental arrangement to measure the long range force between and atom and a surface (Shih & Parsegian, 1975).

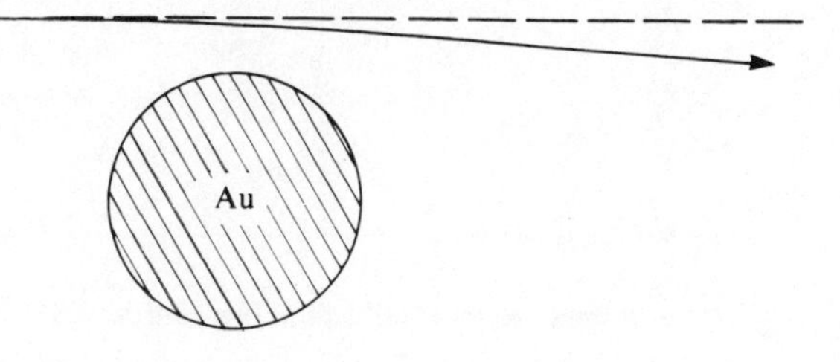

the integrand as a multiplicative factor. This should not be too surprising since $d_\perp$ is a characteristic length that depends on the dielectric properties of the surface (see Chapter 7).

At the closest distances of approach, electrons in the low density spill-out tail of the surface charge distribution begin to 'feel' the presence of a nearby closed-shell atom. Two things occur simultaneously. First, the potential energy of the Bloch electrons is lowered by interaction with the attractive atomic nuclear potential. Second, the kinetic energy of these electrons is raised by the requirement that their wave functions orthogonalize to the atomic valence electron wave functions. At sufficiently short distances, the repulsion always wins and we recover the effective medium result (Fig. 4.7). In fact, the linear relationship between the immersion energy and the host (surface) charge density illustrated there permits us to write the total interaction potential as

$$V(z) = Kn(\mathbf{r}) - \frac{C_V}{|z - z_V|^3}. \tag{8.7}$$

Here, K is a constant read off from plots similar to Fig. 4.7 for other closed-shell atoms and $n(\mathbf{r})$ is the ground state charge density of the surface of interest. As noted in the discussion of inelastic helium atom scattering in Chapter 6, $n(z)$ decays exponentially into the vacuum* so that the

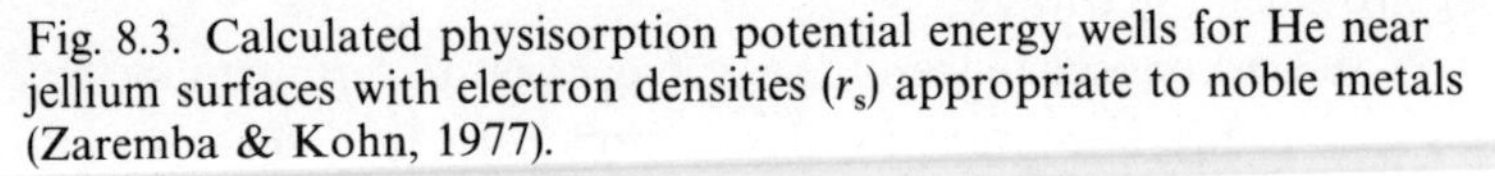

Fig. 8.3. Calculated physisorption potential energy wells for He near jellium surfaces with electron densities (r_s) appropriate to noble metals (Zaremba & Kohn, 1977).

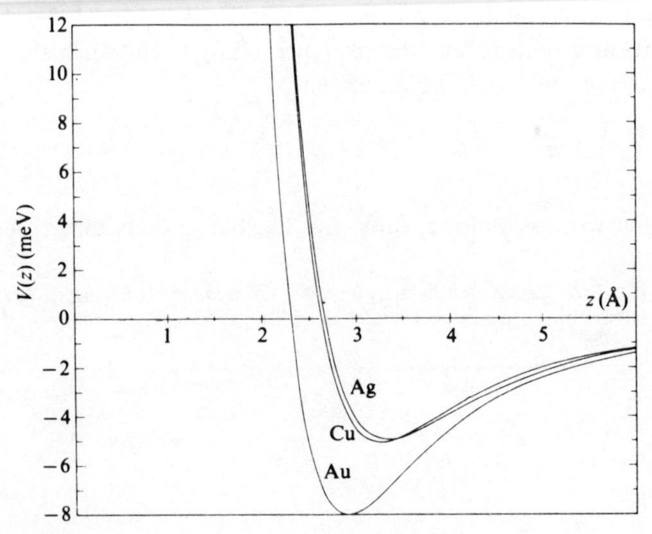

* This is an exact result (Almbladh & von Barth, 1985).

physisorption potential invariably has a shallow minimum a few Ångströms from the surface (Fig. 8.3).

Experimental study of the detailed shape of the physical adsorption well proceeds most easily by analysis of the diffracted beam intensities of *elastic* helium (or other neutral) atom scattering. In the early days of quantum theory, Otto Stern and collaborators used this method to check the de Broglie relation for He and H_2 beams diffracted from alkali halide surfaces. In some of those experiments, Frisch & Stern (1933) noticed considerable structure in the beam intensities as the azimuthal angle of incidence varied. A more recent example of this phenomenon is illustrated in Fig. 8.4. This effect, known as 'selective adsorption', was soon explained by Lennard-Jones & Devonshire (1936).

Surface diffraction of a beam of particles of wave vector $\mathbf{K}$ occurs when $\mathbf{K}(f) = \mathbf{K}(i) + \mathbf{G}_s$, where $\mathbf{G}_s$ is a reciprocal lattice vector of the surface net. In the usual case, an outgoing beam emerges because $K_z^2(f) = K^2(i) - |\mathbf{K}_{\parallel}(f) - \mathbf{G}_s|^2 > 0$. However, the peaks and troughs in Fig. 8.4 appear because of interference from multiple diffraction events where the motion of the atom in the intermediate state is entirely *parallel* to the surface, so that $K_z^2 < 0$ (Fig. 8.5). This situation occurs when the particle can make a transition into (and back out of) one of the quantum mechanical bound states of the physisorption well, E_0, E_1, E_2, etc. The particle gains kinetic

Fig. 8.4. Azimuthal angle dependence of the specular beam intensity of 17 meV He atoms scattered from LiF(100). See text for notation (Derry, Wesner, Krishnaswamy & Frankl, 1978).

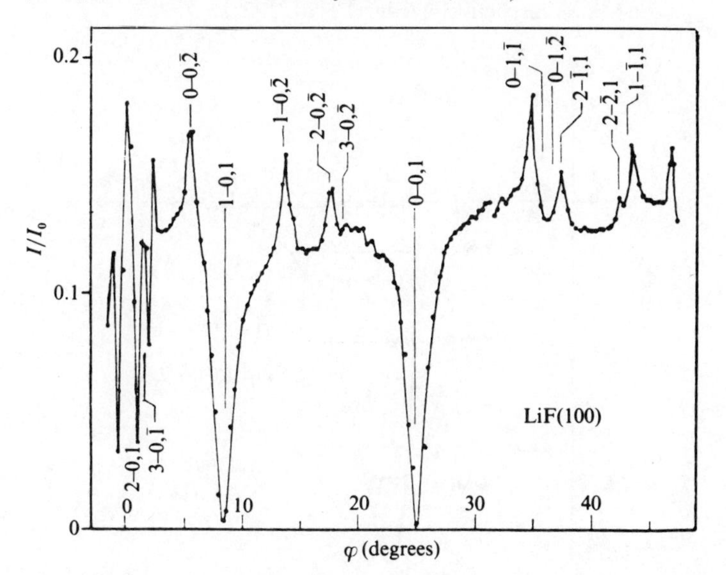

Fig. 8.5. Schematic view of a selective adsorption diffraction event.

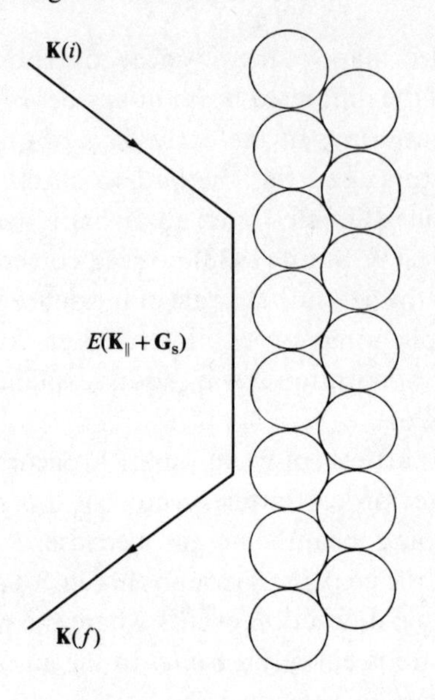

Fig. 8.6. Comparison of experimental physisorption binding energies (hatched lines) determined from H and D scattering from LiF(001) with the calculated bound states of a Morse potential (dashed and solid lines, respectively) (Finzel *et al.*, 1975).

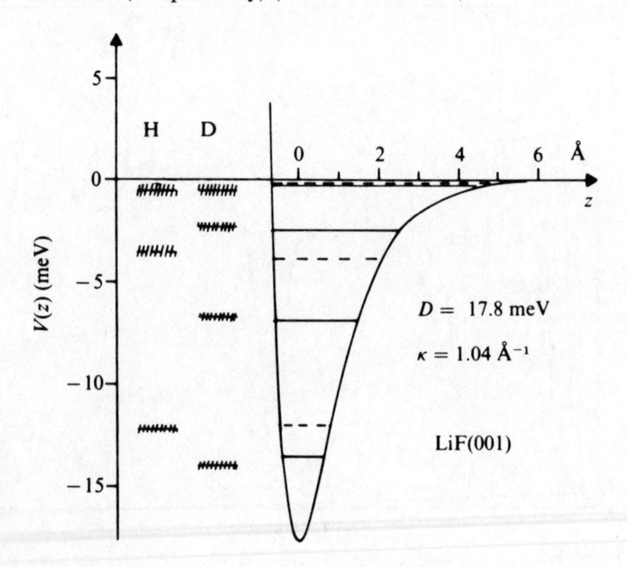

energy at the expense of the potential energy of the well during the time spent trapped within the well:

$$E(\mathbf{K}_{\parallel} + \mathbf{G}_s) = \frac{\hbar^2 K^2}{2m} + |E_n|. \tag{8.8}$$

The selective adsorption resonance condition follows directly from conservation of energy and momentum. This means that one can label the various maxima and minima in Fig. 8.4 ($[n - h, k]$ where n indexes the energy level and (h, k) indexes $\mathbf{G}_s$) without recourse to detailed theory. Of course, a complete picture of the physisorption well *shape* emerges only after one compares experimental binding energies with bound state energies calculated for various model potentials. For example, one can use a parameterized version of the theoretically prescribed exponentially repulsive *plus* inverse cube attractive well of (8.7). Alternatively, practical experience shows that the three-parameter Morse potential, $V(z) = D\{\exp[-2\kappa(z - z_0)] - 2\exp[-\kappa(z - z_0)]\}$, does just as well in many cases (Fig. 8.6). This suggests that the asymptotic van der Waals interaction (8.6), *per se*, does not play a critical role in the equilibrium physisorption 'bond'.

Fig. 8.7. EELS spectrum from physisorbed H_2/Cu(100) (Andersson & Harris, 1983).

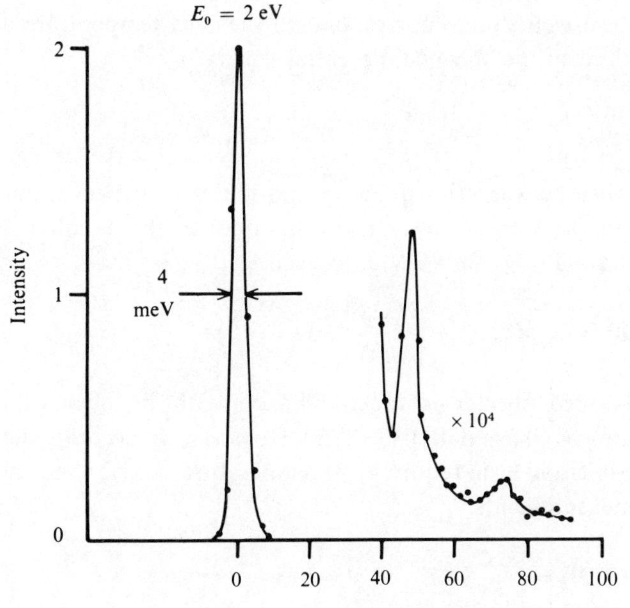

Physisorption well depths are very small compared to typical chemical bond energies ($\sim 1\,\text{eV}$) and thermal energies ($\sim 25\,\text{meV}$). Therefore, it is natural to ask what evidence exists to prove that atoms and/or molecules actually adsorb into such states. One piece of spectroscopic evidence is shown in Fig. 8.7. This is an EELS spectrum of 2 eV electrons scattered from a Cu(100) surface dosed with a small amount of H_2. The spectrum was collected under an ambient pressure of 8×10^{-11} Torr with the sample held at 10 K. At this low temperature, it is certain that the two prominent loss features observed at 45 and 73 meV originate from adsorbed species. However, these energies precisely correspond to the first two rotational excitation energies of *gas phase* H_2. We conclude that rotations of the adsorbed H_2 molecules are essentially unhindered – a view that is consistent with a physisorption picture of very weak adsorbate–substrate binding. Indeed, the reader may wonder whether these excitations actually can kick the molecule *out* of the well by transfer of rotational energy to kinetic energy perpendicular to the surface. The answer to this question is non-trivial and raises a number of interesting dynamical issues to be explored in greater depth in Chapter 14.

Thermodynamics

Thermodynamic measurements provide a view of physisorption that is complementary to the picture drawn from neutral atom and electron scattering spectroscopy. For example, the lowest energy level of the physisorption well should correspond to the zero temperature and zero coverage limit of the chemical potential, viz.,

$$\lim_{\substack{T \to 0 \\ N \to 0}} \mu(T, N) = -|E_0|. \tag{8.9}$$

The difference between this quantity and the chemical potential at high temperature is deduced by integrating one of the familiar Maxwell relations, $(\partial\mu/\partial T)_N = -(\partial S/\partial N)_T$, to yield

$$\mu(0, N) = \mu(T, N) + \int_0^T \left.\frac{\partial S}{\partial N}\right|_{T'} \mathrm{d}T'. \tag{8.10}$$

If the adsorbed species is in equilibrium with its own vapor, high temperature gas phase data fix $\mu(T, N)$. However, an accurate determination of the entropy term requires low temperature, constant coverage heat capacity measurements:

$$S(T, N) = \int_0^T \frac{C_N}{T'} \mathrm{d}T'. \tag{8.11}$$

This procedure has been used to determine the binding energy of He adsorbed onto a polycrystalline graphite substrate (Elgin, Greif & Goodstein, 1978). Values for $C_N(T)$ come from the difference between high-precision calorimetry measurements of a closed cell both with and without the adsorbed species. The resulting binding energy, 12.3 meV, is in excellent agreement with the results of beam scattering experiments.

It will be useful to sharpen up the thermodynamic argument given above and provide a careful derivation of (8.10). First, the phenomenon of adsorption requires a generalization of our previous analysis (Chapter 1) to the case of a two-component system for which both the adsorbate surface particle number (N_s) and the substrate surface particle number (N_s') are legitimate thermodynamic variables. The Gibbs phase rule immediately permits us to set $N_s' = V_s = 0$, so that the surface Gibbs *potential* takes the form

$$G_s(T, P, N_s, A) = H_s - TS_s = U_s - TS_s, \tag{8.12}$$

where H_s is the surface enthalpy. We focus on this particular thermodynamic potential function because its *per particle* value, $\mu_s = (\partial G_s/\partial N_s)_{T,P,A}$, is the adsorbate chemical potential.* Now, because the derivatives of G depend explicitly on the same thermodynamics variables as G itself, we can write

$$\begin{aligned} d\mu_s &= \left(\frac{\partial \mu_s}{\partial T}\right)_{P,N_s,A} dT + \left(\frac{\partial \mu_s}{\partial A}\right)_{T,P,N_s} dA + \left(\frac{\partial \mu_s}{\partial N_s}\right)_{T,P,A} dN_s \\ &= -\left(\frac{\partial S_s}{\partial N_s}\right)_{T,P,A} dT + \left(\frac{\partial \gamma}{\partial N_s}\right)_{T,P,A} dA + \left(\frac{\partial \mu_s}{\partial N_s}\right)_{T,P,A} dN_s \end{aligned} \tag{8.13}$$

where three different Maxwell relations have been used. The differential form (8.10) is thus applicable only to situations where A and N_s are held constant:

$$d\mu_s = -\bar{S}_s \, dT. \tag{8.14}$$

Equation (8.14) suggests an alternate thermodynamic approach that does not require very low temperature experiments. Suppose the above analysis is repeated for the gas phase. Then,

$$d\mu_g = -\bar{S}_g \, dT + \bar{V}_g \, dP. \tag{8.15}$$

* In what follows, a horizontal bar will denote the *per particle* value of other thermodynamics quantities, e.g., $\bar{S}_s \equiv (\partial S_s/\partial N_s)_{T,P,A}$.

In equilibrium, $\mu_s = \mu_g$, so we obtain a Clausius–Clapeyron relation,

$$\left.\frac{\partial P}{\partial T}\right|_{N_s, A} = \frac{\bar{S}_g - \bar{S}_s}{\bar{V}_g} \tag{8.16}$$

which, assuming ideal gas behavior in the vapor phase, commonly is written in the form,

$$R\left.\frac{\partial \ln P}{\partial(1/T)}\right|_{N_s/A} = \bar{H}_s - \bar{H}_g = -q_{st}, \tag{8.17}$$

using the definition of the enthalpy. q_{st} is called the isosteric heat of adsorption and measures an *effective* adsorbate binding energy at finite temperature.

A thermodynamic study of xenon adsorption on a stepped palladium surface illustrates the use of (8.17) to analyze a physisorption system. The Pd(810) substrate can be viewed as a collection of terraces formed from eight rows of atoms normal to [100] separated by monoatomic steps oriented normal to [110] (Fig. 8.8). The experiment consists of measurements of the sample work function, ϕ, and the adsorbate particle number, N_s, as a function of surface temperature at various fixed values of the xenon vapor pressure. The surface particle number, reported in terms of the *coverage*, θ, is derived from the peak-to-peak value of the xenon Auger signal (cf. Fig. 2.2) calibrated against some absolute measurement, e.g., adsorbate mass uptake.*

At fixed pressure, the coverage increases smoothly as the temperature decreases. At the lowest xenon pressure ($\sim 10^{-10}$ Torr), the coverage

Fig. 8.8. Schematic view of the Pd(810) stepped surface.

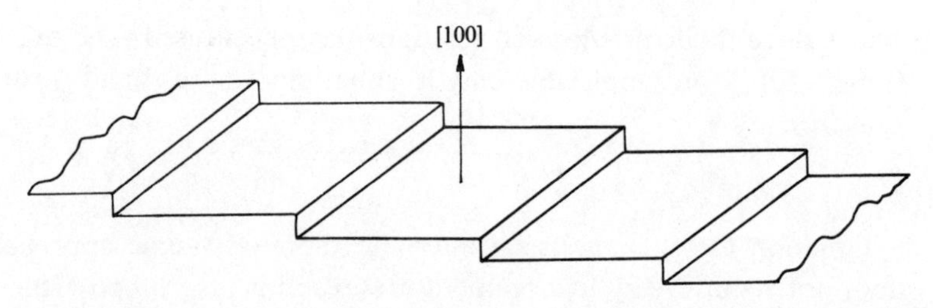

* The coverage usually is defined as the ratio of the adsorbate particle density (atoms/cm^2) to the substrate particle density so that one monolayer ($\theta = 1$) corresponds to one adsorbed atom for each substrate atom. However, in situations like the present, where a large mismatch occurs between the adsorbate/substrate atomic radii or lattice constant, it is more natural to define $\theta = 1$ as the adsorbate density for which a close-packed hard-sphere layer completely covers the available surface area.

saturates at about $\theta = 0.15$, whereas at higher pressure ($> 10^{-8}$ Torr) saturation occurs only at $\theta = 1$. The entropy of the adsorbed layer and the heat of adsorption follow from (8.16) and (8.17) when the isobar data are converted to equilibrium values of P and T at constant coverage, i.e., isosteres. The coverage-dependent values of q_{st}, S_s and ϕ are plotted in Fig. 8.9.

The labelling of Fig. 8.9 suggests a consistent interpretation of the thermodynamic data. Monolayer adsorption of xenon on this stepped surface proceeds in two well-defined stages. Consider first the heat of adsorption. An adatom nestled at the base of a step is better coordinated and feels a greater attractive force than an atom adsorbed in the middle of a terrace. Accordingly, the greater binding energy step sites are filled

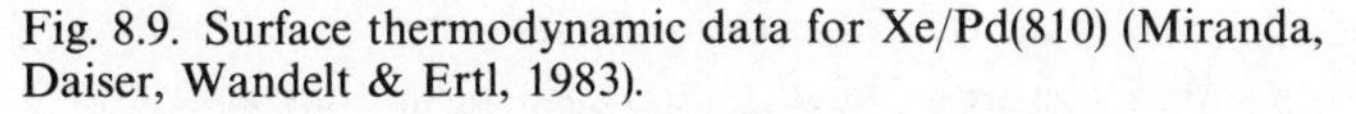

Fig. 8.9. Surface thermodynamic data for Xe/Pd(810) (Miranda, Daiser, Wandelt & Ertl, 1983).

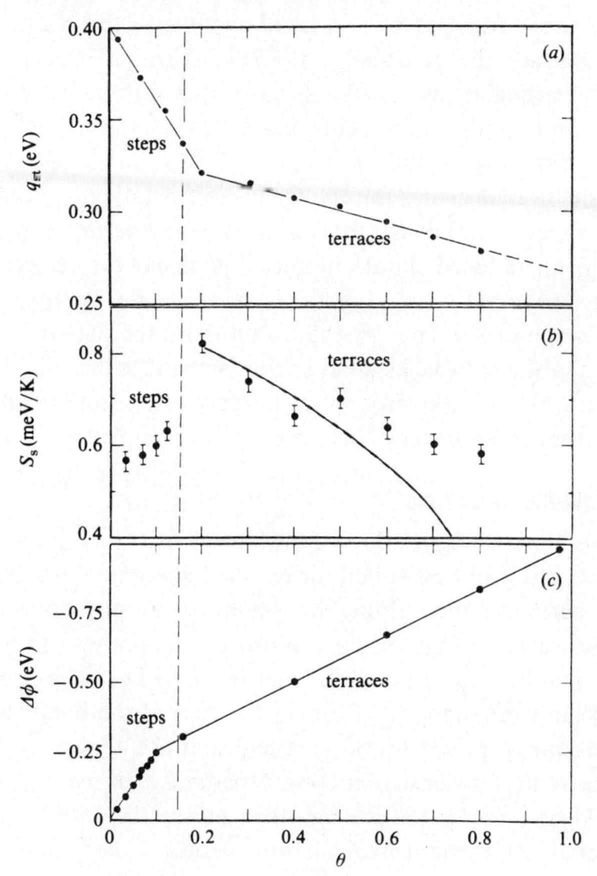

first. The coverage, $\theta = 0.15$, corresponds to complete filling of the step sites. Further adsorption simply fills the terraces.

A specific description of the physical state of the adsorbed layer follows from an analysis of the entropy results. The solid curve in Fig. 8.9(*b*) is the entropy *calculated* for a two-dimensional gas of hard-sphere xenon atoms. The agreement between this parameter-free calculation and the experimental data supports a simple picture of complete adatom mobility across the terraces. If this model is correct, the connection between q_{st} and the adsorbate binding energy follows from equipartition considerations. The enthalpy of the gas phase is $H_g = U_g + PV_g = (5/2)kT$. In the adsorbed phase, free translations are restricted to two dimensions but there is vibrational motion normal to the surface so that $H_s = U_s = -|E_0| + 2kT$. From (8.17), $q_{st} = |E_0| + (1/2)kT$.

The reduced entropy at the step sites is consistent with a further restriction of translational freedom. In fact, it may be more appropriate here to adopt a picture of *localized physisorption* that takes account of the corrugation of the surface potential induced by the truly three-dimensional nature of the surface charge density in (8.7) (cf. Fig. 3.22). Each adatom is 'stuck' to a particular site on the surface with little or no thermally activated motion over the energy barriers that separate adjacent sites. In this case, it is easy to show that $q_{st} = |E_0| - (1/2)kT$.

The work function measures the change in electrostatic potential felt by an electron as it traverses the surface layer. If, as we saw earlier, physisorbed atoms develop an induced dipole moment, $\mathbf{p}$, elementary electrostatics predicts a macroscopic work function change $\Delta\phi = 4\pi N_s p$. By this measure, the slope of the lines in Fig. 8.9(*c*) indicates that the moment induced at a step site is twice as great as the moment induced at a terrace site. We must look to the microscopic charge rearrangement within the physisorbed atom to appreciate this observation completely.

Electronic structure

In this book, we have adopted the local density functional method as the calculational tool best suited for detailed analysis of the electronic structure of surfaces. We adopt the same philosophy for study of adsorbates. However, at first glance, physisorption appears to present an example that manifestly cannot be treated by the LDA. The problem is the van der Waals interaction. Just as in the case of the image force (cf. (4.9)), this long range, power law interaction does not appear in the local approximation to the full density functional theory. Let us see why this is so.

Electrons keep away from one another under the influence of the quantum mechanical, exchange-correlation portion of the Coulomb force.

Qualitatively, we say that each electron digs a positive *exchange-correlation* hole for itself that is totally depleted of charge from neighboring electrons. The hole is large enough to accommodate exactly one unit of charge. As an electron moves through a crystal, it carries along this polarization 'sphere of influence'. The success of the LDA largely is due to its rather good description of electrons in intimate contact with their exchange-correlation holes. However, this 'hole' is left behind (in the form of a polarization charge) when an electron exits a solid through the surface. There is no image or van der Waals interaction in the LDA because a *local* theory cannot account for an electron that is spatially separated from its hole.

It is then perhaps surprising to learn that the local density approximation provides a reasonably good description of the binding and polarization of a closed-shell atom to a simple substrate. Fig. 8.10(*a*) illustrates the ground state charge density obtained by solving the LDA equations (4.3) and (4.4) with an external potential given by (4.5) plus the Coulomb potential from a single charge ($Q = +54$) situated in the vacuum 5 Bohr radii from the positive background. This is Xe/jellium. Notice that there

Fig. 8.10. Contours of constant charge density from an LDA calculation of xenon on jellium ($r_s = 2$): (*a*) total charge density (Lang & Williams, 1982); (*b*) adsorption-induced change in charge density. Solid (dashed) lines indicate a surfeit (depletion) of electrons (Lang, 1981).

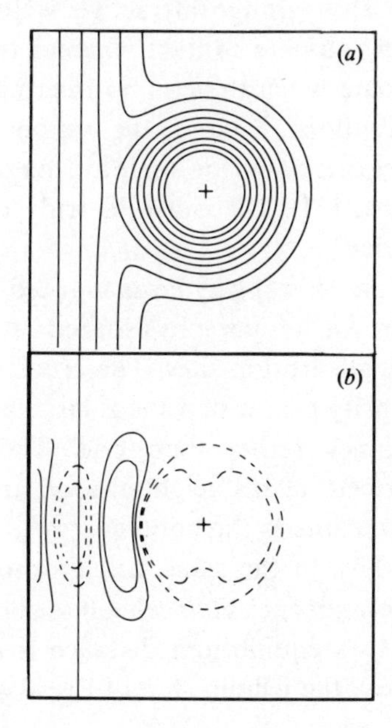

is an intimate intermingling of the xenon valence electron charge density and the metal surface charge density. Considerably more revealing is a contour plot of the *difference* in charge density between Xe/Je and a superposition of the bare-metal and free-atom charge densities (Fig. 8.10(*b*)). This type of plot pinpoints the changes that occur upon physisorption.

The polarization of the physisorbed xenon atom is evident from the charge density difference map. However, the preference of a valence electron to be located on the metal side of the adsorbate does not indicate a build-up of bond charge in the usual chemical sense. Instead, bonding and polarization occur because the valence electrons lower their energy most when surrounded (as much as possible) by their exchange-correlation hole. The interaction energy is simply

$$\Delta = -\int \mathrm{d}\mathbf{r}\,\mathrm{d}\mathbf{r}' \frac{\rho_{xc}(\mathbf{r})\rho_{el}(\mathbf{r}')}{|\mathbf{r}-\mathbf{r}'|} \tag{8.18}$$

where $\rho_{el}(\mathbf{r})$ and $\rho_{xc}(\mathbf{r})$ denote the electronic and exchange-correlation hole charge distributions, respectively. It is important to realize that the van der Waals and local density approach both attribute physisorption effects to the interaction of a valence electron with its 'image'. The former focuses on long range electrostatic forces while the latter emphasizes short range exchange-correlation forces. Support for the LDA point of view comes from two types of spectroscopic evidence. First, it is consistent with the fact, noted earlier, that selective adsorption binding energies often closely correspond to the bound states of short range attractive wells. The adsorbate and substrate are in rather intimate contact whereas the van der Waals language is most appropriate when there is no intermingling of the two interacting charge distributions. Second, the xenon dipole moment and core level binding energies calculated for Xe/Je are in excellent agreement with experiments for Xe/Al(111). Of course, a truly correct picture encompasses both points of view.

One now can identify the microscopic reason why large induced dipole moments and bond energies occur for atoms physisorbed at steps relative to their neighbors on terrace adsorption sites. The crucial point is simply that the substrate charge density near a step site is large because it is bounded by two metallic planes rather than one. Therefore, exchange-correlation holes, $\rho_{xc}(\mathbf{r})$, formed in this neighborhood are very compact and a large induced dipole maximizes the bond energy (8.18) by localizing valence electron charge, $\rho_{el}(\mathbf{r})$, in the same neighborhood. A quantitative illustration of this dipole moment enhancement is shown in Fig. 8.11 for an atom, positioned at its equilibrium distance from the surface, along a continuous path across the jellium step of Fig. 4.4.

Let us return to the quantum mechanical origin of dipole moment formation – a first-order change in the wave function that mixes excited states into the ground state. The nominal atomic configuration of the first excited state of the xenon atom is $5p^5 6s$. However, when physisorbed, the wave function associated with this state admixes with energetically degenerate substrate states. Since the solid lies only to one side of the atom, the induced dipole moment scales with the amount of substrate character mixed into the atomic excited state. To estimate this, denote the energy of the $5p^5 6s$ state relative to the vacuum level as I^* and specialize to the case of a metal substrate. Crudely, if $I^* > \phi$, the work function, the Pauli principle forbids virtually excited electrons from approaching the metal surface; the moment will be small. However, if $I^* < \phi$, the virtually excited electron will have significant wave function weight at the metal surface; the moment will be large.

An indication of the veracity of this simple picture is shown in Fig. 8.12. There, the measured work function change that accompanies xenon physisorption (at monolayer coverage) is plotted for twenty metals arranged by their clean surface work functions. Clearly, the value of I^* plays an important role, although the difference in moment that appears for metals with the same work function indicates that additional factors, such as the magnitude of adsorbate–substrate coupling, are needed to complete the picture. Nonetheless, these results suggest that an enormous moment would develop on high work function substrates if the 6s orbital actually *was* occupied in the atomic ground state, as is the case for the adjacent atom of the periodic table, cesium.

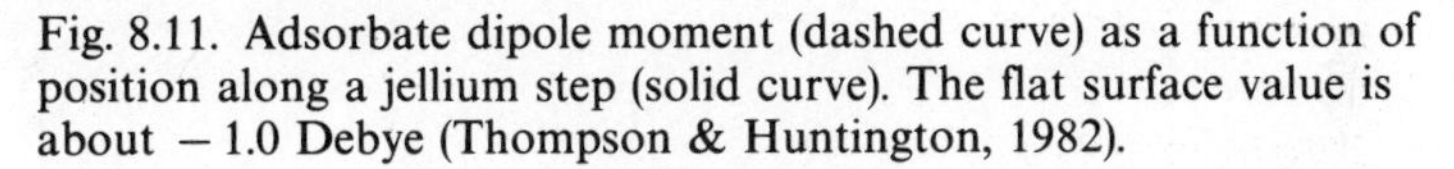
Fig. 8.11. Adsorbate dipole moment (dashed curve) as a function of position along a jellium step (solid curve). The flat surface value is about – 1.0 Debye (Thompson & Huntington, 1982).

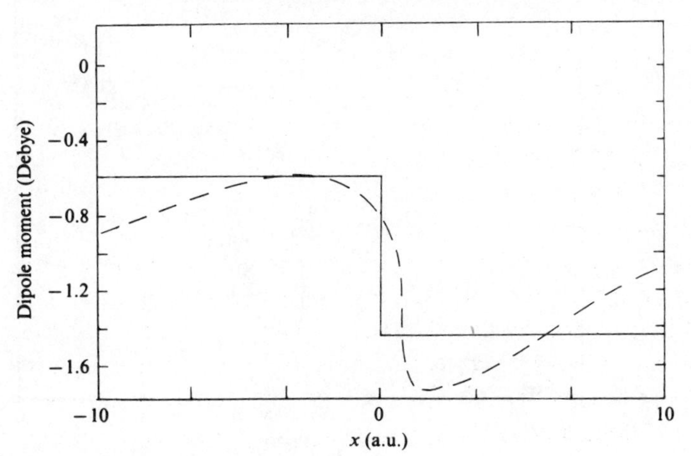

We conclude this chapter with an interesting example of physisorption quite different from the inert gas/metal surface problem examined thus far. In fact, let us reverse the situation entirely. How does a metal adsorbate interact with a nominally 'inert' substrate? Fig. 8.13 shows that this question has significant technical import. This is an electron microscope image of dislocation-induced steps on the (100) cleavage face of a NaCl crystal. The curved steps are one atom layer high and the straight steps are two atoms high. The contrast appears because the crystal has been 'decorated' by gold clusters that preferentially adsorb at step sites after vapor deposition and agglomeration. From our present perspective, the main issue centers on the nature of the interaction between a metal atom and a stepped ionic crystal surface. In particular, why are steps favored?

Fig. 8.12. Experimental work function change for monolayer xenon adsorption on metal surfaces. A vertical arrow denotes the excited state energy of xenon (Chen, Cunningham & Flynn, 1984).

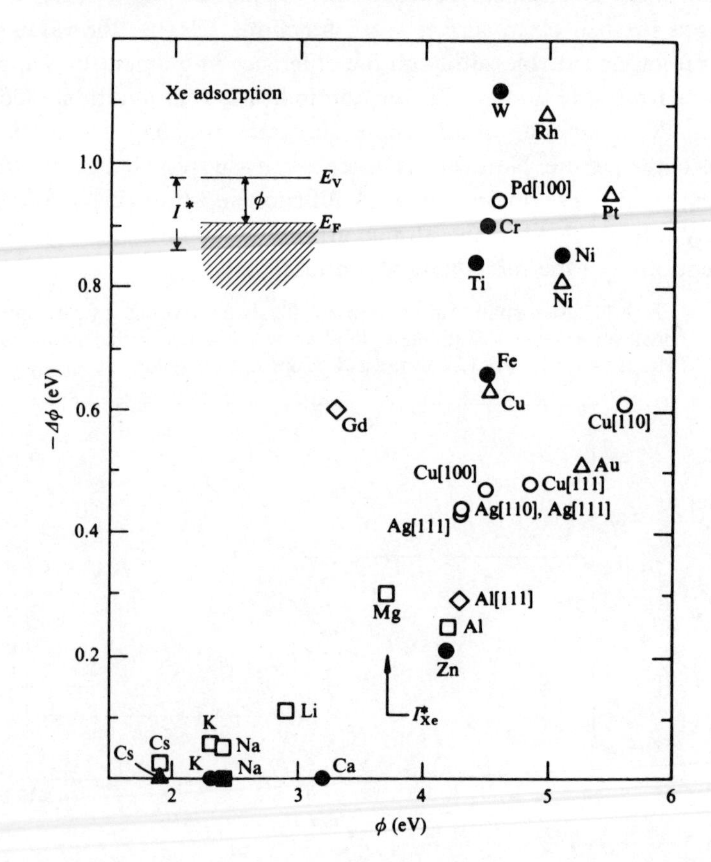

We previously have identified an attractive dispersion force and a Pauli principle–induced repulsive contribution to the physisorption potential energy of an adsorbate above a crystal surface. Here, it is essential to include the electrostatic interaction between the charged substrate ions and the induced dipole moment of the adsorbate. In (8.7) the net effect of the first two terms was written as an effective potential as a function of a single coordinate, $V(z)$. The *corrugation* of the potential energy surface was neglected. We cannot make this approximation now since it is precisely the variations in total energy as the atom traverses the surface that are of interest. Consequently, we write the total potential energy at any adsorbate position ($\mathbf{r}$) as a sum of pairwise interactions from each atom of the substrate ($\mathbf{r}_i$):

$$U(\mathbf{r}) = \sum_i V(\mathbf{r} - \mathbf{r}_i). \tag{8.19}$$

The pair potential $V(\mathbf{r} - \mathbf{r}_i)$ contains the three contributions noted above parameterized in terms of measured values of the polarizability and diamagnetic susceptibility of the adsorbate atom and substrate ions (Chan,

Fig. 8.13. Electron microscope image of a stepped NaCl(100) surface decorated by Au clusters (Bethge, 1982).

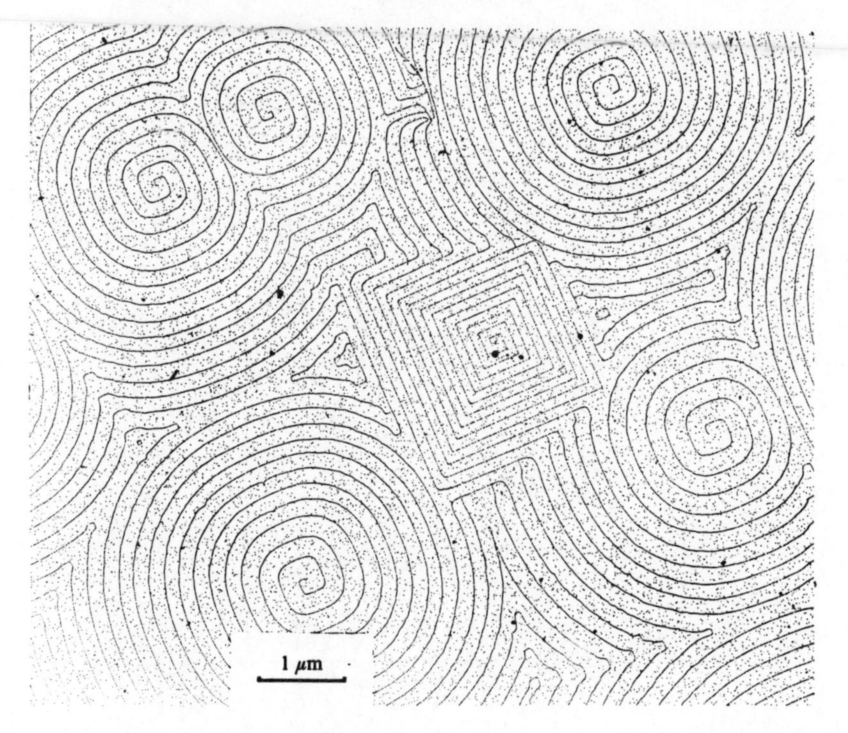

Buckingham & Robins, 1977). This is a semi-empirical scheme but it should be sufficient to reveal characteristic features. The stepped geometry and a contour plot of the adiabatic potential energy surface calculated from (8.19) for a gold atom atop NaCl(100) appear in Fig. 8.14. The sum involves 800 substrate ions. The minimum energy positions (labelled M) are in the vicinity of a sodium ion and do indeed sit at the lower edge of the step. Moreover, the binding energy at a Na^+ *terrace* site (0.7 eV) is in excellent agreement with the measured *desorption* energy of Au from this surface. We defer until Chapters 14 and 15 a detailed discussion of the connection between adsorption and desorption and the significance of the potential energy surface for surface reaction dynamics.

Fig. 8.14. Au physisorption on NaCl(100): (*a*) monoatomic step geometry; (*b*) contour plot of the variation in adatom binding energy as a function of position across the surface. The adatom z-position has been adjusted to its minimum energy value at each point. Contours are drawn at intervals of 0.1 eV. The circles and squares denote Na^+ and Cl^- sites, respectively (Yanagihara & Yamaguchi, 1984).

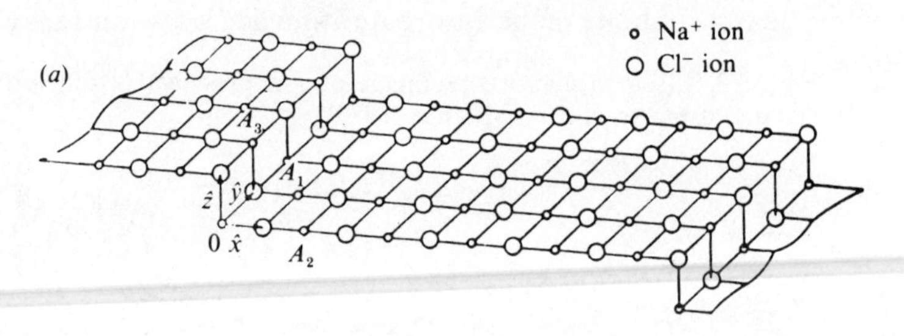

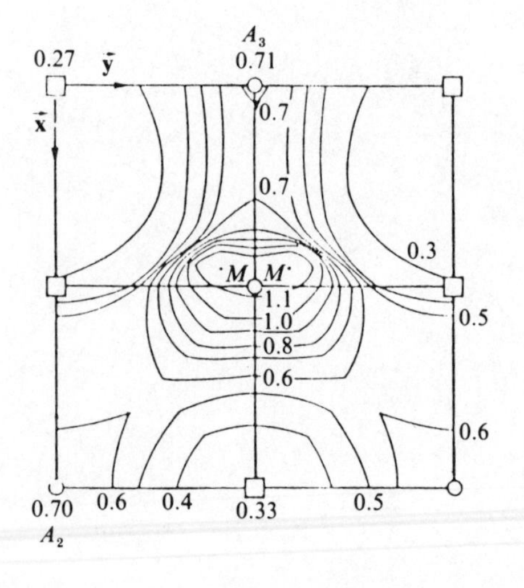

General references

Fundamentals

Landman U. & Kleiman, G.G. (1977). Microscopic Approaches to Physisorption: Theoretical and Experimental Aspects. In *Surface and Defect Properties of Solids*, vol. 6, pp. 1–105. London: The Chemical Society.

Thermodynamics

Cerny, S. (1983). Energy and Entropy of Adsorption. In *The Chemical Physics of Solid Surfaces and Heterogeneous Catalysis* (eds. King & Woodruff), vol. 2, pp. 1–57. Amsterdam: Elsevier.

Electronic structure

Lang, N.D. (1981). Interaction between Closed-Shell Systems and Metal Surfaces. *Physical Review Letters* **46**, 842–5.

9

CHEMISORPTION

Fundamentals

At the beginning of this century, scientists generally believed that all adsorption phenomena were of the sort we have called physisorption. Some (unspecified) long range attractive force drew gas phase matter toward a solid and the increased concentration of the gaseous substance near the surface was thought to be analogous to the retention of the Earth's atmosphere by the gravitational field. The adsorbed layer was viewed as a 'compressed vapor' with little or no interaction with the atoms of the substrate. However, compelling experimental evidence soon accumulated that pointed to another, distinctly different, form of adsorption.

Langmuir (1916) introduced and extensively investigated the idea that there can exist strong, short range forces between adsorbates and a substrate. He regarded the arrangement of atoms at the surface of a solid as a sort of Chinese checkerboard that defines a specific density of potential adsorption sites. Foreign gas atoms that strike the surface may either bounce back into the gas phase or bind to one of these sites through formation of a *surface chemical bond.* The latter process is termed *chemisorption* and, in this view, it is not unreasonable to regard the adsorbate/substrate complex as an enormously large molecule.

Adsorption, and chemisorption in particular, lowers the free energy of any closed system that contains only a free surface and atoms or molecules in the gas phase. This statement may be made more precise by appeal to the Gibbs adsorption equation (1.11):

$$d\gamma + \frac{S_s}{A} dT + \Gamma \, d\mu + \sum_{i,j} (\gamma\delta_{ij} - \sigma_{ij}) \, d\varepsilon_{ij} = 0. \tag{9.1}$$

As in the previous chapter, we have set $V_s = N_s' = 0$ so that only the surface excess number density ($\Gamma \equiv N_s/A$) of the adsorbed species appears explicitly. Now imagine variations in the adsorbate vapor pressure (assumed to behave as an ideal gas) under conditions of constant temperature and constant surface strain:

$$d\gamma = -\Gamma\, d\mu = -kT\Gamma \frac{dP}{P}. \tag{9.2}$$

This equation tells us that a clean surface is thermodynamically unstable to adsorption because the surface tension of the solid decreases with increasing adsorbate particle density.

In a limited number of cases, one can check the Gibbs adsorption isotherm (9.2) directly by experiment. Fig. 9.1 illustrates the variation of γ measured by the zero creep method (see Chapter 1) for oxygen adsorbed onto a polycrystalline copper substrate at 1100 K. At the lowest oxygen pressures ($< 10^{-18}$ Torr) there is evidently no adsorption.* However, the surface tension does indeed fall precipitously thereafter and we calculate an oxygen surface excess of 4.25×10^{14} atoms/cm^2 from (9.2). This corresponds to 1/4 monolayer of oxygen on a copper (111) plane.

Fig. 9.1. Test of the Gibbs adsorption isotherm for oxygen on copper (Bauer, Speiser & Hirth, 1976).

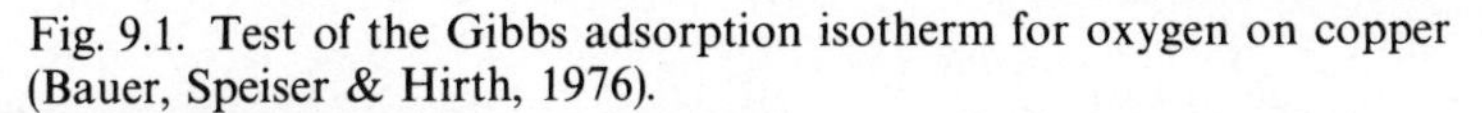

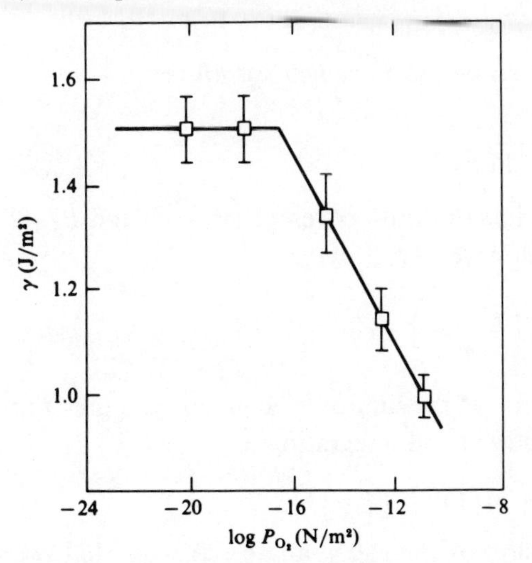

* One obtains and calibrates extremely low partial pressures of oxygen by pumping the gas into the sample chamber through a solid zirconia electrolyte whose conductivity is determined entirely by the oxygen diffusion current. Such a device is used to measure trace concentrations of oxygen in the combustion products of automobiles.

From a microscopic point of view, it is clear that the magnitude of surface tension reduction that accompanies adsorption must be related to the binding energy of the adsorbed species. To establish a precise relation, it is necessary to integrate (9.2) with some choice for the functional form of $\Gamma(P)$. To illustrate the point, we adopt Langmuir's model of non-interacting particles that adsorb onto a checkerboard substrate. Here, a gas phase particle traps at any one of N_0 surface sites with a binding energy ε_b. In addition, we assume that the particle executes small vibrations within the adsorption well characterized by a frequency ν_0.

The grand partition function appropriate to a gas in equilibrium with a substrate modelled in this way is

$$\Xi = 1 + z_{vib}\, e^{\beta(\mu - \varepsilon_b)}, \tag{9.3}$$

where z_{vib} is the vibrational canonical partition function and $\beta \equiv 1/kT$. From Ξ, the equilibrium coverage follows immediately,

$$\Gamma = \Gamma_0 \frac{z_{vib}\, e^{\beta(\mu - \varepsilon_b)}}{1 + z_{vib}\, e^{\beta(\mu - \varepsilon_b)}}. \tag{9.4}$$

If we eliminate the chemical potential by using its gas phase value,

$$e^{\beta\mu} = \beta P \left(\frac{2\pi\beta\hbar^2}{m}\right)^{3/2}, \tag{9.5}$$

(9.4) assumes the form known as *Langmuir's isotherm*:

$$\Gamma = \Gamma_0 \frac{P}{P + P_{1/2}(T)}, \tag{9.6}$$

where Γ_0 corresponds to saturation coverage ($P \to \infty$) and $P_{1/2}(T)$ is the pressure that yields half coverage,

$$P_{1/2}^{-1}(T) = z_{vib}\beta \left(\frac{2\pi\beta\hbar^2}{m}\right)^{3/2} e^{-\beta\varepsilon_b}. \tag{9.7}$$

The predicted variation of the surface tension with gas pressure follows by combining (9.2) and (9.6) and integrating:

$$\gamma = \gamma(0) - \Gamma_0 kT \ln\left[1 + P/P_{1/2}(T)\right]. \tag{9.8}$$

$\gamma(0)$ is the surface tension of the clean surface. We see that even if the requisite surface tension measurements were easy to perform (they are not) it would be difficult to extract the weak dependence of γ on ε_b in this way. Nevertheless, this exercise does have an interesting consequence. Equation (9.8) implies that it may be possible to drive the surface tension of some materials *negative* at sufficiently elevated temperature and pressure. This means that the surface is unstable and a reconstruction must occur. Actually, even if this condition is not met, chemisorption may induce reconstruction anyway if a lower (positive) net surface tension obtains (see Chapter 13).

A straightforward measurement of chemisorption binding energies follows from the same Clausius–Clapeyron analysis of isostere data outlined in connection with physisorption (8.17). For example, the isotherms measured for carbon monoxide adsorbed at low coverage onto a single crystal Pd(111) surface follow the Langmuir isotherm at high temperature (Fig. 9.2). The thermodynamic analysis yields $q_{st} \cong 1.5$ eV and combining (8.17), (9.6) and (9.7) we find that $q_{st} = |\varepsilon_b| - (1/2)kT$*. The thermal energy is negligible at these temperatures so one obtains a binding energy of the order of an electron volt. In fact, since chemisorption is a form of conventional chemical binding we expect that the heat of adsorption will vary on the familiar 1–10 eV scale. Fig. 9.3 makes this point clearly. The scatter in the data reflects both variations in adsorption from substrate to substrate as well as the existence of *inequivalent binding sites* on a given single crystal surface.

The short range chemisorption bond forms only when there is direct intermingling of the substrate and adsorbate charge densities. This observation permits us to set a scale for this phenomenon, independently from the thermodynamic data, by use of the effective medium theory. The

Fig. 9.2. Adsorption isotherms for CO/Pd(111) (Ertl & Koch, 1970).

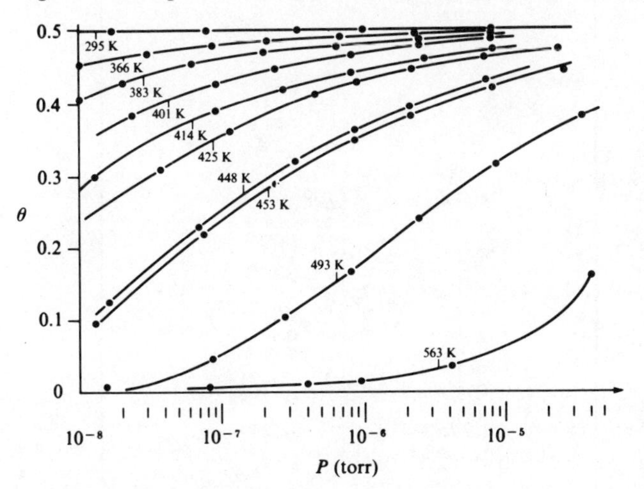

* We suppose that the vibrational partition function in (9.7) takes its high temperature limiting value: $z_{vib} = (1/\beta h\nu)^3$.

oxygen immersion energy curve in Fig. 4.7 clearly demonstrates the existence of a chemisorption potential energy *well* normal to the surface if we identify the electron density scale with the variation of the substrate charge density near the surface. The minimum occurs at a binding energy of about 7 eV, which agrees surprisingly well with the measured *average* heat of adsorption of oxygen on polycrystalline transition metals (Fig. 9.4). The data also exhibit a clear trend not captured by the effective medium theory. The heat of adsorption decreases monotonically as one proceeds from left to right across any transition metal row. The explanation of this striking effect requires additional physics; we will return to it presently.

There is an important qualitative difference between the examples of carbon monoxide and oxygen chemisorption. CO normally adsorbs in molecular form whereas O_2 typically *dissociates* so that bonding occurs with individual oxygen atoms. Neither merely physisorbs except possibly at very low temperatures. A qualitative understanding of the origin of

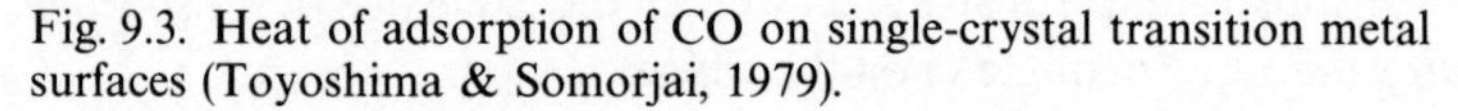

Fig. 9.3. Heat of adsorption of CO on single-crystal transition metal surfaces (Toyoshima & Somorjai, 1979).

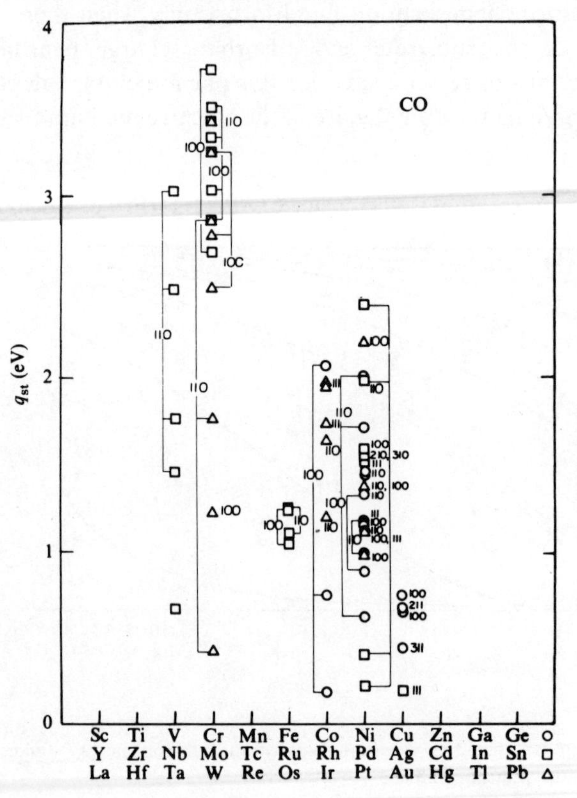

this (and related) behavior follows most easily from a simple model introduced by Lennard–Jones (1932). Let $V(z)$ represent the potential energy of the adsorbate and substrate as a function of their mutual separation (Fig. 9.5). As the molecule approaches the surface, $V(z)$ exhibits the familiar physisorption minimum (dashed curves). Now split the molecule into its atomic constituents. Far from the surface, the separated-atom energy must lie above the energy of the bound molecule (taken as zero). At closer distances, the van der Waals interaction lowers the energy of this configuration as well. More importantly, $V(z)$ can develop a deep minimum if the individual atoms form strong chemisorption bonds to the surface (dashed-dot curves). The energy of the true adsorbate complex evolves along the configurational path of minimum energy (solid curves).

The ground state of the adsorbate depends on the details of the competing potential energy curves and their crossing points. For example, the molecule spontaneously dissociates if the curve crossing occurs *below* the zero of energy (Fig. 9.5(*a*)). By contrast, molecular physisorption results at low temperature if the curve crossing occurs *above* the energy zero (Fig. 9.5(*b*)). The dissociated species ultimately appears in either case at higher temperatures because thermal activation carries the system

Fig. 9.4. Heat of adsorption of CO and O on polycrystalline transition metal surfaces (Toyoshima & Somorjai, 1979).

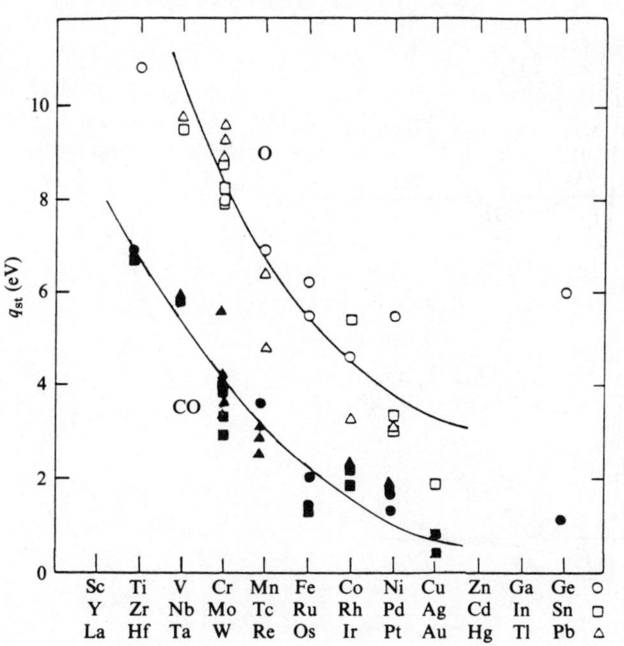

into the lowest energy state over any intervening potential energy barriers. In such cases, one says that the physisorbed state is a *precursor* to the chemisorbed state. Similar considerations apply when dissociation is always unfavorable. Fig. 9.5(*c*) illustrates a case where the adsorbate passes directly into a molecular chemisorbed state. Even within this simplified one-dimensional model it is easy to imagine other possibilities – multiple precursors and direct competition between atomic and molecular chemisorption configurations to name but two.

Fig. 9.5. Schematic diagrams of the potential energy of an absorbate/substrate complex appropriate to three different ground state configurations: (*a*) dissociative chemisorption; (*b*) molecular physisorption; (*c*) molecular chemisorption (Lennard-Jones, 1932).

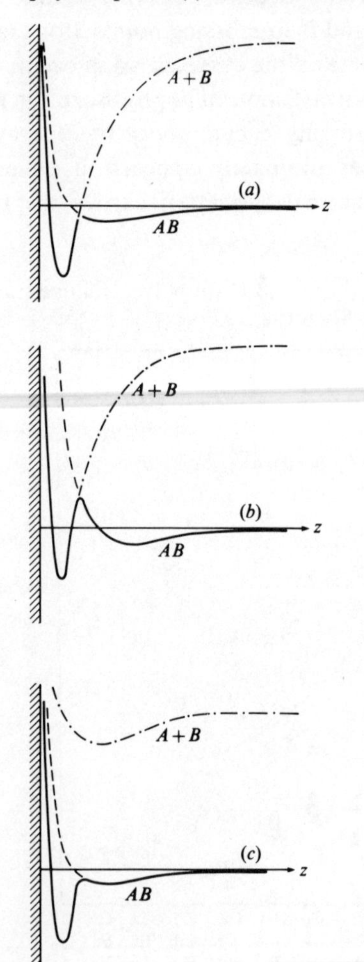

A working knowledge of the bonding and valence concepts of elementary inorganic chemistry is a valuable tool for many chemisorption considerations (see, e.g., Douglas, McDaniel & Alexander, 1983). Consider the molecular orbital diagrams appropriate to the isolated CO and O_2 molecules (Fig. 9.6). In homonuclear oxygen, electrons fill the 5σ and 1π bonding orbitals while two unpaired spins in the unfilled anti-bonding 2π orbital lead to the familiar paramagnetic triplet ground state. The electronic structure of the closed-shell carbon monoxide molecule is rather different. The mismatch between oxygen and carbon atomic levels dictates that bonding occurs only through the 4σ and 1π orbitals whereas the high-lying 5σ molecular orbital is filled by two non-bonding 'lone pair' electrons localized on the carbon end of the molecule.

What do we expect when these molecules are brought near a surface? CO remains in molecular form upon adsorption because its closed electronic shells are quite stable. However, rather than physisorb, the adsorbate orients itself carbon-end down and gains chemisorption energy by lowering the 5σ orbital via a bonding interaction with a localized orbital of the substrate. In chemical language one would say that CO acts as a Lewis base by 'donating' its 5σ electrons to the metal (but see Chapter 12). Oxygen acts rather differently, and for good reason. The intrinsic bond

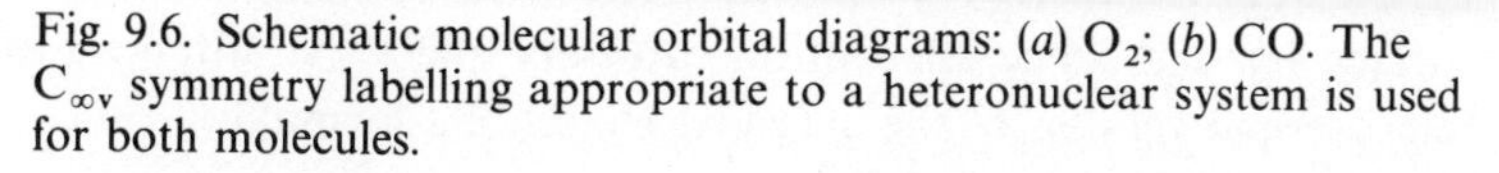

Fig. 9.6. Schematic molecular orbital diagrams: (*a*) O_2; (*b*) CO. The $C_{\infty v}$ symmetry labelling appropriate to a heteronuclear system is used for both molecules.

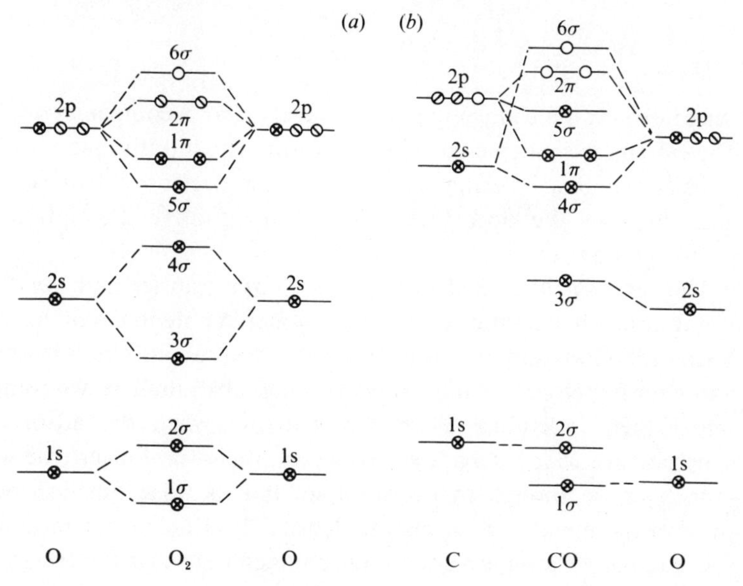

strength of O_2 is less than half that of CO (dissociation energies of 5.2 eV and 11.23 eV, respectively). Therefore, energy lost by breaking the intramolecular bond usually can be more than compensated by local bonding of the highly electronegative oxygen atoms to the substrate. As a result, dissociative chemisorption is the rule for oxygen on most metal and semiconductor surfaces.

The arguments presented above are highly qualitative. Ultimately, it will be necessary to investigate the bonding geometry and electronic structure of adsorbate/substrate complexes in some detail. However, we can develop a considerable amount of insight with a preliminary overview designed to illuminate some general trends of surface chemical bonding. At this stage, we emphasize the role of electron spectroscopy and the use of simple, yet surprisingly rich, *model systems* designed to capture the essential physics of chemisorption.

Metals

Langmuir developed his concept of chemisorption from careful measurements of adsorption-induced changes in the work function of refractory metals – notably Cs/W. In some cases, data of this kind reveal very large changes ($\Delta\phi$) as a function of coverage (Fig. 9.7). How does this come about? From (4.8), we know that the work function is determined by the bulk Fermi energy and the electrostatic dipole associated with the surface barrier. Since adsorption can only affect the latter we look to a microscopic version of the classical Helmholtz formula:

$$\Delta\phi = -4\pi e \int d\mathbf{r}\, z\, \delta n(\mathbf{r}), \tag{9.9}$$

where $\delta n(\mathbf{r})$ is the change in charge density that accompanies adsorption. The extreme cases depicted below conform to an intuitive picture of charge transfer and ionic bonding in accord with the electronegativities of lithium and chlorine. We now seek a more quantitative description of this phenomenon.

The *resonant level model* addresses charge transfer and bonding to a metal surface in the simplest possible manner. A potential well that contains a single bound state at an energy $-|\varepsilon_a|$ represents the adsorbate. We adopt the free electron jellium model for the substrate. The two components retain their individual electronic structure when the adsorbate and substrate are widely separated. However, this is no longer true when the atom is close enough to the solid so that its wave function begins to overlap the metal surface charge density. It is no longer meaningful to speak of purely 'atomic' states or purely 'metal' states in the energy domain near the atomic bound state (Fig. 9.8). The eigenstates of the combined system are a mixture that broadens the sharp atomic level into a resonance that 'leaks' (via quantum mechanical tunnelling) into the metal.

The electronic structure of the resonant level model is very simple to analyze. Label the continuum states of the metal by wave vectors $|k\rangle$ and denote the degree of mixing between the atomic level and a metal state by an overlap matrix element, $V_{a,k}$. The energy levels of the adsorbate/substrate complex follow from diagonalization of the Hamiltonian matrix:

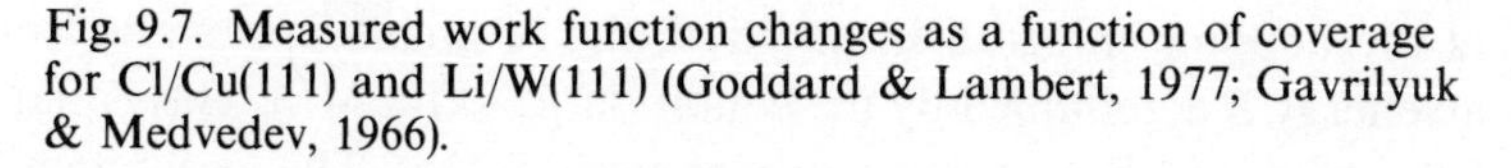

$$\begin{bmatrix} \varepsilon_a - E & V_{a,k_1} & V_{a,k_2} & \cdots \\ V^*_{a,k_1} & \varepsilon_{k_1} - E & 0 & \cdot\cdot 0 \\ V^*_{a,k_2} & 0 & \varepsilon_{k_2} - E & \cdot\cdot 0 \\ \vdots & \vdots & \vdots & \ddots \\ 0 & 0 & 0 & \end{bmatrix}$$

Fig. 9.7. Measured work function changes as a function of coverage for Cl/Cu(111) and Li/W(111) (Goddard & Lambert, 1977; Gavrilyuk & Medvedev, 1966).

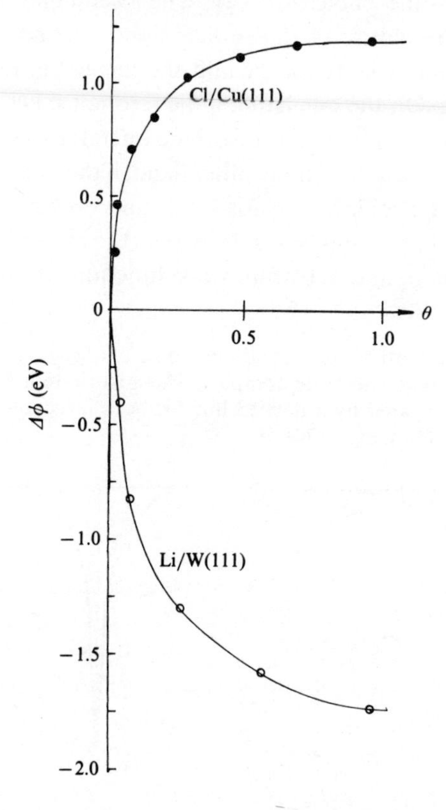

The local density of states (4.25) provides the most natural way to display the results of this diagonalization:

$$\rho(\mathbf{r}|E) = \sum_{\alpha} |\psi_{\alpha}(\mathbf{r})|^2 \delta(E - E_{\alpha}). \tag{9.10}$$

Here, $\rho(\mathbf{r}|E)$ is to be evaluated at the position of the adsorbate. After a bit of mathematics one finds (Newns, 1969):

$$\begin{aligned} \rho_{\mathrm{a}}(E) &= \frac{1}{\pi} \frac{\Delta(E)}{[E - \varepsilon_{\mathrm{a}} - \Lambda(E)]^2 + \Delta^2(E)}, \\ \Delta(E) &= \pi \sum_{k} |V_{a,k}|^2 \delta(E - \varepsilon_k), \\ \Lambda(E) &= \frac{1}{\pi} \int \mathrm{d}E' \frac{\Delta(E')}{E - E'}. \end{aligned} \tag{9.11}$$

The LDOS broadens from a sharp, bound state, delta function into a Lorentzian line shape with a width, $\Delta(E)$, determined by the magnitude of the hopping integral. In addition, the resonant level shifts from ε_{a} to an energy E determined by the condition $E - \varepsilon_{\mathrm{a}} = \Lambda(E)$. This equation can be solved graphically for some choice of $V_{a,k}$. The occupancy of the adsorbate resonance and the nature of the surface bond depend on the relative position of the resonant level energy and the metal Fermi level. There are two limiting cases. On the one hand, if the resonance lies above (below) E_{F}, charge transfer occurs from the adsorbate (metal) to the metal (adsorbate) and an ionic bond results. On the other hand, if the finite-width resonance just straddles the Fermi level the adsorbate and substrate share electrons – a covalent bond forms. Fig. 9.8 is drawn for the case when the overlap between the adsorbate and substrate wave functions is small so

Fig. 9.8. Schematic potential energy and electronic energy level diagram for an adsorbate/substrate complex. The adsorbate local density of states is indicated by a dashed line for both large and small separation distances (Gadzuk, 1974).

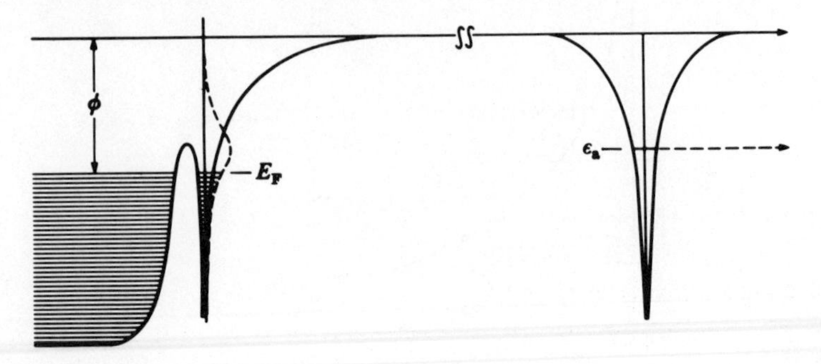

that the level-mixing shift, $\Lambda(E)$, is small – a good approximation for a free electron metal. However, we cannot neglect two additional effects that also tend to shift the position of the resonant level away from $-|\varepsilon_a|$.

Focus more careful attention on the energy of the adatom valence level. In the previous chapter, we computed the total electrostatic energy between an atom and a metal surface as the sum of four image-like terms (8.1). Notice that only the first term in that expression remains if the orbiting electron is not present. Therefore, the change in energy of that electron *alone* is given by the static limit ($r = 0$) of the remaining three terms:

$$\varepsilon_a \to \varepsilon_a + \frac{e^2}{4z}. \tag{9.12}$$

This interaction *raises* the energy of the valence level as the adsorbate approaches the surface. Equation (9.12) is accurate as long as the image approximation is valid ($z > d_\perp(0) \sim 2\,\text{Å}$).

Yet another effect becomes important when the adsorbate is close enough to the solid to sample directly the substrate charge distribution. In that case, the effective potential that constitutes the solid surface barrier (cf. Fig. 4.3) acts as an external field on the adsorbate. First-order perturbation theory is sufficient to estimate the change in the valence level energy. The result is quite simple – the resonant level position simply *tracks* the surface barrier potential:

$$\Delta\varepsilon_a = \langle\psi_a|v_{\text{eff}}(z)|\psi_a\rangle \cong v_{\text{eff}}(z_a), \tag{9.13}$$

where $v_{\text{eff}}(z_a)$ is the magnitude of the substrate effective potential evaluated at the adsorbate position. This effect *lowers* the energy of the adsorbate valence level (Fig. 9.9). Of course, the ultimate energy position of the

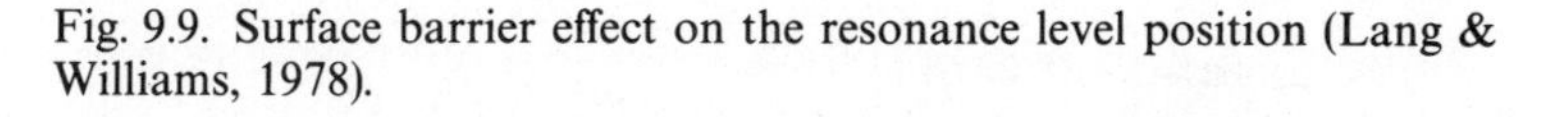

Fig. 9.9. Surface barrier effect on the resonance level position (Lang & Williams, 1978).

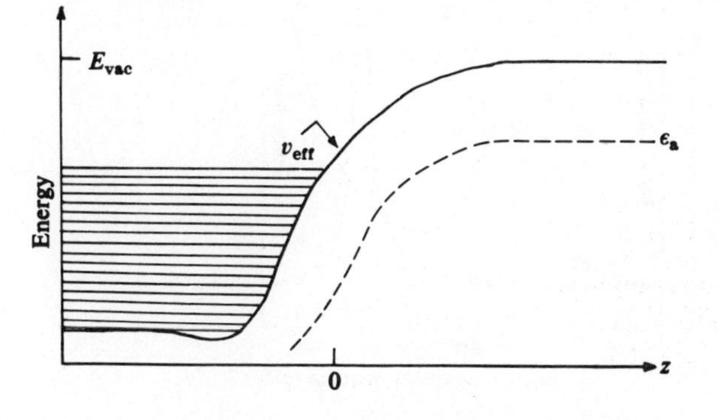

adsorbate level reflects the combined influence of the level-mixing, image and surface barrier shifts evaluated at the equilibrium adsorbate–substrate separation.

The resonant level model completely neglects the detailed electronic structure of the adsorbate. Yet, we must include this aspect of the problem to visualize truly the charge transfer and bonding characteristics of chemisorption. This can be done exactly with the local density functional scheme if we are willing to retain the jellium model of the metal substrate. The relevant calculations are precisely analogous to those discussed in Chapter 8 in connection with xenon physisorption. The adatom–substrate separation is chosen to minimize the total energy of the system and the quantities of interest are the local density of states at the adsorbate site and the charge density distribution.

Fig. 9.10 shows the induced density of states for chlorine, silicon and lithium chemisorbed onto a high density jellium substrate. The Lorentzian-like shapes clearly are consistent with the prediction of the resonant level model. Moreover, the energy positions of the resonances are just as expected from the relative electronegativities of the adatoms. The lithium 2s resonance level and the chlorine 3p resonance level lie above and below the metal Fermi level, respectively. This much could be guessed. Far more interesting are the contour plots of the electron density, which reveal the detailed character of the surface bond (Fig. 9.11). In particular, the lower panel depicts the total charge density *minus* the superposition of the bare-metal plus bare-atom charge densities, i.e., $\delta n(\mathbf{r})$, the quantity that determines the chemisorption-induced work function change (9.9). The

Fig. 9.10. Change in density of states due to chemisorption for Cl, Si and Li atoms adsorbed on jellium ($r_s = 2$). The silicon curve exhibits both 3s-derived and 3p-derived resonances (Lang & Williams, 1978).

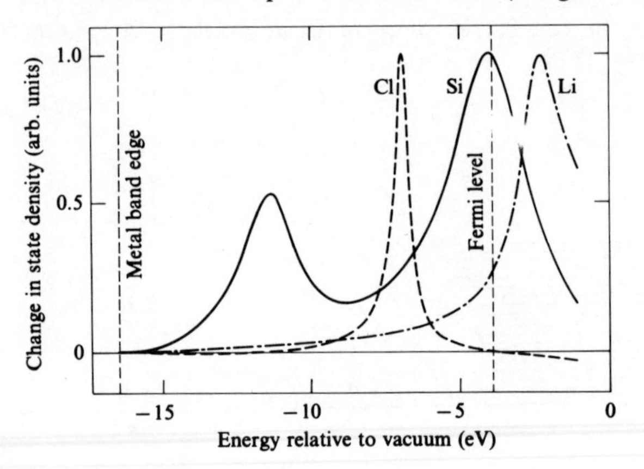

Fig. 9.11. Contours of constant charge density for Cl, Si and Li atoms adsorbed on a jellium substrate: (*a*) total charge; (*b*) induced charge. Solid (dashed) curves denote a surfeit (depletion) of electrons (Lang & Williams, 1978).

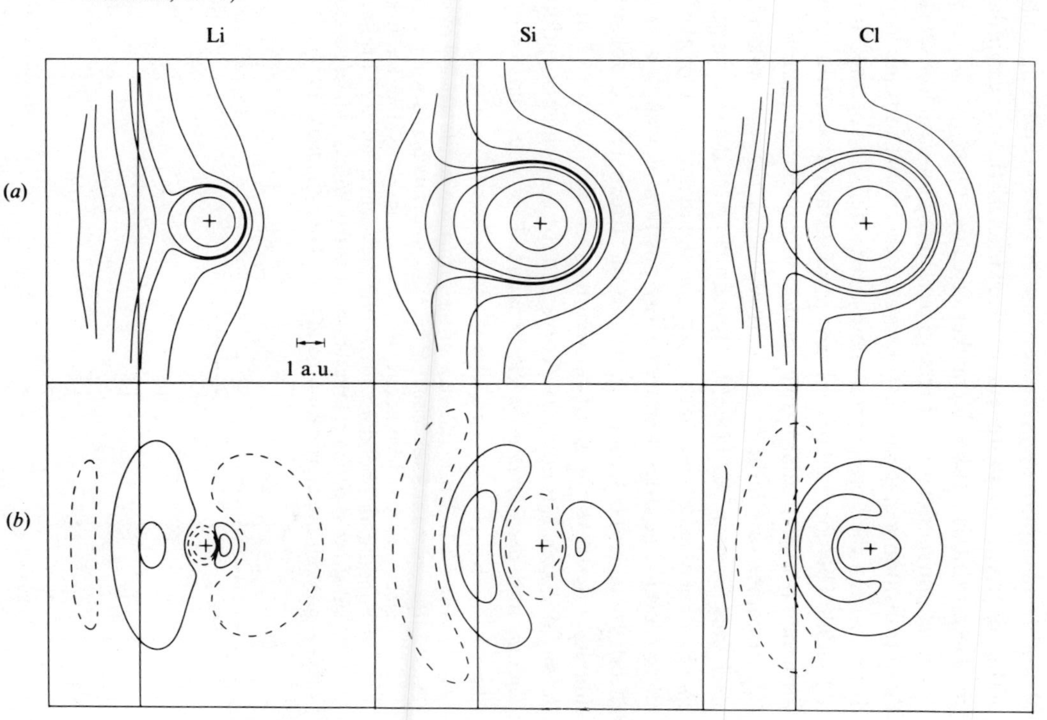

lithium and chlorine plots reflect almost unidirectional charge transfer between adatom and substrate. Dipole moment formation is obvious and the bonding is surely ionic. The charge rearrangement that accompanies silicon chemisorption is much more subtle. Charge builds up on both sides of the adsorbate. This turns out to be characteristic of *covalent* bonds that involve p-orbitals (Bader, Henneker & Cade, 1967).

The jellium model is not directly relevant to non-free electron substrates. For a transition metal, we must generalize the model and embed a localized d-state within the free electron energy levels depicted on the left hand side of Fig. 9.8. This compact orbital can have substantial wave function overlap with the adsorbate charge distribution. If so, the interaction parameter, $V_{a,k}$, becomes very large over a narrow energy range. The corresponding level shift function $\Lambda(E)$ (see (9.11)) acquires considerable magnitude and energy dependence and one expects a significant shift of the resonant energy level. The d-level shifts energy position as well (a sort of 'recoil' effect). In fact, the strong admixture of adsorbate and substrate wave functions destroys the original identities of the two peaks in the local density of states. It is now more correct to speak of chemisorption-induced *bonding* and *anti-bonding* levels broadened into resonances by the free

Fig. 9.12. Schematic view of the formation of the density of states for the resonant level model of transition metal chemisorption. A free electron s-band broadens two localized levels synthesized from the bonding/anti-bonding interaction between an adatom orbital and substrate d orbitals. (*a*) adatom interacts with a single d-state; (*b*) adatom interacts with a band of d-states (Gadzuk, 1974).

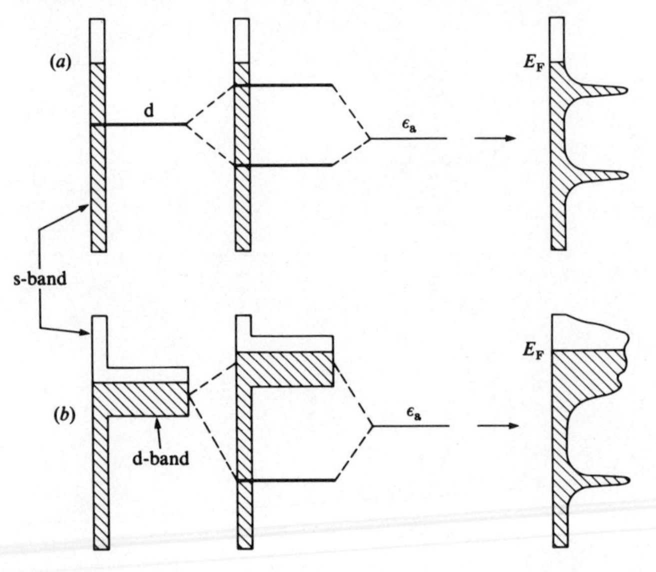

electron s-band (Fig. 9.12*a*). This is the *surface molecule* limit of the resonant level model.

The surface molecule idea is the new ingredient needed to understand the trend noted earlier in the heat of adsorption of oxygen on transition metals (Fig. 9.4). A slight generalization of the argument given above is sufficient to describe the *hybridization* of the adsorbate level in this case. Simply replace the single localized d-orbital by an entire *band* of d-states (Fig. 9.12*b*). To get the trend we need only note two principal changes that occur in a transition metal d-band as the atomic number increases across a row. First, the center of gravity moves down in energy due to increased nuclear attraction. This moves the d-band closer to the oxygen 2p level, which in turn leads to a larger bonding/anti-bonding splitting. Second, the d-band fills to maintain charge neutrality. Therefore, as one proceeds across the row, we occupy more and more anti-bonding states located at progressively higher and higher energies. This explains the net loss of adsorbate–substrate cohesion since, to a first approximation, the total energy of the system is given by the sum of the occupied one-electron energy levels (cf. (4.29)).

We can do more. If the surface molecule picture is truly appropriate, the bonding properties of certain chemisorbed systems should resemble the bonding in a corresponding gas phase free molecule. This notion has

Fig. 9.13. Ball-and-stick model of $Ir_4(CO)_{12}$ (Plummer, Salaneck & Miller, 1978).

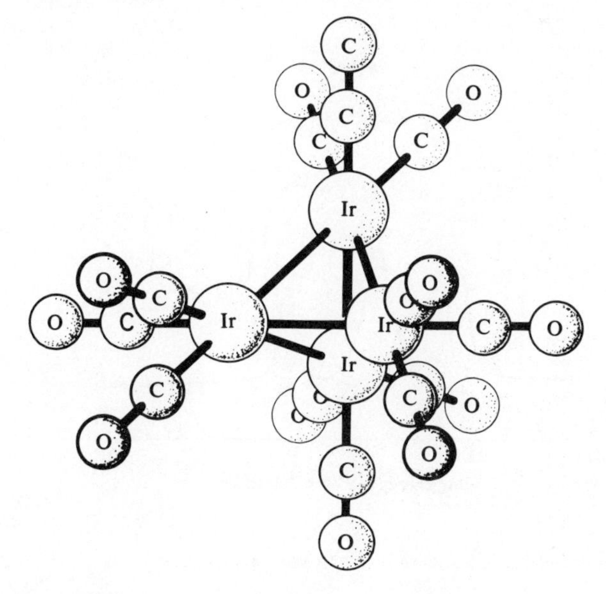

been particularly fruitful for study of CO adsorption on transition metals due to the existence of a class of stable molecules known as *carbonyls* (Cotton & Wilkinson, 1962), which consist of CO molecules terminally bonded to a small cluster of metal atoms (Fig. 9.13). We shall regard the CO 5σ level as the adsorbate 'bound state' of the resonant level model and ask how this level moves when the surface chemical bond forms.

Ultraviolet photoemission is the proper experimental probe for study of orbital binding energies since the energy distribution curve reflects the occupied density of levels (cf. Fig. 4.18).* As Fig. 9.14 shows, the photoelectron spectrum for CO does indeed reflect the 4σ, 1π and 5σ states expected from the molecular orbital diagram (Fig. 9.6). However, the spectrum obtained from $Ir_4(CO)_{12}$ exhibits only *two* broad peaks below the iridium-derived d-levels. The most natural interpretation of the data

Fig. 9.14. UV ($h\nu = 40.8$ eV) photoemission energy distribution curves: (*a*) CO; (*b*) $Ir_4(CO)_{12}$; (*c*) CO/Ir(100) (Plummer, Salaneck & Miller, 1978).

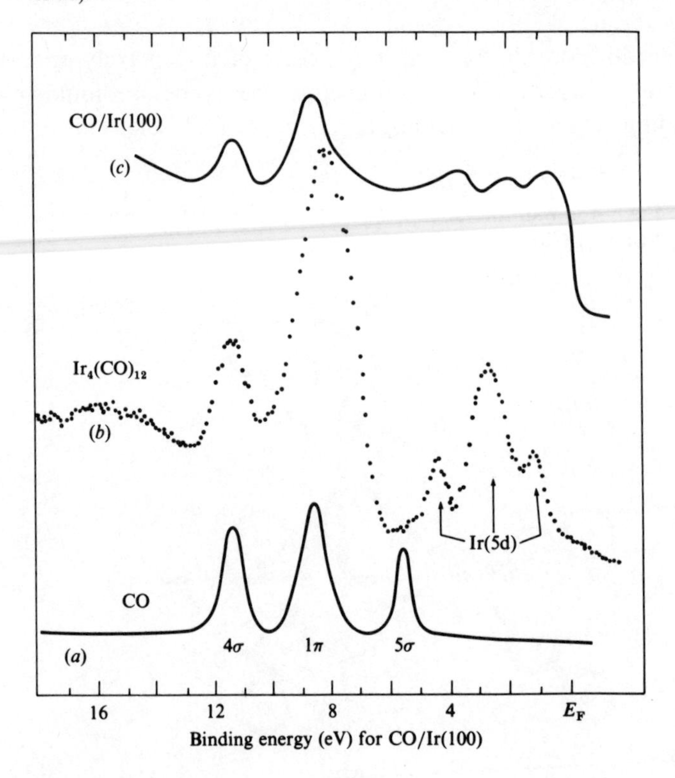

* We ignore so-called relaxation effects; see, e.g., Lang & Williams (1977).

is that a bonding interaction has pulled the 5σ level down to lower energy and broadened all the levels so that the 5σ cannot be resolved from the 1π. More importantly, as a test of the surface molecule concept, the EDC for CO adsorbed onto an Ir(100) surface is remarkably similar to the carbonyl spectrum. We conclude that very localized bonding persists at the semi-infinite iridium surface.

Semiconductors

A general overview of the surface physics of chemisorption on semiconductors is not easy to construct. For one thing, the experimental data base from which trends might be extracted is a bit problematical. Simply put, it is difficult to prepare clean semiconductor surfaces for controlled experimental adsorption studies. Unlike metals, these materials do not anneal well after sputtering and, for compound semiconductors, sputtering and segregation typically lead to non-stoichiometric surfaces. The literature abounds with contradictory results for nominally identical systems. To make matters worse, recall that a semiconductor surface (in equilibrium!) goes to rather extreme (and non-intuitive) lengths to trade off energy gained by local bond formation and energy lost by elastic distortion in search of the lowest free energy geometrical configuration. This suggests that any systematic analysis might be doomed to failure by case-by-case idiosyncrasies. Luckily, the situation is not quite that desperate.

Dangling bonds are undesirable. That was a recurring theme in our discussion of clean semiconductor surfaces (Chapter 4). Most reconstructions had the effect of eliminating or greatly reducing the number of these high energy objects. Therefore, the most natural interaction between a semiconductor and an adsorbate would saturate dangling bonds during the formation of a local surface chemical bond. However, another possibility arises due to the poor screening characteristics of the substrate. Ionic bonding can occur that involves charge transfer between the adsorbate and the *bulk* of the semiconductor. Unfortunately, there is no simple analog to the resonant level model that clearly describes both mechanisms, much less their interplay. Accordingly, we adopt here a more pragmatic approach and appeal directly to experiment to illustrate the range of behavior. To make contact with our previous discussion we retain oxygen as a 'model' adsorbate. Oxygen adsorption on the cleavage faces of silicon, gallium arsenide and zinc oxide – three semiconductors arranged in order of increasing ionicity – will serve to identify a number of characteristic variations on the basic bonding mechanisms sketched above.

In Part 1, we showed that the existence of unfilled electronic states at the surface of Si(111) leads to the phenomenon of Fermi level pinning, i.e., the independence of the sample work function to bulk doping (Fig. 4.39). Fig. 9.15 illustrates the simplest model consistent with the explanation of pinning given in Chapter 4. The undoped surface is characterized by a band of surface states/resonances distributed in the vicinity of the fundamental gap. As discussed in connection with Fig. 4.38, charge neutrality requires that this band be half-full and centered not far from mid-gap. Upon n-type doping of the bulk, the Fermi level rises toward the conduction band minimum. However, a lower total energy results if electrons vacate the high-lying, bulk, donor levels and populate the lower-lying surface states. As charge transfer proceeds, the Fermi level drops and the surface band fills until thermodynamic equilibrium is achieved. Since the surface density of states is large ($\sim 10^{15}/\text{cm}^2$), E_F will be 'pinned' within a fairly narrow energy range independent of the bulk dopant density, N_d. To complete the picture, the Schottky (1939) model posits that the concentration of immobile bulk dopants is uniform right up to the crystal surface. The uncompensated positive donor ions establish an electric field within an adjacent bulk (depletion) layer of thickness z_0. Within this *space-charge layer* Poisson's equation takes the form

$$\frac{d^2V}{dz^2} = -\frac{4\pi n}{\varepsilon} = -\frac{4\pi e N_d}{\varepsilon}, \tag{9.14}$$

where ε is the dielectric constant of the semiconductor. Integrating twice,

$$V(z) = V_{\text{bulk}} - \frac{2\pi e N_d}{\varepsilon}(z - z_0)^2. \tag{9.15}$$

Fig. 9.15. Electron energy levels near the surface of a clean semiconductor: (*a*) undoped sample; (*b*) disequilibrium between n-type bulk and its surface; (*c*) band bending and Fermi level pinning at equilibrium.

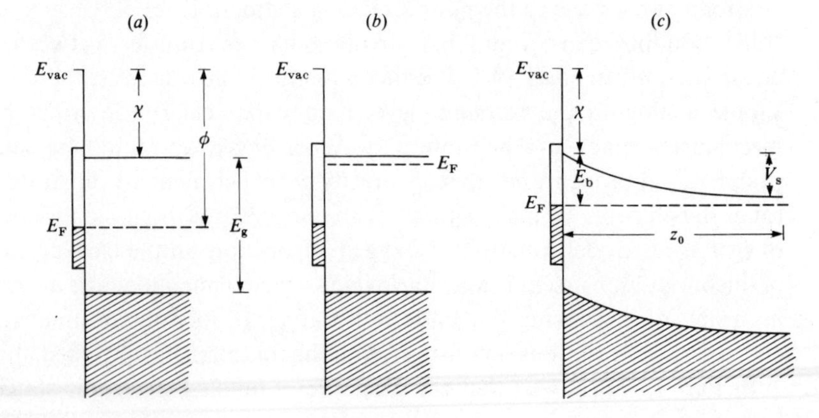

The change in electron energy levels near the surface of the crystal in response to this parabolic electrostatic potential is known as *band bending*. Therefore, in addition to the work function, ϕ, the surface is characterized by the band bending, V_s, and its electron affinity, χ.

What happens when oxygen is permitted to adsorb onto this silicon surface? Fig. 9.16 compares photoelectron electron energy distribution curves obtained from heavily doped ($N_d \sim 10^{20}/cm^3$) Si(111) both before and after saturation coverage with oxygen. The clean spectrum differs from the oxygen-covered results in three significant ways. First, oxygen removes electronic states from within the forbidden gap (small hatched region). This is another example of the 'crud effect' observed earlier in connection with surface excitons (Fig. 6.1). Adsorbates disrupt the subtle boundary conditions that permit surface states to exist. In this case, adsorption removes electronic surface states that pin E_F in the clean sample. Second, sharp features in the density of states associated with bulk transitions

Fig. 9.16. UPS electron energy distribution curves for clean and oxygen-covered Si(111) 2 × 1. Surface states and oxygen-induced features are shown cross-hatched (Wagner & Spicer, 1974).

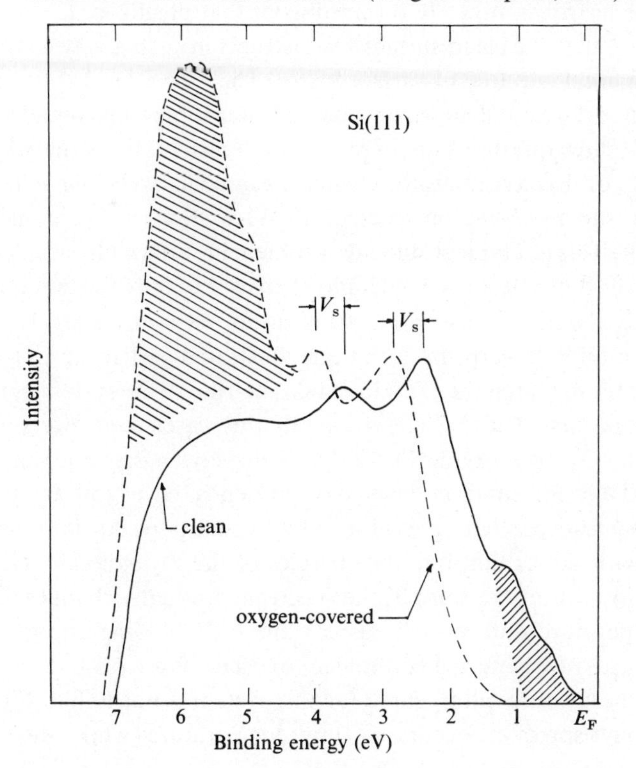

uniformly shift to greater binding energy. This reflects the 'unbending' of the bands. Third, a new broad feature about 6 eV below the Fermi level appears in the spectrum (large hatched region). This is direct photoemission from the 2p states of atomic oxygen. Our identification follows from the fact that the 2p orbital binding energy in a free oxygen atom is 13.6 eV while the electron affinity and band gap of Si(111) are 4.25 eV and 1.1 eV, respectively.

All three observations are connected in the following way. Suppose an occupied band gap surface state releases an electron it acquired from a dopant impurity site in the bulk. The liberated electron returns to the bulk. This narrows the space charge layer and helps unbend the bulk bands. Meanwhile, the newly dangling bond surface state now is free to form a covalent bond with an oxygen 2p orbital directed perpendicular to the surface. The old surface state vanishes from the gap and is replaced by a new surface (bonding) state far down in energy in the vicinity of the oxygen 2p energy. In fact, this level lies about 1 eV above the free atom binding energy because the chemisorption orbital now contains more charge (covalently donated by a silicon atom) to screen the oxygen nucleus.

Consider next the case of the initial oxidation of GaAs(110). The situation is nearly completely reversed from that of silicon. It was shown in Chapter 4 that the clean surface reconstructs in such a way as to sweep the fundamental gap free of surface states. The Fermi level is not pinned at the clean surface but *does* become pinned after submonolayer adsorption of oxygen! Three questions are of primary interest: (*a*) Does the adsorbate bond as O_2 or do oxygen atoms chemisorb dissociatively? (*b*) What is the adsorption site and bonding scheme? (*c*) What states are responsible for the pinning? This is a typical 'difficult' problem in surface physics. A decade of intense effort has produced a definitive answer to only the first question.

We study the initial adsorption scenario for oxygen/GaAs(110) by use of the same UPS 'fingerprint' technique illustrated earlier for CO/Ir(100). That is, we try to match the experimental spectrum of the system of interest to known results of a model system. In this case, there are only two candidates – O_2 and atomic O. Fig. 9.17 displays a sequence of photoelectron EDC's for submonolayer oxygen deposited at low temperature (45 K). The as-deposited spectrum exhibits sharp peaks that correlate precisely with the orbital binding energies of the O_2 *molecule*. However, as one slowly warms the sample, the spectrum gradually changes until the room temperature data clearly display the 6 eV binding energy feature characteristic of chemisorbed *atomic* oxygen. We conclude that the potential energy adsorption curve for this system is not unlike Fig. 9.5(*b*) so that physisorption occurs at low temperature while dissociative chemisorption characterizes room temperature adsorption.

At a more microscopic level, we wish to know exactly where the oxygen atom binds. A simple chemical argument might suggest that bonding occurs to the raised arsenic surface atoms of GaAs(110) (see Fig. 4.43) by donation of the As 4s lone pair electrons to the oxygen. Unfortunately, it does not appear to be as simple as that (Landgren *et. al.*, 1984). Experiments designed to address this question rely heavily on the fact that subtle changes in chemical environment affect the binding energies of the deep core levels of both the adsorbate and the substrate atoms. As outlined in

Fig. 9.17. UPS energy distribution curves obtained for clean GaAs(110) (bottom panel), submonolayer O_2/GaAs(110) deposited at 45 K and during subsequent warming to room temperature (Frankel, Anderson & Lapeyre, 1983).

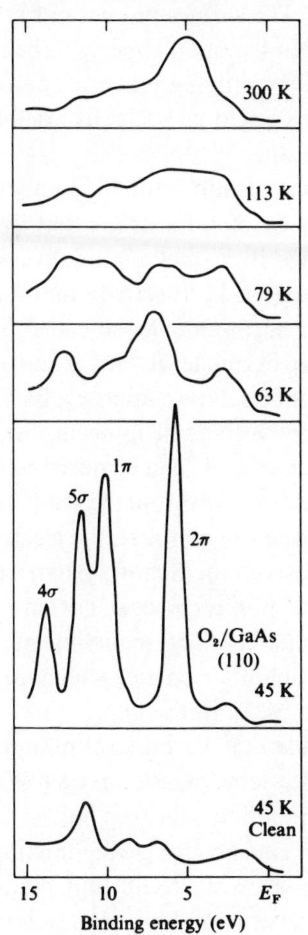

Chapter 2, this information is accessible with an XPS measurement (also known, in this context, as electron spectroscopy for chemical analysis (ESCA)). Difficulties arise because the interpretation crucially depends on a rather subtle analysis of the photoemission core level line shapes. For example, surface core level shifts (Chapter 4) must be distinguished from true chemical shifts. This is not an unambiguous procedure. Accordingly, a conservative observer would conclude that, at present, we simply do not know to which GaAs(110) surface atoms the first few oxygen atoms choose to bind.

The gloomy conclusion of the preceding paragraph does not mean that nothing further can be learned about O/GaAs(110). As the silicon example demonstrated, photoelectron spectroscopy readily can track the relative position of the Fermi level and valence band maximum. The ability to measure and adequately interpret this particular electronic property is of more importance than one might think. The surface physics of Fermi level pinning in the submonolayer chemisorption system appears to be relevant to the electrical properties of *commercial* rectifying junctions constructed from the same materials. To see how this might be so, let us briefly review the microscopic mechanism of rectification.

Suppose one places a piece of bulk metal into intimate contact with a semiconductor surface pinned as in Fig. 9.15(*c*). Electrons that flow from the semiconductor conduction band toward the metal must surmount (by thermal activation) the band-bending barrier, V_s. Electrons that flow from the metal into the semiconductor must surmount the so-called *Schottky barrier*, $E_b = \phi - \chi$. In equilibrium, the Fermi levels align and the two electron flows are equal. No net current flows. Now suppose a bias voltage is applied across the junction. The entire voltage drop occurs within the highly resistive depletion layer. This means that V_s can be raised or lowered at will so that one can vary the rate of electron flow from the semiconductor to the metal by many orders of magnitude. By contrast, the electron flow in the opposite direction is unchanged since the Schottky barrier height is unaffected by the bias voltage. The non-reciprocal current–voltage characteristics of a diode follow immediately. The rectifying junction is characterized by the magnitude of the Schottky barrier which, in turn, is determined by the equilibrium position of E_F in the gap.

In the present context, it is remarkable that Fermi level pinning at the GaAs(110) surface is completed at *very* low coverages of oxygen (Fig. 9.18). Moreover, the magnitude of the corresponding Schottky barrier height is identical to the barrier height found in *commercial* grade junctions built from GaAs(110)! The mystery deepens when we recall that pinning and band bending involve charge transfer between bulk donor levels and surface states in the semiconductor gap. Where do these surface states come from? They must be adsorbate-induced since we know they do not exist on the clean surface. Conventional photoelectron spectroscopy is not sensitive enough to reveal directly the presence of these pinning states (as opposed to indirect band-bending measurements) since no more than 0.1% of the surface sites need be involved. Nevertheless, a more sophisticated spectroscopic experiment *can* see them.

Fig. 9.19 illustrates photoelectron spectra collected from *laser excited* GaAs(110) surfaces. These samples are heavily p-doped so that the Fermi

Fig. 9.18. Fermi level pinning of heavily n-doped GaAs(110) as a function of adsorbed oxygen coverage as determined by photoelectric band-bending measurements. The coverage scale* is logarithmic and E_F is measured relative to the valence band maximum. The unpinning of the Fermi level for n-type Si(111) is shown for comparison (Spicer *et al.*, 1979).

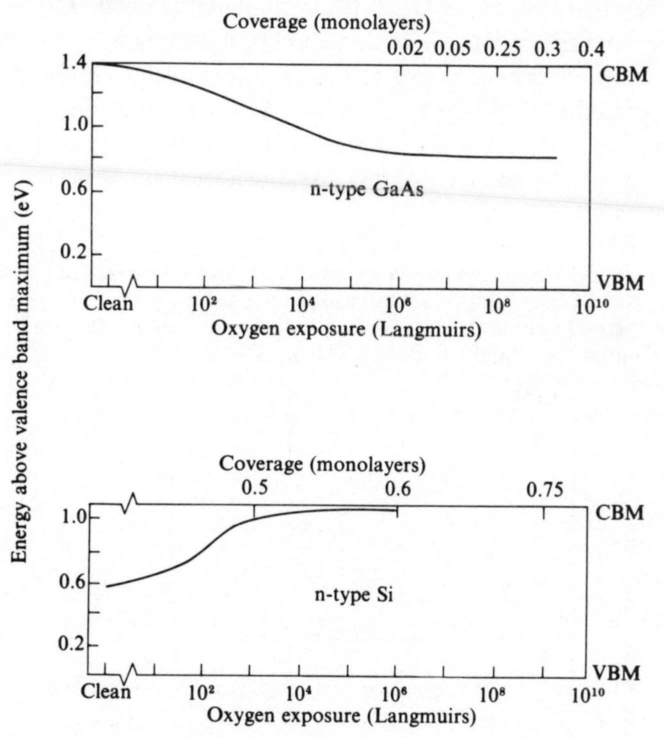

* The length of time that a surface is exposed to a foreign gas at a fixed ambient pressure often is reported in units of 10^{-6} Torr s, i.e., one Langmuir unit. From (2.1), this quantity provides an estimate of the total amount of gas permitted to strike the surface.

level lies just above the valence band maximum (VBM). Under these conditions, nearly all band gap surface states are unoccupied in the ground state. A short (80 ps) laser pulse (2.3 eV) now excites electrons from the valence band into any states that might exist in the gap. Simultaneous UV photoemission reveals the presence of these transiently occupied states. It is evident that submonolayer oxygen chemisorption induces potential pinning surface states and resonances throughout the fundamental gap. A discussion of the possible microscopic origin of these states appears in Chapter 12.

Covalent bonding considerations become increasingly less important as the ionicity of a semiconductor increases and crystal stability is dominated by electrostatic Madelung effects. Zinc oxide ($E_g \equiv 3.2$ eV) will serve as a prototype borderline case where ionic effects begin to play a significant role. For example, there is good evidence that oxygen adsorbs to this surface by a charge transfer mechanism. Electron spin resonance experiments clearly indicate that O_2^- is the dominant adsorbed species on both the polar and non-polar surfaces of ZnO (a naturally n-type material) as well as many other ionic oxides and insulators (Lunsford, 1974). A sensible model postulates that oxygen molecules flop onto their sides and attract electrons from the bulk conduction band, leaving behind a positively charged space-charge layer. In this way, the adsorbate forms a long range

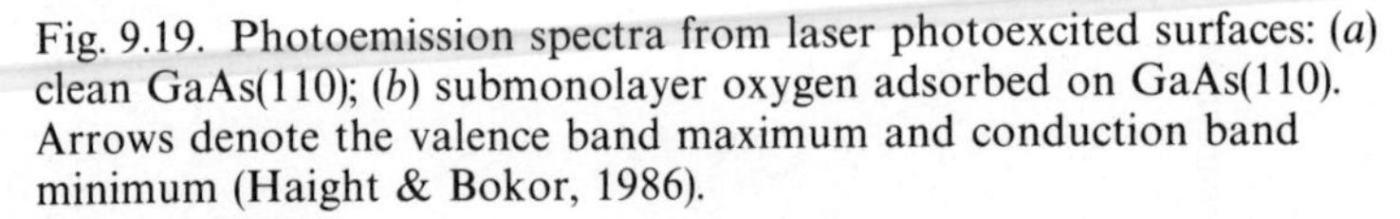

Fig. 9.19. Photoemission spectra from laser photoexcited surfaces: (*a*) clean GaAs(110); (*b*) submonolayer oxygen adsorbed on GaAs(110). Arrows denote the valence band maximum and conduction band minimum (Haight & Bokor, 1986).

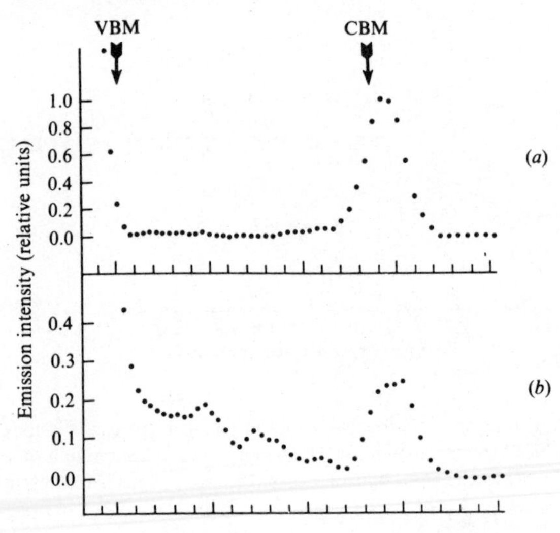

ionic bond with the substrate and still benefits from the bond energy of the stable O_2^- ion.

Observe that this model of ionoadsorption is intrinsically *self-limiting*. We can regard the oxygen 'acceptor level' as playing the role of an unoccupied surface state in Fig. 9.15(*b*). The key point is simply that one electron must be drawn from the bulk for each gas phase molecule that attempts to adsorb. This charge transfer increases the width of the depletion layer and hence increases the magnitude of band bending (9.15). The increased barrier height impedes subsequent electron transport and ultimately shuts off further adsorption. In detail, one typically observes an exponential decay in the adsorption rate – a result known as Elovich's law (Somorjai, 1972).

Spectroscopic evidence for the presence of molecular oxygen on the Zn-terminated face of ZnO is shown in Fig. 9.20. In particular, the central panel displays a *difference curve* obtained by subtracting the spectrum of

Fig. 9.20. UV photoelectron energy distribution curves for O_2/ZnO(0001): (*a*) clean (dashed curve) and oxygen-covered (solid curve); (*b*) difference between the two curves in (*a*); (*c*) gas phase O_2. The gas phase spectrum is aligned so that all spectra share a common vacuum level (Dorn, Luth & Buchel, 1977).

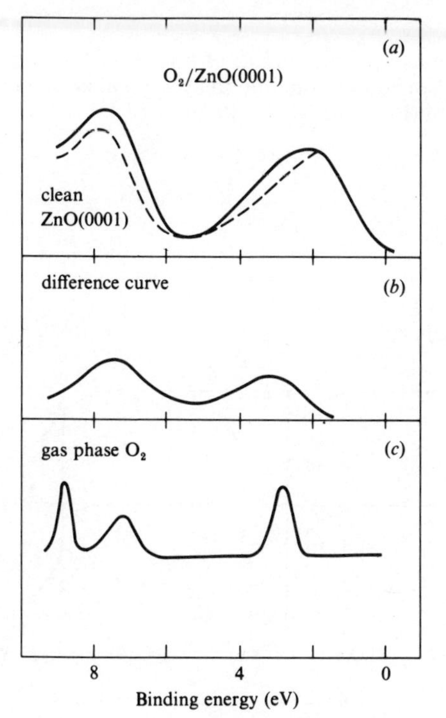

the adsorbate-covered surface from the spectrum of the clean surface. This is a very common technique used to reveal small spectral changes that accompany adsorption. Here, the difference curve corresponds reasonably well to the gas phase O_2 spectrum and there is no trace of dissociated oxygen emission near 6 eV.

We have seen that both local chemical bond formation and long range, electrostatic, adsorbate–bulk substrate interactions play a role in chemisorption on semiconductors. Unfortunately, it is not generally straightforward to cleanly separate the two in many experimental arrangements. For example, suppose interest centers on the change in electron affinity that accompanies adsorption. A measurement of the adsorbate-induced work function is insufficient since $\Delta\phi$ can reflect the effect of charge transfer to both the bulk (ΔV_s) and surface ($\Delta\chi$) of the substrate. Luckily, one can determine $\Delta\chi$ straightforwardly by subtracting the measured band gap from the threshold energy for photoelectric emission (see Fig. 9.15). In this way, it has been established that cesium adsorption lowers the electron affinity of the (110) cleavage face of p-type GaAs(110) from 4 eV to nearly zero. In fact, adsorption of Cs_2O actually lowers the vacuum level *below* the bottom of the (empty) conduction band (Fig. 9.21). The large surface dipole responsible for this effect is formed by

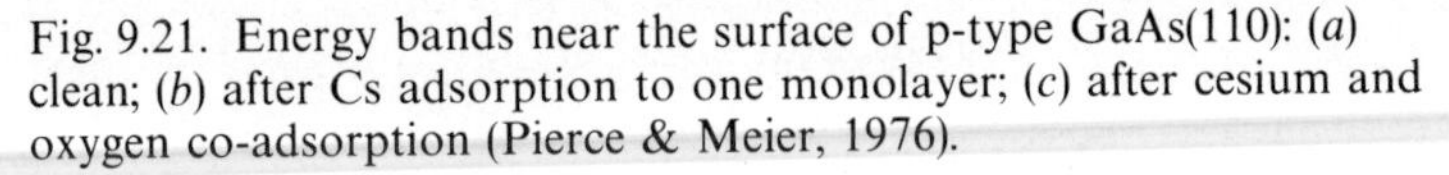

Fig. 9.21. Energy bands near the surface of p-type GaAs(110): (*a*) clean; (*b*) after Cs adsorption to one monolayer; (*c*) after cesium and oxygen co-adsorption (Pierce & Meier, 1976).

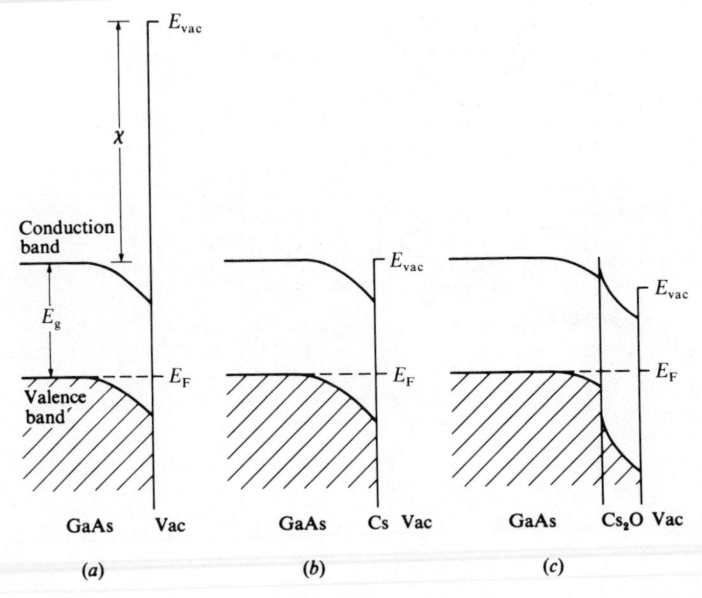

exactly the same ionic electron transfer mechanism that lowers the work function of Li/W. Accordingly, electrons excited up into the conduction band simply 'roll' out of the solid. This fact has a significant consequence because GaAs is a direct gap semiconductor for which the spin-orbit interaction splits the electronic states at the top of the valence band. Circularly polarized radiation with an energy just above E_g pumps electrons out of the crystal with a net spin polarization (Pierce & Meier, 1976). The negative electron affinity Cs_2O/GaAs(110) surface is the source of polarized electrons used in most of the spin-polarized LEED measurements discussed in Chapter 5.

In this chapter, it has been possible to overview only the most qualitative features of chemisorption on metals and semiconductors. Even at this level, we find a striking range of behavior that reflects both the diversity of the corresponding clean surfaces and the specific nature of the adsorbates. However, our general approach dictates that qualitative behavior be understood at the microscopic level. It comes as no surprise that this more detailed information only follows from an accurate knowledge of the precise locations of the adsorbed atoms and/or molecules. Unfortunately, since the adsorbate is, by our definition, a different species from the substrate, size mismatch, chemical bonding requirements, steric constraints and even adsorption-induced reconstruction conspire to considerably expand the range of possible adsorption geometries. We cannot proceed with our analysis of the surface physics of adsorption without an overview and assessment of the methods of adsorbate crystallography.

General references

Fundamentals

Rhodin, T.N. & Adams, D.L. (1976). Adsorption of Gases on Solids. In *Treatise on Solid State Chemistry* (ed. Hannay), vol. 6*A*, pp. 343–484. New York: Plenum.

Metals

Lundqvist, B.I. (1984). Chemisorption and Reactivity of Metals. In *Many-Body Phenomena at Surfaces* (eds. Langreth & Suhl), pp. 93–144. Orlando: Academic.

Semiconductors

Williams, R.H. & McGovern, I.T. (1984). Adsorption on Semiconductors. In *The Chemical Physics of Solid Surfaces and Heterogeneous Catalysis* (eds. King & Woodruff), vol. 3, pp. 267–309. Amsterdam: Elsevier.

10

CRYSTAL STRUCTURE

Introduction

The physics of chemisorption hinges on the static and dynamic properties of the surface chemical bond. Bond formation, bond stability and bond dissolution all are crucial to our subject. In this chapter we begin our investigation motivated by the well-known intimate connection between bonding and structure (see, e.g., O'Keeffe & Navrotsky (1981)). The spatial distribution, strength and reactivity of the electronic bonds within a chemisorption complex depend sensitively on the relative position of the adsorbate and substrate nuclei. In the best case, all bond distances and bond angles will be known to any desired numerical accuracy. Minimally, we should know the local symmetry of the adsorption site, the gross orientation of the adsorbate with respect to the surface and the nature of any structure within the adsorption layer itself. Unfortunately, in the world of surface science, none of these quantities reveals itself in a straightforward, mechanical fashion.

What is the geometrical arrangement of atoms in the surface region of a clean crystal after adsorption by a foreign species? Consider first the case of a single adatom. As discussed in Chapter 8 for the case of Au/NaCl(100), the most probable binding sites occur at substrate positions where the total adatom/substrate potential energy of interaction has minima. For an ordered substrate this defines Langmuir's checkerboard. Substrates of different symmetry exhibit different 'checkerboards' and, as more atoms adsorb, the spatial distribution of occupied sites becomes an issue. One might suppose that particles occupy the available sites randomly. For example, the statistical adsorption model developed in the previous chapter implicitly assumes this scenario. When it occurs, one says that the overlayer is in its structurally *disordered* phase. More typically, and certainly at low temperature, interactions among the adparticles lead to *ordered* arrangements of the adsorbing species. The ordered regions may be restricted to finite-sized *islands*, or, under other conditions, occupy the entire available surface area.

The crystallographic description of ordered overlayer structures follows the notation introduced in Chapter 3. One specifies the ratios of the lengths of the adsorbate (A) and substrate (S) unit mesh translation vectors and (if needed) their relative rotational orientation. Hence, the overlayer depicted in Fig. 10.1(a) is designated $S(110)\ (6 \times 2)$-A relative to the underlying BCC lattice. Fig. 10.1(b) shows ordered adsorption on a

Fig. 10.1. Examples of commensurate ordered overlayers. Open and closed circles denote substrate and adsorbate atoms, respectively. Both the substrate (dashed lines) and overlayer (solid lines) primitive translation vectors are indicated: (a) BCC(110) (6 × 2); (b) FCC(100) ($\sqrt{2} \times \sqrt{2}$–45°); (c) HCP(0001) ($\sqrt{3} \times \sqrt{3}$–30°).

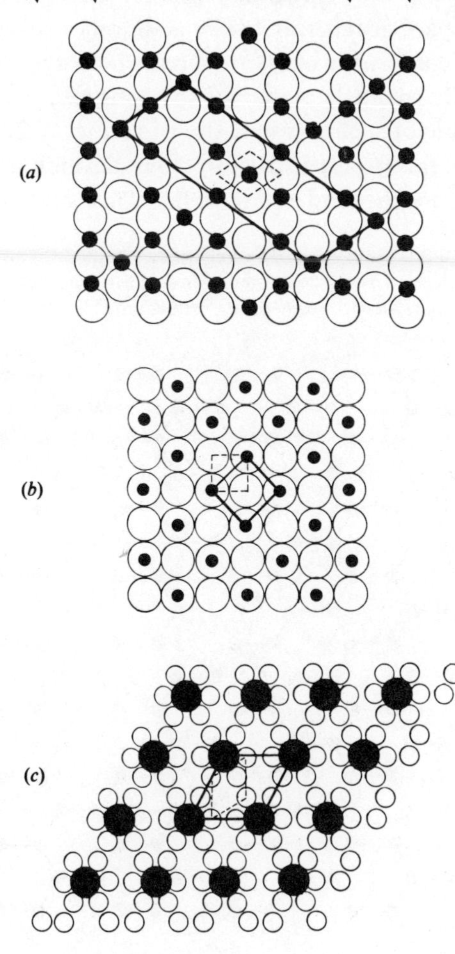

FCC(100) crystal face. The adatom arrangement displayed there commonly is denoted $S(100)$ c(2×2)-A although, strictly speaking, there is no centered square primitive lattice (see Fig. 3.3). The correct designation for this structure is $S(100)$ $(\sqrt{2} \times \sqrt{2}$–$45°)$-$A$. Adsorption onto the basal plane of graphite requires yet a different symmetry description – that of a hexagonal crystal. In our example (Fig. 10.1(*c*)), the appropriate overlayer designation is $S(0001)$ $(\sqrt{3} \times \sqrt{3}$–$30°)$-$A$.

It often is convenient to define the adsorbate coverage (θ) so that $\theta = 1$ occurs when the adsorbate occupies all equivalent adsorption sites (cf. footnote on p. 194). In this case, the three structures illustrated in Fig. 10.1 correspond to coverages of 2/3, 1/2 and 1/3, respectively. We shall say that an overlayer structure is *commensurate* with the substrate whenever θ is a rational number. More precisely, a commensurate adsorbate lattice is characterized by a two-dimensional space group that differs from the substrate space group only by the addition or subtraction of specific symmetry elements. Adsorbed structures whose symmetry is not, or is only accidentally, related to that of the substrate are called *incommensurate*. An example of such a case is shown as Fig. 10.2. Notice that the overlayer atoms in this ordered structure do not exhibit a unique binding site with respect to the underlying solid. This situation can

Fig. 10.2. An incommensurate ordered overlayer. The (small) substrate atoms sit at the vertices of the hexagonal network and the heavy circles denote the (large) adsorbate atoms (Brinkman, Fisher & Moncton, 1982).

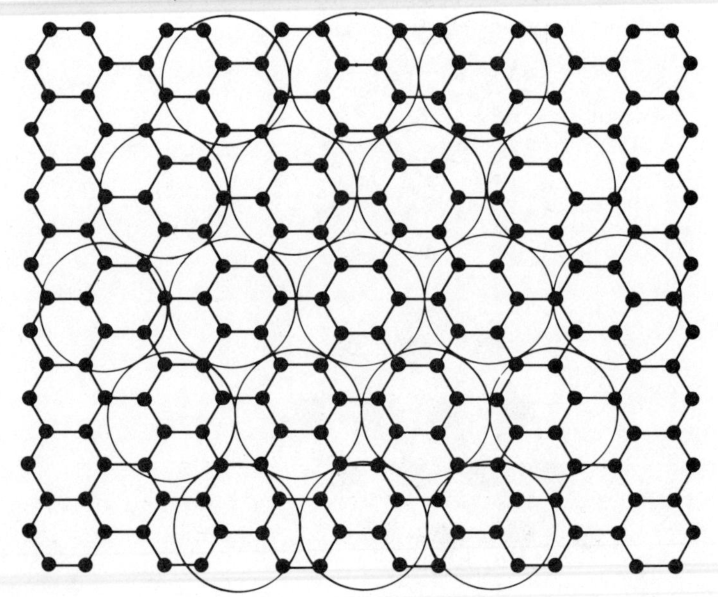

occur for gas coverages near unity if inter-adsorbate interactions that set the overlayer's natural lattice constant successfully compete with the adsorbate–substrate potential energy. A fuller discussion of this and other types of incommensurate overlayer structures appears in Chapter 11.

The crystallographic description of commensurate overlayers outlined above can be obtained from a LEED pattern with little effort (Clarke, 1985). Unfortunately, this analysis tells us nothing about the specific adsorption sites. For example, Fig. 10.1 illustrates three common adsorption geometries: 'bridge' sites (which suggest simultaneous bonding to two substrate atoms), 'on-top' sites (suggestive of head-on bonding to a single surface atom) and 'hollow' sites. The last may reflect either interstitial minima in the substrate corrugation potential (for physisorption) or bonding to multiple surface atoms (for chemisorption). An additional complication arises for adsorbed molecules. Depending on the molecular geometry, these species can stand up, lie down, or bond tilted at some angle with respect to the surface. In principle, dynamical LEED analysis and surface x-ray scattering (see Chapter 3) provide 'complete' solutions that address both long range order and short range bond conformation issues. However, at their present state of development, technical problems (multiple scattering (for LEED) and low signal-to-noise (for x-ray scattering)) effectively preclude either from unsupported *ab initio* surface crystallography on a routine basis. Therefore, as in the clean surface case, no adsorbate structure determination should be trusted until it is shown to be consistent with a battery of surface structural probes. These probes include all those outlined in Chapter 3 as well as others that are either particularly well suited to, or specifically designed for, adsorption studies.

Topography

Scanning tunnelling microscopy provides a detailed topographic view of the geometry of chemisorption. This is both a virtue and a vice. The tunnelling microscope is an extremely local probe and, as such, reveals all the intimate details of the scanned region. A typical image includes steps, defects, pits and other structure in addition to the adsorbed species. This is illustrated graphically by Fig. 10.3, which shows a sequence of STM scans (similar to Fig. 3.22) of an annealed Pt(100) surface after adsorption of a submonolayer amount of carbon. The front central region exhibits a uniform corrugation pattern characteristic of the clean surface. However, sharp irregular spikes and a defected monoatomic step bound this clean area. Needless to say, this is not quite what we want. It is difficult to identify any particular feature of this image with an adsorbed carbon atom.

One obtains a rather more 'defocused' view of surface topography from elastic helium atom scattering measurements. Here, one performs a diffraction experiment so that irregular surface features mostly contribute to a diffuse background. The surface sensitivity of this probe has been amply demonstrated earlier in connection with both surface phonon measurements (Chapter 6) and observations of 'selective adsorption' (Chapter 8). In common with all diffraction techniques, we extract structural information by comparison of experimental diffracted beam intensities with those calculated for a set of model structures. This procedure is straightforward and unambiguous if the scattering process between the sample and the probe is simple and well known (e.g., x-ray scattering) and is difficult and tedious if the scattering process is complicated (e.g., LEED). In the present case, we take advantage of the fact that 60 meV helium atoms barely feel the dispersion force and interact principally with the steeply rising repulsive part of the helium–substrate interaction potential (Fig. 8.3).

Diffraction occurs because this potential is modulated in the surface plane in accordance with the corrugation of the surface charge density (cf. (8.7)). Of course, the corrugation reflects the periodicity of the surface,

$$\xi(\mathbf{R}) = \sum_{\mathbf{G}_s} \xi_{\mathbf{G}_s} e^{i\mathbf{G}_s \cdot \mathbf{R}}, \tag{10.1}$$

where the $\mathbf{G}_s$ are reciprocal lattice vectors of the surface net and $\mathbf{R}$ is a

Fig. 10.3. Scanning tunnelling microscope scans of disordered C/Pt(100) annealed to 1100 K (Hosler, Behm & Ritter, 1986).

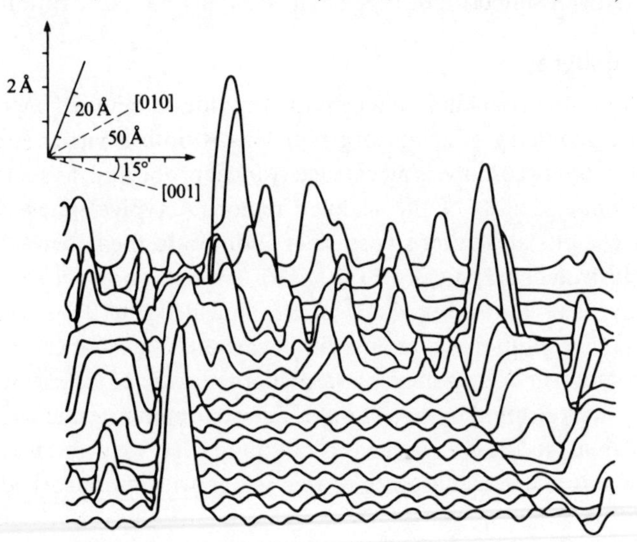

two-dimensional vector in real space. To analyze scattering data, it is simplest to regard the slope of the repulsive wall as infinite, i.e.,

$$\begin{aligned} V(\mathbf{R}, z) &= 0 \quad \text{for } z > \xi(\mathbf{R}), \\ V(\mathbf{R}, z) &= \infty \quad \text{for } z \leqslant \xi(\mathbf{R}). \end{aligned} \tag{10.2}$$

This is the so-called *corrugated hard-wall model.* It is possible to solve the associated scattering problem exactly for the diffracted beam intensities both classically and quantum mechanically. In practice, one normally makes some simplifying assumptions (appropriate to weak corrugation) to reduce the computational effort (Engel & Rieder, 1982). The end result is a set of amplitudes and Fourier components from which the corrugation function (10.1) is constructed.

The principal advantage of this technique is that it is equally sensitive to all elements of the periodic table and can be applied equally well to metals, semiconductors and insulators. The principal disadvantage is that the corrugation topograph does not quantitatively determine the absolute position of the adsorbate with respect to the substrate. As an example, we consider a series of five ordered structures that appear at different coverages following dissociative chemisorption of H_2/Ni(110) at 100 K (Fig. 10.4). The three-dimensional images clearly indicate the overlayer symmetry and corrugation – particularly when compared to the clean Ni(110) result. The right hand side of each panel depicts a *proposed* surface structure that is consistent with the topograph and independent theoretical evidence. Even if the specific zig-zag bridging adsorption site is not precisely correct, the corrugation function itself points directly to highly anisotropic interactions among the adsorbed hydrogen atoms.

Site symmetry

Detailed adsorbate site symmetry analysis requires a local probe that is capable of chemical specificity. Vibrational spectroscopy is almost uniquely qualified in this respect. For chemical identification, the characteristic internal vibrational mode frequencies of gas phase molecules serve as excellent 'fingerprints' that can be sought in the signal derived from a solid surface covered with unknown chemisorbed species. The sensitivity of this type of measurement is about ~ 0.005 monolayer when the vibrations are probed by infrared absorption spectroscopy (IRAS) or electron energy loss spectroscopy (EELS). In the present context, useful structural information can be obtained due to the specific symmetry properties of vibrational excitations.

An N-atomic molecule has $3N$ degrees of freedom, of which 3 are translational, 3 are rotational (2 for a diatomic molecule) and the remainder are vibrational. However, free translation and rotation cannot occur if a chemical bond forms between this molecule and a solid surface. This means that a total of $3N$ local vibrational modes are associated with an isolated adsorption complex. Note carefully that this does *not* imply that $3N$ *distinct* vibrational frequencies are present. The symmetry of the adsorption site – the surface molecule point group – determines the number of non-degenerate vibrational frequencies. In simple cases, the degeneracy can be

Fig. 10.4. Best-fit corrugation surfaces and hard-sphere models for atomic hydrogen adsorption on Ni(110) at different coverages as determined by He atom scattering (Rieder & Engel, 1980; Rieder, 1983).

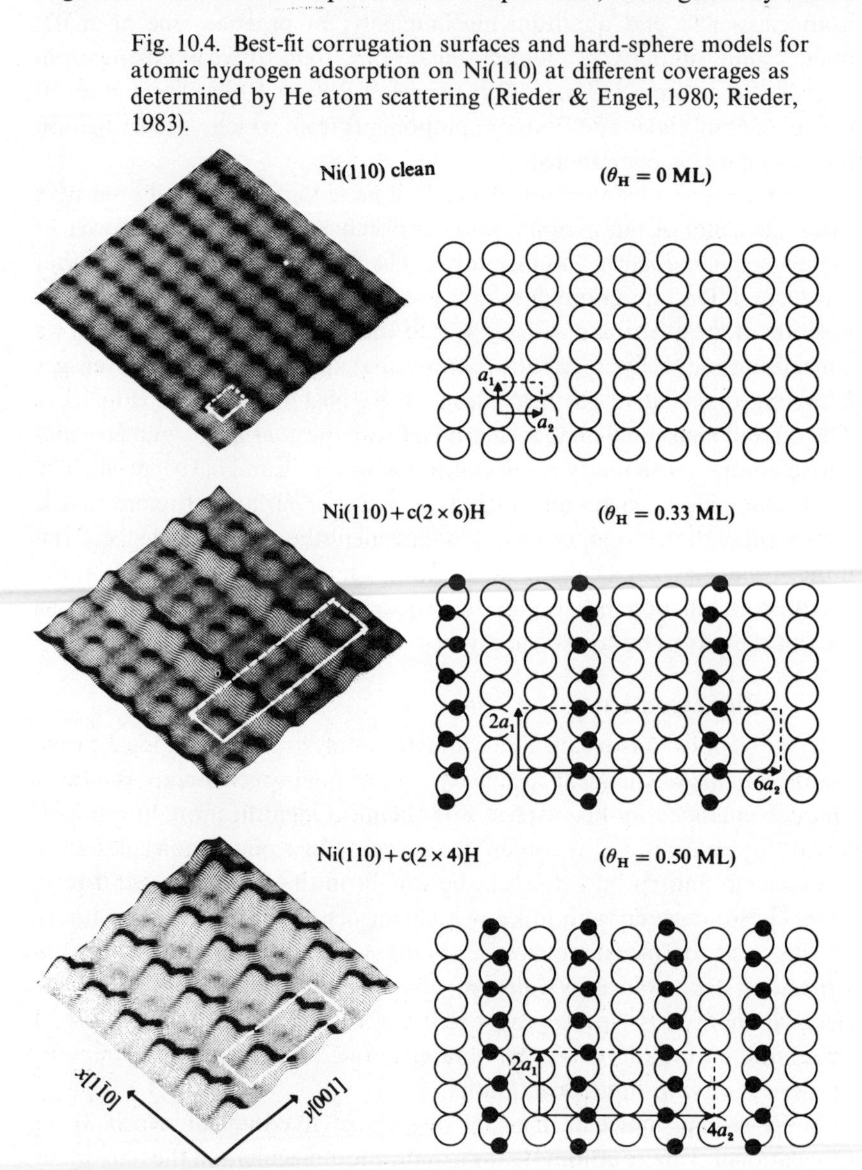

determined by inspection and intuition. More generally, one harnesses the full power of group theory to determine the number of distinct modes to be expected for particular adsorption site symmetries (Ibach & Mills, 1982). In the following we limit ourselves to a few examples of the simple variety.

Fig. 10.5 illustrates vibrational mode degeneracy for adsorption into three common bonding configurations. In the first case, an isolated adatom (H) occupies a three-fold hollow site (C_{3v} symmetry) (Fig. 10.5(*a*)). One mode is associated with frustrated translational motion normal to the surface. Two modes remain but the symmetry of a triangle demands that these bending modes (frustrated translation in the plane) share a common frequency. The latter degeneracy is broken if the adatom occupies a bridge site (C_{2v} symmetry) since the bridging atoms define a preferred

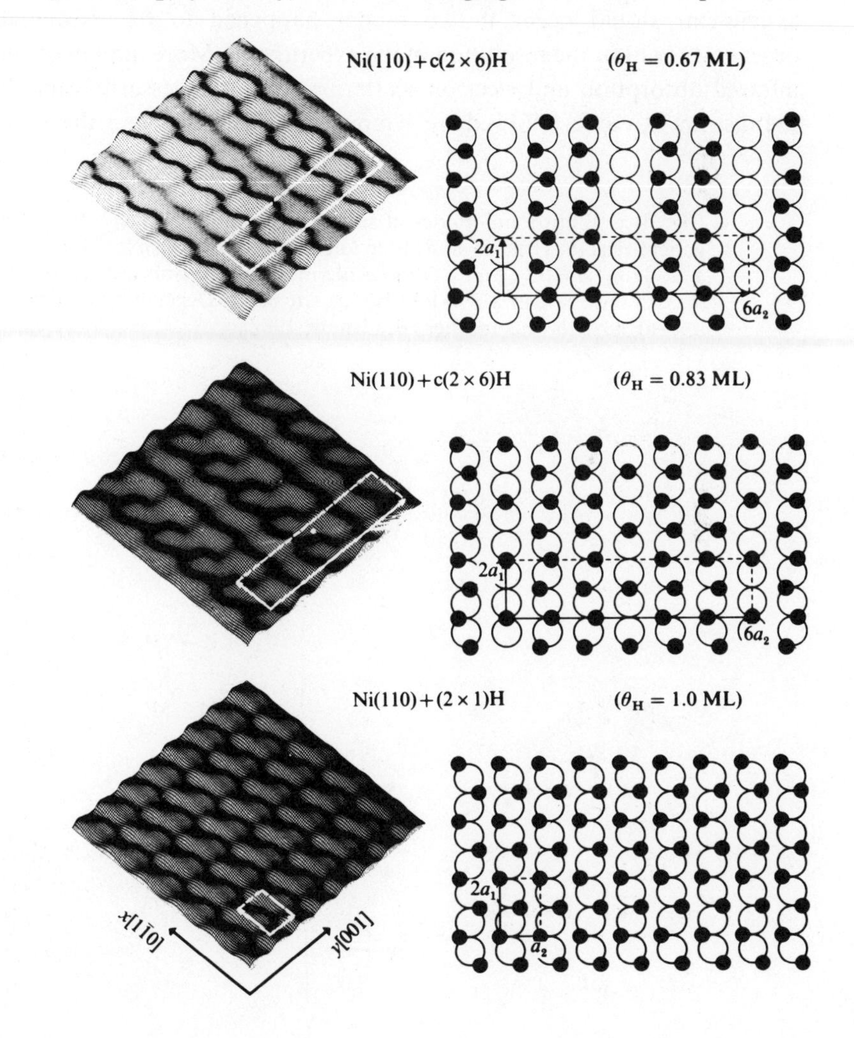

direction in the plane (Fig. 10.5(*b*)). In principle, three distinct stretch frequencies can appear in the experimental spectrum. Figs. 10.5(*c*) and (*d*) pertain to the case of a diatomic molecule (CO) which possesses six normal modes. Adsorption onto a high (C_{4v}) symmetry on-top site results in degeneracies among the possible bending modes so that only two distinct frequencies occur. Two non-equivalent vertical 'beating' modes bring the total count to four. As in the previous example, the full mode structure of the adsorbate can only appear if adsorption occurs onto a sufficiently low symmetry site, e.g., a C_{2v} bridge position.

The foregoing suggests that the simplest 'mode-counting' argument suffices to identify adsorbate site symmetry. However, this is true only if one actually *observes* all the distinct vibrations. For example, an incorrect assignment would occur if two modes happened to be accidentally degenerate (within the resolution of the experiment). More fundamentally, infrared absorption and electron scattering do not necessarily excite all the distinct stretches allowed by symmetry. Both IRAS and the strong

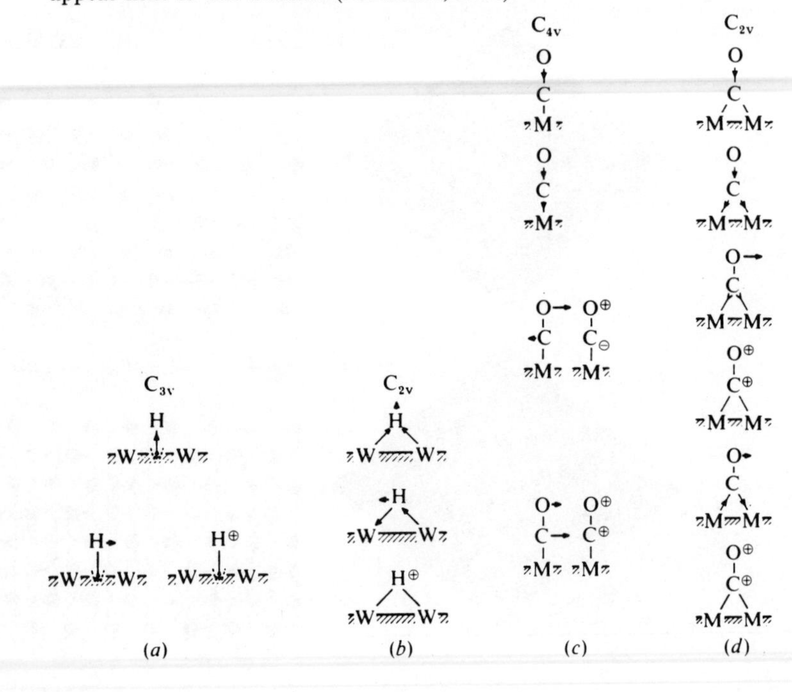

Fig. 10.5. Vibrational modes of simple adsorbates in different bonding geometries: (*a*) H atom in a three-fold hollow site (C_{3v}); (*b*) H atom in a bridging site (C_{2v}); (*c*) CO molecule in a four-fold on-top site (C_{4v}); (*d*) CO molecule in a two-fold bridge site (C_{2v}). Degenerate modes appear next to one another (Bradshaw, 1982).

dipole scattering of EELS excite such vibrations through a Golden Rule matrix element of the form

$$|\langle F|\boldsymbol{\mu}\cdot\mathbf{E}|I\rangle|^2, \tag{10.3}$$

where $|I\rangle$ is (typically) the ground vibrational state of a particular mode and $\langle F|$ is the first excited state. The coupling operator, $\boldsymbol{\mu}\cdot\mathbf{E}$, involves the local electric field at the adsorbate site and the dipole moment established by the intra-adsorbate nuclear motion associated with the vibrational mode.

The principles of group theory establish a strict selection rule: the matrix element in (10.3) can be non-zero only when the mode in question belongs to the same irreducible representation of the adsorbate point group as at least one Cartesian component of $\boldsymbol{\mu}$ (Heine, 1960). Such modes are termed 'dipole active'. Additional 'pseudo-selection rules' come into play if specific adsorption conditions fix the direction of **E**. For example, metals reflect below the plasma edge due to their good screening properties. The same property produces fields outside the solid that can be described in the language of electrical images. Therefore, if some stretching mode produces an adsorbate dipole moment aligned normal to the metal surface, the induced image dipole reinforces the local electric field and strong vibrational excitation may be expected (Fig. 10.6). By contrast, a mode that creates a dipole parallel to the surface induces an image dipole that largely cancels the local dipolar field leaving only a weak quadrupole field. The resulting excitation may be undetectable experimentally. Although no such 'selection rule' holds for materials that are transparent in the appropriate frequency regime, the excitation matrix can always be small 'accidentally', even for nominally dipole-active modes.

Infrared absorption and electron energy loss have contrasting strengths and weaknesses. As in any optical technique, the energy resolution of IRAS is excellent – typically 0.05 meV. Unfortunately, sources and detectors exist only over a limited spectral range. By contrast, an EELS experiment is sensitive to inelastic losses over a spectral range of several electron volts. The price one pays for this versatility is a much degraded resolution (~ 5 meV). However, if high resolution is not required, EELS

Fig. 10.6. Image dipoles near a metal surface.

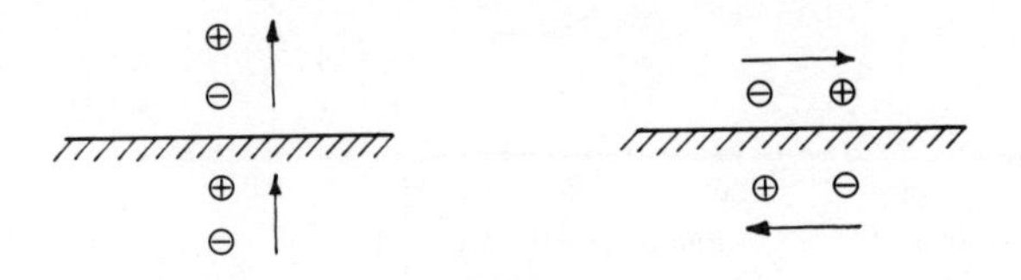

Fig. 10.7. EELS spectra of W(100) c(2 × 2)-H for an electron beam of 5.5 eV incident at a polar angle of 60° at various angles of detection relative to the specular direction. The elastic peak and mode assignments are indicated at left and right, respectively (Barnes & Willis, 1978).

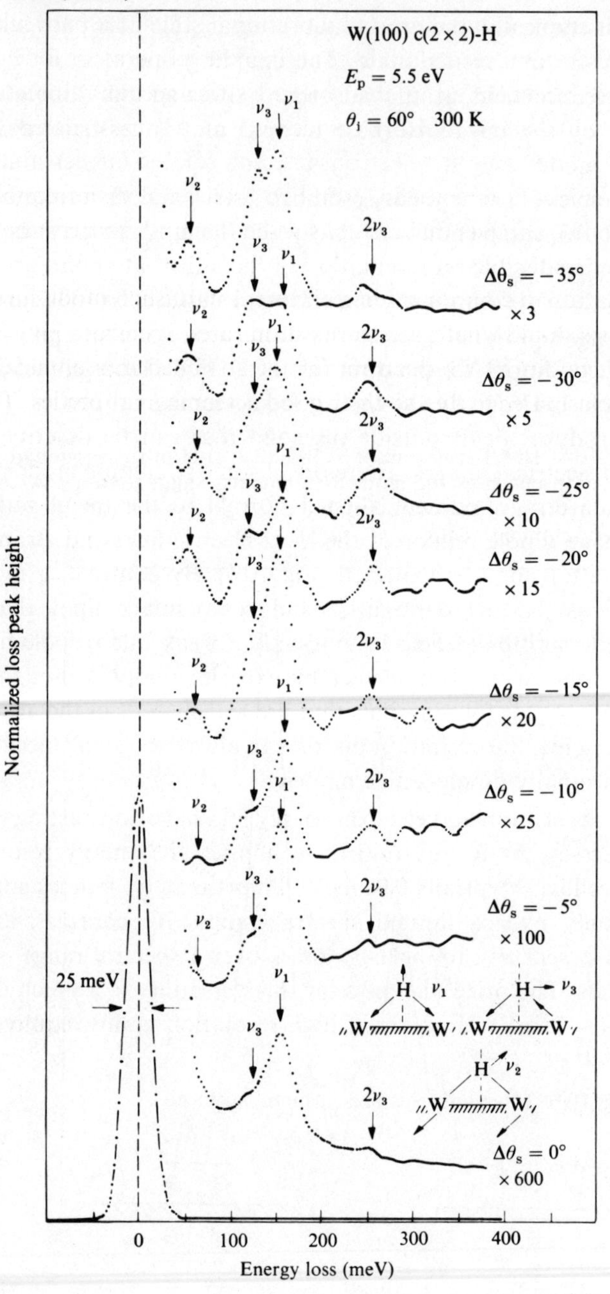

possesses one significant advantage: short range 'impact' scattering of the incoming electron from the local adsorbate potential can excite non-dipole-active modes. The cross section for this excitation mechanism is rather weak so one must look to electrons that scatter away from the forward (specular) direction to avoid the signal-swamping effect of dipole scattering. Fig. 10.7 illustrates a direct mode-counting site assignment for hydrogen adsorbed on tungsten using this trick. In the specular direction, the dipole-active hydrogen 'beating' mode (ν_1) at 130 meV dominates the spectrum. However, as the electron detector rotates farther and farther from the specular direction ($\Delta\theta_s \neq 0$), two additional distinct modes and their overtones appear that establish the local symmetry as C_{2v} – a bridge-bonded site.

Hydrogen adsorption on a semiconductor surface, Si(100), furnishes a contrasting example where the merits of infrared spectroscopy are particularly evident. An IRAS spectrum for the so-called monohydride phase of Si(100) (2 × 1)-H clearly shows two local vibrational modes (Fig. 10.8)

Fig. 10.8. IRAS spectrum of Si(100) (2 × 1)-H in the monohydride phase. Insets show the mode assignments. Small solid arrows indicate the direction of atom motion and open arrows indicate the induced dipole moment (Tully *et al.*, 1985).

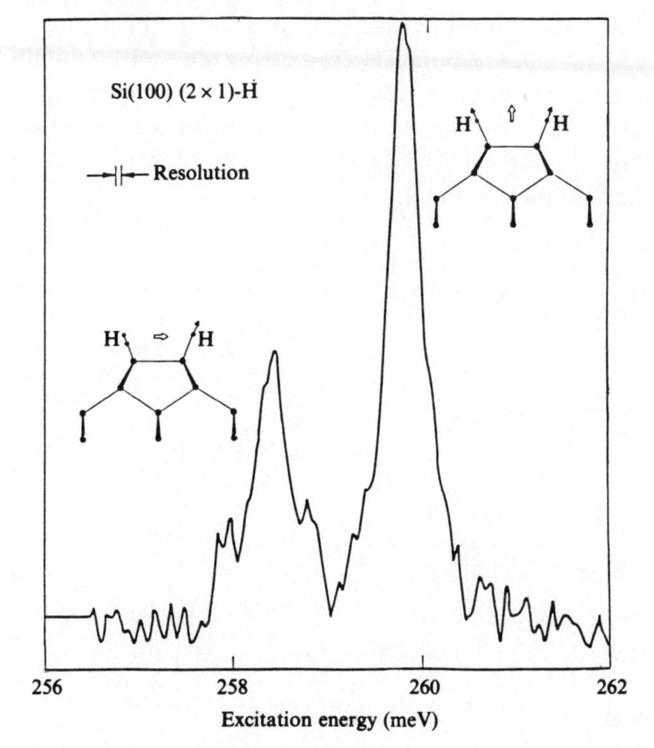

that would not be resolved in an EELS experiment. We observe a high frequency symmetric mode that produces a dipole moment normal to the surface and an antisymmetric mode (adjacent hydrogen atoms move out of phase) that produces a dipole moment parallel to the surface. Although an EELS experiment would be sufficient to conclude that the H–Si bond is stronger than the W–H–W bond (based on the relative adsorbate– substrate stretching frequencies), quantitative results for the vibrational linewidth and mode splitting require the resolution of IRAS. In conjunction with theoretical calculations, the latter quantities provide both structural discrimination and dynamical information (see Chapter 13).

Bond lengths

We now turn to a number of experimental techniques designed to extract quantitative information about the surface chemical bond. Two electron spectroscopic methods have been developed that provide adsorbate-specific geometrical parameters (bond lengths, bond angles, coordination, etc.) which do not require the complicated calculations that must accompany a complete LEED analysis. Both techniques employ x-ray photons to eject electrons from the deep core levels of adsorbate atoms. The elemental specificity of core level binding energies guarantees that the adsorbate signal always can be distinguished from substrate emission. X-ray photoelectron diffraction (XPD) monitors the direct photoelectron current while surface-extended x-ray absorption fine structure (SEXAFS) measures the Auger or fluorescence yield that follows deep core-hole production. These quantities reflect local surface structure in a remarkably simple manner.

Fig. 10.9. Decomposition of the outgoing photoelectron wave field from a surface atom in terms of elementary scattering amplitudes.

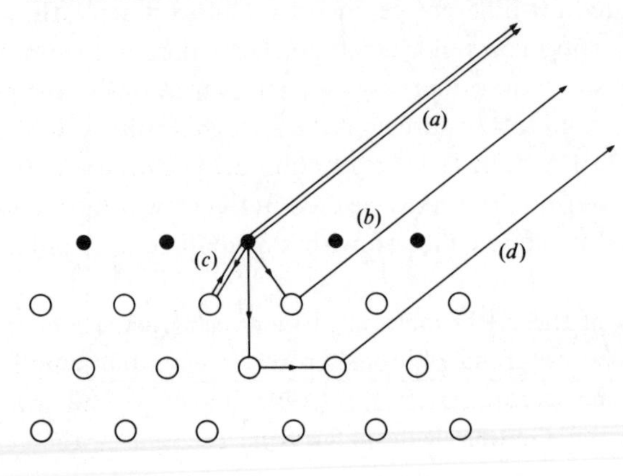

Consider the matrix element for photoemission from an initial core state to a final continuum state that propagates in the direction **k**: $|\langle \psi_f(\mathbf{k})|\boldsymbol{\varepsilon}\cdot\mathbf{r}|\psi_{core}\rangle|^2$. Here, $\boldsymbol{\varepsilon}$ is the polarization vector of the incident radiation. Now decompose the outgoing photoelectron wave field into a sum of amplitudes, each of which leads to the same final state (Fig. 10.9). The direct beam propagates freely away from the central absorption site (*a*). However, this beam interferes with those beams that propagate first to a near neighbor atom and then elastically scatter into the **k** direction (*b*). Second-order contributions to $\psi_f(\mathbf{k})$ include beams that *backscatter* from a nearby atom and enter the final state after scattering from the central atom (*c*) and beams that scatter from two different neighboring atoms before leaving the crystal (*d*). Clearly, bond length information resides in the phase shift introduced by non-direct beams that traverse a greater path length than the direct beam. XPD and SEXAFS provide access to this information in different ways.

We restrict our attention to only those photoelectric events for which the final state electron has kinetic energy in excess of 200 eV. At these energies the electron–atom elastic scattering cross section is small. Multiple scattering events are improbable so that only processes (*a*) and (*b*) of Fig. 10.9 contribute appreciably to the final state wave function of the XPD photoelectron. A key point is that inelastic effects diminish the influence of single scattering from atoms outside the immediate vicinity of the central atom. Therefore, the outgoing wave field images a 'diffraction grating' formed by the absorbing atom's coordination shell of nearest neighbors. Plots of the azimuthal (ϕ) dependence of the photoemission intensity clearly display characteristic interference maxima and minima consistent with the symmetry of the substrate (Fig. 10.10).

In practice, the location of an adsorbate relative to a substrate is determined by comparing measured XPD polar and azimuthal intensities with those calculated for different geometries. Unfortunately, the method is restricted to adsorption on single crystal surfaces and, more importantly, the diffracted intensities are rather insensitive to variations in the position of an adsorbate normal to the surface if the adsorbate sits well above the substrate surface (> 1 Å). The latter restriction arises because high energy elastic electron scattering is strongly peaked in the forward direction. The experimental signal must be collected within a small cone of polar angles near grazing exit.

The limitations of the XPD method all derive from the fact that one explicitly detects the outgoing photoelectron. An essential simplification results if, instead, one *averages* over all possible final state emission angles. Keeping in mind the Golden Rule (4.34), one recognizes the resulting expression as proportional to the total absorption coefficient, $\mu(\omega)$. Moreover, it is easy to show that the contributions from all scattering paths that do not return to the central site before exiting the crystal *cancel* from the sum (Lee, Citrin, Eisenberger & Kincaid, 1981). Therefore, the absorption coefficient several hundred eV above a core level edge is dominated by the interference between processes (*a*) and (*c*) in Fig. 10.9. This effect manifests itself as small amplitude oscillations in $\mu(\omega)$ above the absorption edge which directly reflect a phase shift of $2kR$ – the product of the electron wave vector and the extra length of path (*c*) relative to path (*a*). This is the basis of all SEXAFS experiments.

The qualitative power of this technique becomes evident by comparison of raw SEXAFS absorption data for sulfur adsorbed onto a Ni(100) substrate with bulk adsorption data for nickel silicide (Fig. 10.11). We can conclude that the nickel–sulfur bond length for the chemisorbed species is smaller than the corresponding distance in the bulk compound merely from the observation that the oscillation wavelength derived from the bulk sample is shorter than that observed in the Auger signal.

Quantitative analysis of SEXAFS data to determine adsorbate geome-

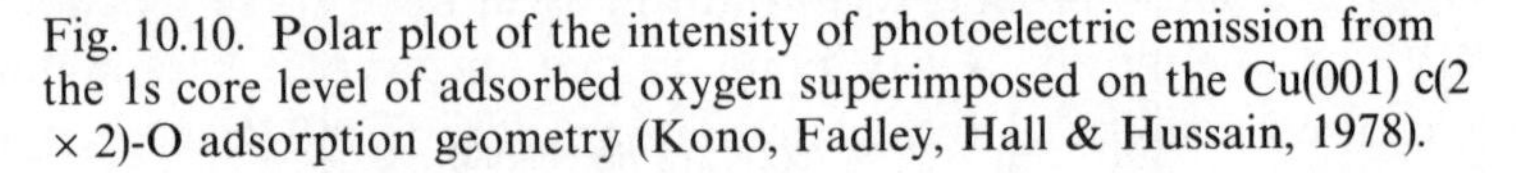

Fig. 10.10. Polar plot of the intensity of photoelectric emission from the 1s core level of adsorbed oxygen superimposed on the Cu(001) c(2 × 2)-O adsorption geometry (Kono, Fadley, Hall & Hussain, 1978).

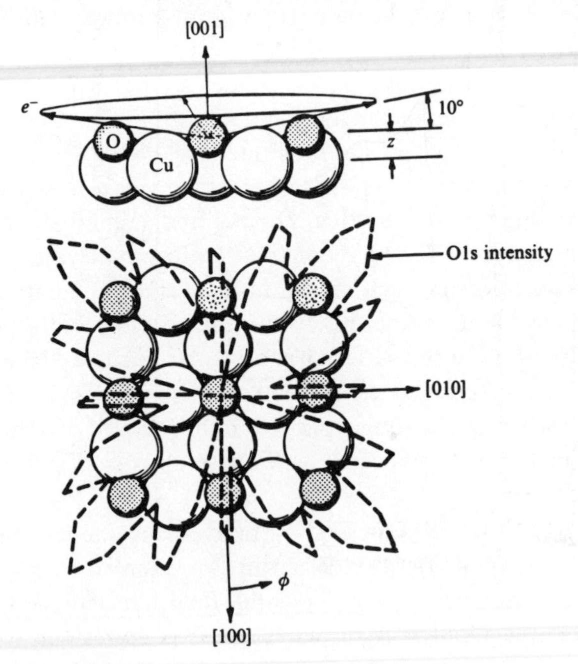

trical parameters is relatively straightforward. First note that the oscillation fine structure apparent in Fig. 10.11 occurs as a function of incident photon energy and hence photoelectron wave vector. This suggests that a simple Fourier inversion of the raw data would directly yield the nearest neighbor bond distance, R. This is not quite the case. Although one does employ a Fourier inversion scheme, it is necessary to be cognizant of the fact that the oscillations generally contain a contribution from the second coordination shell and, more importantly, the backscattering process itself introduces an additional energy dependent phase shift that must be removed from the data. Nevertheless, adsorption bond lengths can be routinely extracted with a precision of 0.01 Å. The *accuracy* of these determinations (when independent results are available) appear to be good to about 0.01 Å as well.

The principal disadvantage of SEXAFS experiments is their expense – data must be collected at synchrotron radiation sources.* On the other hand, the fact that synchrotron radiation is highly polarized permits one

Fig. 10.11. SEXAFS Auger yield from Ni(100) c(2 × 2)-S and total yield (absorption) from bulk NiS. Both spectra obtained above the sulfur K-edge (Stohr, Jaeger & Brennan, 1982).

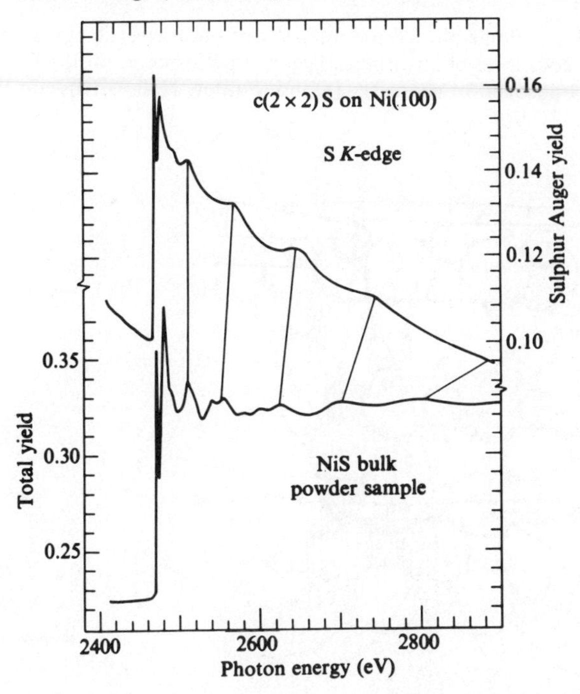

* XPD experiments require only an x-ray tube and an electron spectrometer.

to extract bond site as well as bond length information. To see this, observe that scattering process (*c*) is the most significant structure sensitive contribution to the experimental signal. This means that the absorption matrix element operator $\boldsymbol{\varepsilon}\cdot\mathbf{r} \cong \boldsymbol{\varepsilon}\cdot\mathbf{R}$. Hence, SEXAFS oscillations

Fig. 10.12. SEXAFS data for Ge(111) (2 × 8)-Cl for two extreme orientations of the incident radiation polarization vector. Both the raw data (top) and background-subtracted data (bottom) are shown. The smooth solid curves result from a Fourier filtering analysis (Citrin, Rowe & Eisenberger, 1983).

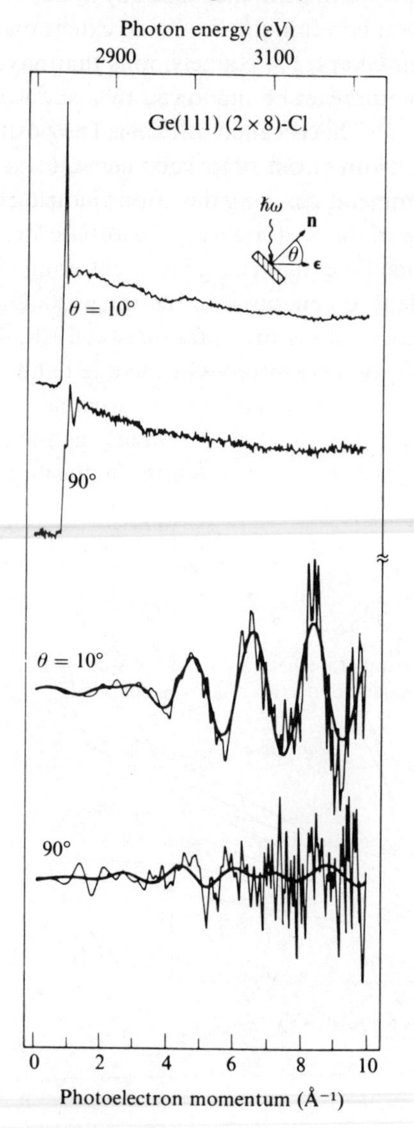

vanish if the polarization vector is perpendicular to the chemisorption bond axis. We illustrate this application of polarization-dependent SEXAFS for the case of Ge(111) (2 × 8)-Cl. Auger yield data collected above the chlorine *K*-edge show intense oscillations only when the light polarization vector points normal to the surface (Fig. 10.12). This strongly suggests that chlorine atoms adsorb directly atop germanium atoms of the substrate.

Ion scattering spectroscopy is the final tool we will discuss that is capable of measurements of bond lengths and bond angles. The key to quantitative adsorbate crystallography with this technique lies in an extension of the shadow cone idea introduced in Chapter 3. Namely, ions that have been backscattered from a substrate atom can be shadowed by a surface atom as they exit the crystal (Fig. 10.13). This is called *blocking*. The position of adatoms relative to a substrate follows from simple geometrical triangulation if one chooses the shadowing and blocking directions judiciously. To this point, the most effective use of this method as an adsorbate structural tool has been for what was termed medium energy ion scattering (MEIS) in Chapter 3. That is, the incident ion energy is in the range 20–200 keV. One reason for this is that the necessary instrumentation is well established. More importantly a subtle trade-off is involved. On the one hand, low-*Z* elements often are the most interesting adsorbates whereas the ion scattering cross section in the HEIS range falls off rapidly as the atomic number of the target decreases. On the other hand, the requisite quantitative Monte Carlo trajectory analysis (see footnote on p. 45) requires a very well-known ion–atom scattering potential and this quantity is somewhat uncertain in the LEIS range.

Fig. 10.13. Typical scattering trajectories that show shadowing and blocking of ions at normal incidence (Williams & Yarmoff, 1983).

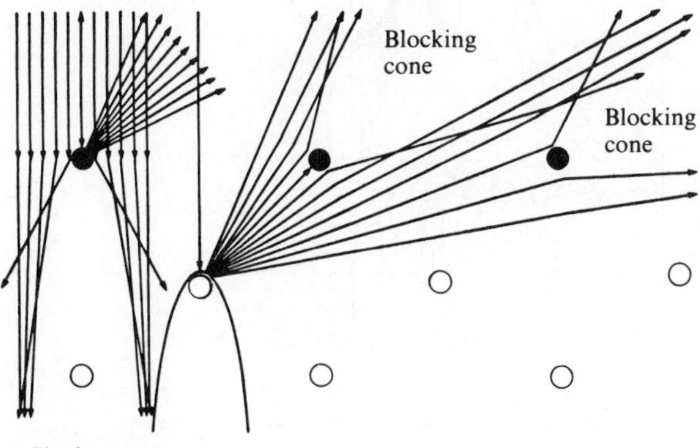

We illustrate the MEIS shadowing and blocking approach to adsorbate structure determination with the example of 0.5 monolayer adsorption of sulfur on Ni(110). Fig. 10.14 shows the intensity of the surface peak as a function of scattering angle in the vicinity of the $[01\bar{1}]$ exit blocking direction. Note first that the minimum in the clean surface blocking dip (dashed curve) is shifted slightly to smaller angles than the bulk $[01\bar{1}]$ direction. This indicates the usual first layer contraction at the surface of a clean metal.* Significant changes occur upon sulfur adsorption. First, the principal dip shifts to the opposite side of the blocking direction by an amount $\Delta\alpha$. Second, an additional blocking dip appears at smaller scattering angles. The first observation implies that the adsorbate induces an *expansion* of the nickel surface layer relative to its bulk value. The second observation determines the z-component of the sulfur–nickel bond distance (about 0.9 Å in this case). Additional analysis of the scattering data strongly suggests that the adsorbates occupy four-fold hollow sites.

Orientation

A new structural variable enters the crystallography when molecular (as opposed to atomic) adsorption occurs. Now one must know

Fig. 10.14. Angular dependence of MEIS scattering intensity using shadowing along [101] and blocking along $[01\bar{1}]$ of a nickel (110) surface. 100 keV protons are incident upon both a clean surface and a sulfur-covered surface. See text for discussion (Van der Veen, Tromp, Smeenck & Saris, 1979).

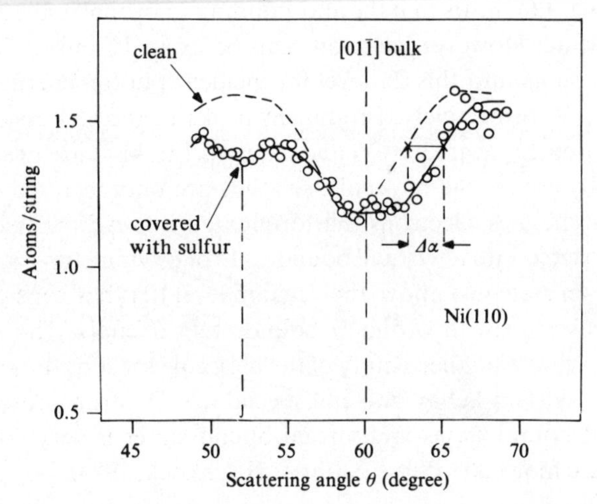

* The angular scan in Fig. 3.17 uses a horizontal axis labelling opposite to that of Fig. 10.14. Both show clean surface relaxation.

the *orientation* of the adsorbate with respect to the substrate. This information is crucial to a proper description of the chemisorption bond. More importantly, it determines the nature of inter-adsorbate interactions and can control possible energy flow and reaction pathways. It is easy to see that the most useful experiments are directly sensitive to the angle between a molecular symmetry axis and some preferred, yet controllable, direction in space. In fact, expressions that exhibit the requisite control parameter already have appeared twice in this chapter. An angle-dependent coupling between the external probe and the adsorption complex exists in both the matrix element for dipolar vibrational excitation (10.3) and the matrix element for photoemission or adsorption described in the previous section. Both can be used effectively in the present context.

Near-edge x-ray absorption fine structure (NEXAFS) spectroscopy focuses on the absorption properties of adsorbed molecules within about 50 eV of a deep core absorption edge.* This is precisely the spectral region eschewed in XPD and SEXAFS analyses because explicit calculation of $\mu(\omega)$ requires multiple scattering considerations. However, here we will require no explicit calculations of the absorption coefficient at all – only the symmetry properties of the matrix element will enter the analysis. Furthermore, the substrate plays no essential role except in so far as it anchors the adsorbate to the solid. This simplicity arises from the fact that core level near-edge absorption from adsorbates is dominated by particularly simple intra-molecular resonant transitions.

Consider the case of carbon monoxide. The energy level diagram for this molecule (Fig. 9.6) shows that the anti-bonding 2π orbital is unoccupied in the ground state. However, electrons can be excited from the carbon or oxygen 1s orbitals into this 2π level for incident photon energies near 285 eV and 531 eV, respectively. Prominent peaks in the CO absorption spectrum also appear near 305 eV and 550 eV. In this latter case, the electrons actually escape the molecule, i.e., they are photoemitted. But at these particular energies, electrons that originate from carbon or oxygen 1s orbitals are excited into a 'virtual' bound state of σ symmetry – a special type of continuum state just above the vacuum level that has considerable wave function overlap with ordinary bound state orbitals. The excited electrons 'rattle around' in the vicinity of the molecule for some time before escaping to infinity. This behavior is not special to CO. Strong absorption into unoccupied bound states and virtual bound states is very common in small gas phase molecules (Sette, Stohr & Hitchkock, 1984). For present

* Surface science is plagued by acronyms. This particular technique is sometimes called XANES: x-ray absorption near-edge-spectroscopy.

application, take particular note of the fact that these two absorption final states have *different* symmetry.

The relative intensity of x-ray absorption into the two final states described above depends on the dipole selection rules embodied in the $\boldsymbol{\varepsilon}\cdot\mathbf{r}$ transition operator. The rules are simply stated for excitation from the deep core σ initial states of a diatomic molecule (Stohr & Jaeger, 1982). If the vector $\mathbf{M}$ denotes the molecular symmetry axis, only σ-type final states can be reached if $\boldsymbol{\varepsilon} \parallel \mathbf{M}$, whereas if $\boldsymbol{\varepsilon} \perp \mathbf{M}$, transitions will occur only to final states of π symmetry. As an example, Fig. 10.15 shows the yield of Auger electrons that follow the production of core holes in the oxygen 1s level of CO/Ni(100). At grazing incidence, the electric

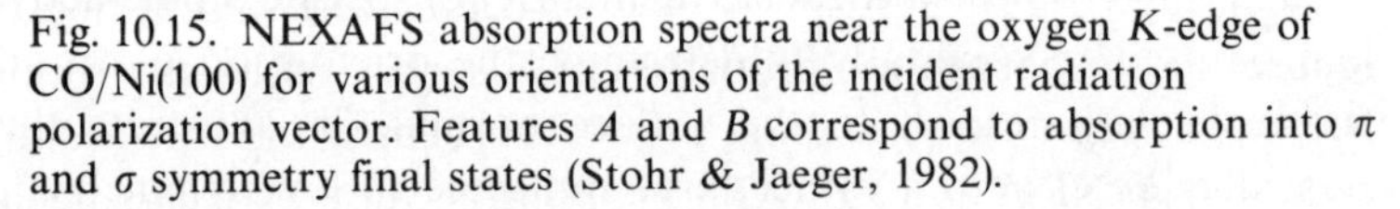

Fig. 10.15. NEXAFS absorption spectra near the oxygen *K*-edge of CO/Ni(100) for various orientations of the incident radiation polarization vector. Features *A* and *B* correspond to absorption into π and σ symmetry final states (Stohr & Jaeger, 1982).

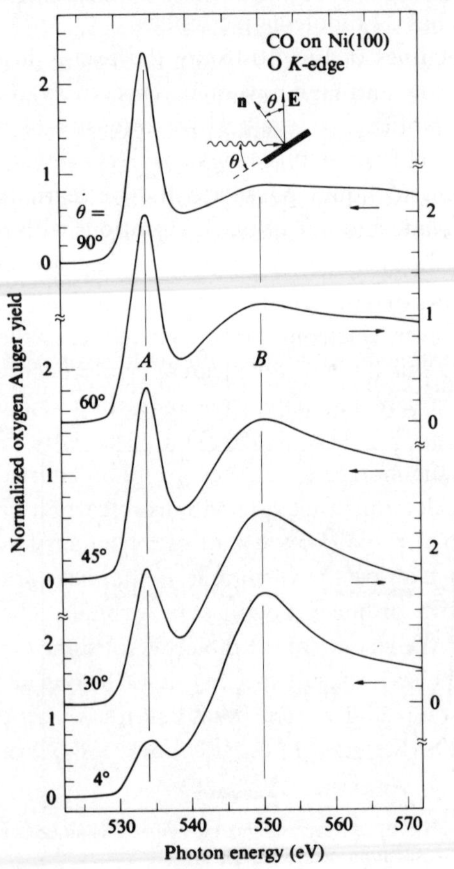

vector is nearly aligned with the surface normal and most of the absorption strength appears in the $1s \rightarrow \sigma$ resonance channel (*B*). At normal incidence, the incident electric field lies in the plane of the surface and absorption into the 2π level (*A*) dominates the spectrum. Both observations are consistent with the suggestion that the CO molecule stands upright with its molecular axis oriented no more than about 10° from the normal.

The (by now) familiar EELS technique is another electron-based spectroscopy that is sensitive to molecular orientation. In this case, we take advantage of the 'pseudo-selection rule' that holds for metal surfaces. As we have seen, the dipole excitation matrix element (10.3) is very small unless the molecular stretch mode in question induces a dipole moment normal to the surface. Hence, if 'impact' excitations are presumed to scatter electrons isotropically, the relative intensity of specific dipole-active modes can be used to qualitatively determine the orientation of an adsorbate molecule with respect to the surface normal. To illustrate this point, consider the case of a rather large molecule of low symmetry, pyridine (C_5H_5N). This molecule has 27 dipole-active modes.

The solid and dotted curves of Fig. 10.16 are the energy loss spectra for pyridine adsorbed at low and high coverage, respectively, on a single crystal Ag(111) surface. Notice that the relative intensity of modes at $610\,cm^{-1}$ and $700\,cm^{-1}$ ($1\,cm^{-1} = 0.124\,meV$) is *reversed* at the two coverages. By comparison to liquid phase results, we learn that these modes correspond to in-plane and out-of-plane vibrations with respect to the ring structure of the molecule. By the argument sketched above, we

Fig. 10.16. EELS vibrational loss spectra of pyridine on Ag(111) at $T = 140\,K$ for coverages below (solid curve) and above (dotted curve) the orientational transition (Demuth *et al.*, 1982).

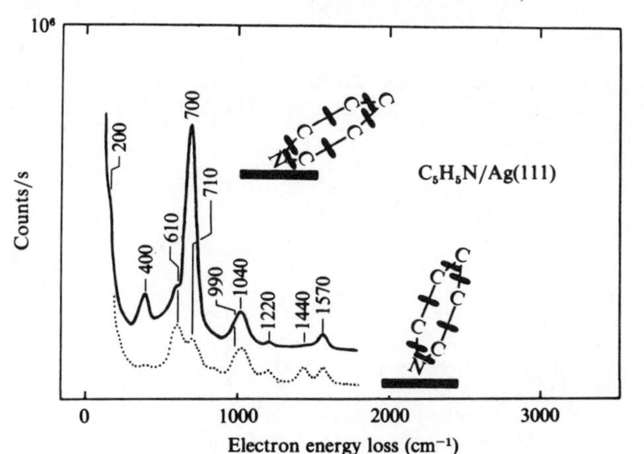

conclude that the C_5H_5N molecule is tilted away from the surface normal significantly more in the low-coverage phase than in the high-coverage phase. This orientational phase transition permits more pyridine molecules to squeeze onto the surface and gain chemisorption bond energy at high coverage.

The EELS data for pyridine cited above suggest that adsorbate–adsorbate interactions can profoundly affect the orientation of a chemisorbed molecule with respect to the substrate. Accordingly, it is natural to inquire how such interactions affect the orientation of adsorbates with respect to one another. This information is not readily accessible using the spectroscopic techniques outlined to this point. Instead, a real-space imaging technique, preferably not limited to ordered overlayers, would be desirable. The scanning tunnelling microscope furnishes one such view, projected onto the direction normal to the surface. Another view, somewhat more sensitive to the polar angle of orientation, derives from electron stimulated desorption ion angular distributions (ESDIAD).

Desorption is a rather complicated process that is not well understood (see Chapter 14). Nevertheless, it is well known that chemisorption bonds can rupture upon bombardment by focused electron or photon beams. If this occurs, it is not unreasonable to suppose that the desorbing particles (positive ions, negative ions and neutrals) exit the surface in collimated cones of emission directed along the original surface bond (Fig. 10.17). A projected image of the adsorbate layer (not unlike an FIM image (Fig. 3.18)) forms if one intercepts the ejected particles with a position sensitive detector. Typical experimental arrangements discriminate in favor of one of the three final charge states and, coupled with mass analysis, display the angular distribution data in digital or analog form.

ESDIAD has been particularly useful for providing qualitative guidance to the construction of structural models of co-adsorption, i.e., the situation

Fig. 10.17. Schematic view of the relationship between surface bond angles and ion desorption angles in ESDIAD (Madey, Doering, Bertel & Stockbauer, 1983).

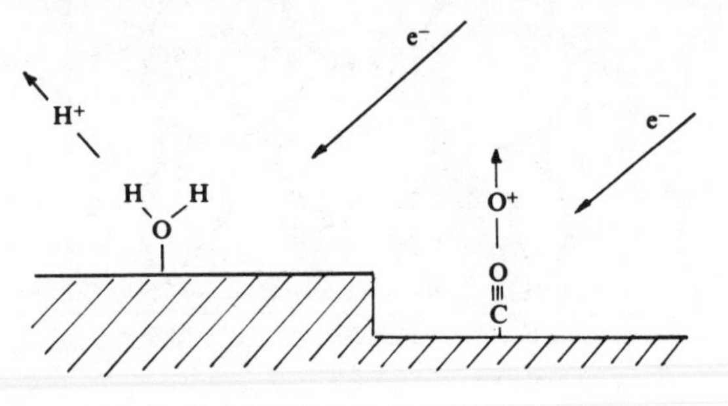

when two different gas phase species adsorb onto a single surface. For example, Fig. 10.18(*a*) illustrates an ESDIAD pattern for the emission of H^+ ions from a Ni(111) surface exposed to a submonolayer dosage of water. The isotropic pattern with a dim center is consistent with H_2O molecules bonded oxygen-end down with a random distribution of azimuthal and polar tilt angles. But, the ion angular distribution is *very* different if the nickel surface is pre-dosed with a very small amount of oxygen before the water is admitted into the sample chamber. A distinct anisotropic pattern appears which exhibits intense emission along $[\bar{1}\bar{1}2]$

Fig. 10.18. ESDIAD study of oxygen and water co-adsorption on Ni(111) at 80 K: (*a*) H^+ ion angular distribution for H_2O on a clean nickel surface; (*b*) H^+ ESDIAD pattern for water adsorption following submonolayer adsorption of oxygen; (*c*) proposed structural model of $H_2O/O/Ni(111)$ co-adsorption (Madey & Netzer, 1980).

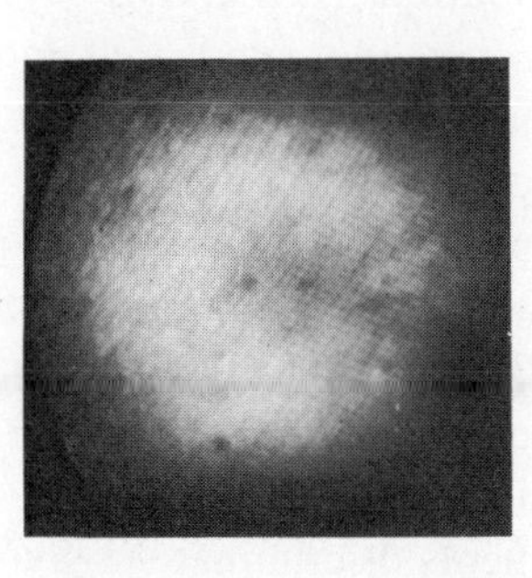

(*a*)

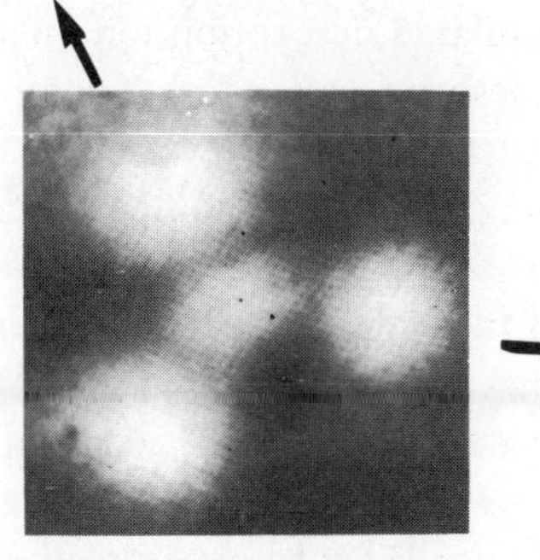

(*b*)

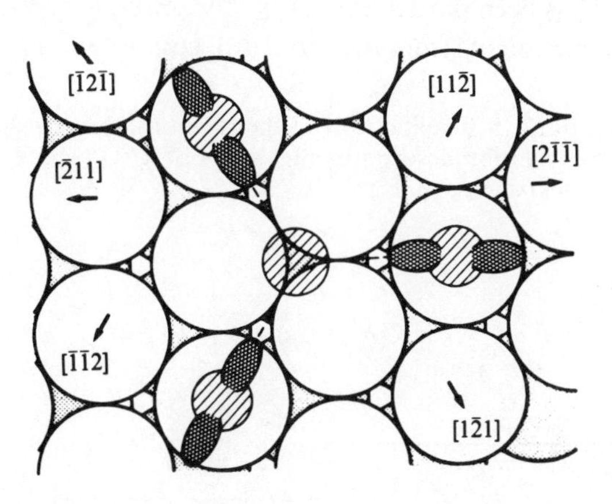

(*c*)

azimuths (Fig. 10.18(*b*)). A structural model consistent with this result places the oxygen atoms in three-fold hollow sites so that three water molecules can hydrogen bond to each O as shown (Fig. 10.18(*c*)). The resulting surface O–O bond distance is quite similar to that found in hydrogen-bonded ice. In this picture, one assumes that it is the hydrogen ligands not involved in the O–H_2O bond that contribute to the ESDIAD signal.

General references

Topography

Engel, T. & Rieder, K.H. (1982). Structural Studies of Surfaces with Atomic and Molecular Beam Diffraction. In *Structural Studies of Surfaces* (ed. Hohler), pp. 55–180. Berlin: Springer-Verlag.

Site symmetry

Willis, R.F., Lucas, A.A. & Mahan, G.D. (1983). Vibrational Properties of Adsorbed Molecules. In *The Chemical Physics of Solid Surfaces and Heterogeneous Catalysis* (eds. King & Woodruff), vol. 2, pp. 59–163. Amsterdam: Elsevier.

Bond lengths and orientation

Margoninski, Y. (1986). Photoelectron Diffraction and Surface Science. *Contemporary Physics* **27**, 203–40.

Stohr, J. (1984). Surface Crystallography by Means of SEXAFS and NEXAFS. In *Chemistry and Physics of Solid Surfaces* (eds. Vanselow & Howe), vol. 5, pp. 231–55. Berlin: Springer-Verlag.

Van der Veen, J.F. (1985). Ion Beam Crystallography of Surfaces and Interfaces. *Surface Science Reports* **5**, 199–288.

Madey, T.E., Doering, D.L., Bertel, E. & Stockbauer, R. (1983). Electron and Photon Stimulated Desorption: Benefits and Pitfalls. *Ultramicroscopy* **11**, 187–98.

11

PHASE TRANSITIONS

Introduction

The phenomenon of condensation is one of the most familiar properties of bulk matter and so has attracted the attention of physicists for decades. The basic questions are straightforward to pose but remarkably difficult to answer. In fact, a conceptual revolution was required before truly rapid progress was achieved (Wilson, 1979). As we have indicated earlier (Chapter 5), dimensionality plays a crucial role in this modern theory of phase transitions (Ma, 1976). It then is natural to ask how much (if any) of our common three-dimensional experience and intuition carry over to the two-dimensional problem. Typical questions might be: What is the nature of the adsorbate phase diagram? How does a surface species pass from an ordered crystallographic state to a disordered state? What microscopic mechanisms are involved? How does an overlayer freeze and/or melt? Are any properties unique to two dimensions?

From the thermodynamic point of view, we have learned that clean surface critical phenomena and melting do indeed both differ from their three-dimensional bulk counterparts. We further quantify this notion here and examine the *universality* hypothesis, which states that only symmetry considerations and (in some cases) the range of adsorbate interactions determine the intrinsic nature of overlayer phase changes on a Langmuir checkerboard. Both static and dynamic issues will receive attention. From the crystallographic point of view, a synergistic interplay between theory and experiment has uncovered and elucidated a fascinating new structural transformation in adsorbed films: the commensurate–incommensurate transition. Although elegant mathematical theories exist, we outline instead a phenomenological approach that most readily exposes the essential physics. Melting reappears as an important feature. Surprisingly perhaps, the same conceptual framework permits, as well, a brief account of contemporary studies of superfluidity and superconductivity in adsorbed layers.

Let us approach our problem in stages. Suppose just a small fraction of gas phase material adsorbs onto a solid surface. In some situations this small perturbation is enough to trigger a structural transformation of the substrate surface layer. An excellent example of this phenomenon appeared in Fig. 10.3. Recall that this is a scanning tunnelling microscope image of C/Pt(100). The corrugated central foreground region is a patch of clean Pt(100) which, like Ir(100) (see Chapter 5) reconstructs to form a close-packed surface layer. Look now at the right side of the scan just behind the spikes in the foreground. We observe a sizeable flat region that does not show the prominent corrugation of the foreground patch. This is not due to poor perspective in the figure.

The smooth region of Fig. 10.3 corresponds to a portion of Pt(100) that exhibits an ideal 1×1 bulk-like surface termination. Evidently, adsorbed carbon atoms (not imaged) have *undone* the original reconstructive phase transition! This is not so surprising if we recall that adsorption quite generally reduces surface tension (Fig. 9.1). The issue then is the *relative* reduction in γ for the two possible surface structures compared to their clean surface difference in free energy. In this case, the ideal surface wins.

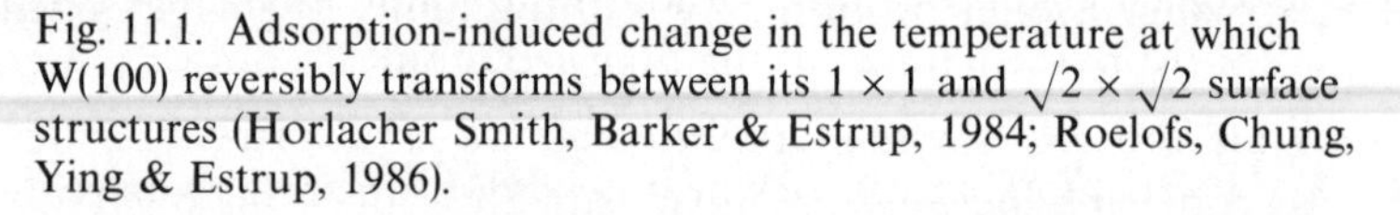
Fig. 11.1. Adsorption-induced change in the temperature at which W(100) reversibly transforms between its 1×1 and $\sqrt{2} \times \sqrt{2}$ surface structures (Horlacher Smith, Barker & Estrup, 1984; Roelofs, Chung, Ying & Estrup, 1986).

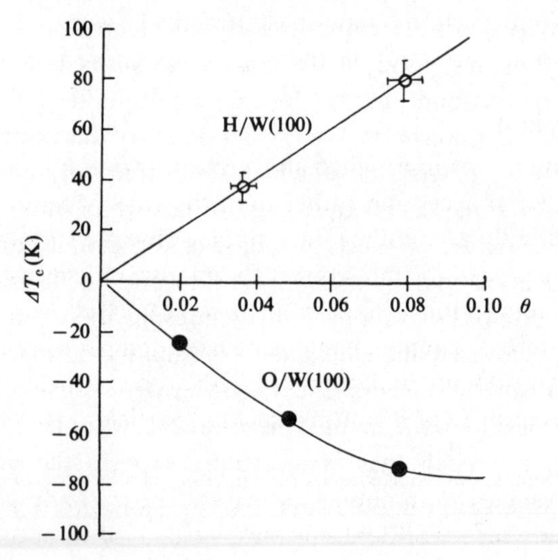

Moreover, the unreconstruction helps explain the monoatomic step surrounding the reconstructed clean patch. The close-packed surface is 20% denser than the 1×1 surface. Hence, when the phase transition occurs, surplus atoms are forced upward into the next layer creating a step relative to the reconstructed clean surface.

A more microscopic interpretation of adsorbate-induced structural sensitivity is possible for the W(100) surface. In that case (Fig. 5.3), the clean surface transforms from 1×1 to $\sqrt{2} \times \sqrt{2}$ just below room temperature. The interesting observation is that small amounts (< 0.1 monolayer) of adsorbed hydrogen *increase* the transition temperature while equally small amounts of oxygen *decrease* the transition temperature* (Fig. 11.1). In other words, H adatoms stabilize the zig-zag reconstructed structure whereas O atoms destabilize the $\sqrt{2} \times \sqrt{2}$ structure. The simplest explanation relies on the following observation: H atoms occupy bridge sites while O atoms occupy four-fold hollow sites.

Suppose that each adsorbate exerts a force on the nearest neighbor substrate atoms. For the hydrogen case, notice that the zig-zag reconstruction (exaggerated in Fig. 5.3) possesses two inequivalent bridge sites – a 'long' bridge and a 'short' bridge. This means that adatoms preferentially adsorb onto a long (short) bridge site if the H–W interaction is repulsive (attractive). In either case, transformation to the 1×1 phase is energetically unfavorable so that hydrogen adsorption retards the transition. Now consider oxygen chemisorption. All four-fold hollow sites are equivalent in the $\sqrt{2} \times \sqrt{2}$ phase. However, the tungsten atoms do not sit symmetrically with respect to the adsorption site. Adsorption costs energy (relative to the 1×1 structure) because one near neighbor substrate atom is always too close (far) if the O–W interaction is repulsive (attractive). Hence, oxygen adsorption encourages transformation to the high temperature phase.

Phase equilibrium

Notwithstanding the examples cited above, chemisorption-induced surface reconstruction is the exception rather than the rule. Most of the time, gas phase atoms simply accumulate on a stable substrate surface as the ambient pressure increases. The most direct thermodynamic measurement of this process is adsorption volumetry (Thomy, Duval & Regnier, 1981). One admits gas into a sample chamber at fixed temperature and measures the equilibrium pressure and number of adsorbed particles. We expect a Langmuir isotherm (9.6) if non-interacting particles randomly

* One monitors the temperature dependence of the intensity of superlattice reflections, i.e., extra diffraction spots (LEED in this case) associated with the reconstructed phase relative to the 1×1 structure.

occupy the surface (cf. Fig. 9.2). However, sometimes one finds isotherms similar to those shown in Fig. 11.2. At high temperature one does indeed find curves similar to the Langmuir form. But at reduced temperature, the isotherm acquires a vertical step. The chemical potential (pressure) does not change as the adsorbate density increases. This behavior is familiar from bulk thermodynamics and signals the presence of two-phase coexistence – a first-order phase transition has occured (Callen, 1985).

Many years before data of this sort were available, Fowler (1936) realized the proper way to modify Langmuir's assumptions so that two-phase coexistence can occur. The idea is quite simple; a condensed phase appears if one permits attractive interactions amongst the adsorbates. To see this we shall not reproduce Fowler's analytic results but, instead, introduce a very useful concept – the lattice gas model. Imagine that the surface of the substrate is broken up into N identical square cells, each roughly the size of a single adsorbate. Define a local occupation variable c_i so that $c_i = 1$ if an ad-particle occupies the ith cell and $c_i = 0$ otherwise. The coverage θ is the thermal average summed over all sites:*

$$\theta = \frac{1}{N}\sum_{i=1}^{N} \langle c_i \rangle. \tag{11.1}$$

Fig. 11.2. Adsorption isotherms of CH_4 on NaF. The dashed curve is a suggested coexistence curve (Morishige, Kittaka & Morimoto, 1984).

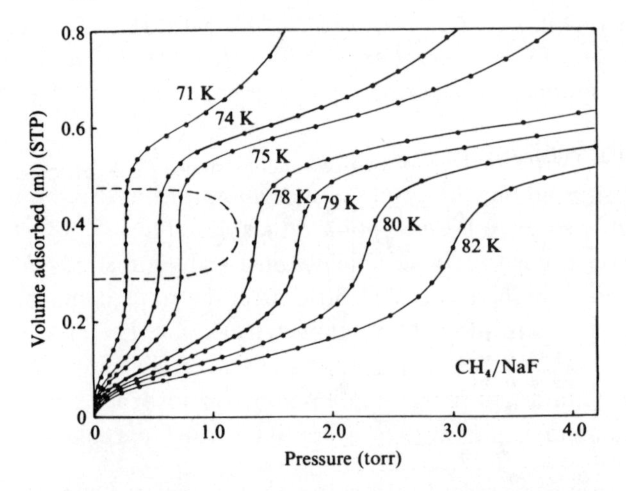

* We suppose that the surface corrugation is weak (or the temperature high) so that an adsorbate effectively samples the entire available surface.

Each particle gains an energy ε upon adsorption and every pair of particles at sites i and j gain an interaction energy ϕ_{ij}. The total energy $\mathscr{H}$ of the system then is

$$\mathscr{H} - \mu N_a = -(\varepsilon + \mu) \sum_{i=1}^{N} c_i - \sum_{i \neq j} \phi_{ij} c_i c_j. \tag{11.2}$$

A factor of $\mu N_a = \mu \sum c_i$ has been subtracted from both sides of (11.2) to remind us that the chemical potential μ rather than the coverage is the appropriate independent variable when the adsorbate is in equilibrium

Fig. 11.3. Adsorption isotherms for a square lattice gas with nearest neighbor adsorbate attraction only (Binder & Landau, 1981).

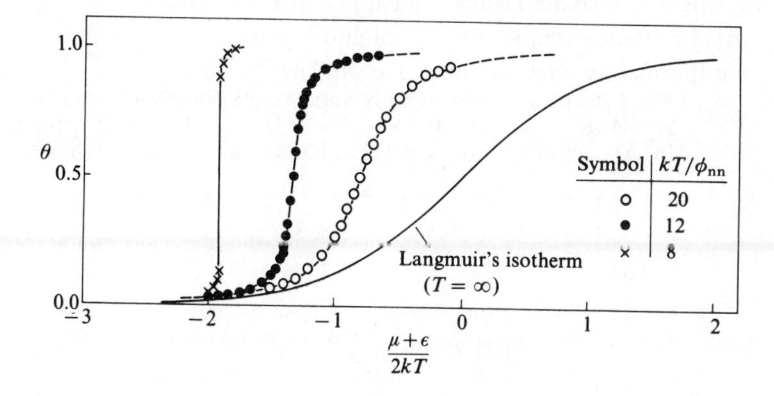

Fig. 11.4. Schematic view of two-phase coexistence for a lattice gas.

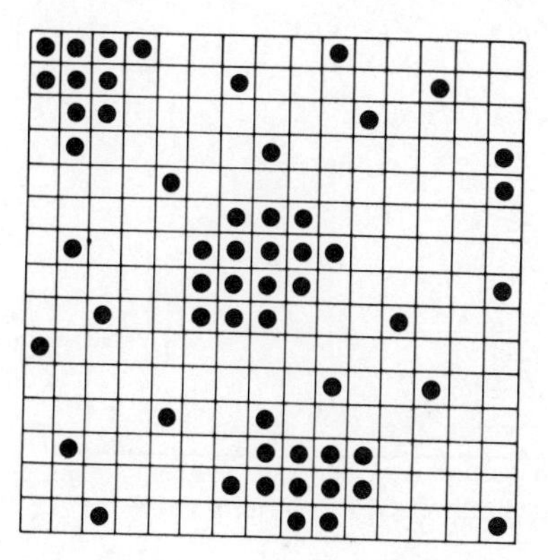

with its own vapor. The relationship between the two is

$$\mu = \frac{1}{N}\frac{\partial F}{\partial \theta}\bigg|_T, \tag{11.3}$$

where

$$F = -kT \ln \mathrm{Tr}_{\{c_i\}} \exp(-\mathscr{H}/kT) \tag{11.4}$$

is the Helmholtz free energy of the lattice gas system.

Equation (11.2) defines a problem in statistical mechanics. We are interested in the properties of this system when the $\{c_i\}$ are characteristic of an equilibrium ensemble. One way to do this is by Monte Carlo simulation (Roelofs, 1982). This is a numerical technique by which one computes sums like that in (11.4) by sampling the configurations $\{c_i\}$ according to a Boltzmann probability distribution. For simplicity, let all

Fig. 11.5. Coexistence curve between single-phase and two-phase regions of Au on W(110). Solid line is the van der Waals prediction for $T_c = 1130\,\mathrm{K}$ and $\theta_c = 0.26$ (Kolaczkiewicz & Bauer, 1984).

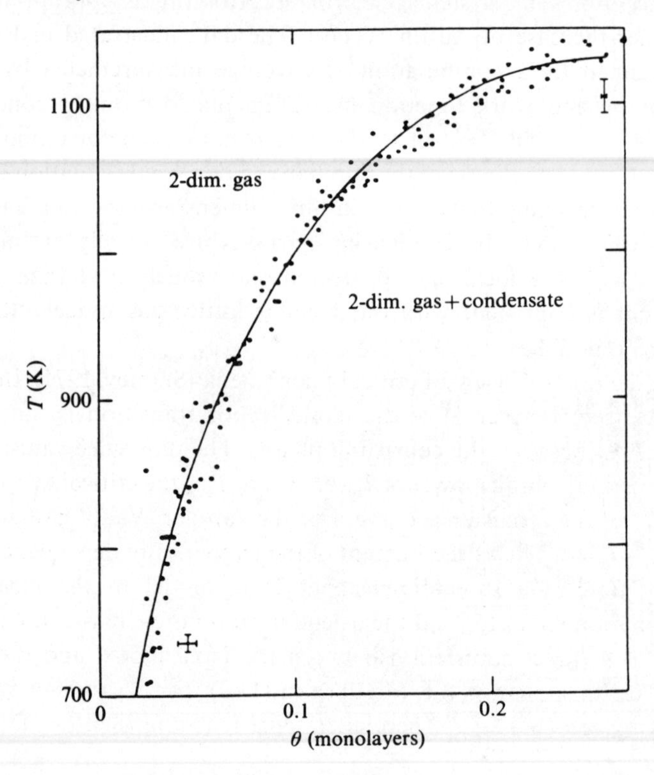

the ϕ_{ij} be zero except when i and j are adjacent cells. The presence of attractive interactions restricted to nearest neighbors (nn) in this way is sufficient to produce condensation. We see this from a sequence of adsorption isotherms for the lattice gas model computed from (11.3) by the Monte Carlo method (Fig. 11.3). Quite reasonably, the effect depends on the ratio kT/ϕ_{nn} rather than on the interaction strength itself. On the lattice, the two-phase region resembles Fig. 11.4. Groups of islands (the high density condensed phase) coexist with a random distribution of occupied sites (the low density gaseous phase).

A two-dimensional gas system is characterized by an equation of state which expresses the relationship between the pressure, the coverage and the temperature. Different assumptions about the nature of inter-adsorbate interactions lead to different predictions – usually expressed in the form of virial coefficients (Steele, 1974). Here we appeal directly to experiment and examine the shape of the coexistence curve. This is straightforward for physisorption systems for which one readily establishes equilibrium between the sample and a particle reservoir above the surface. Isotherm data are easy to obtain. By contrast, the ambient vapor pressure can be immeasurably low for chemisorption systems and one must turn to other techniques. For some systems, electron spectroscopy is appropriate.

Consider the case of Au on W(110). The data illustrated in Fig. 11.5 were obtained by a combination of coverage measurements by Auger spectroscopy and work function measurements to monitor condensate island formation. The best fit to the data (solid curve) corresponds to a two-dimensional analogue of the familiar van der Waals equation of state. It is worth remarking that, just as in three dimensions, the van der Waals model assumes that the condensing particles are entirely mobile. This contrasts with the localized absorption site models of Langmuir and Fowler but is consistent with our cellular lattice gas model – the condensate is a liquid.

From the point of view of critical phenomena (Stanley, 1971), the *order parameter* (see Chapter 5) of the liquid–vapor transition is the density difference $\Delta\rho$ between the coexisting phases. The difference vanishes as a characteristic algebraic power of $T_c - T$ where T_c is the critical temperature at the top of the coexistence curve. For the van der Waals case one finds $\Delta\rho \propto (T - T_c)^{1/2}$. This is reminiscent of the temperature dependence of the magnetization of a three-dimensional Ising model in the mean field approximation (cf. (5.15) and the attendant footnote). This is no accident as there is a direct connection between the Ising model and the lattice gas. To see this, change the variables in (11.2) from c_i to spin variables $S_i = \pm 1$ by $c_i = (1 - S_i)/2$. Then, after some rearrangement we find

$$\mathscr{H} = -H \sum_{i=1}^{N} S_i - \sum_{i \neq j} J_{ij} S_i S_j + \text{const.}, \qquad (11.5)$$

where $J_{ij} = (1/4)\phi_{ij}$ and $H = -(\varepsilon + \mu)/2 - (1/4)\sum\phi_{ij}$. This is precisely the spin 1/2 Ising model in a magnetic field (Ma, 1985). The magnetization

$$M = \frac{1}{N} \sum_{i=1}^{N} \langle S_i \rangle, \qquad (11.6)$$

is related to the coverage (11.1) by $\theta = (1 - M)/2$.

We saw in Chapter 5 that mean field theory breaks down as one approaches the critical point and fluctuations in the order parameter become more and more important. In the present case, the Ising correspondence means that the shape of the coexistence curve must deviate from van der Waals behavior very near T_c. Instead, the critical exponent ought to be $\beta = 1/8$ reflecting the exact solution of the two-dimensional Ising model (Fig. 5.4). Although the data in Fig. 11.5 fall a bit short of the critical region ($|1 - T/T_c| \ll 1$), very precise isotherm measurements for $Ar/CdCl_2$ (Larher, 1979) and heat capacity measurements for CH_4 on graphite (Kim & Chan, 1984) confirm the Ising prediction. Heat capacity data represent another traditional experimental approach to phase transition characterization: calorimetry (Marx, 1985). The aforementioned CH_4 data

Fig. 11.6. Experimental specific heat of ^4He on graphite. The observed critical exponent is $\alpha \cong 1/3$ (Ecke & Dash, 1983).

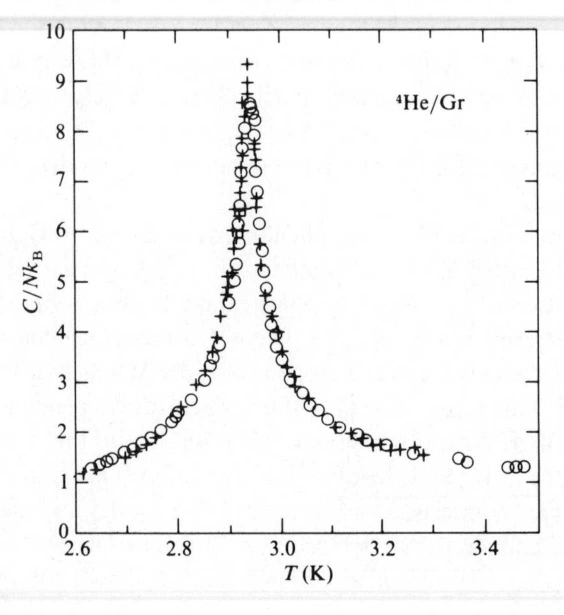

diverge logarithmically at T_c in perfect accord with expectations based on the two-dimensional Ising model. However, not all overlayer phase transitions are of this sort. This is illustrated by Fig. 11.6, which shows a specific heat anomaly near 3 K for ^{4}He adsorbed onto the basal plane of graphite. In this case, the data are well fitted to a form $C = A|T - T_c|^\alpha + B$ where A and B are constants and $\alpha \cong 1/3$. It is necessary to reintroduce the adsorbate crystallography to understand this result.

Fig. 11.7. Equilibrium temperature–coverage phase diagrams for inert gases on graphite: (*a*) argon (Migone, Li & Chan, 1984); (*b*) krypton (Butler, Litzinger & Stewart, 1980; Specht *et al.*, 1984); (*c*) xenon (Heiney *et al.*, 1983). See text for notation and discussion.

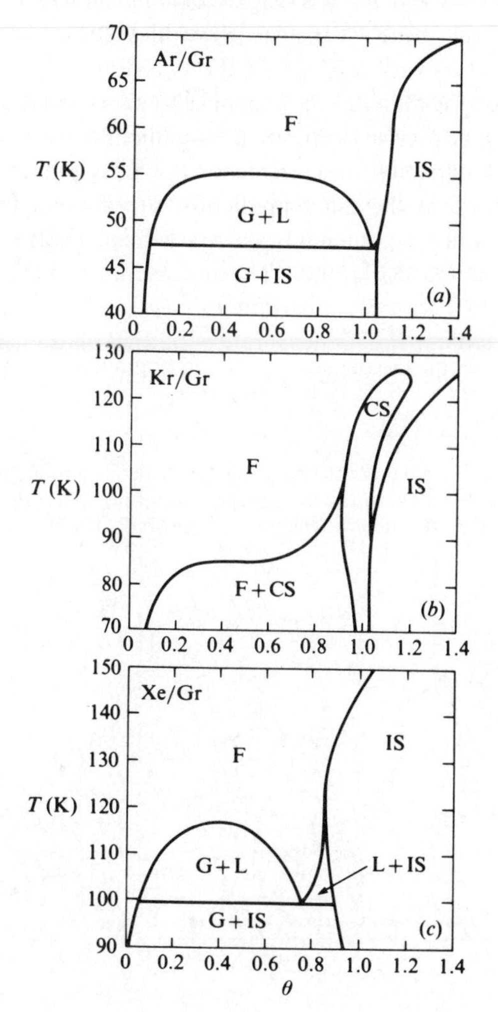

Physisorption systems

Our study of phase transitions into and out of ordered adsorbate arrangements begins with physisorbed overlayers on smooth substrates. We do this because one can analyze the problem completely. The adsorbates interact with the substrate and one another in a completely known fashion: weak attraction from the dispersion force and hard-core repulsion (Fig. 8.3). This simplification does not reduce the problem to a triviality. Quite the contrary. A wealth of phenomena occur precisely because we restrict attention to substrates where the height of the potential energy barriers between adsorption sites, i.e., the corrugations, are comparable to thermal energies. To prove the veracity of this statement we need only exhibit the experimental phase diagrams for three inert gases: argon, krypton and xenon, adsorbed onto the basal plane of graphite (Fig. 11.7).

The dissimilarity of the three phase diagrams in Fig. 11.7 is remarkable – particularly when one notes that the bulk phase diagrams for argon, krypton and xenon are essentially identical. Only the xenon example exhibits a typical 'bulk' topology with regions of two-phase coexistence of gas (G) + liquid (L), gas + solid and liquid + solid in addition to a well-defined triple point and critical point. The solid, however, is crystallographically incommensurate with the substrate (Fig. 10.2 is drawn to scale for exactly this case). A similar incommensurate solid (IS) phase is found for argon although its phase diagram lacks a region of liquid/solid

Fig. 11.8. Krypton monolayer adsorbed on graphite: (*a*) commensurate $\sqrt{3} \times \sqrt{3}$–30° structure where the adatoms occupy one of the three equivalent sublattices; (*b*) incommensurate phase (Bak, 1982).

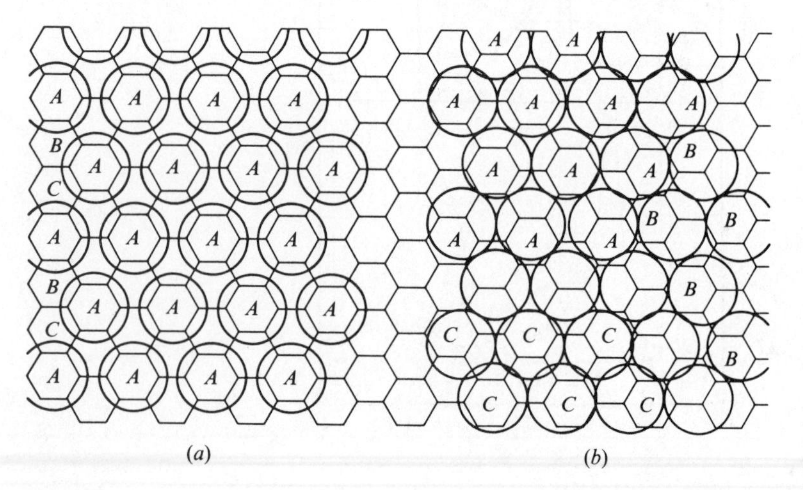

coexistence (or it is very small). Only krypton exhibits a legitimate commensurate solid (CS) phase ($\sqrt{3} \times \sqrt{3}$–30°) at low density (Fig. 11.8(*a*)); but a transition to an incommensurate phase (Fig. 11.8(*b*)) occurs at a higher coverage. Moreover, krypton apparently shows no triple or critical point and the CS melts directly to a fluid (F) phase. Evidently, the substrate has a profound effect on the behavior of these simple gases.

The qualitative difference between the krypton phase diagram and the other two inert gases phase diagrams arises principally from an adsorbate/substrate size mismatch effect. To see this, calculate the fractional difference

Fig. 11.9. X-ray scans across the adsorbate (1,0) Bragg peak for inert rare gases on graphite. Vertical line passes through the position of the first surface reciprocal lattice vector of the substrate: (*a*) argon (McTague, Als-Nielsen, Bohr & Neilsen, 1982); (*b*) krypton; (*c*) xenon (Birgeneau *et al.*, 1980).

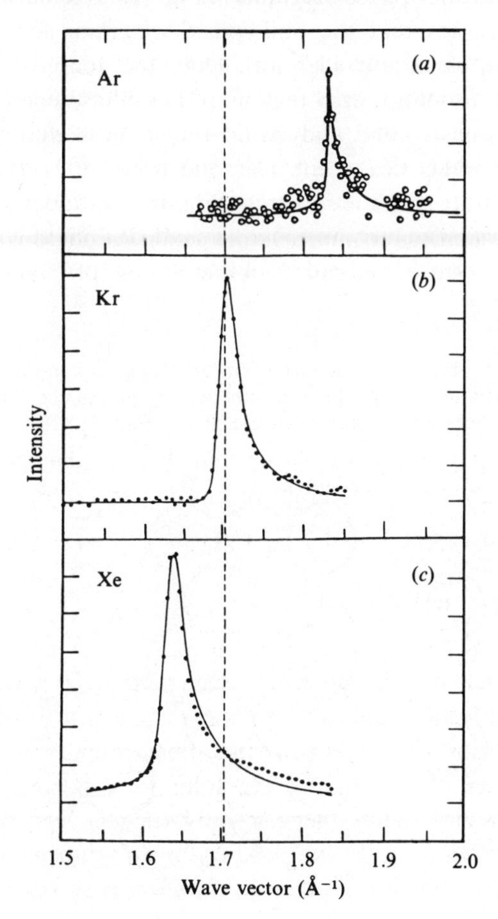

between the in-plane graphite lattice constant and the *bulk* solid phase lattice constant of the gases, viz., $(a_{\text{inert gas}} - a_{\text{graphite}})/a_{\text{graphite}}$. This quantity is called the 'misfit' in the theory of epitaxy (see Chapter 16). The values for argon, krypton and xenon are -8%, -5% and $+8\%$, respectively. Therefore, krypton atoms that form a commensurate overlayer by occupation of the hexagonal hollow sites of the regular graphite lattice very nearly sit at the minimum of the krypton–krypton interaction potential. This condition is not satisfied so well for either argon or xenon. They gain more energy by ignoring the favorable adsorption sites of the substrate and adopting their own 'natural' lattice constant. This conjecture is most easily tested by diffraction. LEED is one obvious approach; another is x-ray scattering. Fig. 11.9 illustrates x-ray diffraction profiles for these systems along the (1, 0) line in reciprocal space. The vertical dashed line denotes the position of the graphite Bragg peak and directly reveals the proclivity of these adsorbates to conform to or ignore the underlying corrugation potential.

In the previous chapter we defined an incommensurate overlayer structure as one whose symmetry is not related to that of the substrate. In the present context this means that there is no choice of integers N and M such that $N\mathbf{G} = M\mathbf{g}$ where $\mathbf{G}$ and $\mathbf{g}$ are the shortest reciprocal lattice vectors of the substrate (in-plane) and the adsorbate, respectively. This fact has an interesting consequence. Suppose we compute the interaction energy between the substrate and a rigid incommensurate overlayer in the usual way,

$$U = \int \mathrm{d}\mathbf{r}\, n(\mathbf{r}) V(\mathbf{r}), \tag{11.7}$$

where $n(\mathbf{r})$ is the density of the overlayer and $V(\mathbf{r})$ is the substrate corrugation potential. To do the integral, we Fourier analyze the pieces of the integrand:

$$n(\mathbf{r}) = \sum_{\mathbf{g}} n_{\mathbf{g}} \mathrm{e}^{\mathrm{i}\mathbf{g}\cdot\mathbf{r}}; \quad V(\mathbf{r}) = \sum_{\mathbf{G}} V_{\mathbf{G}} \mathrm{e}^{\mathrm{i}\mathbf{G}\cdot\mathbf{r}}. \tag{11.8}$$

One sees immediately that $U \propto \delta_{G,g}$ so, by the argument given above, the interaction energy identically vanishes! An incommensurate overlayer simply 'floats' atop its substrate – it is truly two-dimensional.

The unique properties of such a floating solid were introduced in Chapter 5 in connection with the Wigner electron lattice on ^{4}He. We saw there that thermal fluctuations of the ion positions in two dimensions are much larger than in three dimensions. In fact, the effect is so strong that the 'crystal' does not exhibit broadened delta-function peaks in a diffraction experiment. Instead, a power-law singularity occurs when the Bragg condition is satisfied (Jancovici, 1967), although it is difficult to discern visually in Fig. 11.9. Be that as it may, perhaps the most interesting aspect of a two-dimensional solid is its melting. In the electron/^{4}He case, the dislocation unbinding mechanism of Kosterlitz & Thouless (1973) clearly was implicated. Does the same occur here?

It turns out that the thermal destruction of order within an adsorbed film can follow a number of different scenarios. It is simplest to begin with the case of a commensurate overlayer. In that case, we can adapt the Landau argument given in Chapter 5 (in connection with reconstruction) to the case of a lattice gas. Now, however, the cells of the lattice gas correspond precisely to the known adsorption sites of the substrate. Let this lattice be described by a space group G_0. The cells are randomly occupied in the high temperature, disordered phase so that the symmetry of the overlayer is identical to the symmetry of the substrate. At low temperature, the ordered phase preferentially picks out certain sites for occupation and the mass density of the overlayer exhibits a lower symmetry group G. As in the reconstructive case, the order–disorder transition within an adsorbed layer *can* be continuous only if G is a subgroup of G_0 and the mass density transforms in an appropriate manner.*

If the transition is continuous one can make definite statements about the critical exponents one should see in experiment (Schick, 1981). Below the transition temperature, the overlayer density contains non-zero Fourier components at wave vectors $\mathbf{K}$ characteristic of the group G. Hence, an appropriate order parameter is

$$\psi_{\mathbf{K}} = \sum_{\mathbf{r}} e^{i\mathbf{K}\cdot\mathbf{r}} \langle n(\mathbf{r}) \rangle. \tag{11.9}$$

This quantity correctly vanishes above T_c because $\langle n(\mathbf{r}) \rangle$ is equal to a constant, independent of $\mathbf{r}$, in the disordered phase. We now invoke the concept of universality and construct a free energy functional – a so-called Landau–Ginzburg–Wilson (LGW) Hamiltonian (Mukamel & Krinsky, 1976) – which contains combinations of the $\psi_{\mathbf{K}}$ and their gradients (cf. (5.14)) that are invariant under the symmetry operations of G. Now the problem is reduced to its essence independent of all material properties of the overlayer and substrate. One now looks for simple mathematical models, e.g., the Ising model, that exhibit the same symmetry properties as the LGW Hamiltonian just obtained. Universality claims that the critical

* It is always possible that an 'accidental' crossing of free energy curves leads to a first-order transition that preempts the continuous transition.

exponents for these models (often known) are identical to those of the original physical system. Consider the disordering of a commensurate $\sqrt{3} \times \sqrt{3}$ overlayer structure on the triangular lattice of adsorption sites of graphite. As Fig. 11.8(*a*) shows, there actually are three equivalent adsorption sub-lattices (labelled *A*, *B* and *C*) on this surface. The perfect ordered structure can occupy any one of them. An analysis of the sort sketched above shows that this problem is equivalent to a mathematical model (called the 3-state Potts model) that is similar to an Ising model except that the 'spin' variable takes on three distinct values instead of two. We now look to the krypton phase diagram for a place where the commensurate solid melts directly to a fluid phase. This occurs at the top of the CS 'tongue' near $\theta = 1$. The specific heat, as one crosses this phase boundary, looks just like that shown in Fig. 11.6 (helium also condenses into a commensurate $\sqrt{3} \times \sqrt{3}$ phase on graphite). The Potts model value for the specific heat critical exponent is 1/3.

Turn now to the melting of an incommensurate overlayer. As noted earlier, the submonolayer portion of the argon and xenon phase diagrams resemble bulk phase diagrams. Melting proceeds by a conventional first-order process that evidently preempts any continuous behavior. However, the dislocation unbinding melting mechanism does appear to be operative at coverages near and just above one monolayer. Moreover, the present level of experiments (Rosenbaum, Nagler, Horn & Clarke, 1983) supports a sophisticated extension of this theory whereby the truly disordered fluid phase is preceded by a liquid-like phase that retains some of the bond-orientational order of the solid (Nelson & Halperin, 1979).

Commensurate–incommensurate transitions

The remarkably rich krypton phase diagram of Fig. 11.7(*b*) exhibits yet another fascinating bit of two-dimensional surface physics. A structural transition occurs with increasing adsorbate density between a commensurate solid phase and an incommensurate solid phase. This phenomenon is by no means unique to krypton/graphite; it appears in many physisorption *and* chemisorption systems. The pertinent question is: How does one pass from Fig. 11.8(*a*) to 11.8(*b*)? As so often happens, the basic physics has been rediscovered many times. The original discussion (in the context of plastic deformation of crystals) was by Frenkel' & Kontorova (1939).

Consider a linear chain of atoms, harmonically bound together, which sit atop a corrugated 'washboard' potential. This model is characterized by two *competing* interactions. The interatomic springs prefer an overlayer lattice constant a while the substrate favors an adatom separation of b. We take the case of relatively weak springs with a only slightly less than b. The problem is to determine the position of the atoms with respect to the substrate as a function of external 'pressure' – applied here by pulling or pushing on the ends of the chain.

At low pressure, it is reasonable that all the atoms sit above potential energy minima with their connecting springs slightly stretched. This is the commensurate, or *registered*, phase (Fig. 11.10(a)). Up to a point, this configuration is stable as the pressure increases. Eventually however, the average separation between adatoms must decrease to some value midway between a and b. A possible solution is to uniformly space the adatoms by precisely this amount (Fig. 11.10(b)). But this configuration costs both spring energy and substrate potential energy for every atom of the chain. Instead, the system achieves a much lower energy (with the same *average* adatom separation) by holding most of the atoms at the well minima and squeezing just a few atoms close together (Fig. 11.10(c)). The localized region of high density is called a *domain wall* or a *soliton*. More walls form as the pressure increases further and, since solitons repel one another (Pokrovsky & Talapov, 1984), they arrange themselves on the chain with uniform spacing. This is usually called the 'weakly' incommensurate phase. Ultimately, the chain becomes 'all domain walls' as the average interparticle spacing approaches a. The structure then is completely incommensurate (Fig. 11.10(d)).

It is necessary to generalize this one-dimensional model for application to a real two-dimensional system like krypton/graphite. Nevertheless, the essential point remains. We should view the overlayer crystal structure in the vicinity of the C–I transition as a collection of commensurate regions separated by domain walls. For graphite, the existence of three equivalent adsorption sublattices leads to the possibility of quite complex wall structure (Fig. 11.11). A particularly graphic illustration of this point comes from molecular dynamics simulations (see discussion accompanying Fig. 5.6) of Kr/Gr at a density just slightly in excess of the C–I phase boundary (Fig. 11.12). The white regions of the figures correspond to areas where the krypton atoms are well associated with a particular graphite sublattice. The dark regions are the domain walls where the adatoms mostly straddle the maxima of the graphite corrugation potential. At low temperature the commensurate domains are recognizably hexagonal although their size and shape slowly change due to thermal fluctuations. By contrast, at higher temperature the domain walls are 'frayed' and meander significantly. This is the 'fluid' phase that creeps in between the

Fig. 11.10. The Frenkel′–Kontorova model of the C–I transition: (*a*) commensurate phase at low pressure; (*b*) a high energy configuration at intermediate pressure; (*c*) ground state configuration at intermediate pressure containing one domain wall; (*d*) incommensurate phase at high pressure.

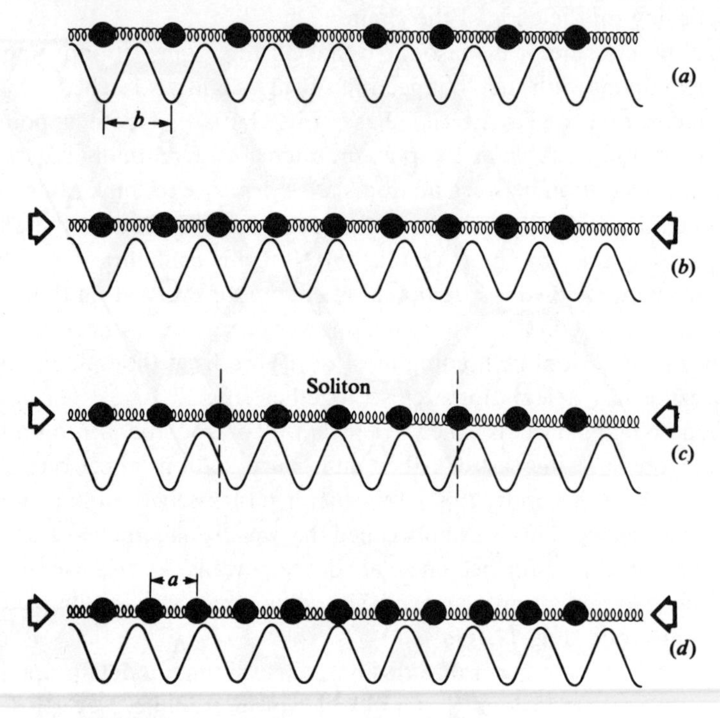

Fig. 11.11. A wall intersection where three commensurate domains meet (Bak, 1982).

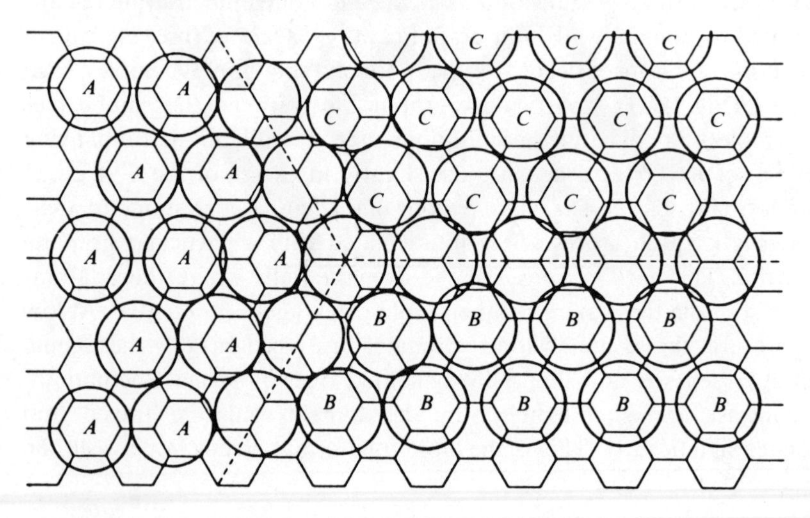

Fig. 11.12. Pictorial representations of the crystallographic structure of two-dimensional krypton on graphite obtained by molecular dynamics simulations just above one monolayer coverage. The depicted rhombus is 620 Å on a side. See text for discussion: (*a*) 17 K; (*b*) 95 K (Koch, Rudge & Abraham, 1984).

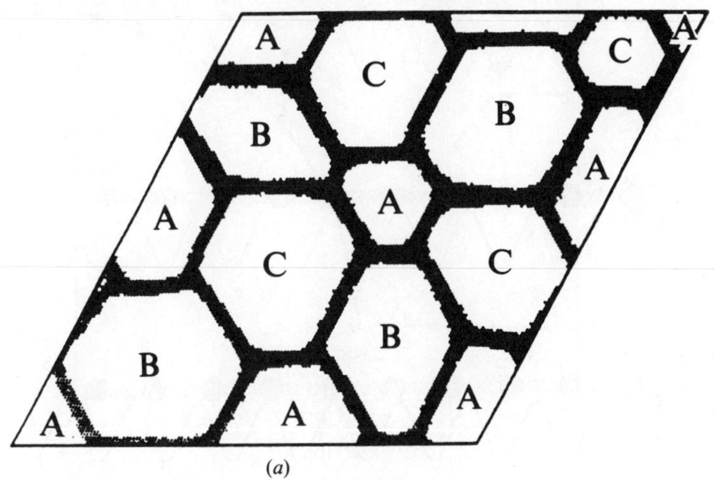

(*a*)

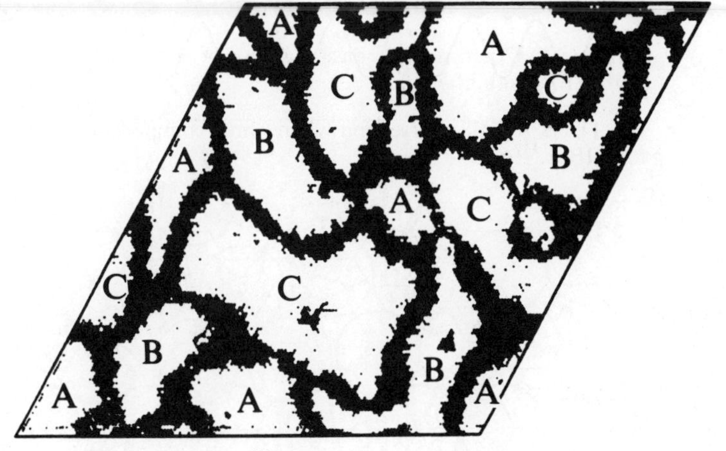

(*b*)

Fig. 11.13. Pressure dependence of the order parameter at 57 K near the commensurate–incommensurate transition of krypton adsorbed on graphite (Chinn & Fain, 1977).

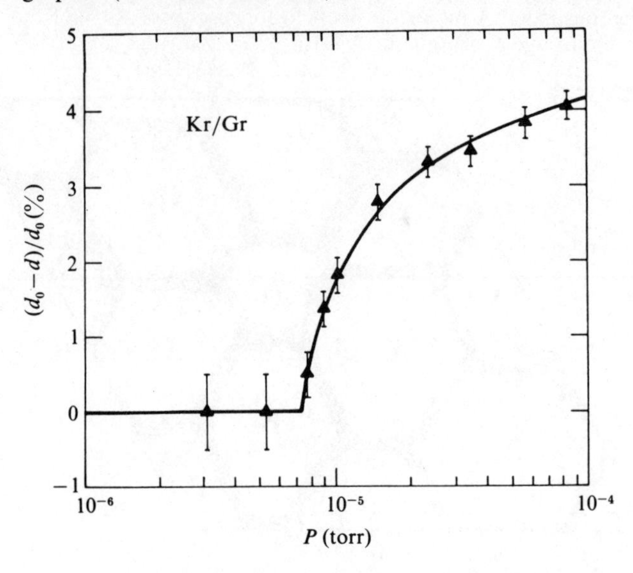

Fig. 11.14. Experimental phase diagram of H/Fe(110) as determined by LEED (Imbihl *et al.*, 1982).

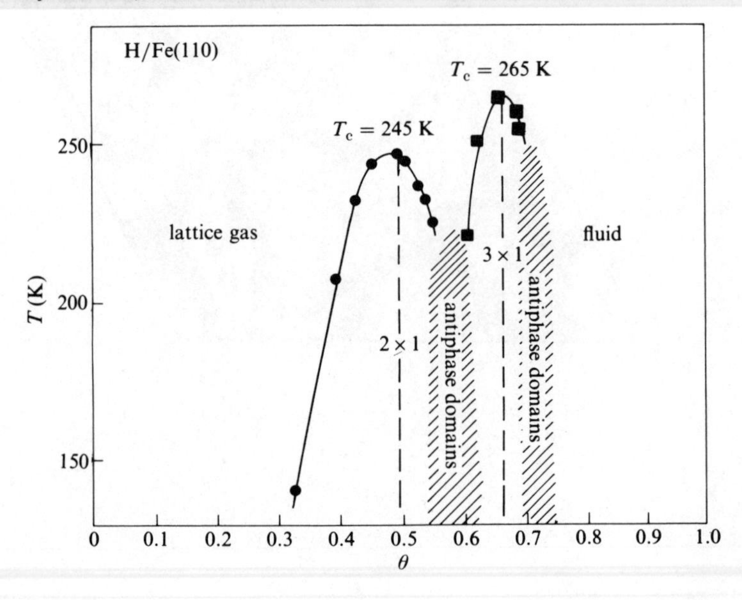

CS and IS phases in Fig. 11.7(*b*).

Very near the C–I transition it is possible to analyze the thermodynamics of the situation qualitatively entirely in terms of the walls (Villain, 1980). More precisely, the internal energy contains a positive term proportional to the total wall length (essentially a surface tension) and a term proportional to the number of wall intersections. If this 'crossing energy' is negative, the system will spontaneously fill the system with walls – the transition is first order. If the crossing energy is negative, the C–I transition ought to proceed in a gradual continuous manner with power-law critical behavior of the order parameter as a function of the chemical potential (pressure). In the present case, the appropriate order parameter is the inverse separation between domain walls. Equivalently, and more conveniently for diffraction experiments, the order parameter is proportional to the mean deviation of the overlayer lattice constant from its value in the commensurate phase. LEED measurements of this quantity for krypton

Fig. 11.15. An anti-phase domain structure formed from adjacent occupancy of the two equivalent adsorption sublattices of an ordered 2×1 overlayer.

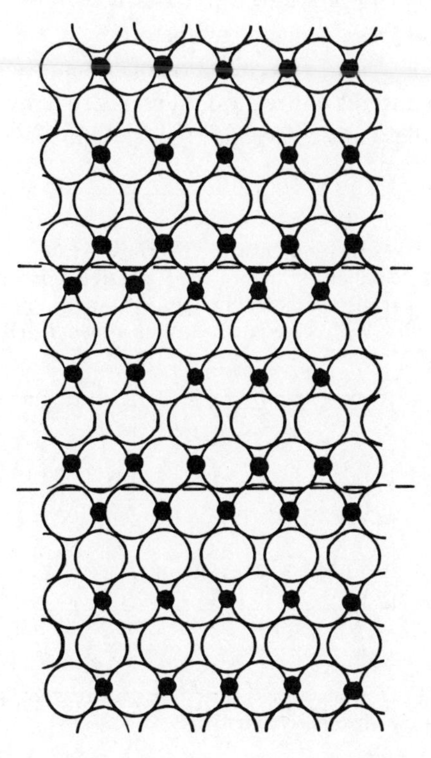

adsorbed on graphite at 57 K are shown in Fig. 11.13. At this temperature at least, the data indicate a continuous transition with a critical exponent $\beta \cong 1/3$.

The domain wall concept now permits us to be more precise about the nature of the topological defects that unbind in the Kosterlitz–Thouless melting transition of an incommensurate overlayer (Chapter 5). We do so by turning to a chemisorption system: H/Fe(110). We will have more to say about phase transitions in such systems presently. Here we merely display the experimental phase diagram (Fig. 11.14). First, one finds regions where the adsorbed hydrogen orders into 2×1 and 3×1 commensurate structures. In addition, however, there are regions of so-called 'anti-phase domains'.* An example of an anti-phase domain structure is shown in Fig. 11.15. It consists of three regions of 2×1 adatom arrangement, each one of which occupies one of the two equivalent 2×1 adsorption sub-lattices. The average adatom density of this defected structure is greater than the pure phase. The domain walls (dashed lines) here are analogous to the domain walls that form in the physisorption case. It is perfectly consistent to refer to Fig. 11.15 as a weakly incommensurate 2×1 phase of the H/Fe(110) chemisorption system.

In general, an overlayer that orders into a $p \times 1$ structure possesses p equivalent adsorption sublattices. At zero temperature, a weakly incommensurate phase can be described as a collection of domains separated by parallel walls perpendicular to the p-direction. At non-zero temperature, thermal fluctuations cause the walls to wiggle (Fig. 11.16(*a*)). Moreover,

Fig. 11.16. Walls in a $p \times 1$ incommensurate overlayer structure. Numbers label regions occupied by one of the p possible adsorption sublattices: (*a*) roughly parallel walls at low temperature; (*b*) bound pairs of dislocations below T_m; (*c*) free dislocations above T_m (Bak, 1984).

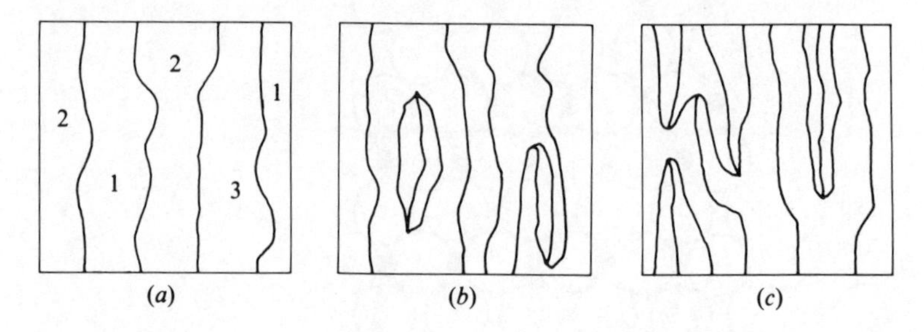

* The name comes from LEED. A phase shift enters the scattering amplitude for electrons that originate from adjacent, symmetry-related domains.

with finite energy cost, the system can support grain boundaries or bound pairs of topological dislocations (Fig. 11.16(*b*)). Such dislocations occur at the junction of p domain walls. This is so because a single wall introduces a phase shift of $2\pi/p$ with respect to atomic positions in the p-direction and a complete circuit around the dislocation must yield 2π, cf. Fig. 5.8. As we have seen, melting of an incommensurate phase can occur via the unbinding of these pairs into free dislocations that disorder the solid (Fig. 11.16(*c*)). Topological domain wall dislocations appropriate to a triangular lattice can be observed in Fig. 11.12(*b*). At present, it is not known if the high temperature state of the anti-phase domains in H/Fe(110) (or any other chemisorption system) constitutes a domain wall liquid.

Chemisorption systems

The phase diagram and critical behavior of a typical chemisorbed species is considerably more complicated to obtain and interpret than that of a physisorption system. The additional complexity arises principally from the fact that chemisorption (by definition) involves strong bonding to the substrate. From the experimental point of view, the existence of a deep adsorption well means that the equilibrium vapor pressure above the sample is extremely tiny. Effectively, the chemical potential is not an adjustable parameter so that one varies the coverage by varying the dosage, i.e., the amount of gas exposed to the sample. Experimenters typically quote only the *relative* coverage as determined by, for example, Auger spectroscopy. High temperatures must be avoided to retard diffusion into the bulk. Perhaps worst, one has to cope with the fact that both limited substrate quality and LEED* instrumental restrictions introduce undesirable finite-size effects into the diffraction process that cannot be ignored (Lagally, 1982). Nevertheless, a significant body of experimental results is available. A survey of these data reveals an important point.

Unlike the physisorption examples, chemisorbed species typically exhibit more than one ordered structure in their phase diagram. We have seen this already for hydrogen (Figs. 10.4 and 11.14) but the phenomena persist for much larger adsorbates as well (Fig. 11.17). This richness directly reflects the effect of complex interactions among the adatoms. To see this, return to the lattice gas model introduced earlier (11.2). For present purposes it is simpler to work with the equivalent Ising spin model with pairwise interactions (11.5). Notice that this expression is invariant under the substitution $H \to -H$ and $\{S_i\} \to \{-S_i\}$. Since $\{S_i\} \to \{-S_i\}$ means that

* Signal-to-noise problems have so far restricted the use of surface x-ray scattering for this purpose. However, only *kinematic* LEED theory (Chapter 3) is needed since detailed atom positions are not required.

$M \to -M$ we have $\theta \to 1-\theta$ as well (11.6). As a result, the phase diagram is *symmetric* around $\theta = 1/2$ – a fact that is certainly *not* true of the experimental phase diagrams seen up till now. We conclude that a lattice gas model can only reproduce realistic phase diagrams if *many-body* interactions are added to the model. The simplest such choice is called a 'trio interaction' whereby one adds a term to (11.2) of the form

$$\mathscr{H}_t = -\sum_{i \neq j \neq k} \phi_t c_i c_j c_k. \tag{11.10}$$

This term breaks the up/down spin symmetry of the Ising model and thus admits non-symmetrical phase diagrams.

The foregoing analysis demonstrates that the interplay between theory and experiment in the study of phase transitions in chemisorbed layers addresses two rather distinct pieces of physics. On the one hand, one hopes to learn about the detailed interactions among absorbates (ϕ_{ij}, ϕ_t, etc.) by fitting Monte Carlo lattice gas simulations to experiment. On the other hand, critical phenomena are sensitive to only the *range* of such interactions so one can test the universality hypothesis in the chemisorption context as well. We pursue both avenues below. But first, it is useful to have some feeling for what to expect. Luckily, there is quite a bit of independent information about the nature of inter-adsorbate forces.

The Coulomb interaction accounts for the majority of the interaction energy among *ionically* bonded chemisorbed species. Fig. 9.11 clearly shows that charge transfer bonding can lead to a substantial dipole moment on a highly electropositive or electronegative adsorbate. This is true, for example, for the strontium case of Fig. 11.17. Although the magnitude of this dipole decreases as the coverage increases (see Chapter 12), long range (R^{-3}) repulsive dipole–dipole forces dominate the problem. Remarkably, one can prove that a one-dimensional version of this problem exhibits an *infinite* number of stable commensurate phases as a function of coverage (Bak & Bruinsma, 1982)! The corresponding phase diagram in two dimensions is unknown but undoubtedly complex.

A second contribution to the mutual interaction energy of an overlayer arises from elastic distortions of the substrate. This energy is positive so the effective interaction between two adsorbates is repulsive. To see this, suppose that two separated adsorbates each exert a net attractive (repulsive) force on neighboring substrate atoms. The total energy rises because each adatom must do positive work to overcome the expansion (compression) of the deformable substrate lattice induced by the other. Detailed calculations show that the interaction energy varies as the inverse cube of the adatom separation although the absolute magnitude of the effect depends on the Poisson ratio and shear modulus of the substrate (Lau & Kohn, 1977).

The elastic interaction just discussed is mediated by the phonons of the substrate; it is an indirect interaction. There exists another type of indirect adsorbate interaction that is significant for covalent chemisorption on metals where the coupling is effected by the electrons of the substrate. The physics is best understood by use of the resonant level model of chemisorption (Chapter 9). Consider the following process (Fig. 11.18). An electron hops off one of the adsorbates and enters the metal at the Fermi level. Simultaneously, an electron from below E_F hops out of the metal onto the adsorbate. This creates an electron–hole pair in the metal. The pair propagates through the metal until it encounters a neighboring adsorbate. The hopping process occurs again at the second adatom, but this time in reverse order. Clearly, this establishes some communication between the adsorbates. However, the existence of a Fermi cutoff in momentum space means that not all wave vectors contribute to the process. As a result, the interaction potential has Friedel oscillations (cf. the discussion of Fig. 4.2 and Einstein (1978)).

Fig. 11.17. Phase diagram for Sr/W(110). The incommensurate solid phase at high coverage has hexagonal symmetry. Also, several more large unit cell commensurate phases appear at very low coverage (Kanash, Naumovets & Fedorus, 1975).

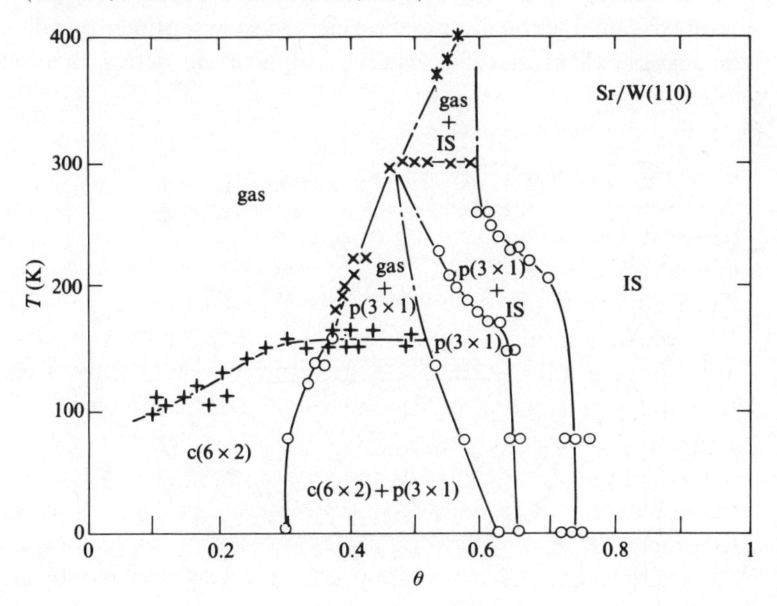

$$U(R) \sim \frac{\cos(2k_F R)}{R^5}. \tag{11.11}$$

This form is reminiscent of the RKKY interaction between dilute impurities in a metal (Ziman, 1972). In that case, the interaction decays as R^{-3}. Roughly speaking, the surface interaction falls off more rapidly because fewer wave functions have finite amplitude near the surface. There are fewer open channels of communication.

The indirect electronic interaction between adsorbates has been observed by a clever application of field ion microscopy (Chapter 3). The idea is the following. Imagine a *gedankenexperiment* wherein one randomly throws two non-interacting atoms down onto a well-characterized surface and records the distance between them. After many such trials we may construct a distribution function for these spacings; call it $p_0(R)$. For a real system, interactions between the adatoms lead to a different equilibrium distribution function, $p_{exp}(R)$. It is this latter quantity that can be obtained from FIM images. The pair interaction energy then follows from

$$\frac{p_{exp}(R)}{p_0(R)} = \kappa e^{-U_{exp}(R)/kT}. \tag{11.12}$$

Fig. 11.19 illustrates $U(R)$ extracted in this way for W and Ir atoms on a W(110) surface. Note the order of magnitude of the interaction energy.

No experiments directly provide information about many-body forces like the trio interaction discussed earlier. Quasi-first principles calculations are possible (Muscat, 1984), but it is difficult to assess their validity.

Fig. 11.18. Schematic view of the indirect electronic interaction between two adsorbates in the resonant level model.

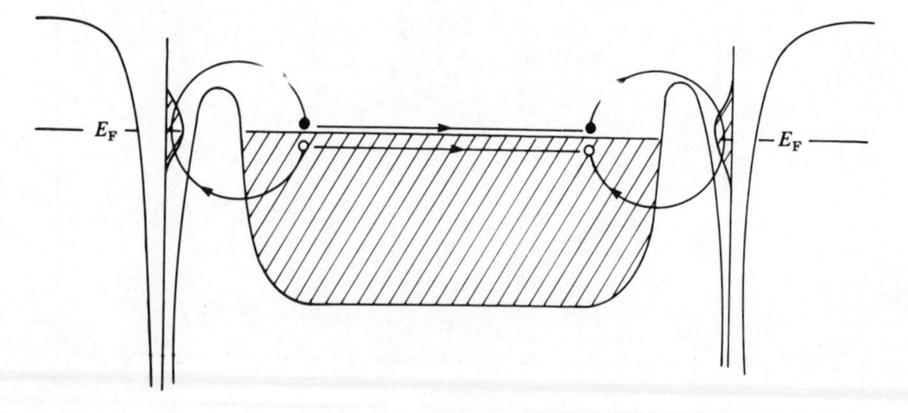

Accordingly, we adopt a pragmatic approach by reverting to the Monte Carlo lattice gas model. The question becomes: what characteristic values for the pair and trio interactions are needed to reproduce experiment?

Extensive lattice gas studies exist for H/Fe(110) assuming that hydrogen adsorbs onto a bridge site. The Hamiltonian is (11.2) including nearest, next-nearest and third-nearest neighbor pair interactions and a trio interaction (11.10) (Fig. 11.20(*a*)). The gross topology of the experimental phase diagram (Fig. 11.14) constrains the search to a relatively small part of parameter space. The best fit (Fig. 11.20(*b*)) uses the values $\phi_{nn} = -13$ meV, $\phi_{2n} = -83$ meV, $\phi_{3n} = -22$ meV and $\phi_t = +35$ meV. Notice that all the pair interactions are *repulsive* here. Moreover, the trio interaction is not strong enough to produce net attraction. Then, in accordance with expectation, first-order condensation and phase coexistence do not occur. Instead, the phase transitions are second order (continuous) in agreement with experiment.

This brings us back to critical phenomena and, in particular, the measurement of critical exponents. This requires a considerable amount of delicacy due to the aforementioned limitations of LEED. A particularly

Fig. 11.19. Distance dependence of the pair interaction energy between a tungsten atom and an iridium atom on a W(110) surface (Tsong & Casanova, 1981).

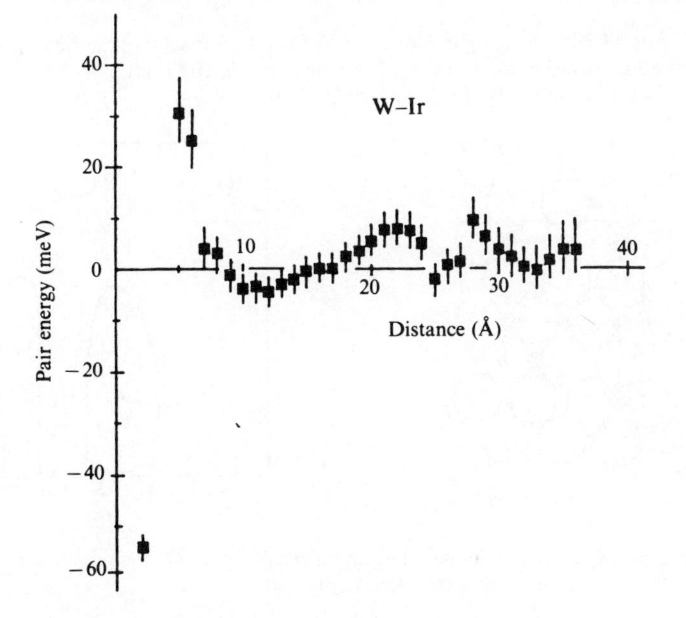

carefully studied system is W(112) p(2 × 1)-O.* The phase diagram is quite simple. It looks just like Fig. 11.20(*b*) without the region of 3 × 1 order. A line of critical points separates the ordered region from the disordered region. In addition, a symmetry analysis suggests that the transition belongs to the Ising universality class. Experiments confirm this prediction in striking fashion.

Fig. 11.21 shows the temperature dependence of the Bragg peak intensity and full-width at half-maximum (FWHM) of the (1/2 0) superlattice LEED beam for $\theta = 0.5$ monolayer of oxygen. This reflection corresponds to the shortest reciprocal lattice vector that accompanies the one-dimensional doubling of the surface unit cell upon adsorption. The data contain *three* critical exponents. The long range order parameter exponent β comes directly from the temperature dependence of the (1/2 0) beam intensity just below T_c. Superlattice intensity persists above T_c due to the existence of diffuse scattering from critical fluctuations, i.e., short range order. This quantity diverges near T_c with a characteristic 'susceptibility' exponent γ. Finally, the temperature dependence of the FWHM above T_c reflects the critical behavior of the correlation length (Chapter 5) through an exponent ν. These three exponents and the value of T_c are treated as fitting parameters. The result is: $\beta = 0.13 \pm 0.01$, $\gamma = 1.79 \pm 0.14$ and $\nu = 1.09 \pm 0.11$. All three values agree well with the exact values for the two-dimensional Ising model: 1/8, 7/4 and 1, respectively.

Surface physics is a uniquely flexible laboratory for the study of critical

Fig. 11.20. Lattice gas model of H/Fe(110): (*a*) schematic view of the interactions retained in the Hamiltonian; (*b*) the final calculated phase diagram (Selke, Binder & Kinzel, 1983).

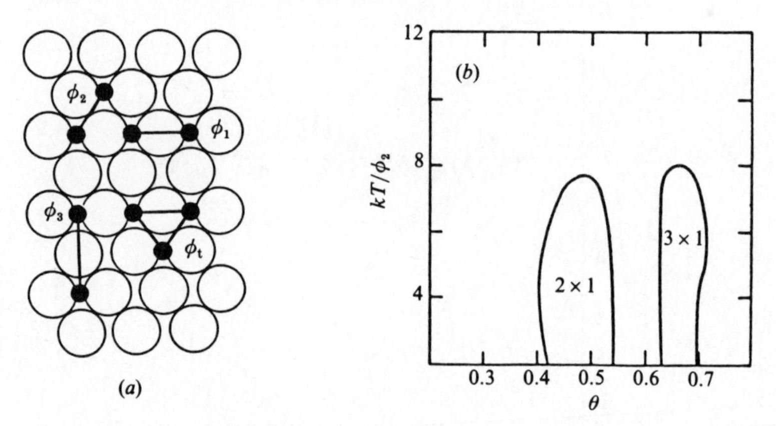

* One frequently sees this notation. The 'p' in p(2 × 1) stands for 'primitive' to denote the smallest unit cell of the correct symmetry.

phenomena because it is so easy to move between different universality classes and hence between vastly different models in statistical mechanics. For example, let us retain exactly the same chemical system discussed above – oxygen adsorbed on tungsten – but simply change the substrate to the (110) surface. According to Landau (Domany, Schick, Walker & Griffiths, 1978), adsorption on this surface of a BCC crystal also can support a continuous transition from a lattice gas to an ordered 2×1 structure. However, the relevant LGW Hamiltonian corresponds to a model where a spin variable can rotate freely in a two-dimensional plane (rather than just point in two or three directions as in the Ising or 3-state Potts models) with a slight preference for alignment along the orthogonal coordinate axes. The experimental phase diagram exhibits not only the continuous transition but a genuine first-order transition as well (Fig. 11.22).

Both experiment and theory agree that the O/W(110) system exhibits a so-called *multicritical point* at low coverage. This occurs when a line of continuous transitions meets a line of first-order transitions. It appears here at about 460 K because the continuous phase boundary between the $p(2 \times 1)$ structure and the lattice gas must hit the first-order coexistence

Fig. 11.21. Normalized peak intensity (open squares) and FWHM (closed circles) vs. temperature for the (1/2 0) LEED beam from W(112) $p(2 \times 1)$-O at half coverage (Wang & Lu, 1985).

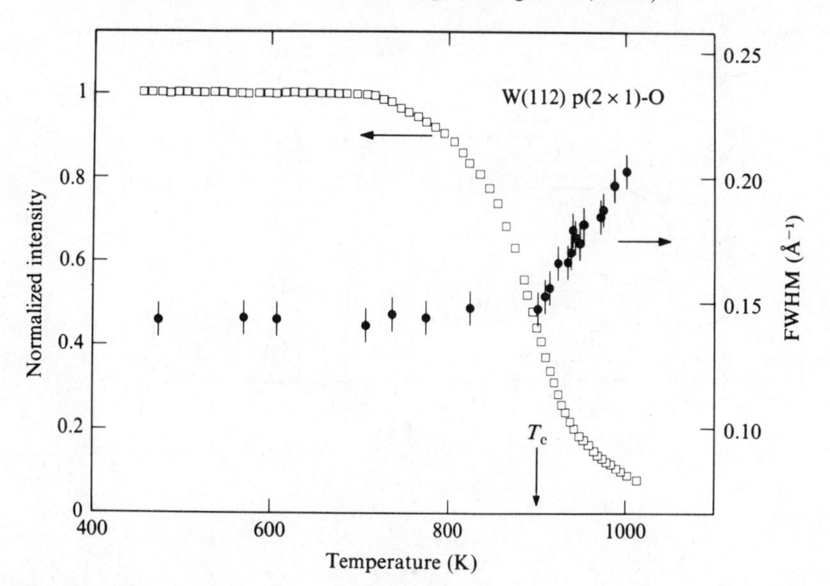

Fig. 11.22. Phase diagram for O/W(110): (*a*) experimental (LEED) phase boundaries (symbols) and proposed phase boundaries (solid curves (Lagally, Lu & Wang, 1980); (*b*) results for a lattice gas model that includes attractive first and *fifth* neighbor interactions and repulsive second neighbor, third neighbor and trio interactions (Rikvold, 1985).

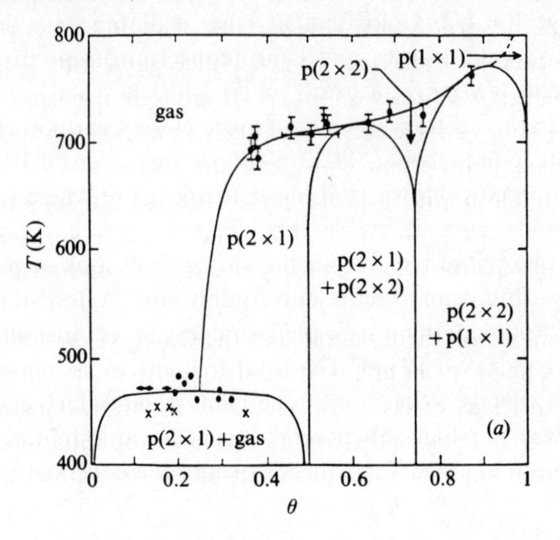

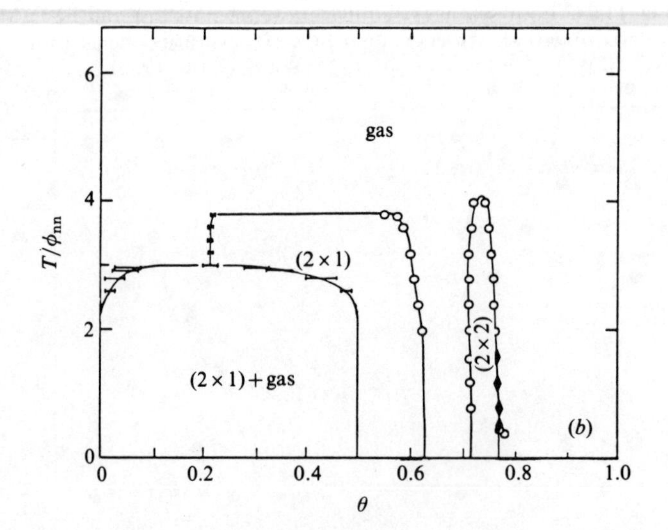

curve somewhere. The sensitivity of the lattice gas model is particularly clear in this case. An attractive fifth neighbor interaction is used which is only 10% as strong as the first neighbor interaction. However, if the fifth neighbor term is neglected, the coexistence region and the multicritical point disappear! Observe also that the two proposed phase diagrams disagree above $\theta = 1/2$. Unfortunately, the available data do not discriminate between first-order and continuous transitions involving the p(2 × 2) phase at low temperature.

We conclude this section by looking more closely at another aspect of the p(2 × 1) structures in Fig. 11.22. Suppose one quenched this system from high temperature down to a lower temperature where the ordered phase is thermodynamically stable. If this occurs at low coverage ($\theta < 0.2$) we enter the two-phase region and the surface will look something like Fig. 11.4. But this cannot represent equilibrium. A (one-dimensional) surface energy/unit length must be paid for every bit of circumference that bounds each condensed island. The total free energy is minimized only when the little islands coalesce into one giant island which coexists with the gas. The way in which this process occurs is a question of kinetics.

A similar question arises if the quench occurs at a coverage greater than

Fig. 11.23. Four symmetry-equivalent realizations of a p(2 × 1) ordered overlayer structure on a BCC(110) surface.

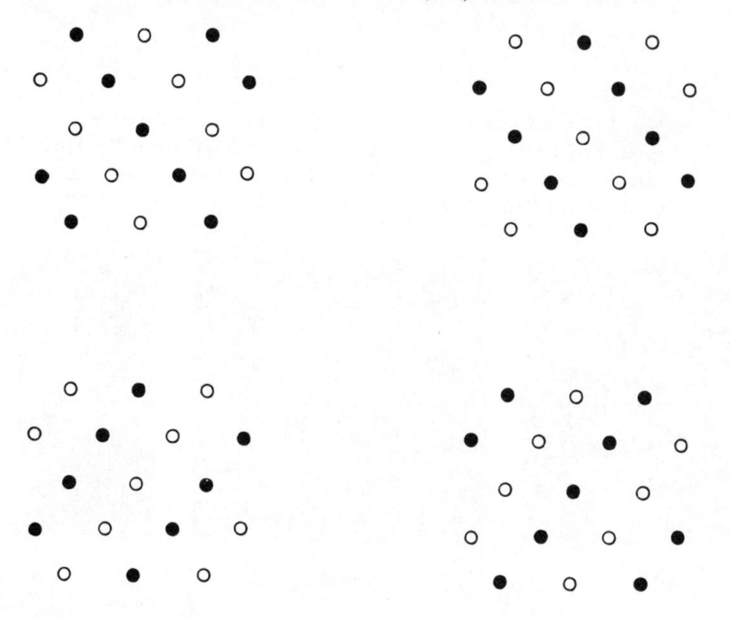

that of the multicritical point. In that case, the system is supposed to convert completely to the ordered phase at once. But there is a problem. If there is one adsorption site per unit cell, the initial quench will produce numerous small islands of each of the four equivalent p(2 × 1) arrangements: two translational and two rotational anti-phase domains (Fig. 11.23). Every boundary between any two domains costs energy; ultimately, the system must get rid of them. The Monte Carlo method provides a means to visualize this process.

Fig. 11.24 presents four 'snapshots' in the time-evolution of the adsorbate arrangement of a quenched lattice gas system that undergoes a continuous order–disorder transition into a p(2 × 1) ordered state. The coverage is near saturation and the four different symbols denote atoms that belong to each of the four domains. No symbols are drawn for atoms that belong to walls which separate domains. The system clearly rearranges in a manner designed to rid itself of domain walls. The simulation data can be quantified if one defines an average domain size $L(t)$. This quantity ought to increase as time proceeds. Indeed, after long times, the linear dimension $L(t)$ is much larger than any microscopic length scale. This suggests an analogy with critical phenomena where the correlation length similarly exceeds all microscopic lengths near T_c. If this is correct, we guess that $L(t) \propto t^x$ where x is a 'universal' exponent. For the system under consideration, the Monte Carlo data fit this form well (at long times) with an exponent $x \cong 1/3$.

Experiments on the real O/W(110) system confirm the 'kinetic critical

Fig. 11.24. Monte Carlo snapshots of the time evolution of a quenched lattice gas with the symmetry of W(110) p(2 × 1)-O. The 'times' listed are not simply related to laboratory time. See text for discussion (Sadiq & Binder, 1984).

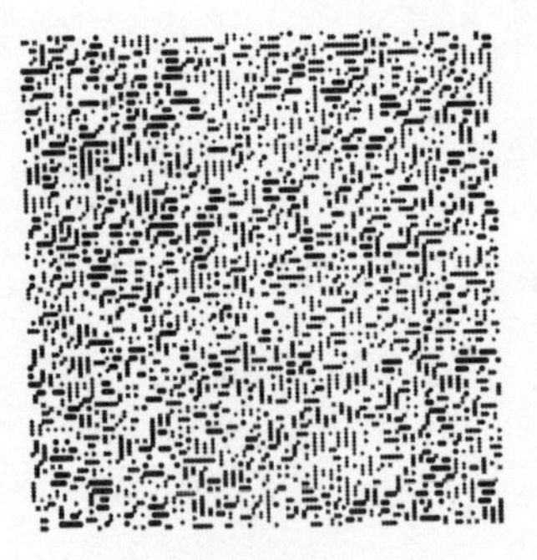

phenomena' observed in the lattice gas simulations. Moreover, they set the time scale for the phenomena. The appropriate (LEED) measurement simply monitors the time dependence of the intensity of an overlayer superlattice reflection after a quench into the one-phase region. Fig. 11.25 results if a reasonable assumption is made about the relationship between the observed intensity and the average domain size. One finds a 1/3 power growth law for three independent quenches from the disordered phase – each to a different low temperature.

Fig. 11.25. Plot of the cube of the average linear dimension of a p(2 × 1) domain of O/W(110) vs. time extracted from LEED superlattice intensity data. Three trials, all at saturation coverage, are shown (Tringides, Wu, Moritz & Lagally, 1986).

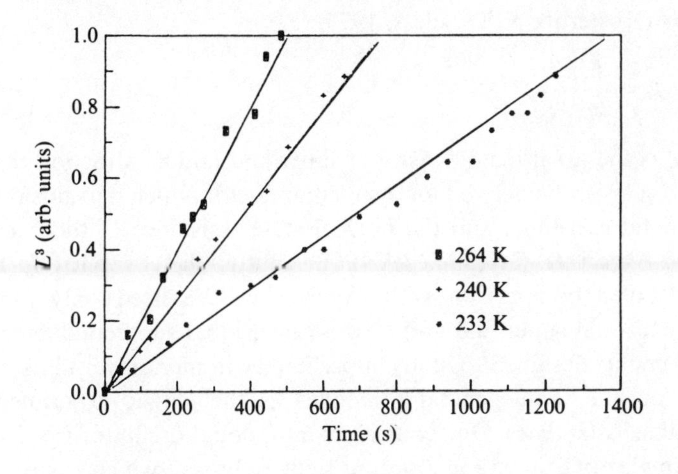

Superfluidity and superconductivity

The condensed phase that characterizes a superfluid or a superconductor (macroscopic numbers of ions or Cooper pairs in a single quantum state) appears to have little in common with the structural phases discussed so far in this chapter (March & Parrinello, 1985). However, in two dimensions, one can imagine a mechanism for the *destruction* of these ordered states that is quite familiar by now: the Kosterlitz–Thouless (KT) process. The trick is to identify an appropriate topological defect for each system. The energy of a single such defect should diverge logarithmically with the size of the system. This prevents the existence of free defects at low temperature. However, the energy of a bound *pair* of defects should be finite so that an equilibrium population of pairs is present at any temperature. Then, if the defects are localized, the entropy also is proportional to the logarithm of the system size (cf. Chapter 5 on the KT melting of the Wigner lattice) and the system is unstable to disordering via defect pair unbinding at a temperature T_m.

The requisite excitations for both a superfluid and a superconductor are well known and studied. They are called *vortices* and, in both cases, involve quantized circulation of some kind. In quantum liquids, the vortex is a circulating flow of superfluid around a core of normal fluid. The flow velocity is restricted to take on discrete values (Feynman, 1972). In a superconductor, the vortex is a filament of magnetic flux that penetrates the sample and is screened by a circulating flow of supercurrent. The flux enclosed by the flow is quantized in units of $hc/2e$ (Ashcroft & Mermin, 1976). A straightforward application of the theory leads to explicit predictions for the transition temperature. For example, for superfluid ^{4}He one finds (Kosterlitz & Thouless, 1973)

$$T_c = \frac{\hbar^2 \pi}{2m^2 k} n_s, \tag{11.13}$$

where n_s is the areal mass density of superfluid and m is the ion mass. A similar result can be derived for a superconductor which depends only on fundamental constants and the normal state resistance of the overlayer (Beasley, Mooij & Orlando, 1979). Evidently, these explicit formulae invite detailed comparison with experiment. We need only locate a two-dimensional superfluid and a two-dimensional superconductor.

Experiments designed to study superfluidity in monolayer films of ^{4}He operate on the same general principles as the classic experiments of Andronikashvili (1946). One constructs a torsional oscillator from a long strip of mylar plastic wrapped into a tight 'jelly-roll' which is suspended from a rigid rod. The period of the oscillator depends on the moment of inertia of the mylar roll. Normal state helium adsorbed onto the mylar within the roll is dragged along and contributes to the moment of inertia. However, in the superfluid state, the liquid decouples from the substrate. The mass density of the superfluid is calculable directly from the change in period of the oscillator. Fig. 11.26 illustrates the relationship between the measured superfluid mass density and transition temperature for submonolayer coverage of helium. The solid line is the prediction of (11.13). We conclude that the unbinding of vortex–antivortex pairs is indeed the dominant mechanism for the destruction of superfluidity in helium adlayers.

An interesting set of experimental and theoretical considerations apply to any proposed study of superconductivity in two dimensions. On the one hand, it is not necessary to look to the properties of a single adsorbed monolayer. The characteristic length scale in the condensed state is the coherence length, i.e., the spatial separation between electrons in a Cooper pair, which is of the order of 1000 Å. Therefore, metal films with thickness less than, say, 100 Å (25–30 atomic layers) will be effectively two-dimensional from the point of view of superconductivity. On the other hand, it has proven very difficult to prepare crystalline metal films of this thickness on non-conducting substrates. Deposited metal typically does not adsorb onto such surfaces in a regular layer-by-layer manner. Instead, the metal clumps into three-dimensional islands that coalesce into amorphous films. This is a well-known problem in epitaxy (Chapter 16). As a result, all experiments to date work with rather 'dirty' disordered metal overlayers whose resistivity in the normal state is at least an order of magnitude greater than typical 'good metal' values. Nevertheless, this does not mean that no physics emerges from these studies.

Fig. 11.27 shows the variation of superconducting transition temperature with sample thickness (t) for a sequence of amorphous Mo–Ge alloy films deposited on amorphous Ge. Also plotted is the film resistance (usually called the resistance per square, $R_{\square}$) which is related to the resistivity by $R_{\square} = \rho/t$. We see that T_c is suppressed (relative to its bulk value) by much

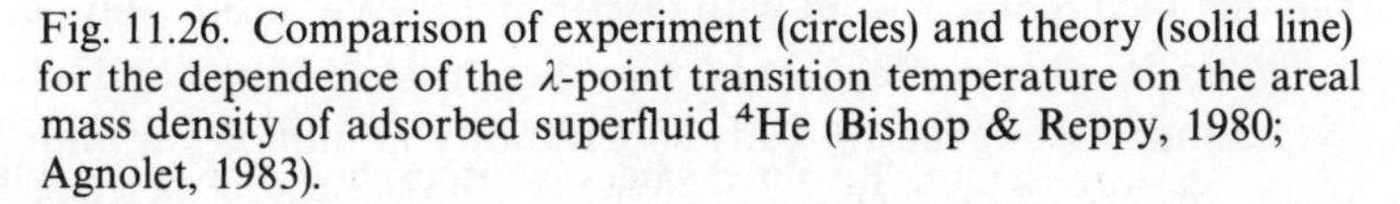

Fig. 11.26. Comparison of experiment (circles) and theory (solid line) for the dependence of the λ-point transition temperature on the areal mass density of adsorbed superfluid ^{4}He (Bishop & Reppy, 1980; Agnolet, 1983).

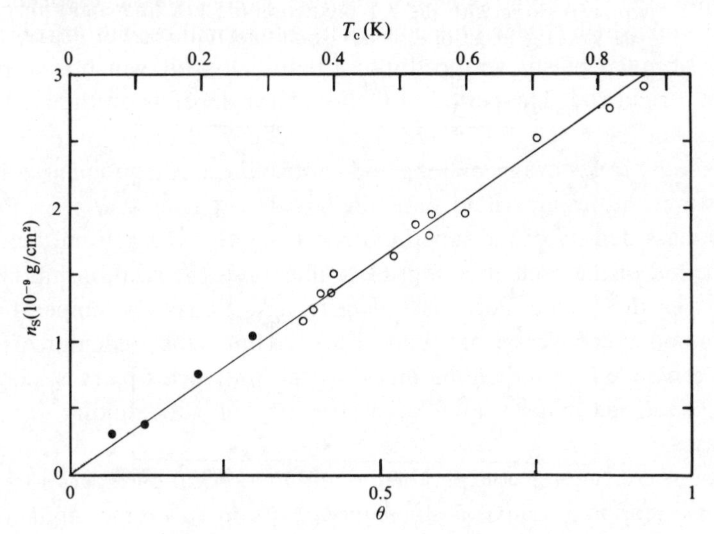

more than KT theory predicts (dashed line). Evidently, some other mechanism preempts the vortex-pair unbinding mechanism. It turns out that the principal effect is precisely the disorder in the film. To understand this crudely, recall that electrons in a disordered metal suffer collisions with impurities and defects which considerably impede their motion through the material. This effect degrades the ability of the metal to screen the Coulomb interaction. The collective state vanishes when the mutual Coulomb repulsion within Cooper pairs exceeds the phonon-mediated attraction that leads to superconductivity in the first place. The solid line in Fig. 11.27 comes from a proper theory of this interplay between disorder and Coulomb interactions in two dimensions (Maekawa & Fukuyama, 1981).

Recent developments in epitaxial growth technique (Chapter 16) permit us to carry this investigation one further step. With proper care, a crystalline monolayer of metal can be made to chemisorb onto a semiconductor substrate. The Ag/Ge(100) system is an example (Burns, Lince, Williams & Chaikin, 1984). As it happens, modern theories (Lee & Ramakrishnan, 1985) suggest that this structural breakthrough is of little help as regards electron transport. In two dimensions, one cannot escape the effects of even the smallest amount of disorder. Nonetheless, this example is interesting. The silver film appears to undergo a superconductive transition at about 1.6 K even though bulk silver itself does not superconduct!

Fig. 11.27. Variation of T_c (relative to its bulk value) as a function of sheet resistance. Measured values for amorphous $Mo_{79}Ge_{21}$ thin films (squares) along with the KT prediction (dashed line) and a theory based on the effect of disorder (solid line) (Graybeal & Beasley, 1984).

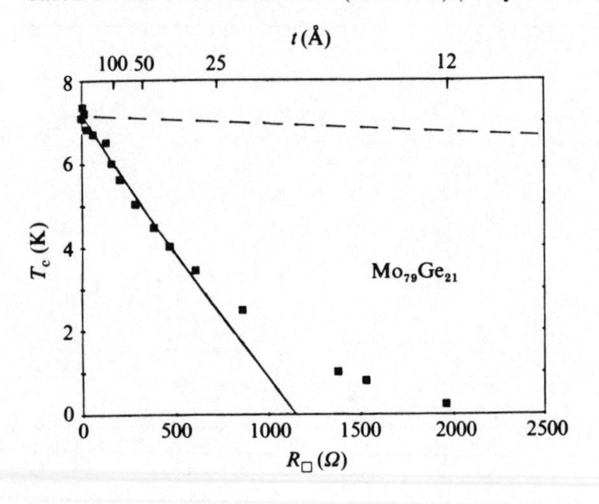

General references

Phase equilibrium

Dash, J.G. (1975). *Films on Solid Surfaces.* New York: Academic.

Physisorption systems

Villain, J. & Gordon, M.B. (1983). Static Properties of Physisorbed Monolayers in Their Solid Phases. *Surface Science* **125**, 1–50.

Commensurate–incommensurate transitions

Pokrovsky, V.L. & Talapov, A.L. (1984). *Theory of Incommensurate Crystals. London: Harwood.*

Chemisorption systems

Roelofs, L.D. & Estrup, P.J. (1983). Two-Dimensional Phases in Chemisorption Systems. *Surface Science* **125**, 51–73.

12

ELECTRONIC STRUCTURE

Introduction

Some of the most venerable work in the history of surface science directly concerns the effect of foreign gas adsorption on the electronic properties of a solid surface. Or, more precisely, the effect of adsorption on the electron emissivity characteristics of a clean metal surface. Long ago, Kingdom & Langmuir (1923) discovered that the current of electrons evaporated from a hot tungsten filament greatly increased after the filament was exposed to various metal vapors. The basic effect, known as *thermionic emission*, involves the thermal activation of electrons over the surface energy barrier (Fig. 4.3) and into the vacuum. This establishes the sample work function ϕ as the relevant material property – a fact quantified by the famous Richardson–Dushman equation (Kittel, 1966):

$$J = \frac{4\pi me}{h^3}(kT)^2\,\mathrm{e}^{-\phi/kT}, \tag{12.1}$$

for the emission current density J. Evidently, the greatest emission occurs for adsorbates that maximally decrease the substrate work function. The early experiments identified cesium and thorium as particularly effective in this regard.

Today, the search for more efficient thermionic emitters focuses more on high power microwave tubes than on the applications Langmuir had in mind (Tuck, 1983). Nevertheless, the original Cs/W system still represents one of the most dramatic examples of the work function lowering phenomenon (Fig. 12.1). We hinted at the underlying mechanism earlier in connection with dipole moment formation in adsorbed xenon layers (Fig. 8.12) and were more explicit for the example of Li/jellium (Fig. 9.11). In the simple resonant level picture, the 6s electron of highly electropositive cesium hops into the metal and remains in the immediate vicinity to screen the adsorbed ion. The large induced dipole moment opposes the clean substrate charge spill-out dipole (Fig. 4.2) with a concomitant decrease in the work function.

But there is more to it than that. The suppression in ϕ saturates and actually reverses as cesium adsorption approaches monolayer coverage. We can understand this simply as follows. Initially, each adsorbate contributes individually to the work function change; the decrease is linear in coverage. Eventually, however, the dipole fields from neighboring atoms begin to feel one another (Fig. 12.2). In particular, each dipole *depolarizes* its neighbors. The adatoms slowly neutralize one another even as the nominal dipole moment increases with coverage.

It is easy to build this effect into the resonant level model. Consider the interaction energy of a collection of parallel dipoles arranged in a planar array:

$$U_{\text{dip-dip}} = \sum_{i,j} \frac{p^2}{|r_i - r_j|^3} = \kappa \frac{p^2}{a^3}. \tag{12.2}$$

Fig. 12.1. Measured work function of several single-crystal surfaces of tungsten as a function of cesium coverage (Kiejna & Wojciechowski, 1981).

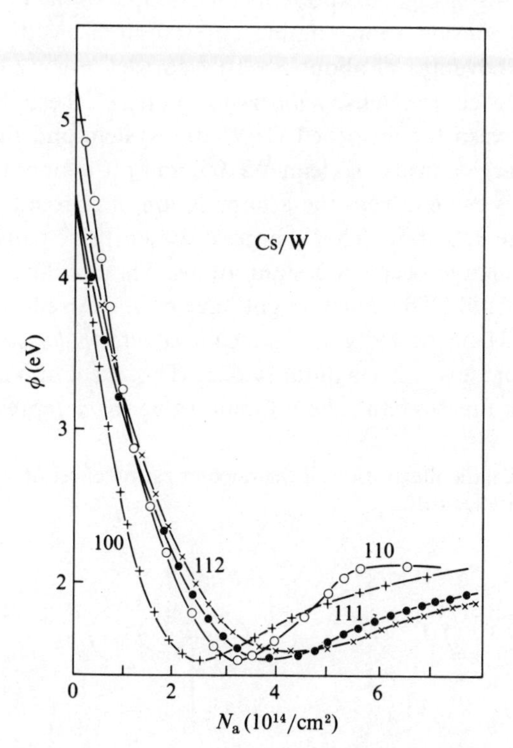

Here, p is the dipole moment, a is the lattice constant and κ is a numerical constant that depends on the adsorption geometry. Then, if N_a denotes the adsorbate coverage/unit area, we readily modify an old result (9.12) to

$$\varepsilon_a'(N_a) = \varepsilon_a + \frac{e^2}{4z} - \kappa p^2 N_a^{2/3}. \tag{12.3}$$

As the number of adatoms increases, the zero-coverage resonant level position (by definition, well above the Fermi level for this case) drops in energy and begins to cross E_F. Electrons flow back to reneutralize the adsorbed ion. The finite width of the resonant level prevents the process from proceeding to completion. Nevertheless, the induced dipole moment diminishes and a more covalent bond replaces the original strongly ionic chemisorption bond.

The scenario sketched above is correct in broad outline. But the real world is rather more complicated than the resonant level model permits. We learn this primarily from detailed electronic structure calculations of the sort first presented in Chapter 4. That is, results from a carefully parameterized tight-binding model, or better, full solutions to the Kohn–Sham equations (4.4) in the local density approximation (LDA) for a slab geometry (Fig. 4.16). Fig. 12.3 presents an example of the latter for the case under present consideration – a tungsten (100) surface with cesium atoms adsorbed to a coverage of about $5 \times 10^{14}/\text{cm}^2$ (cf. Fig. 12.1). The upper panel is again a charge density *difference* plot, i.e., the difference in charge density between the adsorbed Cs/W(100) system and the sum of the superposed charge density of clean W(100) and a Cs monolayer.

Covalent bonding is evident from the accumulation of charge between the adsorbate and the substrate. This charge is drawn principally from the 6s outer valence charge between cesium atoms. The bonding occurs to the substrate 5d orbitals. This much might have been inferred from the resonant level model. Unexpectedly, a significant *counter-polarization* of the cesium 5p near-core level also is quite evident. This effect opposes the reduction of the work function and the ultimate value for ϕ represents a

Fig. 12.2. Schematic illustration of the depolarization effect of neighboring dipolar fields.

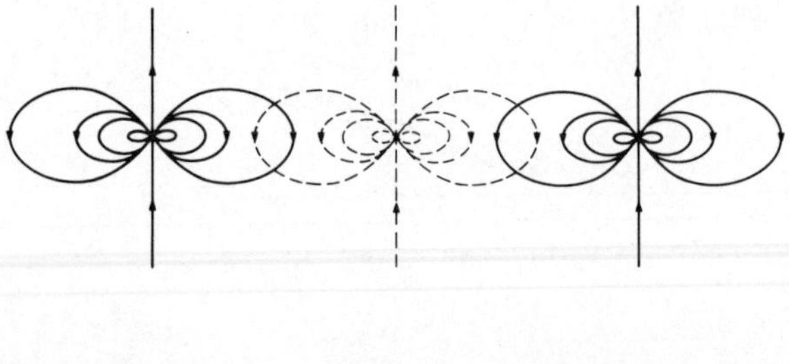

Fig. 12.3. Results from an LDA electronic structure calculation for a c(2 × 2) overlayer of Cs on either side of a five-layer slab of W(100): (*a*) charge density difference contour plot where solid (dashed) lines indicate a surfeit (depletion) of electronic charge; (*b*) difference in the Coulomb potential averaged in planes parallel to the surface (V(Cs/W) − V(clean Cs) − V(Cs monolayer)) (Wimmer, Freeman, Hiskes & Karo, 1983).

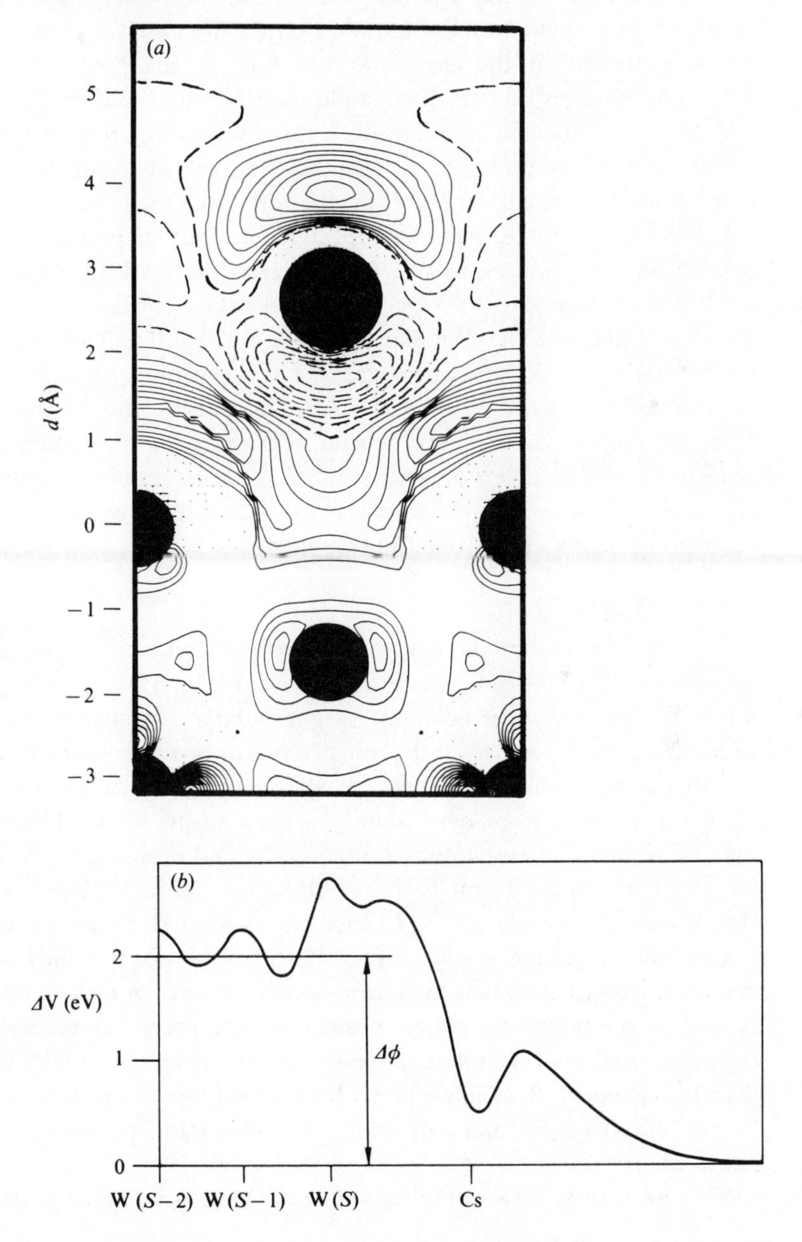

self-consistent compromise between 6s and 5p polarization. The Cs-induced charge rearrangement is complex indeed. An electron propagating toward the surface from deep within the bulk experiences considerable variations in kinetic energy near the surface before gaining exit from the crystal (Fig. 12.3(*b*)).

The remainder of this chapter extends and broadens the analytical approach just applied to the Cs/W example. We analyze experimental results pertinent to the electronic structure of adsorbed atoms and molecules in terms of both the simple models introduced in Chapter 9 and more sophisticated calculations. A three-way comparison is essential because it is important to know when the intuitively appealing (and easily applied) models are correct and when they err. Even then one may require the detailed calculations anyway. Many questions of practical import depend on the numbers, to wit: How large is the work function? How small is the energy barrier? Why does one adsorbate concentration passify a surface reaction while a similar one does not? etc. The basic idea is to build intuition about the changes wrought by adsorption. Already we have emphasized the crucial role played by crystallography. Indeed, the results from all theoretical methods depend sensitively on the geometry of adsorption. Hence, unless otherwise stated, the bond lengths, binding sites and orientations used in the calculations discussed below are chosen to conform as nearly as possible to experiment.

Metals

We begin our inquiry with adsorption onto a simple metal – aluminum. This is a case where the atom-on-jellium model (Chapter 9) might be expected to be adequate. A particularly illuminating sequence of comparisons is possible for the case of oxygen chemisorption on Al(111). For this case, of course, we expect the adsorbate to 'oxidize' the substrate, i.e., grab electrons from the conduction band of the metal. This means that the oxygen 2p resonant level ought to fill and drop below the Fermi level (cf. the chlorine example of Fig. 9.10). One sees just this behavior in the chemisorption-induced local density of states (LDOS) for oxygen on a semi-infinite jellium substrate (Fig. 12.4(*a*)). The positive background density is chosen to match the mean electron density of bulk aluminum. Note that the resonance energy position is in excellent agreement with the position of an oxygen-induced feature in the experimental UPS energy distribution curve. Recalling Fig. 9.9 we see that this energy is set largely by the effective potential created by the aluminum conduction band electrons.

We now turn to three completely different calculations that purport to

Fig. 12.4. Comparison between experiment and theory for adsorbate-induced features in the electron spectrum of O/Al(111): (*a*) EDC from ultraviolet photoemission spectroscopy (solid curve) and LDOS from O/Je ($r_s = 2$) (dashed curve) (Eberhardt & Himpsel, 1979; courtesy of N. Lang, IBM Watson Research Laboratory; (*b*) LDOS from an LDA finite cluster model (dotted curve), a tight-binding slab model (dashed curve) and a self-consistent LDA slab calculation (solid curve). All the theoretical results are normalized to the same integrated area as the experimental curve (Salahub, Roche & Messmer, 1978; Bullett, 1980; Wang, Freeman & Krakauer, 1981).

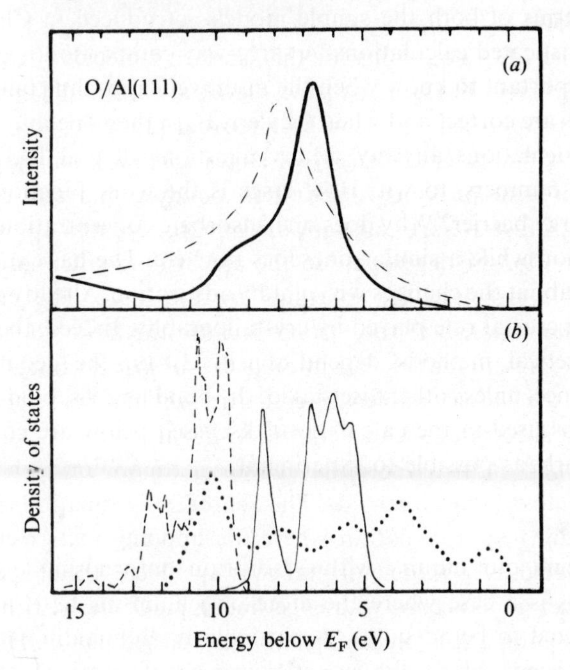

Fig. 12.5. The arrangement of atoms in a 19-atom cluster that locally models the (111) surface of an FCC substrate. Circles, triangles and squares denote atoms in the top, second and third layer, respectively.

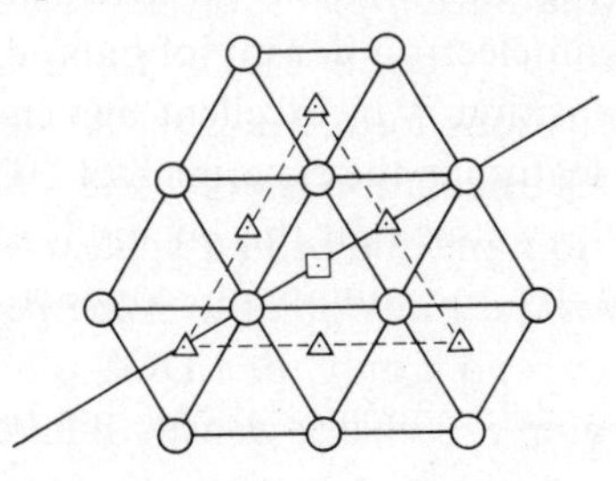

model the substrate more accurately than the jellium model, e.g., they include the aluminum ion cores. The first of these is designed to take advantage of the fact that the interaction of oxygen 2p orbitals with Al 3p orbitals should be quite local. The substrate is modelled by a small cluster of atoms (19 in this case) arranged to reflect the crystal geometry in the immediate neighborhood of a few Al(111) adsorption sites (Fig. 12.5). Note that there are two inequivalent three-fold hollow sites. Guided by experiment, we consider only the sites for which there is no aluminum atom in the second layer directly below the adsorbate.

Solution of the LDA equations for this 'surface molecule' yields a set of discrete energy levels. In Fig. 12.4(*b*), these levels are artificially broadened into an LDOS to represent the semi-infinite solid (dotted curve). We see immediately that the induced density of states spreads over more than 10 eV. The individual peaks correspond to bonding and anti-bonding combinations of substrate orbitals with oxygen $2p_x$, $2p_y$ and $2p_z$ orbitals. The latter are no longer degenerate in the anisotropic potential field of the substrate. Nevertheless, the splittings are too large. The bonding/anti-bonding interactions are too strong because the finite cluster does not adequately represent the delocalization of the *substrate* orbitals. Moral: in most cases, the surface molecule construct should be reserved for localized orbital substrates such as transition metals.

The next calculation uses a seven-layer slab of aluminium with a 1×1 overlayer of aluminium on either side. The Schrödinger equation is solved in the tight-binding approximation where the hopping matrix elements (4.21) are calculated from atomic orbital overlaps. Here, the shape of the LDOS is in rather good accord with experiment, particularly with regard to the splittings among the oxygen 2p orbitals (dashed curve). However, the energy of the oxygen levels is severely misplaced with respect to the Fermi level. The problem is the surface barrier and dipole layer which set the absolute energy scale for the problem. These are determined by self-consistent charge transfer and rearrangement effects which are substantial for highly electronegative adsorbates but completely neglected here. Moral: conventional tight-binding calculations are applied most confidently to covalent bonding situations.

Finally, Fig. 12.4(*b*) also shows the oxygen LDOS for a fully self-consistent solution to the LDA equations for a five-layer Al slab with a 1×1 oxygen layer on each side. This calculation is expensive – but it gets the answer right! The energy position and the orbital splittings are well reproduced. The question arises: was there any point to going beyond the atom-on-jellium model? The answer is no if only the LDOS is of interest. But there is more to learn. For example, a charge density difference plot

(Fig. 12.6) very nicely demonstrates how oxygen atoms attract electrons from aluminium interstitial positions to fill out their 2p charge clouds.

The power of the sophisticated slab calculation becomes clearer when one directs attention to the details of the adsorbate–substrate and adsorbate–adsorbate interactions. To this end, Fig. 12.7 shows the principal oxygen-derived states (at the SBZ center) in three energy windows that span the induced LDOS of O/Al(111). In the central panel (-6 eV to -8 eV) one finds primarily *non-bonding* oxygen $2p_x$ and $2p_y$ orbitals whose charge density is confined entirely to the immediate vicinity of the oxygen atom. By contrast, bonding and anti-bonding combinations of oxygen $2p_z$ and substrate 4s and 4p orbitals appear at the bottom and top of the LDOS, respectively. These states spread charge throughout the surface region. As one moves away from the zone center, the wave functions change character and charge builds up in the region between adjacent oxygen atoms. In other words, the adsorbate states form two-dimensional energy bands that disperse in energy as a function of $\mathbf{k}_{\parallel}$, the electron wave

Fig. 12.6. Charge density difference contour plot for Al(111) p(1 × 1)-O. The plane of the figure is perpendicular to the surface and passes through two inequivalent three-fold sites along the skew line in Fig. 12.5 (Wang, Freeman & Krakauer, 1981).

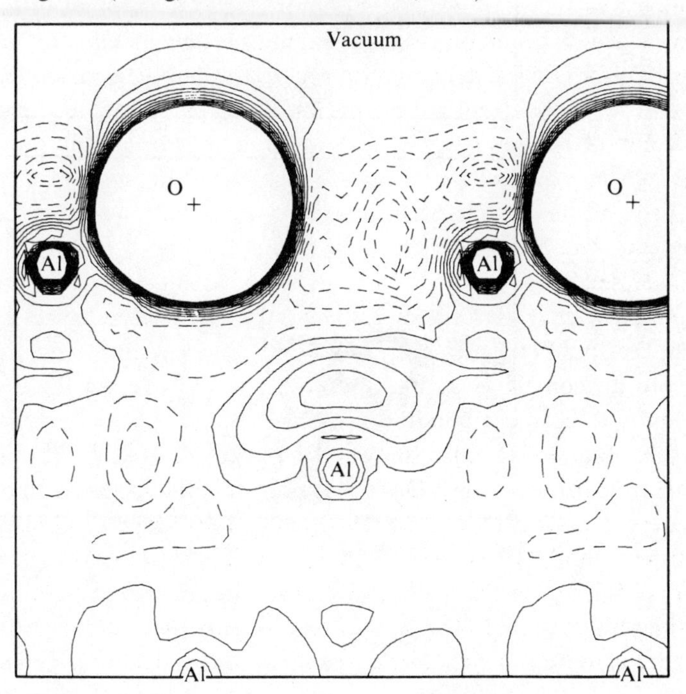

Fig. 12.7. Charge density contours for three oxygen-localized states at the Γ point in the surface Brillouin zone of O/Al(111): (*a*) bonding state at the bottom of the LDOS; (*b*) non-bonding oxygen state; (*c*) anti-bonding state at the top of the LDOS (Wang, Freeman & Krakauer, 1981).

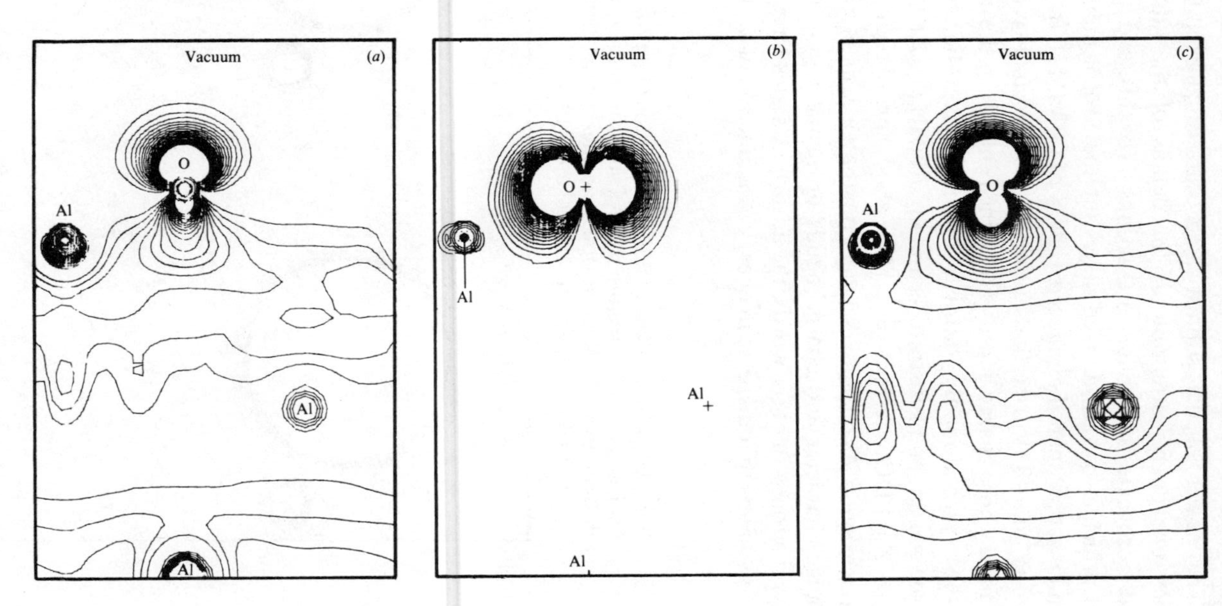

vector in the surface plane. Fig. 12.8 compares the calculated dispersion with the results of polarization-dependent angle-resolved UPS (Chapter 4).

As noted earlier, the surface molecule concept is most appropriate for adsorption on a transition metal. To investigate this claim, consider the case of a clean titanium surface exposed to hydrogen gas. At the outset, it is not obvious whether this gas adsorbs dissociatively or in molecular form. Unlike oxygen, where the large atomic heat of adsorption virtually guarantees dissociative adsorption (Fig. 9.4), the dissociation energy of H_2 is about 2.24 eV/atom while H atom chemisorption energies on metal substrates vary in the 2–3 eV range (Davenport & Estrup, 1987). To decide the issue for the (0001) surface of HCP titanium we can imagine using a slightly enlarged version of the cluster of Fig. 12.5 and compute the total energy of H_2 and H atoms at various sites and separations from the adsorbate complex. In this way (in principle) one could generate potential energy surfaces which are the three-dimensional analogs to Fig. 9.5.

The program sketched above can be pursued with the traditional methods of quantum chemistry (Whitten & Pakkanen, 1980) if the 'supermolecule' constructed above is a reasonable representation of reality. For small clusters, this is a viable alternative to the local density functional approach generally adopted in this book. The main idea is to represent the total wave function of the adsorbate/substrate cluster as a superposition of many Slater determinants built from Hartree–Fock (rather than Kohn–Sham) orbitals. The relative weighting for each determinant and

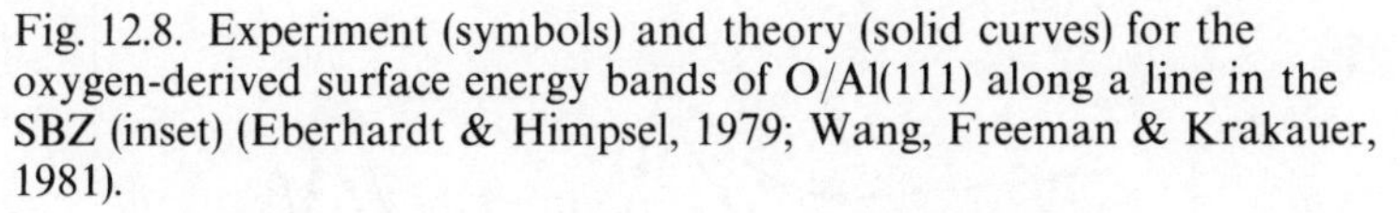

Fig. 12.8. Experiment (symbols) and theory (solid curves) for the oxygen-derived surface energy bands of O/Al(111) along a line in the SBZ (inset) (Eberhardt & Himpsel, 1979; Wang, Freeman & Krakauer, 1981).

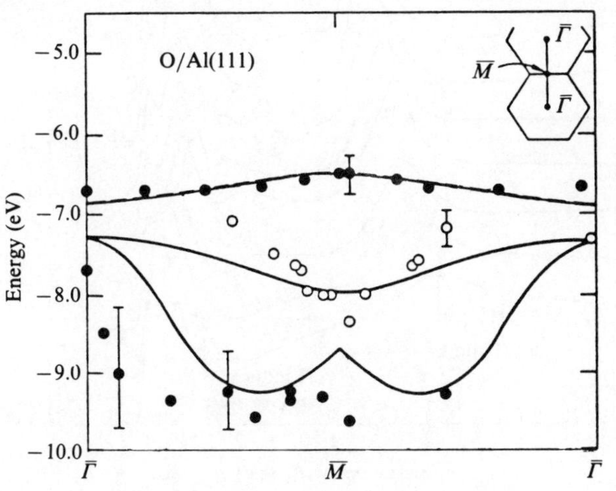

the total energy follow from straightforward diagonalization of the exact Hamiltonian. For titanium, this so-called *configuration interaction* scheme reveals that the energy of an approaching H_2 molecule increases unless the H–H bond distance increases (Cremaschi & Whitten, 1981). In other words, the molecule dissociates. The same three-fold site favored by atomic O/Al(111) emerges as the preferred site for atomic H/Ti(0001).

We next appeal to angle-resolved photoemission experiments to suggest the appropriate way to proceed. Fig. 12.9 shows a sequence of energy distribution curves collected as a function of electron exit angle (cf. Fig. 4.20). A hydrogen-induced feature clearly splits off far below the titanium 3d bands. However, near the center of the SBZ a *second* hydrogen-induced surface state appears within the substrate 3d bands at much higher energy. This is totally unlike the O/Al(111) case and clearly cannot be reproduced by the atom-on-jellium model. But it is completely consistent with the strongly interacting limit of the resonant level model (Fig. 9.12) if the two states correspond to bonding and anti-bonding levels. An LDA slab

Fig. 12.9. UPS photoelectron energy distribution curves for H/Ti(0001) as a function of $k_{\parallel}$. From top to bottom, the curves correspond to emission angles that vary from grazing exit (SBZ edge at K) to normal exit (SBZ center at Γ) (Feibelman, Hamann & Himpsel, 1980).

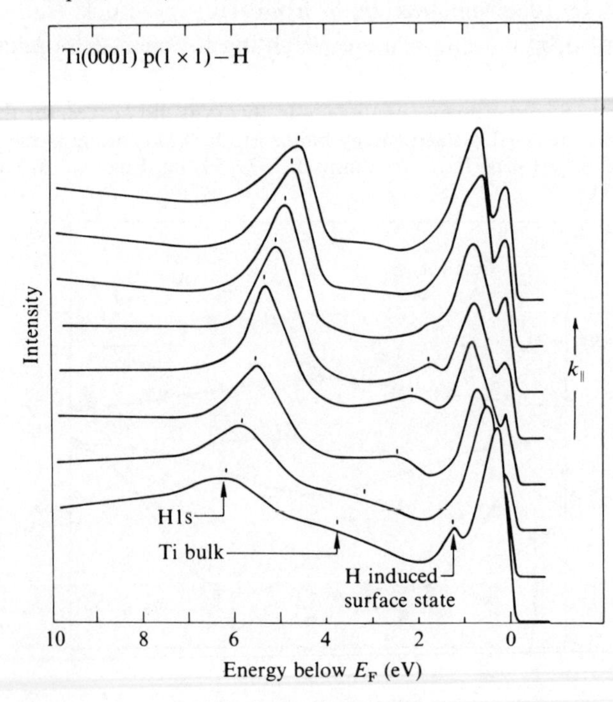

calculation of H/Ti(0001) addresses this issue as well as the dispersion of the surface states.

The calculated surface band structure for this system is shown in Fig. 12.10. Two prominent surface states appear whose energy and dispersion match the experimental results. Analysis of the hydrogen-induced wave functions confirms our previous assignment. This is a strongly covalent system. The surface states are bonding and anti-bonding combinations of the hydrogen 1s orbital and titanium $3d_{3z^2-r^2}$ orbitals. Furthermore, it is physically reasonable (and easily calculable) that the energy separation between the two levels scales with the strength of the coupling matrix element between the orbitals involved. We can use this fact to illustrate a familiar point worth reiterating. Since the H sits in a hollow site, the relevant wave functions appear as in Fig. 12.11. In this configuration, it is clear that the vertical position of the adsorbate greatly affects the magnitude of H–Ti wave function overlap. This, in turn, affects the splitting dramatically. In short, the induced electronic structure reflects the adsorption geometry. This is always true – although the reflection is not also so clear as in the present case.

An interesting experimental observation motivates our example of the

Fig. 12.10. Calculated (solid line) and measured (triangles) dispersion of adsorbate-induced energy levels for Ti(0001) (1 × 1)-H. The dashed lines are states of the 'bulk' slab (Feibelman, Hamann & Himpsel, 1980).

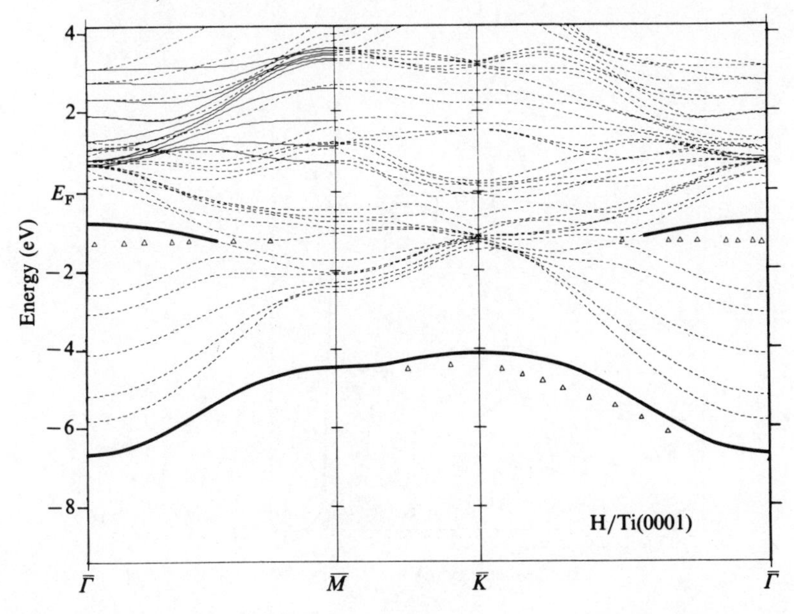

Fig. 12.11. Schematic view of the wave functions involved in H chemisorption on titanium. The large circles represent different vertical positions for the hydrogen 1s orbital.

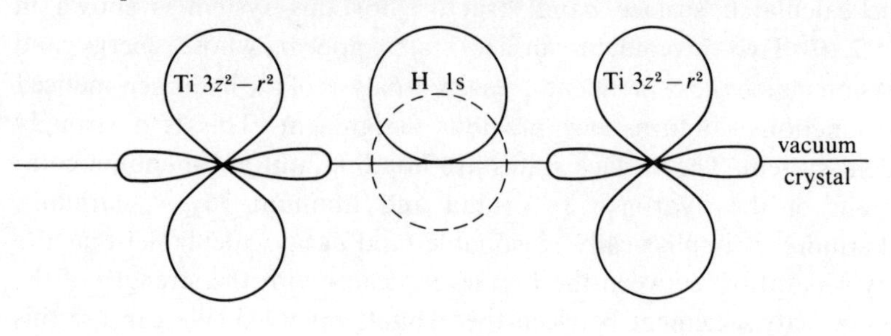

Fig. 12.12. Ultraviolet photoelectron energy distribution curves for monolayer (dashed curve) and greater-than-monolayer (solid curve) coverage of Pd on Nb(110). The dotted curve shows the relative emission from the niobium substrate (El-Batanouny, Strongin, Williams & Colbert, 1981).

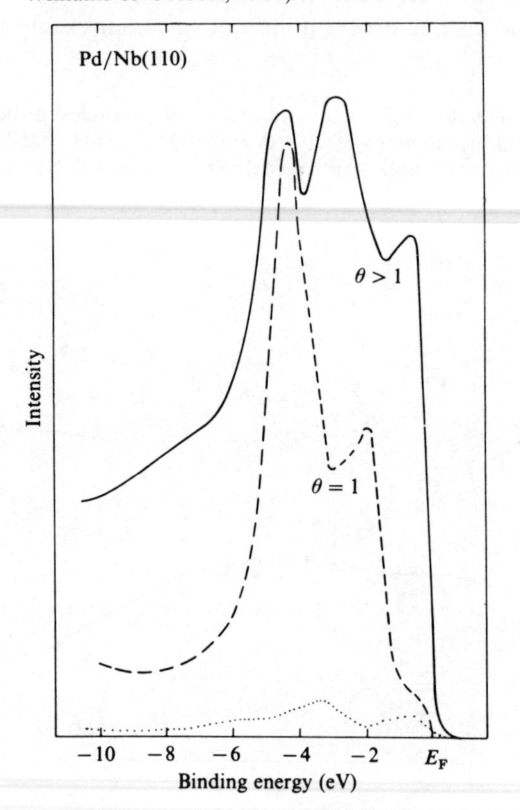

chemisorption of one transition metal onto the surface of a second transition metal. Coordinated LEED and photoemission studies show that a structural/electronic change occurs when palladium atoms adsorb onto a Nb(110) substrate around one monolayer coverage. At $\theta = 1$, one observes a commensurate 1×1 ordered overlayer of Pd/Nb(110). But at slightly higher coverage the diffraction pattern suddenly changes to that of pristine Pd(111). This result may not be too surprising since the surface tension $\gamma(110) > \gamma(111)$ for a typical FCC metal (cf. Fig. 1.8 for lead). What *is* surprising is the accompanying large change in the electronic structure of the palladium overlayer.

Fig. 12.12 illustrates UPS energy distribution curves obtained in the two adsorbate geometries. The photon energy is chosen to take advantage of matrix element effects that suppress emission from the niobium substrate so that the figure essentially reflects the Pd LDOS. Notice that the commensurate Pd(110) overlayer is devoid of density of states at the Fermi level.* By contrast, clean Nb(110) and Pd(111) at $\theta > 1$ exhibit emission spectra with substantial weight at E_F, as one would expect for a normal transition metal. Somehow, adsorption of a single layer of Pd creates a noble metal from two transition metals. How does this occur?

Perhaps a single *unsupported* layer of Pd in the Nb(110) structure already possesses this unusual behavior. To test this idea, we cite the results of calculations that employ a self-consistent tight-binding method designed to take account of charge transfer effects neglected in the O/Al(111) example discussed above. The left panel of Fig. 12.13 illustrates the theoretical DOS for such a free-standing monolayer along with the surface LDOS of Nb(110). A substantial density of states at E_F is apparent for both systems. Evidently, purely two-dimensional Pd(110) is still transition metallic. We conclude that the chemisorption interaction itself is essential to produce noble metal behavior in commensurate Pd/Nb(110).

The right hand panel of Fig. 12.13 shows the theoretical prediction for the local density of states on the top (Pd) and second (Nb) layer of Pd/Nb(110). The curves clearly are reminiscent of the photoemission results. *A posteriori*, we can understand the essential physics by a further generalization of the surface molecule concept as applied in Fig. 9.12. Initially, both the palladium and niobium 4d levels are in the immediate vicinity of the Fermi level. Upon chemisorption they interact and form bonding and anti-bonding *bands* which are driven above and below the Fermi level. The bonding is covalent with little or no charge transfer. Since the center of gravity of the Pd LDOS begins lower in energy than that of

* Presumably, 5s spectral weight exists at E_F that is too weak to observe.

Nb(110) (Fig. 12.13(*a*)), the lower portion of the joint LDOS is primarily Pd-like with a little niobium character mixed in. Conversely, the portion of the LDOS driven up in energy is primarily Nb-like with a bit of palladium mixed in. Additional adsorbed Pd does not interact directly with the niobium substrate and the density of states eventually heals to its bulk value.

It is instructive to connect this discussion with the issue of H_2 adsorption treated earlier. Experiments show that hydrogen dissociatively chemisorbs to Pd/Nb(110) when $\theta > 1$ but that *no* (or little) hydrogen adsorbs when $\theta = 1$ (El-Batanouny *et al.*, 1981). Again, the explanation can be found in the surface molecule limit of the resonant level model – now applied to the H–Pd chemisorption bond. The key observation is simply that the *anti-bonding* level for this system resides somewhere near the top of the d-band (Fig. 9.12). Above one monolayer, palladium behaves like its bulk counterpart. The states near the top of the band lie above E_F and are unoccupied. Chemisorption of H occurs because occupancy of the deep-lying bonding level lowers the energy of the system. However, we have just seen that the Pd 4d band drops below the Fermi level at $\theta = 1$. The

Fig. 12.13. Self-consistent tight-binding results for the electronic structure of Pd/Nb(110): (*a*) surface LDOS of clean Nb(110) and DOS of an unsupported Pd(110) monolayer; (*b*) LDOS of the Pd overlayer and the first Nb substrate layer for commensurate Pd/Nb(110). Vertical dashed lines mark the position of the Fermi level (Kumar & Bennemann, 1983).

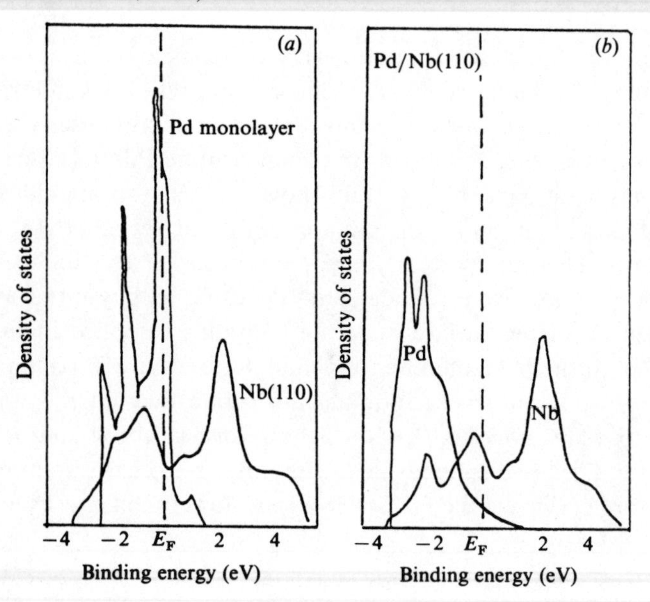

Fig. 12.14. Charge density contour plots appropriate to Ni(100) c(2 × 2)-CO: (*a*) free molecule 5σ orbital; (*b*) free molecule 2π orbital; (*c*) difference between CO/Ni(100) and the superposition of clean Ni(100) and an unsupported CO monolayer. Solid (dashed) lines indicate a gain (loss) of electronic charge (Wimmer, Fu & Freeman, 1985).

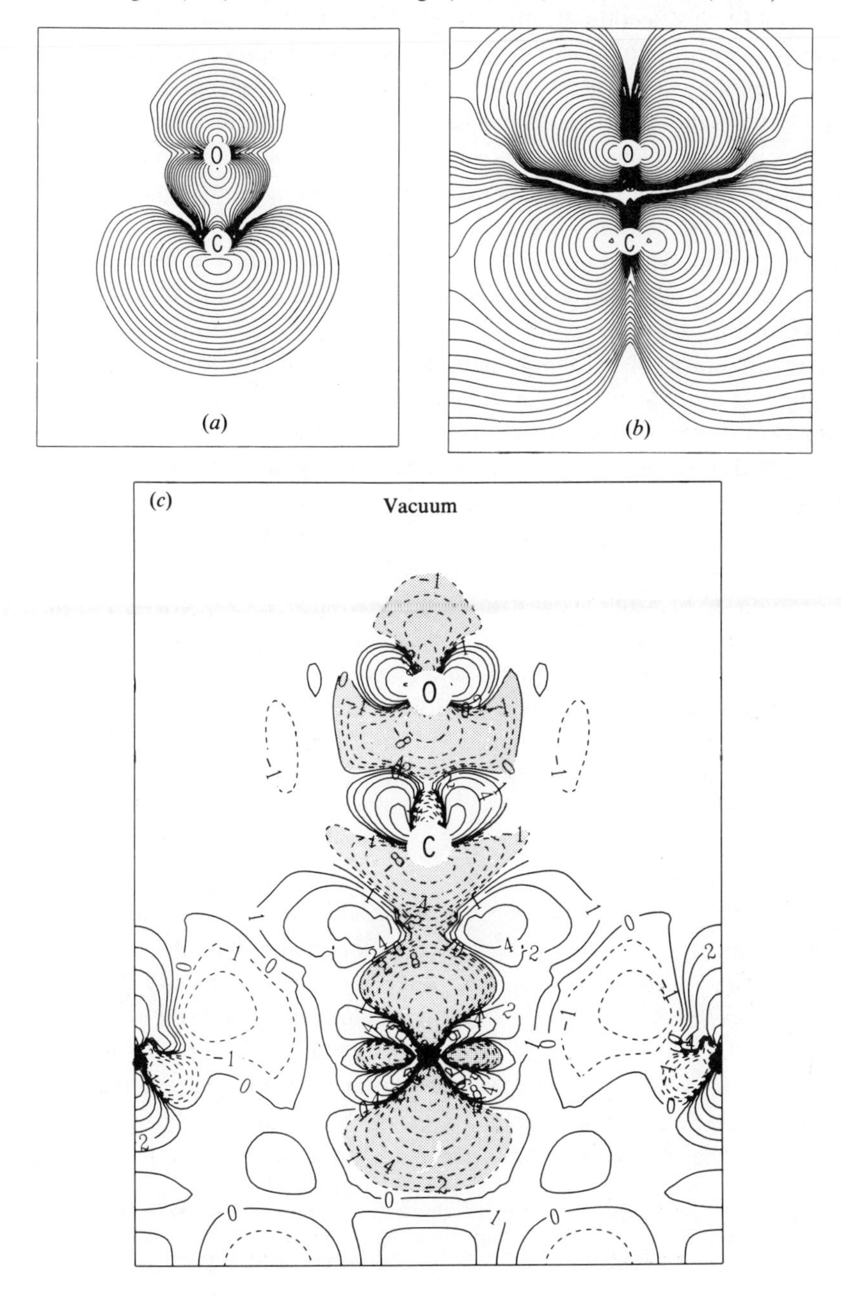

accompanying anti-bonding level drops below E_F as well and, hence, fills with electrons. No adsorption occurs because the hydrogen/palladium complex gains zero net energy if both the bonding *and* anti-bonding states are occupied.

We close this section by turning to a molecular adsorbate – carbon monoxide.* Earlier (Chapter 9) we described the bonding of this molecule to a transition metal in terms of a donation of 5σ electrons from the adsorbate to the substrate. Now we can be more precise. To fix ideas, it is useful to remind oneself of some of the relevant molecular orbitals. Recall in particular that the 5σ state is occupied and the 2π state is unoccupied in the ground state of the free molecule (Fig. 9.6). The spatial character of these wave functions is shown in Figs. 12.14(*a*) and (*b*). The changes that occur when CO chemisorbs to a transition metal (nickel in this case) are shown in Fig. 12.14(*c*).

The 'donation' of 5σ charge actually involves a strong bonding/anti-bonding interaction of this molecular orbital with a $3d_{3z^2-r^2}$ orbital of the nickel substrate. The distinctive shape of both components is seen in the shaded portion of Fig. 12.14(*c*). Bear in mind that this is a covalent bond: some 5σ character is mixed into the metal state and some 3d character is mixed into the CO state. It appears as a net depletion of charge in the difference plot because the final hybrid orbital is less 'pure' than either of the original constituent orbitals. A net gain of charge density that is distinctly '2π' in character also stands out in this figure. This can only mean that chemisorption has dragged the 2π orbital below the Fermi level of the CO/Ni(100) complex. More precisely, a bit of 2π character has been mixed into a previously occupied metal orbital by a second chemisorption interaction. This is called 'back-donation' in the chemical literature.

Magnetism

The foregoing discussion makes clear that considerable charge rearrangement accompanies the formation of chemisorption bonds. This suggests that a corresponding rearrangement of *spin* density probably occurs as well. If so, a broad range of issues immediately beg for attention since we have seen (Chapter 5) that local magnetic moments, ordered magnetic states and magnetic critical phenomena all occur at the clean surface of certain transition and rare-earth metals. We take as a working hypothesis the supposition that all foreign contaminants tend to suppress

* CO is very much the 'hydrogen atom' of surface science. Literally hundreds of experiments and dozens of calculations have been devoted to it. See, for example, Roberts & McKee (1978) and Dunlap, Yu & Antoniewicz (1982).

substrate magnetic effects. After all, substrate electrons recruited to form bonds with adsorbates presumably become less efficient at forming local moments. Moreover, even if substantial moments do form, one guesses that adsorbate interference effects disrupt, rather than enhance, the delicate long range correlations required for magnetic order. Let us be more precise.

The effect of adsorption on local moment formation is best understood for the case of a nickel substrate. This is a relatively simple case since the bulk moment (0.61 μ_B) arises from a very small number of holes at the top of the minority 3d spin band. Just as in the Fe(100) case examined in Chapter 5, a larger moment (0.66 μ_B) obtains on the Ni(100) surface. To better appreciate the effect of chemisorption, it is useful to understand the microscopic origin of this enhancement in terms of a simple model introduced earlier in connection with surface core level shifts (Chapter 4).

Our previous study emphasized local band narrowing as the most important effect that accompanies reduced coordination at surface sites (4.28). But, since there are fewer neighbors available for electron sharing, we also find that excess d-charge builds up in the surface layer. This charge enters the minority spin band because the exchange split majority spin band is virtually full for iron, cobalt and nickel (cf. Fig. 5.9). Hence, at this stage, one finds a *decreased* moment on the surface atoms relative to the bulk. However, Fig. 4.29 showed that (presumably mobile s–p) charge must flow into the surface from the bulk to electrostatically raise the surface d-band and produce a common Fermi level. Reneutralization occurs as (minority) charge flows back out of the rising d-band to compensate for both sources of added charge. The net result is fewer minority electrons than at the start of the process, i.e., an *increase* in the surface magnetic moment.

It is simplest to see the effect of a foreign adsorbate if one adopts the following 'alchemy' approach. Focus attention on the *second* layer of a clean transition metal. These atoms are essentially bulk-like since all their coordination and chemical requirements are met. Now 'transmute' the top (surface) layer to a different chemical species, say, oxygen or hydrogen. The subsurface LDOS must still narrow since the adsorbate cannot provide the same strength of d–d interaction as the self-same transition metal atom. This leads to a decreased moment by the argument given above. However, in contrast to the clean surface case, charge need not flow from the bulk to equilibrate the Fermi levels. Hybridization with the overlayer and adsorbate-induced changes in the density of states are more than sufficient to maintain charge neutrality of the second (metal) layer of the adsorbate/substrate combination. We predict that adsorption 'deadens' the substrate magnetic moment.

Generally (although not universally), experiment and detailed theory bear out these ideas. From LDA electronic structure calculations we learn that adsorbates as diverse as H (Weinert & Davenport, 1985), Cu (Zhu, Huang & Hermanson, 1984), and CO (Raatz & Salahub, 1984) all reduce the magnetic moment of the outermost layer of Ni(100) to substantially below its bulk value. Perhaps surprisingly, the same occurs for oxygen, for which we might have supposed that oxidation of minority spin electrons would lead to a larger surface moment. Instead, spin-polarized *inverse* photoemission clearly shows that the density of unoccupied minority band states decreases upon adsorption (Fig. 12.15).

In this experiment, one directs a spin-polarized beam of electrons towards the target. Some of the electrons undergo radiative transitions from high energy plane wave states to unoccupied conduction band states just above the Fermi level. The measured intensity of emitted photons for up and down spin incident electrons provides a spin-resolved image of the *unoccupied* DOS just as normal photoemission provides an image of the occupied DOS (Dose, 1985). The experimental signal can only pertain to surface magnetism because adsorption cannot affect bulk moments. In accord with our general argument, we conclude that oxygen bonding suppresses moment formation.

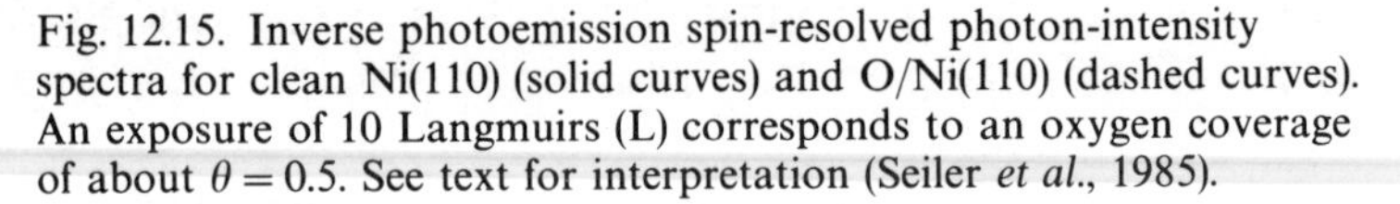
Fig. 12.15. Inverse photoemission spin-resolved photon-intensity spectra for clean Ni(110) (solid curves) and O/Ni(110) (dashed curves). An exposure of 10 Langmuirs (L) corresponds to an oxygen coverage of about $\theta = 0.5$. See text for interpretation (Seiler *et al.*, 1985).

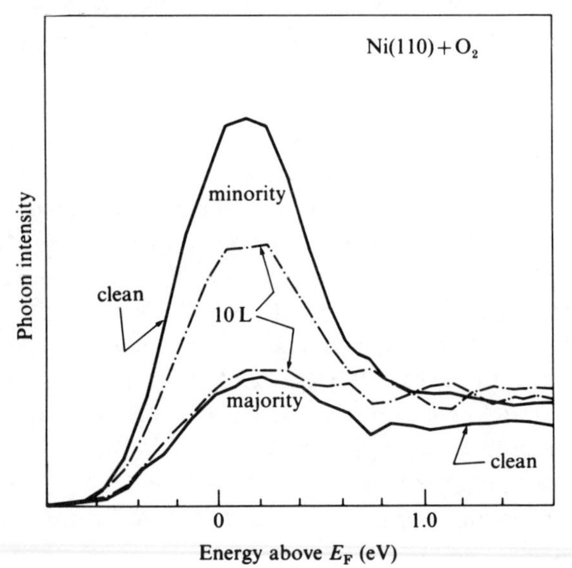

Comparatively little is known about the effect of chemisorption on magnetic ordering and (even less) critical phenomena. We cite a single example. Rare-earth magnets are a suitable laboratory for such studies because the large 4f moments are atomic-like and not directly affected themselves by chemical adsorption effects. This does not mean that the *interactions* between such moments cannot be affected. Fig. 12.16 illustrates this point with data for the spin polarization of photoelectrons ejected from a polycrystalline gadolinium substrate dosed with H_2. Adsorption suppresses both the saturation magnetization and the extrapolated surface critical temperature relative to their bulk values (cf. Fig. 5.18). Similar behavior occurs in bulk magnets when communication between local moments is disrupted by doping with non-magnetic impurities (see e.g., Lagendijk & Huiskamp (1972)). Nevertheless, no theory exists for this effect in the present context.

Semiconductors

It is difficult to frame a simple microscopic picture of the effect of adsorption on the electronic structure of a semiconductor as easily as was possible for the corresponding metal substrate problem. Typically, one doesn't know the position of the ions well enough to speak meaningfully about the electrons. The situation is identical to that encountered for clean semiconductors and occurs for the same reason. Unlike metals, these materials reconstruct rather than strain and their orbitals rehybridize

Fig. 12.16. Temperature dependence of the spin polarization of electrons photoemitted from the surface of H/Gd. H_2 dosage is 0.5 L. Compare with Fig. 5.18 (Cerri, Mauri & Landolt, 1983).

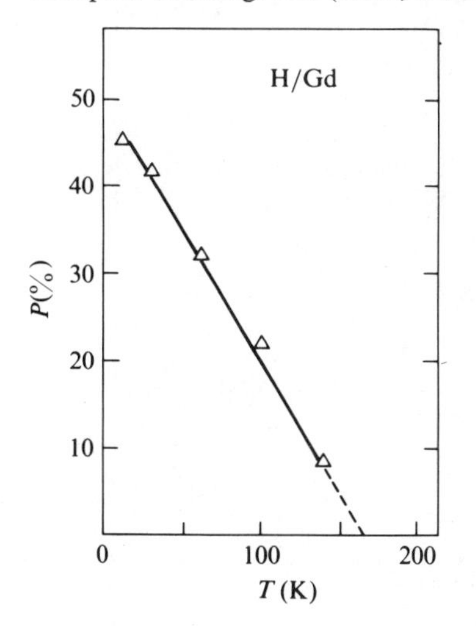

at the first opportunity to lower the total free energy. Hence, one finds adsorbate-induced reconstruction and/or non-transparent adsorbate bonding configurations as often as not. More can be said about a few particularly simple model systems. As a result, we must content ourselves with the identification of a few general (but non-universal) features rather than the extraction of broad trends or unifying model constructs. This is vexing but unavoidable.

We begin simply. The ideal Si(111) surface possesses a clearly defined surface state band within the fundamental gap (Fig. 4.36). The eigenfunctions of this state correspond to dangling $3p_z$ orbitals that stick out into the vacuum. Turning to the simplest adsorbate, hydrogen, it is reasonable to posit that adsorption into a 1 × 1 structure of 'on-top' sites saturates these dangling bonds and produces an energetically stable structure. The half-filled surface states of the clean surface vanish and a surface bonding

Fig. 12.17. Surface energy bands for ideal Si(111) (1 × 1)-H calculated with the tight-binding method. The surface states of both the clean surface (dashed curves) and the adsorbate-covered surface (solid curves) are indicated (Pandey, 1976).

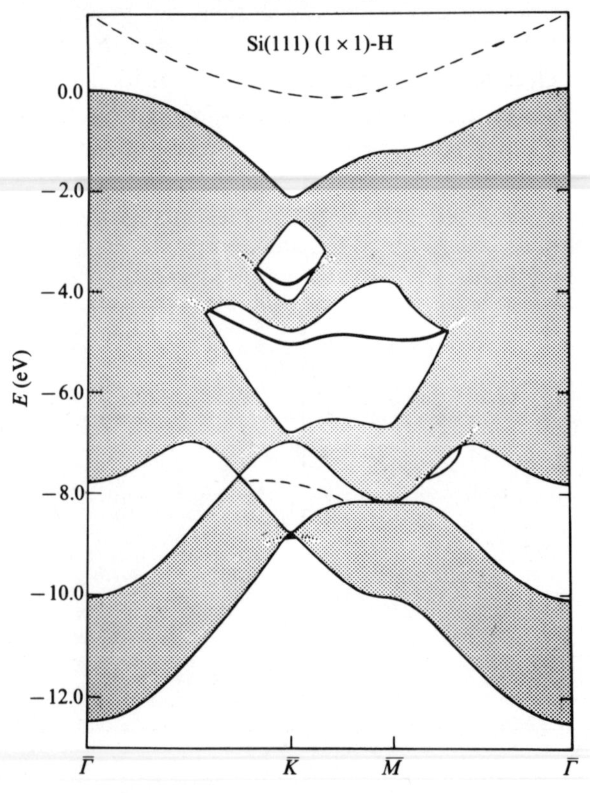

band between silicon and hydrogen forms in its stead. The calculated surface band structure of Fig. 12.17 shows how this occurs. As usual, the shaded region is the projection of the bulk electronic energy bands onto the surface Brillouin zone. The dashed curves denote the dangling bond and back-bond surface states of the clean surface. Three *bona fide* surface states appear in the spectrum of the chemisorption complex. In particular, a covalent bonding interaction with H 1s orbitals lowers the dangling $3p_z$ orbital energy by 5 eV/atom. The resulting hybridized state lives mainly along the periphery of the SBZ (cf. Fig. 4.35).

Heretofore, we have compared the results of calculations such as those presented in Fig. 12.17 with the measured dispersion of surface states obtained by angle-resolved photoemission. But, since the relevant experiments are costly to perform (since synchrotron radiation is required), it is far more common to compare calculated LDOS with UPS energy distribution curves, $N(E)$. Figs. 12.4, 12.12 and 12.13 are examples of this approach. Unfortunately, this is not really correct. The raw intensity of a photoemission spectrum differs from the local density of states due to matrix element effects (4.34), escape depth factors (Fig. 2.1), spectrometer resolution and the presence of secondary electrons (Chapter 2). To address this fact, Fig. 12.18 shows a rare attempt to fold all of these effects into the calculated H/Si(111) LDOS for direct comparison to UPS data.

The agreement between experiment and theory in Fig. 12.18 is rather

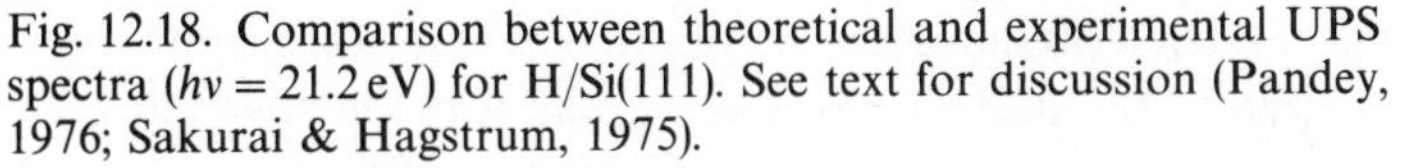
Fig. 12.18. Comparison between theoretical and experimental UPS spectra ($h\nu = 21.2$ eV) for H/Si(111). See text for discussion (Pandey, 1976; Sakurai & Hagstrum, 1975).

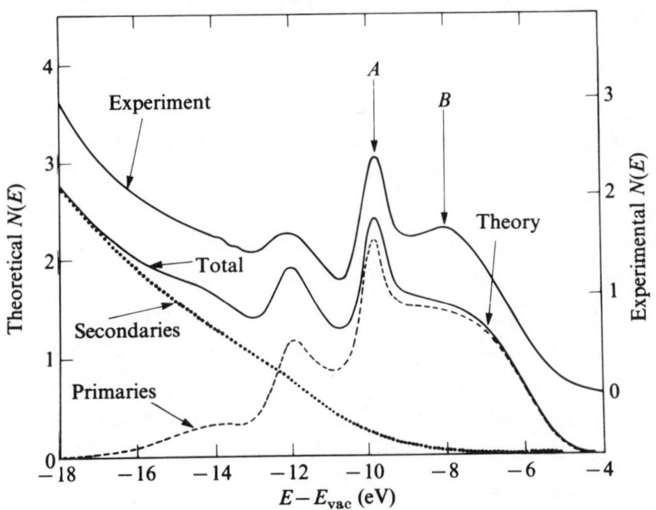

good. However, the calculations pertain to H on ideal Si(111) 1 × 1 whereas the measurements correspond to saturation coverage of hydrogen on Si(111) 7 × 7! The message is clear. If the adsorbate geometrical structure is known, a reasonably accurate band structure in conjunction with laboratory UPS spectra is sufficient to elucidate the electronic structure of the system. However, if the crystal structure is *not* known, agreement between such theory and experiment should be treated with caution. In the present case, we presume that hydrogen saturation of the remaining dangling bonds of the 7 × 7 structure (Fig. 3.20) is indeed responsible for the feature *A* in the spectrum. However, more subtle effects, e.g., feature *B*, presumably arise from details of the reconstruction.

The chemisorption of chlorine on Ge(111) provides a rather different example – from several points of view. First, the adsorption site (on-top) and bond length (2.07 ± 0.03 Å) are known from SEXAFS measurements (Fig. 10.12). Second, an independent (concurring) determination of the geometry is available from total energy LDA slab calculations. The latter, of course, supplies details of the electronic structure. Third, Pauling (1960) provides an alternative discussion of the electronic structure in the context of the properties of the free molecule $GeCl_4$. In particular, he points out (Sec. 9.1) that the Ge–Cl bond length in the molecule (exactly the same as found for Cl/Ge(111)) is much shorter than the sum of the covalent radii of the two constituents. This is reasonable since the electronegativity difference between chlorine and germanium ($\sim$ 1.15) points to considerable ionic character to the bond.

The ionicity of the chemisorption bond appears quite clearly in the calculated LDA total charge density contours for this system (Fig. 12.19). Germanium $4p_z$ charge evidently flows to fill the chlorine 3p shell. However, Pauling goes on to comment that the equilibrium bond length is even shorter than one would estimate based on this ionicity. He suggests that there must be 'partial double-bonding' between the ions. This is interesting because the obvious candidates for such extra bonding, the remaining 4s/4p orbitals of germanium, are tied up in sp^3 hybrids that back bond the surface Ge atom to its substrate neighbors.

Analysis of the calculated induced charge density shows that there is indeed additional bonding in the Cl/Ge(111) system. But, rather than p-orbitals, the bond energy comes from d-orbitals of π symmetry directed along the Cl–Ge bond axis. These states are synthesized principally from 4d states of Ge that are *unoccupied* in the ground state of the clean substrate. This unexpected effect is entirely due to chemisorption. The adsorbate polarizes the 4d orbital and lowers its energy to the point where partial occupancy becomes energetically favorable.

The Al/Ge(100) system represents yet a third electronic archetype of chemisorption on a semiconductor. It provides an introduction to the problem of the metal/semiconductor interface – a central concern of any microscopic theory of Schottky barrier formation. Unfortunately, little is known about this system experimentally near monolayer coverage. Therefore, for the first time, we rely entirely on electronic structure theory to guide our discussion. This is not inappropriate because our main point will remain valid even if it turns out that 'real' Al/Ge(100) is somewhat different. We proceed caveat in hand.

The ideal Ge(100) surface presents a square adsorption lattice to foreign gas particles (Fig. 4.35). We consider two coverages of aluminum atoms adsorbed onto the energetically favored bridge sites. Figs. 12.20(*a*) and (*c*) illustrate top views of the calculated charge density contours for $\theta = 0.5$ and $\theta = 1$, respectively. Figs. 12.20(*b*) and (*d*) are the corresponding surface corrugation topographies. The key point is that the calculated equilibrium spacing (z_0) of the aluminum overlayer from the germanium substrate is vastly different in the two cases. At $\theta = 0.5$ one finds $z_0 = 1.2$ Å while at $\theta = 1$ the spacing $z_0 = 1.8$ Å. This geometric fact profoundly affects the electronic structure.

At half-coverage Al/Ge(100) resembles a typical chemisorption system. The overlap of germanium dangling bonds with aluminium sp^2 orbitals leads to bridge bonding and a highly anisotropic surface charge distribu-

Fig. 12.19. Calculated charge density contours for Cl/Ge(111) in the equilibrium on-top site. The contours are shown in a (110) plane intersecting the adsorption site (Bachelet & Schluter, 1983).

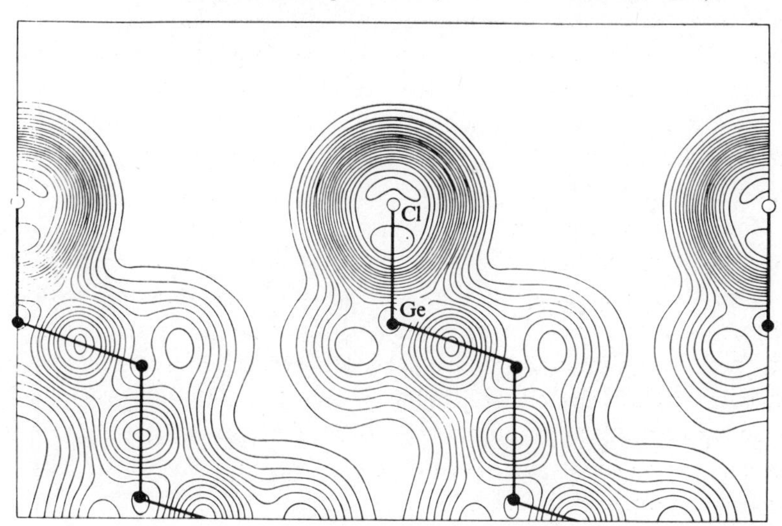

tion. However, the situation changes entirely when aluminum atoms are placed on the remaining bridge sites. The overlayer pulls away from the substrate, chemisorption bond charge practically vanishes and the overlayer charge density resembles that of metallic aluminum (cf. Fig. 4.23). This drastic change is equally obvious from the calculated LDOS (Fig. 12.21). At $\theta = 0.5$, chemisorption-induced bonding and surface states appear 3 eV below E_F and at the Fermi level, respectively. By contrast, the state density at $\theta = 1$ increases in a modulated stepwise manner as the energy approaches E_F. It is easy to see that this is consistent with the 'metallization' of a two-dimensional overlayer that is only weakly coupled to the underlying substrate.

Consider the eigenstates of an electron trapped within a monolayer of simple metal atoms. The confining potential *perpendicular* to the crystal plane is essentially the Coulomb field of a single ion. The spectrum is a discrete set of 'atomic' energy levels. However, each of these levels ('sub-bands') is highly degenerate since a continuous set of nearly-free electron states describes the motion *parallel* to the monolayer. In other

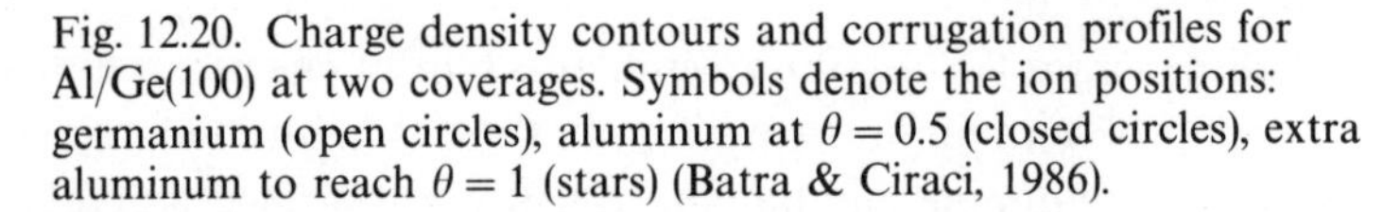
Fig. 12.20. Charge density contours and corrugation profiles for Al/Ge(100) at two coverages. Symbols denote the ion positions: germanium (open circles), aluminum at $\theta = 0.5$ (closed circles), extra aluminum to reach $\theta = 1$ (stars) (Batra & Ciraci, 1986).

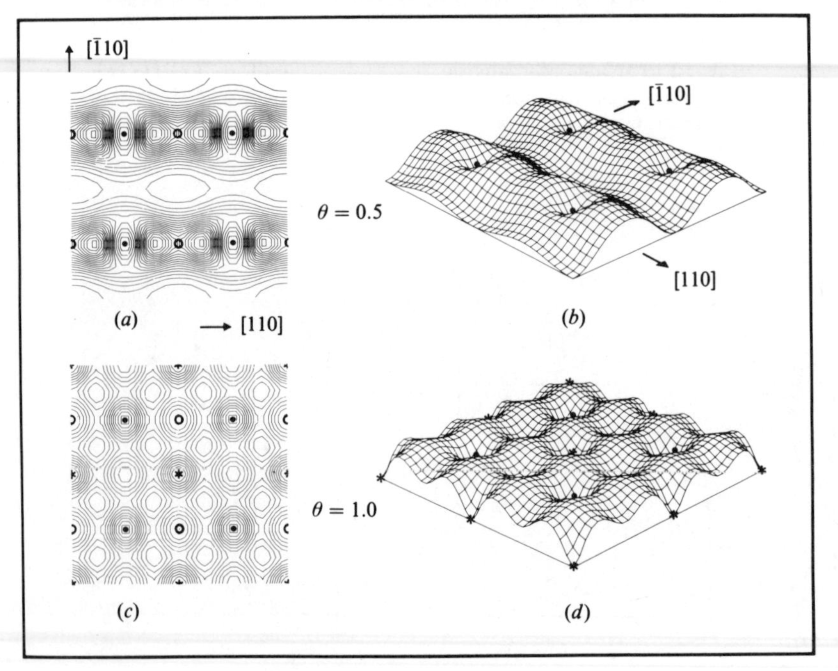

words, there is a two-dimensional free electron gas associated with each of the discrete levels of the perpendicular 'box'. Recall next that the density of states for a two-dimensional free-electron gas is independent of energy: $N(E) = 4\pi m/h^2$. Therefore, in the absence of any potential modulations along the plane, the total DOS for a monolayer increases stepwise each time the Fermi level rises past a new sub-band.

Our central conclusion is striking. Even at the monolayer level, some metals can gain binding energy by loosening their adsorptive attachment to a semiconductor substrate. A lower energy results from the exchange of chemisorption bond energy for overlayer metallic cohesive energy.

Schottky barriers and band offsets

The needs of the microelectronics industry define many of the problems associated with adsorption on semiconductors. For example, empirically one knows that device performance can depend sensitively on the quality of the interface between a semiconductor and a second deposited metal, semiconductor, or oxide (Milnes & Feucht, 1972). Commercial fabrication facilities routinely employ many of the experimental characterization techniques discussed throughout this volume. At a deeper level, microminiaturization has driven us to the point where one might expect microscopic investigation of the initial stages of semiconductor interface formation to have direct relevance to device

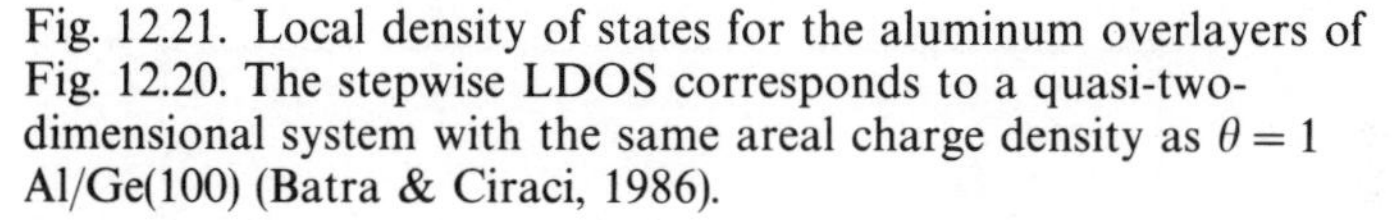

Fig. 12.21. Local density of states for the aluminum overlayers of Fig. 12.20. The stepwise LDOS corresponds to a quasi-two-dimensional system with the same areal charge density as $\theta = 1$ Al/Ge(100) (Batra & Ciraci, 1986).

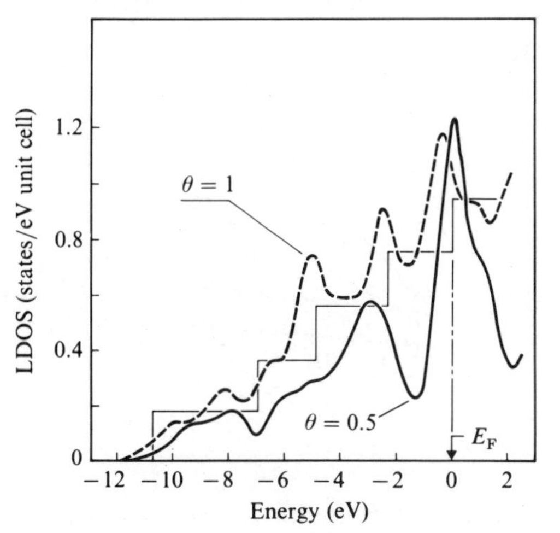

physics. For the most part, this idea has been pursued with regard to the evolution of two interface *energy* parameters: the Schottky barrier at a metal/semiconductor interface and the discontinuity of the valence band maximum at a semiconductor/semiconductor interface. We briefly address both problems.

The magnitude of the Schottky barrier at a metal/semiconductor interface largely determines the rectification properties of the contact. With reference to Fig. 9.15(*c*), we can express this quantity as

$$E_b = E_g - E_F, \tag{12.4}$$

where the zero of energy is taken as the valence band maximum. Equation (12.4) suggests the possibility of continuous control of the barrier height by doping the semiconductor to vary E_F. However, experiments show instead that the magnitude of the barrier E_b is *independent* of bulk doping and varies (if at all) with the electronegativity of the deposited metal (Fig. 12.22). Clearly, this is reminiscent of the phenomenon of Fermi level 'pinning' at a clean surface (Fig. 4.39). As described in Chapters 4 and 9, pinning occurs due to the presence of surface states. More precisely, E_F is pinned because the density of surface states is much greater than the

Fig. 12.22. Measured values of the Schottky barrier height for junctions formed from four common semiconductors with elemental metals. The metals are arranged according to their (Pauling) electronegativity (Louie, Chelikowsky & Cohen, 1977).

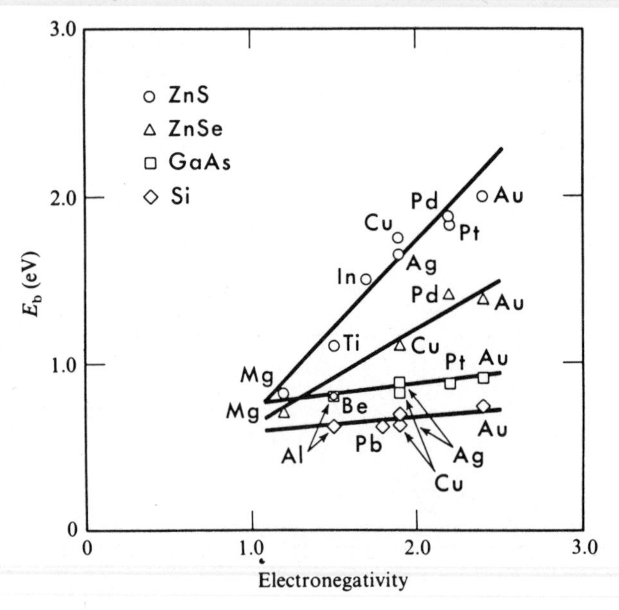

(areal) density of bulk donor/acceptor states. The Fermi level at the surface need not change by very much to accommodate any charge flow needed to equilibrate the surface to the bulk. Bardeen (1947) presumed that intrinsic localized states persist at an interface and pin the Schottky barrier similarly.

Bardeen's original idea is at least a possibility for barriers formed from Si, Ge and ZnSe. However, as we have seen, reconstruction sweeps the fundamental gap free of intrinsic surface states at the clean surface of materials such as GaAs, GaSb and InP. In these cases, pinning occurs only after deposition of a foreign species; sometimes, at the submonolayer level (Fig. 9.18). The question naturally arises: what is the origin and electronic nature of the states that fix the value of the Schottky barrier at a metal/semiconductor interface? For simplicity, we ignore any interdiffusion or reactivity of the metal and semiconductor atoms.

Consider first covalent substrates. Fig. 12.22 shows that E_b is remarkably insensitive to the type of metal used to form the Schottky barrier. In fact, for the cleavage face of GaAs(110), one finds that E_F is pinned at the same point in the gap no matter what adsorbate one uses (Fig. 12.23)! This strongly suggests that pinning is related to some intrinsic property of the semiconductor – despite the fact that no intrinsic electronic states exist on the clean surface. The most natural idea (that is independent of the chemical identity of the adsorbate) is an adsorption-induced structural change. For example, perhaps the energy released by chemisorption is great enough to locally induce surface defects, e.g., vacancies, interstitials, or antisites (anions on cation sites or vice versa). Surface electronic states associated

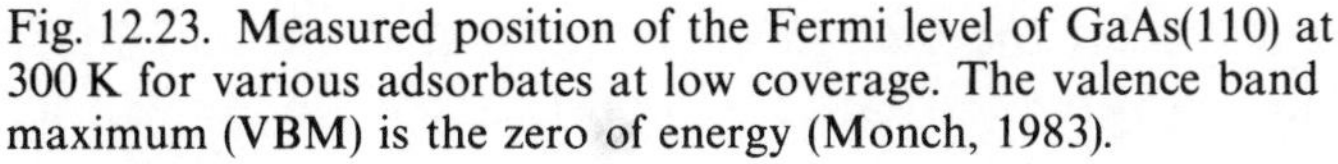

Fig. 12.23. Measured position of the Fermi level of GaAs(110) at 300 K for various adsorbates at low coverage. The valence band maximum (VBM) is the zero of energy (Monch, 1983).

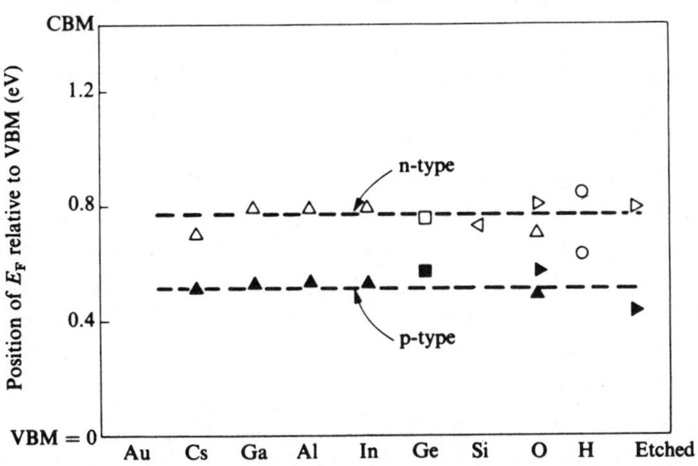

with these defects lie in the fundamental gap (Dow, Allen & Sankey, 1984). Another possibility is that adsorption locally 'unreconstructs' the surface as discussed earlier for the case of C/Pt(100) (Fig. 10.3). In the present case, this would reintroduce dangling bond states into the gap available to trap (donate) charge and pin E_F. It is very likely that one of these mechanisms is operative – at least for very low adsorbate coverage.

A different possibility, more appropriate to monolayer (and greater) adsorbate coverage, was suggested by Heine (1965). We make the key idea explicit by reference to an LDA slab calculation of the electronic structure of Al/GaAs(110). The adsorption geometry for this system is not known. Consequently, an FCC aluminium lattice was placed atop both the ideal GaAs(110) surface and the reconstructed GaAs(110) surface (Fig. 4.43). The results show that adsorption on the *ideal* surface is energetically favorable by about 0.35 eV/surface atom. Aluminum atoms form strongly bonding sp^3 hybrids with the surface As atoms. This prevents the transfer of charge from Ga to As required to form the gallium sp^2 orbitals of the reconstructed surface. More importantly, the layer-by-layer LDOS reveals a very interesting occurrence. The top layer of GaAs becomes 'metallized'.

Fig. 12.24 illustrates the local density of states for several planes parallel to the Al/GaAs(110) ideal interface. The top and bottom panels are the LDOS for the metal overlayer and the third (bulk) layer of the semiconductor. The fundamental gap is evident in the latter around zero energy (E_F). The middle panel is the LDOS for the first layer of GaAs. Notice that there is no gap! It has been filled in by metal-induced gap states (MIGS) that derive from the wave function tails of the immediately adjacent aluminum states. The latter, of course, exist at all energies. Moreover, the calculated density of MIGS is $\sim 10^{14}$ states/eV – more than sufficient to pin the Fermi level.

It is no accident that the Fermi level is 'pinned' in Fig. 12.24 very near the middle of the bulk gap of GaAs(110). This is a consequence of charge neutrality. Metal-induced states, like any states in the gap of a semiconductor, are synthesized from exponentially decaying wave functions that derive their spectral weight from bulk states (cf. Fig. 4.38). MIGS near the bottom (top) of the gap are mostly valence (conduction) band in character. Therefore, if the decay length is long (so that the interface layer is well metallized), the local Fermi level ought to be pinned very near the point where these states cross over from valence band-type to conduction band-type. Sensibly, this occurs somewhere near mid-gap (Tersoff, 1984a). In any event, none of this has anything to do with the adsorbate. Hence, we reproduce the insensitivity of E_b to overlayer chemical identity.

The band gap is larger for more ionic semiconductors. Compared to the covalent case, metal-induced states near midgap are farther away (in energy) from their parent bulk states and the wave function amplitudes decay to zero more rapidly. As a result, these MIGS screen charge transfer at the interface less efficiently. It then is unsurprising that, as the data show (Fig. 12.22), electronegativity effects begin to appear in the value of E_b.

As to the evolution of the Schottky barrier, the electronic structure calculations exhibit an unmistakeable trend: the density of MIGS is inversely proportional to the strength of the metal–semiconductor chemisorption bond. Simply put, chemical bond formation between adsorbate and substrate transfers spectral weight from the gap region into deeper-lying bonding levels. Therefore, we can produce the following (speculative) scenario. At low coverage, conventional chemisorption occurs and Fermi level pinning (if it occurs) is related to local structural rearrangements of the substrate. As metal builds up, either layer-by-layer or in three-dimensional clumps, chemisorption weakens (cf. Fig. 12.20) and metallic behavior increases. This promotes MIGS. Localized levels in the gap broaden into resonances and pinning persists at about the same energy

Fig. 12.24. Planar local density of states near an ideal Al/GaAs(110) interface: (*a*) aluminum overlayer; (*b*) top layer of GaAs; (*c*) bulk GaAs. The zero of energy is the Fermi level (Zhang, Cohen & Louie, 1986).

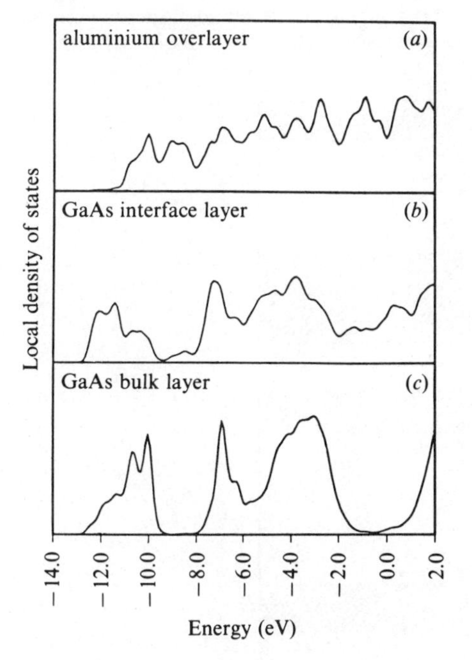

as before – although the precise nature of the pinning states is rather different.

Similar physical considerations apply to the study of semiconductor/semiconductor junctions. This type of interface occurs in so-called quantum well devices (see, e.g., Chemla (1985)) and other commercially useful heterostructures. The most important material parameter in these systems is the relative position of the two constituent semiconductor band gaps at the interface. This is called the band offset. Together with the relative band bending, the band offset determines the electrostatic potential barrier to electron transport across the junction. We use the MIGS idea to estimate its magnitude.

Consider first the case of two identical semiconductors whose bands are artificially offset by an amount ΔV (Fig. 12.25(*a*)). From the discussion above, we expect propagating states from each material to exponentially tail off into the band gap of the adjacent material. These are the analog of MIGS – call them induced gap states (IGS). The charge neutrality condition directs us to occupy the predominantly valence band-derived IGS. The predominantly conduction band-derived IGS remain empty. Hence, an electrostatic dipole forms at the interface. The induced dipole potential opposes ΔV and, in fact, screens it to a value near $\Delta V/\varepsilon$, where

Fig. 12.25. Two examples of the relation between band alignment and the induced interfacial dipole: (*a*) single semiconductor with an imposed offset; (*b*) the interface between two dissimilar semiconductors (Tersoff, 1984b).

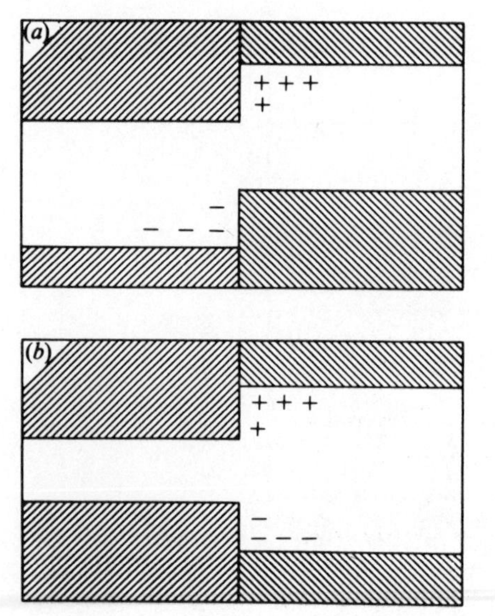

ε is the interfacial dielectric constant. But this practically wipes out ΔV because $\varepsilon \sim 10$ (or greater) for typical semiconductors. In other words, the induced dipole (which costs energy) drives the system back to its original zero-offset condition. Now apply this result to the more realistic example of an interface between two dissimilar semiconductors (Fig. 12.25(*b*)). It is easy to see that charge neutrality again favors one particular value for the band offset – the one that most nearly produces zero net induced dipole. Roughly speaking, this is equivalent to aligning the midgap points of the two components to the junction.

Surface sensitive UPS permits a straightforward test of this simple prediction. At low overlayer coverage, the discontinuity in the valence band maximum ΔE_v can be read off directly from a photoemission energy distribution curve (Fig. 12.26). Core level shifts provide the same information. Unbiased comparison with experiment reveals that the 'zero-dipole' condition predicts observed band offsets with an accuracy of about 0.15 eV (Margaritondo, 1985).

Fig. 12.26. Valence band EDC's of a cleaved CdS crystal both before and after Si deposition. Values in Ångströms indicate the overlayer thickness. The VBM of CdS lies at about 2.5 eV below E_F while the VBM of Si begins to grow in at about -0.9 eV (Katnani & Margaritondo, 1983).

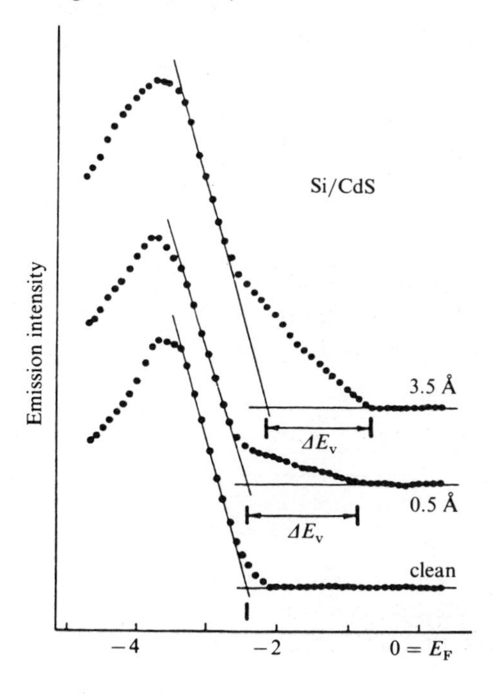

The success of the band offset model does not mean that the microscopic physics at a semiconductor heterojunction is a completed story. For example, Fig. 12.27 illustrates midgap band line-up as observed at a Ge/GaAs(100) interface. The figure shows both the measured ΔE_v and the position of the Fermi level for germanium adsorption onto three different reconstructed surfaces of the non-cleavage (100) face of GaAs. Happily, ΔE_v is independent of the initial reconstruction of the substrate. This confirms the idea that charge neutrality involves IGS derived from wave functions of the bulk. However, the pinned position of E_F evidently *does* depend on more intimate details of the surface. We note an interesting correlation. Surface chemical analysis reveals (Bachrach, Bauer, Chiaradia & Hansson, 1981) that the atomic stoichiometry differs for these three reconstructions. The 4×6, $c(8 \times 2)$ and $c(4 \times 4)$ surfaces exhibit, in sequence, increasingly more arsenic atoms/surface unit cell. Unfortunately, no present theories suggest an intimate connection between surface atom excess and pinning position. Significant work remains to be done.

Insulators

The study of adsorption-induced changes in the electronic structure of wide band gap materials is largely *terra incognita*. Both electron

Fig. 12.27. Schematic view of the energy bands and Fermi level position for Ge/GaAs(100) as determined by photoemission spectroscopy. Band bending is not visible on the scale of the figure (~ 10 Å) (Chiaradia, Katnani, Sang & Bauer, 1984).

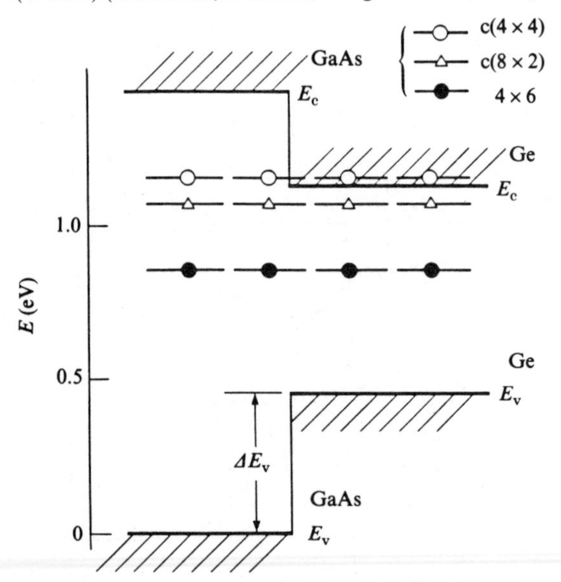

spectroscopy and surface band calculations are tricky to perform when charging or charge transfer effects are large. Consequently, we limit ourselves to one illustrative example of each. We do this mainly for the sake of completeness. As will become clear, it is not possible to draw firm conclusions from the limited information available.

We begin with a photoemission experiment designed to study the adsorption of formic acid (HCOOH) on the non-polar $(1\bar{1}00)$ surface of zinc oxide. This is a marginal insulator with a band gap of 3 eV. As usual, one uses the UPS 'fingerprint' technique discussed in Chapter 9. Unfortunately, only the most subtle changes occur in the measured EDC after adsorption. This occurred also for O_2/ZnO (Fig. 9.20). Therefore, as in that case, we turn to the *difference* between the measured adsorbate-covered surface spectrum and the clean surface spectrum (Fig. 12.28). As it happens, the difference curve bears no particular resemblance to the photoelectron spectrum of molecular HCOOH. However, it is not unlike the experimental spectrum of CO and H_2 co-adsorbed onto ZnO at the same temperature. Moreover, if one identifies the extra peak in the adsorption spectrum at about 10 eV binding energy with free oxygen emission, it is plausible that the original EDC reflects a decomposition of formic acid in CO, H_2 and a free oxygen atom.

Let us pursue this scenario a bit further. Quite generally, the 4σ orbital of carbon monoxide does not participate in surface bonding (Chapter 9). This suggests that one can align the 4σ photoemission peak of CO with the lowest peak in the HCOOH/ZnO spectrum. With this assumption,

Fig. 12.28. Three UPS energy distribution curves which pertain to HCOOH/ZnO$(1\bar{1}00)$. Difference spectra $\Delta N(E)$ for HCOOH (upper solid curve) and CO + H_2 (dashed curve) adsorbed onto ZnO and a conventional EDC $N(E)$ for gas phase CO (lower solid curve) (Luth, Rubloff & Grobman, 1976).

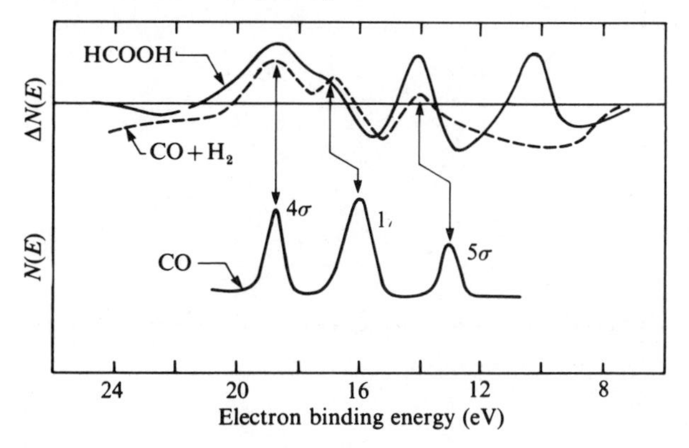

we interpret the downshifts in the adsorption spectrum to mean that, unlike CO bonding to transition metals, CO bonding to ZnO involves both the 5σ and 1π orbitals. Fig. 12.29 is an artist's conception of how this might occur if the free oxygen atom is imagined to be part of the chemisorption complex. It is important to stress that this figure represents a combination of chemical intuition and wild speculation.

In closing, we consider adsorption of a metal onto a wide-gap insulator,

Fig. 12.29. Model for chemisorbed CO and O on ZnO($1\bar{1}00$). One oxygen orbital is not involved in any bond (Luth, Rubloff & Grobman, 1976).

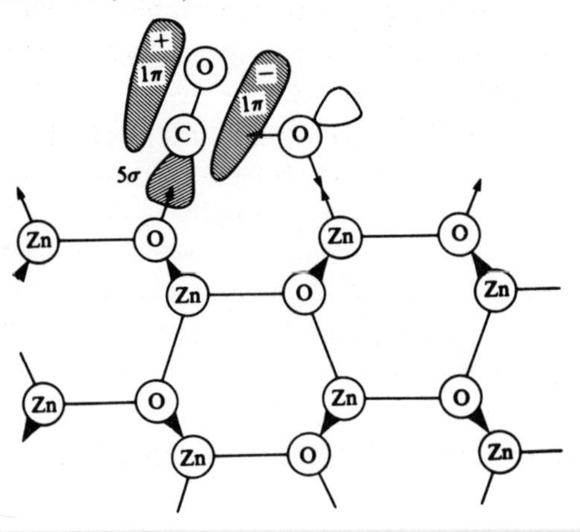

Fig. 12.30. Contour map of the Au–Cl bonding orbital of Au/NaCl(001). Solid and dashed curves refer to positive and negative values for the wave function, respectively (Fuwa, Fujima, Adachi & Osaka, 1984).

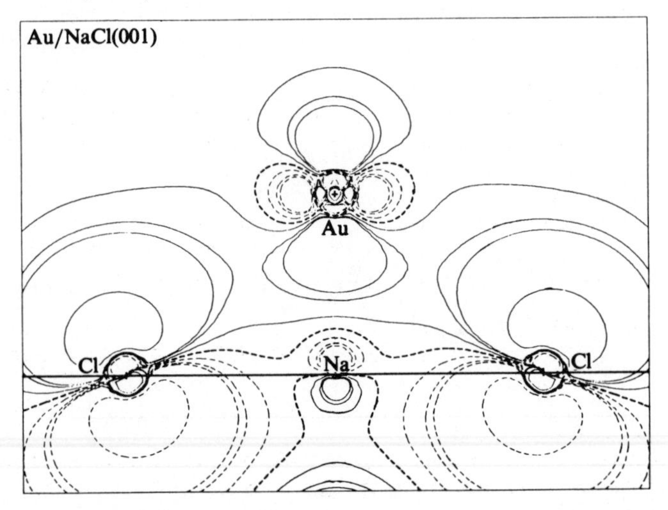

specifically, gold on rocksalt. This choice is interesting because we previously have discussed the potential energy surface of Au/NaCl(100) (Fig. 8.14) in terms of a physisorption interaction between the constituents. However, we also have seen that the local density approximation provides an alternative description of dispersion effects that differs from classical van der Waals theory (cf. the discussion of Fig. 8.10). The latter is appropriate at large adsorbate/substrate separations whereas the LDA may be more appropriate at very short distances. To examine this, we cite the results of an LDA calculation of the electronic structure of a single Au atom situated atop a Na site in a $Na_{17}Cl_{17}$ cluster. The Au–Na bond distance is fixed at the sum of the Na ionic radius and the Au atomic radius.

The calculation shows that there is negligible interaction of the adsorbate with the orbitals of the underlying sodium ion. This occurs because the gold 5d levels fall within a huge energy gap that separates the 2p and 3s sodium-derived energy bands of bulk NaCl. By contrast, the chlorine 3p levels are rather close in energy. A true chemical bond forms between the Au $5d_{3z^2-r^2}$ orbital and $3p_z$ orbitals on the four nearest neighbor Cl ions. A contour map of the principal bonding wave function appears in Fig. 12.30. Electrons that occupy the associated energy level contribute to the net heat of adsorption. Unfortunately, it is not known whether this contribution, or an intrinsically non-local contribution to (8.19) omitted by the LDA, dominates the bond energy and determines the adsorption geometry.

General references

Metals

Scheffler, M. & Bradshaw, A.M. (1983). The Electronic Structure of Adsorbed Layers. In *The Chemical Physics of Solid Surfaces and Heterogeneous Catalysis* (eds. King & Woodruff), vol. 2, pp. 165–257. Amsterdam: Elsevier.

Magnetism

Falicov, L.M., Victoria, R.H. & Tersoff, J. (1985). Electronic and Magnetic Properties of Transition Metal Surfaces, Interfaces and Overlayers. In *The Structure of Surfaces* (eds. Van Hove & Tong), pp. 12–17. Berlin: Springer.

Semiconductors

Williams, R.H. & McGovern, I.T. (1984). Adsorption on Semiconductors. In *The Chemical Physics of Solid Surfaces and Heterogeneous Catalysis* (eds. King & Woodruff), vol. 3, pp. 267–309. Amsterdam: Elsevier.

Schottky barriers and band offsets

Margaritondo, G. (1983). Microscopic Investigations of Semiconductor Interfaces. *Solid State Electronics* **26**, 499–513.

Insulators

Kiselev, V.F. & Krylov, O.V. (1985). *Adsorption Processes on Semiconductor and Dielectric Surfaces.* Berlin: Springer.

13

ENERGY TRANSFER

Introduction

This chapter begins our exploration of the physics of dynamical processes at solid surfaces – adsorption, diffusion, reaction and desorption. To do so, we must leave the ground state problem and concentrate on the excited states of adsorbed atoms and molecules. One way to proceed focuses on the excitation spectrum. As we know, this spectrum comes in two parts: single particle excitations and collective excitations. Our earlier discussion for clean surfaces (Chapter 5) dwelt primarily with the latter and it is possible to duplicate that effort here. For example, Fig. 13.1 illustrates the calculated and measured (by EELS) dispersion of two types of collective excitations for two vastly different adsorbate/substrate combinations. The left panel pertains to vibrations localized in an oxygen adlayer on Ni(100), i.e. overlayer *phonons*. Theory and experiment are in good accord for this system. Both exhibit three dispersive branches. The low frequency acoustic excitation is the (oxygen-modified) Rayleigh mode of the nickel substrate.

A metallized adsorbate layer can support collective excitations of its charge density in addition to the more familiar phonon modes. The right panel of Fig. 13.1 compares theory and experiment for the dispersion of two-dimensional *plasmons* in an ordered potassium overlayer adsorbed on a dimerized Si(100)2×1 surface. Theory predicts three branches (one intraband and two interband excitation channels) whereas the experiment clearly resolves only two. The plasmon spectrum is more complicated than that of a free electron system (cf. Fig. 6.3) due to band structure effects in the potassium chains.

We do not pursue this taxonomy further. Instead, it is the single-particle excitations which require the most careful attention. This is so because it is precisely the low-lying electronic, vibrational, rotational and translational states of *individual* adsorbates that control the interesting physical processes listed at the outset. Quantitative study of these phenomena requires a reasonably detailed description of the participating excited states. Luckily, this is not necessary (at least initially) if only the essential qualitative features are of interest.

The basic issue is energy flow. An adsorbate enters into and exits from various excited states as energy flows between it and its environment. There are many components to this environment. Energy exchange can occur between a specified atom or molecule and some external source, neighboring adsorbates, the substrate, or some combination of these. This remains a complex subject even with the simplification noted above. Consequently, our general approach remains unchanged: we use the simplest models to sort out the physics. Hopefully, the explication of more complex (i.e., realistic) situations then requires only a judicious juxtaposition of our accumulated wisdom.

Electronic and vibrational states

Consider first an external radiation source (e.g., a photon or electron beam) that promotes an adsorbate from its ground state to an excited electronic or vibrational state. This is the basis to many of the

Fig. 13.1. Calculated (solid curves) and measured (symbols) dispersion of overlayer phonons for Ni(100) c(2 × 2)-O (left panel) and overlayer plasmons for Si(100) (2 × 1)-K (right panel). Note the break in the vertical scale for the latter. The central panel shows the substrate (open circles) and absorbate (closed circles) atoms approximately to scale for the two structures (Rahman *et al.*, 1984; Aruga, Tochihara & Murata, 1984).

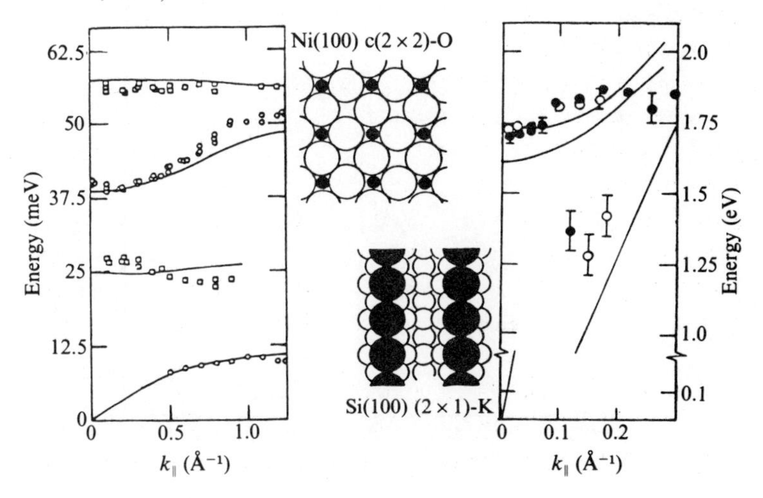

experimental techniques discussed heretofore: EELS, IRAS, NEXAFS, etc. Usually, there is little bottleneck to energy transfer from the incident beam to an adsorbed species and meaningful surface spectroscopy is possible. The well-known technique of Raman scattering is a famous counter-example. Vibrational Raman spectra are used widely in gases, liquids and solids as a complement to infrared absorption spectroscopy for quantitative studies of structure and bonding. Unfortunately, the inelastic Raman amplitude is a second-order process (see, e.g., Baym (1974)) and the corresponding scattering cross section is quite small ($\sim 10^{-30}$ cm^2/molecule-steradian typically). Even with the use of special detection techniques, the observed counting rate for adsorbed nitrobenzene (an extremely strong Raman scatterer: $d\sigma/d\Omega \sim 10^{-28}$ cm^2/steradian) does not bode well for routine surface studies (Fig. 13.2).

Raman spectroscopy was considered essentially useless for surface studies until the discovery (about a decade ago) that the scattering intensity from pyridine molecules adsorbed onto metal electrodes is enhanced over normal values by a factor of 10^5–10^8. By now, surface-enhanced Raman scattering (SERS) has been observed for about 100 different molecules on a dozen different metal substrates (Moskovits, 1985). Where does this tremendous enhancement come from? One possibility is a special case of the conventional resonance Raman effect, i.e., there exists an electronic excitation of the sample that matches the incident (laser) beam energy so that a nearly vanishing energy denominator drives the scattering through

Fig. 13.2. Raman spectra of submonolayer $C_6H_5NO_2$ deposited on Ni(111) at 100 K (Campion, Brown & Grizzle, 1982).

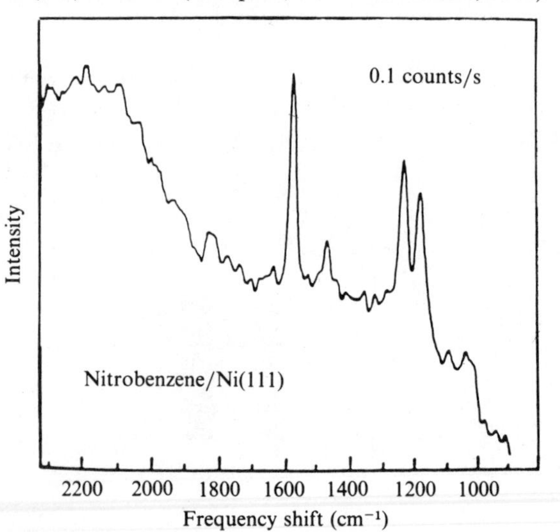

a resonance. Unfortunately, this mechanism cannot be general – it depends on the detailed electronic structure of the molecule/surface complex.

We turn to a second possibility. It is well established that the largest Raman enhancements occur when the metal surface is *roughened.* The roughness must be short on the length scale of the radiation (~ 5000 Å) but typically is long on an atomic length scale. A particularly simple model of such roughness might be a collection of hemispherical protrusions and depressions on an otherwise flat surface (Fig. 13.3). These small structures support local plasmon-like resonances, i.e., collective oscillations of charge confined to the hemispheres whose frequency is determined by the bulk metal dielectric constant and geometrical boundary conditions. If the external field can excite one of these modes, the electric field in the vicinity of the hemisphere is greatly enhanced over its flat surface value.

Completely classical electromagnetic calculations show that the field enhancement near the tip of an ellipsoidal protuberance can approach a factor of 10^2 (Fig. 13.4). This 'lightning rod' effect is sufficient to reproduce the observed enhancements. Recall two facts. First, like any second-order optical process, the Raman cross section is proportional to the square of the product of two dipole matrix elements. Second, each constituent matrix element is itself proportional to the magnitude of the *local* electric field at the adsorbate position (cf. (10.3)). Bearing in mind the local field enhancements, we see that a rough substrate acts as an *amplifier* for both the incoming (ω) and scattered (ω') wave fields. The total Raman cross section scales like $|A(\omega)A(\omega')|^2$.

The notion that surface roughness is responsible for SERS is difficult to check because a real metal surface exhibits roughness on many length

Fig. 13.3. Schematic model of a molecule (not to scale) above a rough metal surface. $A(\omega)$ is the electromagnetic enhancement factor. See Fig. 13.4.

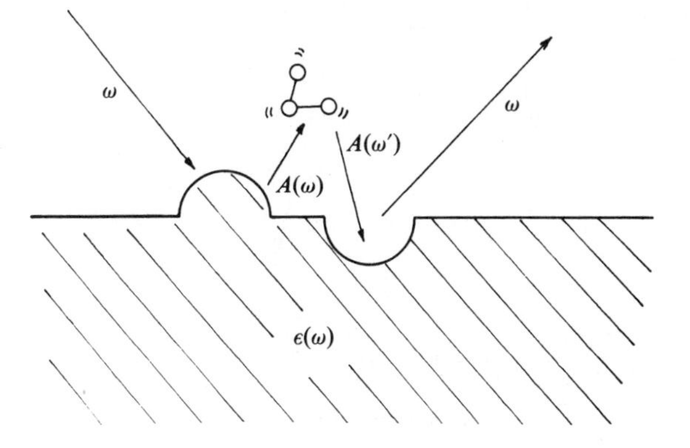

Fig. 13.4. The modulus of the field amplification factor $A(\omega)$ ($E(\omega) = A(\omega)E_{ext}(\omega)$) at the tip of a silver ellipsoid as a function of photon energy for several different aspect ratios. The semi-major axis is taken to be 200 Å (Gersten & Nitzan, 1982).

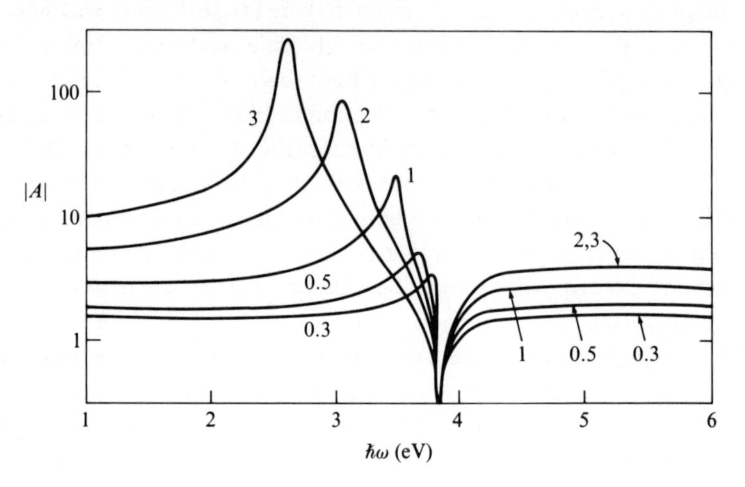

Fig. 13.5. SERS from cyanide molecules adsorbed onto silver ellipsoids (aspects ratios of 3:1 and 2:1) deposited onto a microlithographed SiO_2 substrate. Smooth curves are theory (Liao *et al.*, 1981).

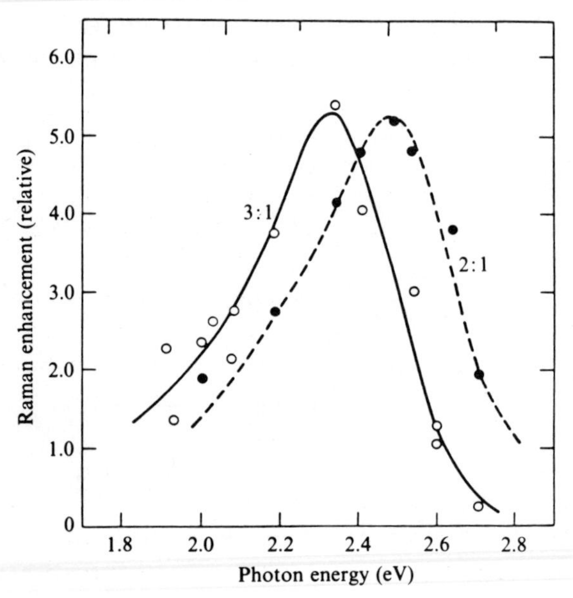

scales. Typical experiments show enhancements over a rather broader energy range than Fig. 13.4 would imply. To circumvent this problem, consider instead an 'artificial' surface created by microlithography. The substrate consists of regular silver ellipsoids evaporated onto posts (5000 Å in height and uniformly separated by 3200 Å) etched onto a SiO_2 wafer. Fig. 13.5 displays the relative Raman enhancement obtained from cyanide molecules deposited onto this surface. The results (for two different ellipsoid aspect ratios) are in excellent accord with classical electromagnetic theory.

The material properties of the substrate entered the foregoing example only to the extent that they set the frequency scale for roughness-induced field resonances. However, the substrate dielectric response profoundly affects the *decay* of an excited electronic or vibrational adsorbate state even if the surface is perfectly flat and there is no external source of radiation. This occurs because one can associate a dipole moment (and an associated dipolar field) with the transition matrix element that connects an excited state $|N\rangle$ to the ground state $|0\rangle$:

$$\boldsymbol{\mu} = |\langle N|e\mathbf{R}|0\rangle|. \tag{13.1}$$

For an electronic excitation **R** is just the position operator while for a vibrational excitation **R** is to be identified with the normal mode coordinate of the ion motion. We address the following question: at what rate does this oscillating dipole transfer energy to its surroundings. Or, equivalently, what is the lifetime of the excited state in the presence of a solid surface?

Experiments show that a metal surface strongly modulates the fluorescent decay rate of an excited molecule even when it is separated from the dipole by distances in excess of the emission wavelength. This remarkable result is illustrated in Fig. 13.6, which shows the measured lifetime of a fluorescent dye molecule containing Eu^{+3} ions as a function of its separation from a silver substrate. At very large metal–molecule separation, the decay rate exhibits its free space value, τ_0. The dramatic oscillations about this value occur because the dipole field of the molecular oscillator reflects from the metal surface. As a function of distance, the reflected wave arrives back at the molecule either in phase or out of phase with the original signal. This is again classical electromagnetism; the oscillations are an interference effect. Note that the Fresnel equations determine the amplitude and phase of the reflected wave so that only the bulk metal dielectric function enters the analysis.

Interestingly, the surface physics of this example occurs in the design of the experiment. The measurements require that the fluorescent molecules be situated at precisely controlled distances from the metal substrate. This is done by adsorbing perfect monolayers of fatty acid molecules by the Langmuir–Blodgett technique – one after another – until any desired 'spacer' distance is achieved (see inset to Fig. 13.6). Only the top adsorbed layer contains the fluorescent species.

The excited molecules decay via emission of photons at most of the distances displayed in Fig. 13.6 There is energy transfer directly to the ambient radiation field. However, at distances of about 100–200 Å, the observed lifetime drops precipitously: a new channel for energy transfer has opened. The photon emission rate remains unchanged but *non-radiative* energy transfer to the substrate begins to occur. Qualitatively, the near (or induction) field of the emitting system induces currents in the substrate. Energy flows from the dipole field of the oscillator to the solid and dissipates via Joule heating as electrons in the metal scatter from phonons, impurities and other electrons. A simple calculation permits us be more quantitative.

Consider first the fluorescent oscillator in free space. The polarizability $\alpha(\omega)$ relates the induced dipole moment to the 'bare' electric field at the molecule (Atkins, 1983):

$$\mathbf{p} = \alpha(\omega)\mathbf{E}, \tag{13.2}$$

Fig. 13.6. Comparison of experiment (dots) and electromagnetic theory (solid curve) for the fluorescent lifetime of an excited Eu^{+3} complex ($h\nu = 2\,\text{eV}$) as a function of the emitter molecule distance above a silver substrate (Chance, Prock & Silbey, 1978). Inset: construction of a spacer layer of fatty acid by the Langmuir–Blodgett technique (Kuhn, 1983).

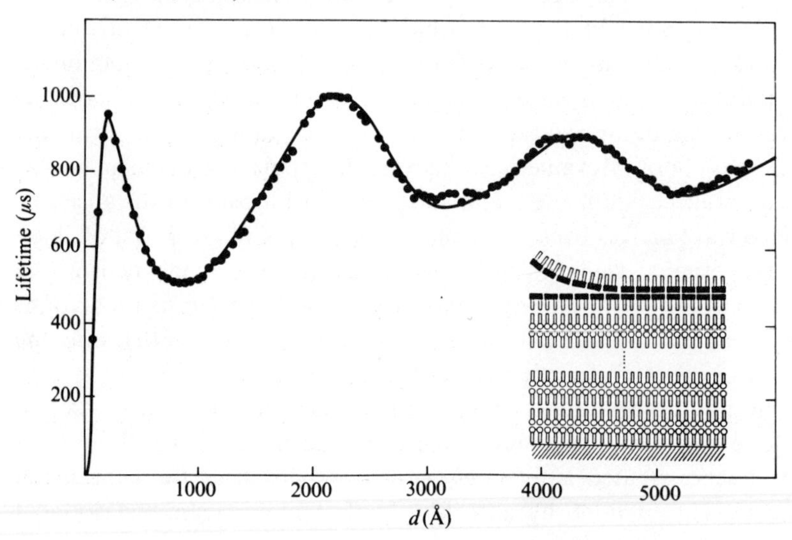

where

$$\alpha(\omega) = \frac{2\mu^2}{\hbar\Omega(1 - \omega^2/\Omega^2)} \equiv \frac{\alpha_0}{1 - \omega^2/\Omega^2}. \tag{13.3}$$

In this expression, μ is given by (13.1) and $\hbar\Omega$ is the transition energy between the ground and excited states. As usual, one includes the lifetime of the excited state by the replacement $\omega \to \omega + i/\tau_0$. We seek a change in the imaginary part of ω due the influence of the substrate. In the present case, the molecule polarizes the metal and the electric field induced at the emitter site is that of an electrodynamic image. Hence,

$$\mathbf{p} = \alpha(\mathbf{E}_{\text{ext}} + \mathbf{E}_{\text{image}}) = \alpha\mathbf{E}_{\text{ext}} + \frac{\alpha\mathbf{p}}{4d^3}\frac{\varepsilon(\omega) - 1}{\varepsilon(\omega) + 1}, \tag{13.4}$$

so that

$$\mathbf{p} = \frac{\alpha}{1 - \dfrac{\alpha}{4d^3}\dfrac{\varepsilon(\omega) - 1)}{\varepsilon(\omega) + 1}}\mathbf{E}_{\text{ext}}. \tag{13.5}$$

The denominator in (13.5) vanishes when $\omega = \omega_{\text{R}}$, the renormalized resonant frequency of the oscillator. Using the expression for α from (13.3) we find

$$\omega_{\text{R}}^2 = \left[1 - \frac{\alpha_0}{4d^3}\frac{\varepsilon(\omega_{\text{R}}) - 1}{\varepsilon(\omega_{\text{R}}) + 1}\right]\Omega^2. \tag{13.6}$$

Then, assuming that $\alpha_0/d^3 \ll 1$,

$$\frac{1}{\tau} = \text{Im}\,\omega_{\text{R}} \cong \frac{\mu^2}{4\hbar d^3}\,\text{Im}\,\frac{\varepsilon(\omega_{\text{R}}) - 1}{\varepsilon(\omega_{\text{R}}) + 1}. \tag{13.7}$$

The dipole oscillator damps with a characteristic inverse cube dependence on the metal–molecule separation.* Dissipative effects enter through the imaginary part of the dielectric function. In particular, if $\varepsilon(\omega_{\text{R}}) + 1 \cong 0$, most of the oscillator energy transfers first to the substrate surface plasmon mode (cf. (6.5)). Nevertheless, only the bulk properties of the substrate enter the problem (again through $\varepsilon(\omega)$).

The qualitative argument given earlier suggests when one might expect an explicit *surface* damping mechanism to become important: the distances d must be less than the inelastic mean free path in the metal. It is only at these distances (< 100 Å for silver) that non-radiative energy transfer to electron–hole pairs in the surface region becomes competitive with bulk loss processes. But note, creation of an electron–hole pair requires

* The energy transfer is proportional to the square of the dipole moment p so that the true damping rate is twice that given by (13.7).

momentum as well as energy. Where does the momentum come from? One source is the surface itself! Broken translational invariance at the surface provides momentum here just as it did in Chapter 4 to render $\mathbf{k}_\perp$ indeterminate for an ejected photoelectron. Accordingly, only electrons which propagate nearly perpendicularly to the surface plane can be excited by this mechanism.

It is easy to find a correction to (13.7) that properly accounts for this surface damping process. Notice that it is not possible to evaluate d in (13.4) for small distances unless we specify where the crystal terminates. But this is precisely the quantity $d_\perp(\omega)$ introduced long ago in a similar context (cf. (4.9) and (7.10)). Therefore, we let $d \to d - d_\perp(\omega)$ in (13.7) and expand to first order in $d_\perp(\omega)/d$. This determines the signature of surface damping: $1/\tau \sim d^{-4}$. Indeed, careful experiments similar to those described above confirm this behavior for biacetyl ($CH_3COCOCH_3$) molecules suspended above a Ag(111) surface by ammonia spacer layers (Fig. 13.7).

Fig. 13.7. Comparison of the measured lifetime of the first triplet state of biacetyl as a function of distance from Ag(111) with surface damping theory (solid line) ($\tau \sim d^4$) (Alivisatos, Waldeck & Harris, 1985).

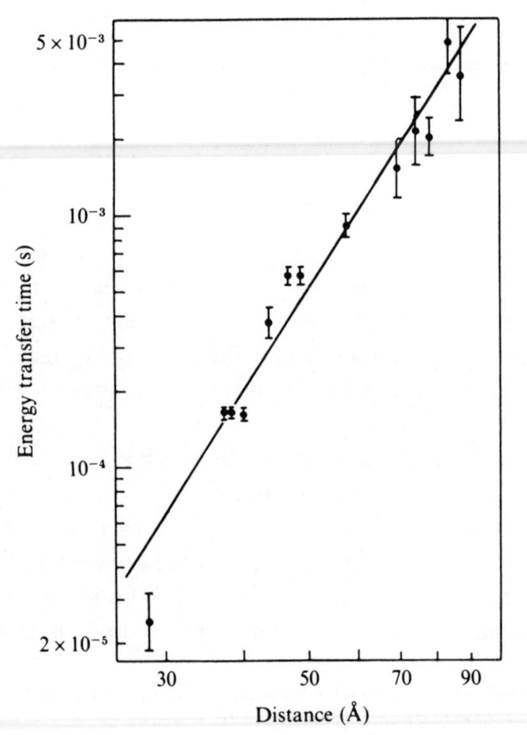

At still smaller values of d, detailed theory suggests that significant damping can occur by yet another mechanism (Persson & Andersson, 1984). Here, the Fourier components of the dipole electromagnetic field itself supply the necessary momentum to electron-hole pairs. However, this process dominates only when $d \ll \omega_F/\Omega k_F(\sim 10$ Å for typical parameter values) so that it becomes questionable to consider the dielectric response of the substrate without including the effect of the adsorbate from the beginning. Certainly a different approach is necessary if the excited state species is in direct contact with the substrate, i.e., an adsorbate.

Adsorbate vibrational modes are infinitely long-lived if one makes both the adiabatic and harmonic approximations to the ion motion. We observe a finite lifetime only because one of these simplifications breaks down. For example, anharmonic effects couple such vibrations to other (lower frequency) modes of the system such as the phonons of the substrate. As it happens, typical adsorbate–substrate and intramolecular vibrational frequencies (50–200 meV) are far greater than typical bulk phonon frequencies. An adsorbate mode can decay only by simultaneous creation of a great many substrate phonons – a very improbable event unless the anharmonicity is very large.

A failure of the Born–Oppenheimer approximation also can lead to vibrational state decay. This is easiest to see in the context of the resonant level model of an atomic adsorbate. An adatom located at a distance z from the substrate induces a local contribution to the density of electronic states in the vicinity of the energy ε_a (Fig. 9.8). All states below E_F are occupied and all states above E_F are unoccupied. But, according the Fig. 9.9, the resonant level position rises and falls as the adatom vibrates back and forth. If the ion motion is slow enough, electrons flow in and out of the resonant level to maintain a common Fermi level with the bulk. However, for sufficiently swift ion motion, the electrons lag behind and states that should be occupied in the adiabatic ground state remain empty. An electron–hole pair appears in the induced DOS, i.e., energy has been transferred from the vibrating ion to the electronic system.

Let us compute the non-adiabatic electronic damping rate from the Golden Rule. It seems reasonable to take the Coulomb field of the vibrating adsorbate as the perturbation which scatters electrons from below E_F to above E_F. Denote the deviation of the vibrational normal mode coordinate from its equilibrium value ($\mathbf{R}_{eq}$) by $\mathbf{R}(t)$. It is sufficient to expand the perturbation to lowest order in $\mathbf{R}(t)$:

$$\frac{Ze}{|\mathbf{r}-\mathbf{R}_{eq}-\mathbf{R}(t)|} \cong \frac{Ze}{|\mathbf{r}-\mathbf{R}_{eq}|} + \mathbf{R}(\mathbf{t})\cdot\frac{\partial}{\partial\mathbf{R}}\frac{Ze}{|\mathbf{r}-\mathbf{R}|}\bigg|_{\mathbf{R}=\mathbf{R}_{eq}}. \tag{13.8}$$

But this cannot be quite right. Surely we must include metallic screening from the substrate. Better still, the true force on the adsorbate arises from the complete self-consistent electron response to all the ions in the problem. In other words, the total effective potential $v_{eff}(x)$ as given, for example, by density functional theory (cf. 4.4).

We proceed as follows. The initial state is just the vibrational level $|N\rangle$ in the presence of a quiescent Fermi sea. The final state contains an excited electron–hole pair (kk') in the presence of a quiescent oscillator $|0\rangle$. Therefore,

$$\frac{1}{\tau} = \frac{2\pi}{\hbar} \sum_{\substack{k<E_F \\ k'>E_F}} |\langle kk',0|\mathbf{R}\cdot\frac{\partial v_{eff}}{\partial \mathbf{R}}|N\rangle|^2 \delta(\hbar\Omega - \varepsilon_{k'} + \varepsilon_k)$$

$$= \frac{2\pi}{M\Omega} \sum_{k,k'} \left| \int d\mathbf{r} \psi_{k'}^*(\mathbf{r}) \frac{\partial v_{eff}(\mathbf{r})}{\partial \mathbf{R}} \psi_k(\mathbf{r}) \right|^2 \delta(\hbar\Omega - \varepsilon_{k'} + \varepsilon_k), \quad (13.9)$$

where we have taken $R = (\hbar/2M\Omega)^{1/2}(b + b^\dagger)$ and included an explicit factor of 2 for spin. Now observe that $\hbar\Omega \ll E_F$ so that the perturbation excites electrons only in the immediate vicinity of the Fermi level. We use this fact to rewrite (13.9) in a form which shows that $1/\tau$ does not depend explicitly on the adsorbate vibrational frequency:

$$\frac{1}{\tau} = \frac{2\pi\hbar}{M} \sum_{k,k'} \left| \int d\mathbf{r} \psi_{k'}^*(\mathbf{r}) \psi_k(\mathbf{r}) \frac{\partial v_{eff}(\mathbf{r})}{\partial \mathbf{R}} \right|^2 \delta(E_F - \varepsilon_k)\delta(E_F - \varepsilon_{k'}). \quad (13.10)$$

The structure and content of (13.10) suggests that the largest electronic damping rates occur for light adsorbates on substrates with loosely bound electrons. Therefore, to set the scale, the top panel of Fig. 13.8 shows the results of an evaluation of this formula for the intramolecular stretching mode of a hydrogen molecule adsorbed upright onto a flat jellium surface. This figure also displays (lower panels) the LDOS at the adsorbate site as a function of adsorbate–substrate separation. Notice that the H_2 bonding level (horizontal arrow) is split off below the substrate band edge and persists as a sharp bound state. By contrast, the *anti-bonding* level overlaps the conduction band and broadens into a resonance. Both levels drop in energy as the center of mass of the molecule approaches the surface. The calculated damping rate (expressed as a vibrational line width $\Gamma = h/\tau$) rises rapidly as the LDOS of the unstretched molecule cuts across the system Fermi level. This is in complete accord with the argument given above generalized to a resonant anti-bonding level. Moreover, the damping rate rapidly *decreases* if the molecule is pushed in to greater (negative) values of Δd since then, the anti-bonding resonance drops completely below E_F.

At present, there are no measurements of the vibrational lifetime of H_2 adsorbed on simple metals. If we are willing to simply scale the adsorbate masses, indirect support for the correctness of the electronic damping scenario comes from the fact that the calculated damping rates are of the same order of magnitude (0.5–2 meV) as those observed for more complex systems such as CO, N_2 and CH_3O on noble and transition metal surfaces. However, by far the most convincing evidence appears in IRAS measurements of the *lineshape* of a vibrational mode of atomic hydrogen bridge-bonded to W(100). The experimental spectrum for W(100)(1 × 1)-H (Fig. 13.9) reveals two distinct features. First, a symmetric absorption feature (v_1) which corresponds to a beating mode of hydrogen against its bridge adsorption site (cf. top panel of Fig. 10.5(*b*)). Second, a highly asymmetric absorption feature associated with the first

Fig. 13.8. LDA calculations of the intramolecular vibrational linewidth (top panel) and induced local density of states (bottom panel) for molecular hydrogen on jellium ($r_s = 4.0$) as a function of adsorbate–substrate separation. The latter is given in terms of $\Delta d = d - d_{eq}$, where d_{eq} is the equilibrium separation (Hellsing & Persson, 1984; Johansson, 1981).

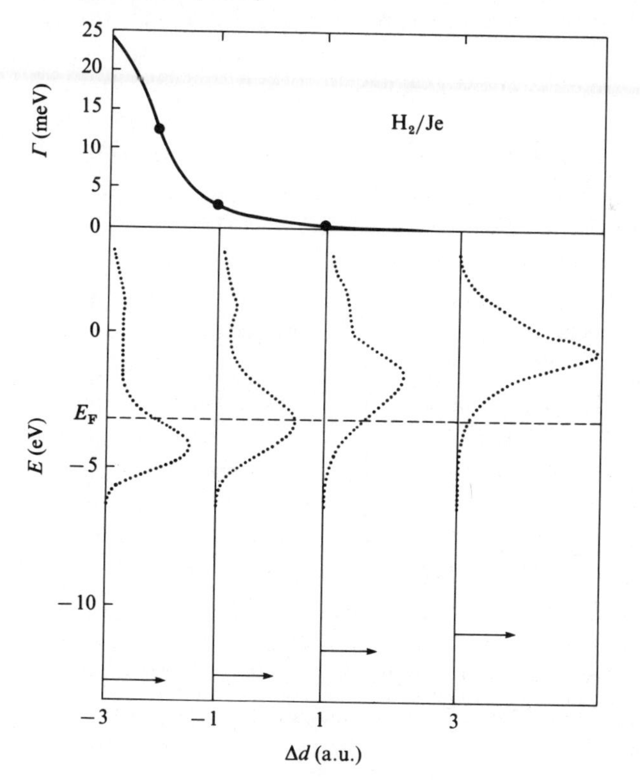

overtone of a low frequency 'wagging' mode (cf. bottom panel of Fig. 10.5(*b*)).* Why are these lineshapes so different?

Generally speaking, vibrational absorption spectra exhibit a Lorentzian lineshape. This form follows from our identification of the mode lifetime with the imaginary part of the mode frequency. The absorption is proportional to $\mathrm{Im}[\mu^2/(\omega - \omega_R - i/\tau)]$. Presumably this is correct for the ν_1 mode. However, the non-adiabatic damping mechanism is different. Charge sloshes into and out of an electronic level at E_F in a manner which is manifestly out-of-phase with the ion motion. Hence, the adsorbate dipole moment itself acquires an imaginary part and the lineshape deviates from a Lorentzian form. Detailed calculations (Langreth, 1985) show that the experimental signature is precisely the asymmetric lineshape seen in Fig. 13.9. We conclude that the wagging mode couples to near-Fermi level states far better than the beating mode.

Translational and rotational states

The question of the exchange of energy between a moving particle and a solid surface arose around the turn of the century in connection with experimental tests of the kinetic theory of gases. The earliest quantitative discussion is due to Baule (1914). He treated the interaction of a gas phase atom (m) with a stationary surface

Fig. 13.9. Infrared absorption spectrum of W(100) (1 × 1)-H. The dashed line is a fit to an asymmetric lineshape discussed in the text (Chabal, 1985).

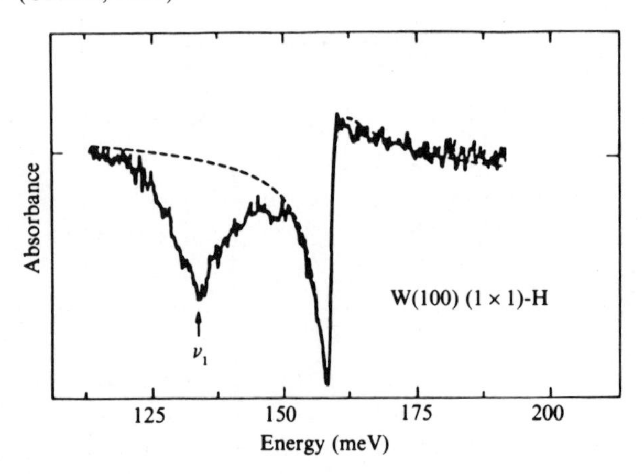

* Notice that the H vibrational frequencies in this high coverage ($\theta = 1$) 1 × 1 phase differ from those observed in the low coverage ($\theta = 0.25$) c(2 × 2) phase (Fig. 10.7). The frequency shifts are related to the reconstructive phase transition of the W(100) surface (Fig. 11.1). See Barnes & Willis (1978).

atom (M) as a binary elastic collision. Elementary conservation of momentum and energy immediately yields the energy transfer $\Delta = E_f - E_i$ as

$$\Delta = \frac{4\mu}{(1+\mu)^2} E_i, \tag{13.11}$$

where $\mu = m/M$ and E_i is the initial kinetic energy of the gas particle. This collision sets the surface 'atom' into motion and one readily can imagine a sequence of subsequent collisions between lattice atoms which dissipates the deposited energy deeper and deeper into the crystal. Although simple, this result provides a first estimate to the so-called *energy accommodation coefficient*, $\alpha = \Delta/E_i$.

A slightly better calculation takes account of the temperature of the surface (T_s) by assigning a velocity V to the surface atom before the original collision occurs. Identical considerations show that

$$\Delta = \frac{4\mu(E_i - \frac{1}{2}MV^2) + 2mv(1-\mu)V}{(1+\mu)^2} \tag{13.12}$$

where v is the velocity of the incoming particle. We average the velocity V over a Boltzmann distribution with $\langle V \rangle = 0$ and $\langle V^2 \rangle = kT_s/M$ to obtain the final result:

$$\Delta(T_s) \cong \frac{4\mu}{(1+\mu)^2}(E_i - \tfrac{1}{2}kT_s). \tag{13.13}$$

The vibrating lattice loses energy to the gas. An alternative approach takes $T_s = 0$ but considers energy transfer to the long wavelength acoustic modes of the solid by a gas whose particle velocities exhibit a Maxwell distribution. Under these conditions, one can show that the mean energy transfer is proportional to V^3 so that $\alpha \sim T^{3/2}$ (Landau, 1935). Let us put this prediction to the test.

Fig. 13.10 illustrates measurements of the energy accommodation coefficient for neon gas in equilibrium with a clean tungsten surface. The experiment follows a classical design pioneered by Knudsen (1934). We observe first that energy transfer is inefficient for a light particle such as Ne. This is consistent with (13.11). However, the (gas) temperature dependence is non-monotonic with a distinct minimum near 200 K. A $T^{3/2}$ law obtains only well above room temperature. The source of this disagreement is the neglect of an essential bit of physics – namely, the existence of a physisorption well. Crudely speaking, the effective kinetic energy of a gas particle that approaches the surface can never drop below U_0, the depth of the adsorption well. Replacing E_i by $E_i + U_0$ in (13.11) immediately implies that α is non-zero as $T \to 0$. More elaborate theories reproduce the upturn at low temperature and show that $kT_{\text{min}} \cong U_0$ (Trilling, 1970).

The calculation presented in Fig. 13.10 is surprisingly crude. The incoming particle interacts through a Morse potential with a solid modelled as an elastic continuum. Successful as it appears, this approach is somewhat at odds with our intuitive notion that an incoming particle collides with a specific surface atom and sets up local vibrations within the solid. The molecular dynamics technique (Chapter 5) provides a bridge between these two points of view. Therein, one treats the solid as a collection of individual atoms which interact via harmonic forces with their immediate neighbors. The motion of individual atoms is constrained only so that the average kinetic energy of the entire crystal reflects a specified temperature T. We also specify a specific gas atom/surface atom force law, such as that derived from a Morse potential.

The simulation begins with a single gas atom moving towards the crystal with a fixed initial velocity $\mathbf{v}$. The time evolution of the entire system follows by solution of Newton's second law for each particle, i.e., by integration of a set of coupled differential equations. In particular, we obtain the trajectory of the gas particle as it transfers energy with the crystal and scatters back into the gas phase. However, the 'solid' must contain hundreds (or thousands) of atoms so that the acoustic phonons are properly described. Simultaneous solution of a great many coupled equations is necessary to obtain even a single trajectory. Worse yet, the

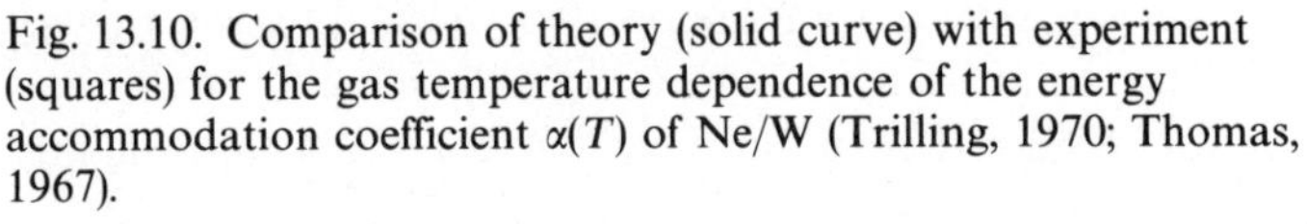

Fig. 13.10. Comparison of theory (solid curve) with experiment (squares) for the gas temperature dependence of the energy accommodation coefficient $\alpha(T)$ of Ne/W (Trilling, 1970; Thomas, 1967).

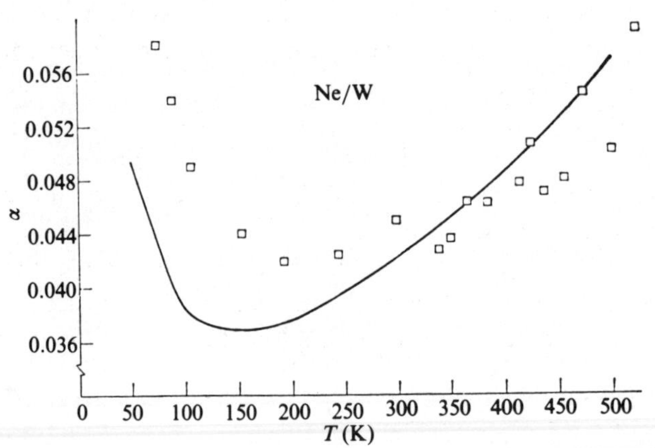

present problem requires that one average the resulting energy transfer over a great number of trajectories computed for gas particles whose initial velocities (and possibly rotations and vibrations) are chosen from a thermal distribution at temperature T. This is an impractical scheme, even for a supercomputer. To make progress, we resort to a trick.

An essential simplification results from the fact that the incident gas particle interacts with no more than a few substrate atoms at a time. There is no question that we must integrate the equations of motion for these 'primary' atoms explicitly. By contrast, the influence of the more distant 'secondary' atoms is indirect and manifests itself in two distinct ways. Both involve energy transfer. First, energy pumped into the system by the gas particle dissipates from the primary zone to the secondary zone in a manner which depends in a complex fashion on the interactions among the lattice atoms. Second, thermal fluctuations of the secondary atoms impart random impulsive forces (and hence energy) to the primary atoms. Guided by this intuition, let us ignore the equations of motion for particles in the secondary zone altogether. To compensate, it is necessary to modify Newton's law for the primary atoms to take account of the influence of the secondary atoms. The result is known as a generalized Langevin equation (Adelman & Doll, 1976):

$$m\ddot{\mathbf{r}} = -\Omega^2\mathbf{r} + \mathbf{f}(\mathbf{r}) - \int_0^t \mathrm{d}t'\Lambda(t-t')\dot{\mathbf{r}}(t') + \xi(t). \tag{13.14}$$

In this expression, $\mathbf{r}$ is an N-component column vector which contains the positions of the N atoms retained in the primary zone including the gas atom. The first two terms on the right hand side are due to Newton. Each primary particle couples harmonically to neighboring primary atoms and is subject to the external gas-surface force $\mathbf{f}(\mathbf{r})$. The second two terms reflect the presence of the rest of the solid. $\xi(t)$ is a random fluctuating force which provides local heating of the primary zone. Most of the complexity of the original problem winds up in the so-called 'memory' function $\Lambda(t)$. Notice that the integral depends on the *velocities* of the primary particles so that it plays the role of a friction force.

It is important to realize that $\Lambda(t)$ and $\xi(t)$ are not completely independent. Energy flow into and out of the primary zone must be regulated (on average) in such a way as to maintain the entire lattice at its fixed temperature T. This is guaranteed if one chooses the memory function proportional to the autocorrelation function of the random force,

$$\Lambda(t) = \frac{1}{kT}\langle \xi(t)\cdot\xi(0)\rangle, \tag{13.15}$$

a result known as the (second) fluctuation–dissipation theorem (see, e.g., Kubo, Toda & Hashitsume, 1985). The art of this method involves a realistic choice of $\mathbf{f}(\mathbf{r})$ and a clever choice of $\Lambda(t)$. Fig. 13.11 shows an application to the case of translational energy transfer between NO molecules and a Ag(111) surface. The simulation involves an average over 1500 trajectories computed under the same conditions as a corresponding molecular beam scattering experiment (see next section below). Good agreement is found for both the average velocity $\langle v \rangle$ and the root-mean-square velocity spread $\sigma = [\langle v^2 \rangle - \langle v \rangle^2]^{1/2}$ for molecules scattered through different angles by the silver surface.

Perhaps a question has entered the reader's mind. What has become of quantum mechanics? The foregoing classical analyses cannot possibly account for an intrinsically quantum mechanical effect such as elastic diffraction of the incoming particle. Furthermore, note that the energy transfers observed in the above examples are less than 10 meV, i.e., well below a typical Debye energy. A classical treatment of the lattice presumes

Fig. 13.11. Comparison of theory (boxes indicate statistical uncertainty) and experiment (points) for final velocity parameters of monoenergetic ($E_i = 85$ meV) NO molecules scattered from Ag(111) ($T_s = 500$ K). The molecules are incident at 50° from the surface normal. θ_f is the scattering angle and the rotational temperature of the beam is 20 K: (*a*) mean velocity; (*b*) rms velocity spread, both normalized to the incident velocity (Muhlhausen, Williams & Tully, 1985; Asada & Matsui, 1982).

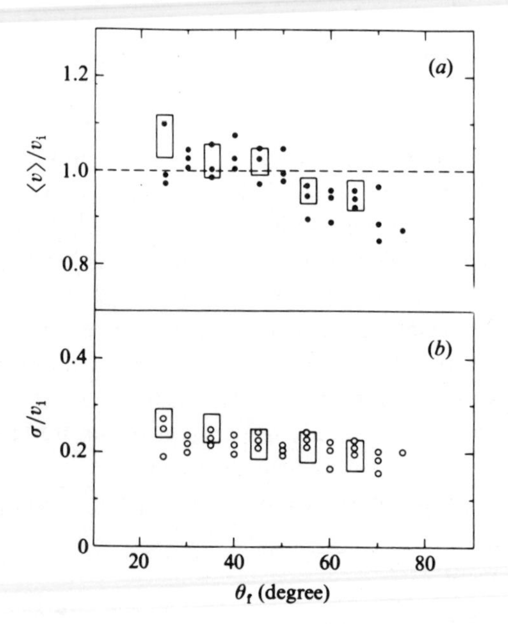

that this energy dissipates to many acoustic phonons of lower energy. What if energy transfer occurs instead by excitation of a single phonon? The mapping of surface phonon dispersion curves by inelastic atom scattering is predicated on precisely this assumption (Chapter 6). One might well expect explicit quantum effects in this regime. Indeed, the very notion of a classical trajectory is problematical for very light 'quantum' particles such as He or H_2. On the other hand, the success of classical theories supports the idea that thermal averaging washes out significant quantum effects. Clearly, the problem deserves further study.

To begin, we restrict ourselves to massive atoms and molecules. In that case, it is meaningful to speak of a gas phase particle which follows a classical trajectory $\mathbf{R}(t)$. The moving particle exerts a time-dependent force $\mathbf{f}[\mathbf{R}(t)]$ on the solid – modelled as a collection of harmonic oscillators. This force is, of course, the same as the one that entered the Langevin equation (13.14). The energy of the solid changes because the gas particle does work by setting lattice atoms into motion. Hence, an appropriate Hamiltonian for the system is:

$$\mathscr{H} = \sum_n \varepsilon_n a_n^\dagger a_n + \mathbf{f}[\mathbf{R}(t)]\cdot\mathbf{u}, \tag{13.16}$$

where $\mathbf{u}$ is the displacement of the surface atom struck by the particle and the subscript n labels the phonon energy eigenstates. To simplify notation, suppose further that $\mathbf{f}(t)\cdot\mathbf{u} = f(t)z$.

We require the amplitude that the solid winds up in oscillator state $|n\rangle$ at a time long after the particle has scattered back into the gas phase. This follows from conventional first-order time-dependent perturbation theory as

$$\langle n|\psi(t\to\infty)\rangle \equiv c_n(\infty) = -\frac{\mathrm{i}}{\hbar}\sqrt{\frac{\hbar}{2M\omega_n}} f(\varepsilon_n), \tag{13.17}$$

where

$$f(\varepsilon) = \int_{-\infty}^{+\infty} \mathrm{d}t\, f(t)\mathrm{e}^{\mathrm{i}\varepsilon t/\hbar}, \tag{13.18}$$

is the Fourier transform of the driving force. Equation (13.17) leads immediately to the probability that the lattice absorbs energy in the range between ε and $\varepsilon + \mathrm{d}\varepsilon$ due to *one-phonon* processes,

$$P_1(\varepsilon) = \sum_n |c_n(\infty)|^2 \delta(\varepsilon - \varepsilon_n) = \frac{|f(\varepsilon)|^2}{2M\varepsilon} g(\varepsilon) \tag{13.19}$$

and the average energy transfer,

$$\Delta_1 = \int_{-\infty}^{+\infty} \mathrm{d}\varepsilon\, \varepsilon P_1(\varepsilon) = \frac{1}{2M}\int_{-\infty}^{+\infty} \mathrm{d}\varepsilon\, g(\varepsilon)|f(\varepsilon)|^2. \tag{13.20}$$

In the last formulae, $g(\varepsilon) = \sum \delta(\varepsilon - \varepsilon_n)$ is the density of phonon states.

Equation (13.20) is a nice tidy expression. Unfortunately, it cannot describe the classical limit because, as noted above, that regime manifestly involves *multi-phonon* excitation. To do so, we must go beyond first-order perturbation theory. Luckily, it is possible to sum the perturbation series to all orders and diagonalize the Hamiltonian (13.16) exactly. One finds that the exact energy loss distribution function $P(\varepsilon)$ can be written in the form*

$$P(\varepsilon) = \int_{-\infty}^{+\infty} \mathrm{d}t \exp\left[\mathrm{i}\varepsilon t/\hbar + P_1(t)\right] \tag{13.21}$$

where

$$P_1(t) = \int_{-\infty}^{+\infty} \mathrm{d}\varepsilon \left[\mathrm{e}^{-\mathrm{i}\varepsilon t/\hbar} - 1\right] P_1(\varepsilon). \tag{13.22}$$

From this it is easy to verify that

$$\Delta = \int_{-\infty}^{+\infty} \mathrm{d}\varepsilon\, \varepsilon P(\varepsilon) = \int_{-\infty}^{+\infty} \mathrm{d}\varepsilon\, \varepsilon P_1(\varepsilon) = \Delta_1. \tag{13.23}$$

In other words, the average energy transfer is insensitive to multiphonon corrections to $P(\varepsilon)$. Moreover, an intrinsically quantum mechanical effect appears. There is a non-zero probability that the gas particle scatters elastically from the solid, i.e.,

$$\lim_{\varepsilon \to 0} P(\varepsilon) = \mathrm{e}^{-2W} \delta(\varepsilon), \tag{13.24}$$

where the 'Debye–Waller' factor is given by

$$2W = \int_{-\infty}^{+\infty} \mathrm{d}\varepsilon\, P_1(\varepsilon). \tag{13.25}$$

The energy loss distribution function exhibits two interesting limits for a crystal in equilibrium at temperature T (Brako, 1982). These are defined with respect to the characteristic energy of the problem: the upper limit to the phonon density of states – call it ε_0. In the extreme quantum limit, both Δ and kT are small compared to ε_0 and we recover the results of perturbation theory. $P(\varepsilon)$ is proportional to $g(\varepsilon)$ and a Bose–Einstein statistical factor. In the extreme classical limit, $\Delta \gg \varepsilon_0$, the no-loss line at zero energy is negligible and $P(\varepsilon)$ has a Gaussian shape centered at the Baule value of Δ (13.11) with a width of $2(kT\Delta)^{1/2}$. The two cases are

* Equation (13.21) is derived clearly, for example, by Sunjic & Lucas (1971), albeit in the context of a different problem.

sketched in Fig. 13.12. Most experiments appear to operate in the regime between 'marginally' classical and 'extremely' classical.

So far we have presumed that adsorbate translational energy dissipates exclusively to substrate lattice vibrations. However, for metal surfaces, the discussion of vibrational damping in the previous section clearly suggests a second possibility: energy transfer to electron–hole pairs. In that case, non-adiabatic effects particularly favor this mechanism if an adsorbate electronic level lies in the immediate vicinity of the Fermi level. We imagine a related process here. The only difference is that translational motion (normal to the surface) during the scattering process plays the role of vibrational motion in the previous case.

To be more quantitative we again adopt the forced oscillator model to evaluate the corresponding energy loss distribution function. This is possible because a coherent electron–hole pair (two fermions) can rigorously be regarded as a single bosonic entity (Gunnarsson & Schonhammer, 1982). But the creation and annihilation operators for the harmonic oscillator are also bosons! Consequently, for this problem, the quantum mechanics of phonon excitation and electron–hole pair production are identical. The preceding analysis and the qualitative results sketched in Fig. 13.12 may be carried over practically intact. The main difference is that the precise form of the forcing function $f(\varepsilon)$ is more uncertain. Parameterization of experiment could provide some information about this quantity. Unfortunately, this is a theory awaiting confirmation. At present, no measurements unambiguously identify electronic friction as a significant damping mechanism to adsorbate translational motion.

With minor modifications, the problem of *rotational* energy transfer at a solid surface can be treated along the lines sketched above. Substrate phonons and electronic excitations presumably damp rotations of an

Fig. 13.12. Schematic view of the energy loss distribution function of the forced oscillator model: (*a*) quantum limit; (*b*) classical limit. (Courtesy of J. Harris, IFF/KFA Jülich.)

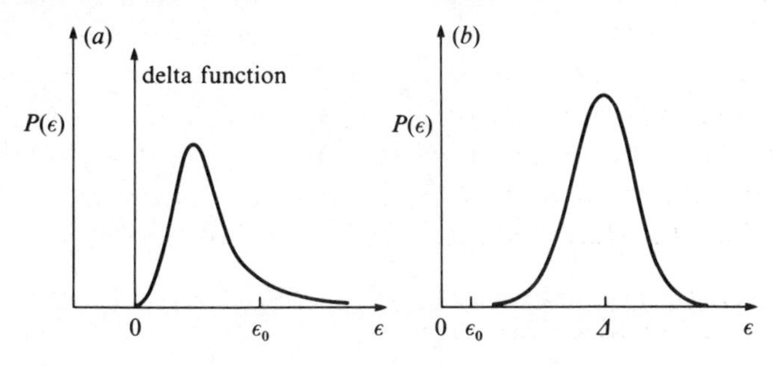

adsorbed molecule in a manner similar to that suggested for the other single-particle excitations. There is no need to repeat that discussion here. Instead, we explore a different question: the role of the surface in the *conversion* of energy between rotational and translational states of the adsorbate itself. The relative importance of quantum mechanical effects again becomes an issue – this time for the scattering problem. Is classical mechanics sufficient or is a wave mechanical approach necessary? For most diatomic molecules the de Broglie wavelength is small compared to the variations in the gas–surface interaction potential. Even so, gas phase studies show that certain kinds of quantum mechanical interference phenomena are quite sensitive to the details of the potential. *A priori*, it is difficult to know when such effects are negligible.

Fig. 13.13. Schematic representation of the collision of a molecule with a solid surface: (*a*) no initial molecular rotation; (*b*) moderate initial state rotation; (*c*) large initial state rotation.

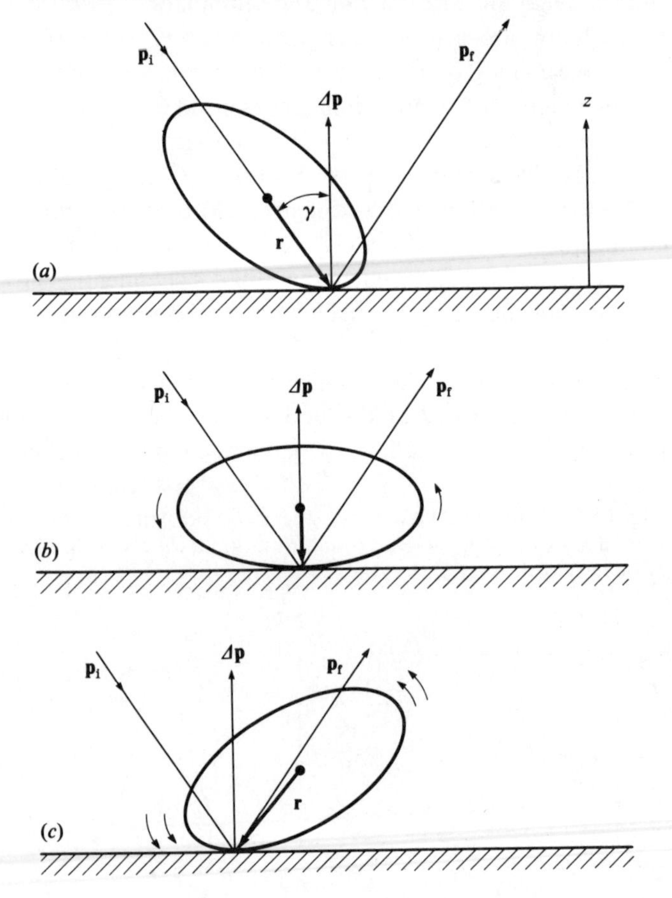

Most of the characteristics of translational to rotational (T → R) energy transfer at a solid surface emerge from the simplest classical considerations – notwithstanding the caveat of the previous paragraph. We treat the surface as a flat (no corrugation), rigid (no phonons) barrier and require only that the molecule not be spherically symmetric. In other words, throw a non-rotating, oblong object (e.g. a football) at a wall and ask what happens. The answer is obvious; the football bounces off spinning because the collision imparts a torque to it. Fig. 13.13(*a*) depicts the process. A rigid ellipsoidal shell (initial linear momentum $\mathbf{p}_i$) strikes the wall with its semi-major axis inclined at some angle γ from the surface normal. The impulse imparts a momentum transfer $\Delta\mathbf{p}$ and hence a torque $\mathbf{r} \times \Delta\mathbf{p}$ to the ellipse. Energy conservation demands that the ensuing translation ($\mathbf{p}_f$) and rotation share the kinetic energy of the initial state.

Now repeat the experiment with a football already set into (counter-clockwise) rotation before it strikes the surface. We represent the situation by turning the ellipsoid through part of this motion at the moment of impact (Fig. 13.13(*b*)). The collision produces no torque whatsoever in this configuration; there is no energy transfer. In our crude model, a more rapidly rotating football turns through an even greater angle by the time the collision occurs. As Fig. 13.13(*c*) shows, the induced torque now *opposes* the original rotational motion. The particle scatters from the surface with less (more) rotational (translational) kinetic energy than it had in the initial state. This is R → T energy transfer.

The results obtained from these simple considerations are borne out in fully quantal calculations. In their simplest form, one again treats the substrate as smooth and rigid, but now it is necessary to solve the Schrödinger equation for the wave function $\psi_\varepsilon(\mathbf{r})$ of the gas phase particle:

$$\left[-\frac{\hbar^2}{2M}\nabla^2 + V(\mathbf{r}) - \varepsilon\right]\psi_\varepsilon(\mathbf{r}) = 0. \tag{13.26}$$

In this expression, ε is the kinetic energy of the molecule and $V(\mathbf{r})$ is the gas–surface interaction potential. The scattering anisotropy evident in Fig. 13.13 enters here if $V(\mathbf{r})$ contains short range repulsion (or attraction) which depends explicitly on the orientation angle γ. One gets a reasonably realistic description with a simple Morse-like choice:*

$$V(\mathbf{r}) = A(\gamma)e^{-2z} - Be^{-z}. \tag{13.27}$$

In the language of scattering theory, T → R energy transfer occurs

* Recall that an equally simple, yet completely different, choice of gas–surface interaction potential was used to extract surface *structural* information from the *elastic* scattering channel (cf. (10.2)).

because the γ dependence of the potential permits mixing among the angular momentum components of the molecular wave function. In particular, l-states with no amplitude before impact can acquire non-zero amplitude after the collision. This is shown in Fig. 13.14 for a case where $A(\gamma)$ has been represented by a three-term expansion in Legendre polynomials. As advertised, the direction of energy transfer between translation and rotation changes sign as one increases the rotational quantum number (J) of the initial state.

The internal dynamics of the energy transfer process becomes clearer if one probes the distribution of *final* rotational states as a function of initial conditions. For simplicity, consider the case of an initially non-rotating ($J = 0$) molecule scattered from a potential (13.27) similar to the previous example. Fig. 13.15 shows the relative probability that the molecule winds up in a final state of rotational quantum number J. If one treats the scattering classically (solid curve) the distribution exhibits two prominent singularities. These are called 'rotational rainbows'. The name arises from an analogy to the physics of a conventional optical rainbow. In that case, one finds that light scatters particularly strongly from small water droplets through an angle θ_R because classical trajectories converge to it from both large and small impact parameters (Fig. 13.16). In the

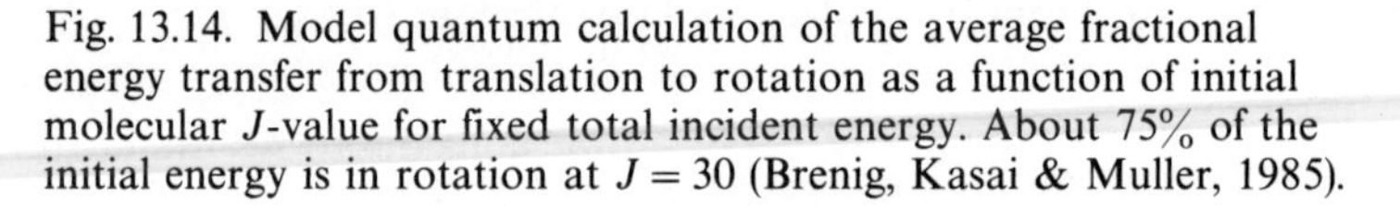

Fig. 13.14. Model quantum calculation of the average fractional energy transfer from translation to rotation as a function of initial molecular J-value for fixed total incident energy. About 75% of the initial energy is in rotation at $J = 30$ (Brenig, Kasai & Muller, 1985).

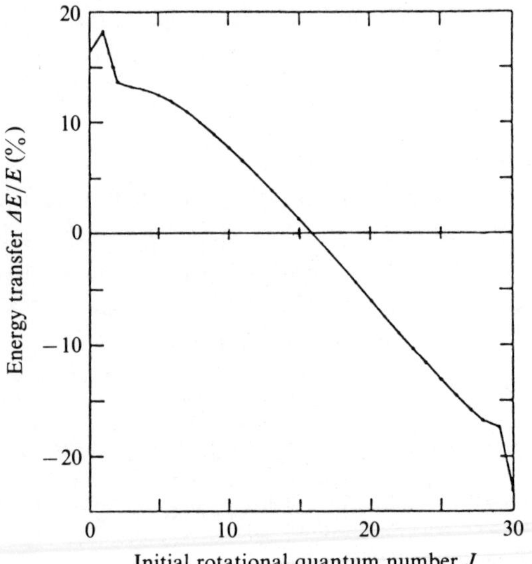

Fig. 13.15. Model potential calculations of the final rotational state distribution for the scattering of a non-rotating molecule from a flat surface. Solid curve is based on classical trajectories while dashed curve is the full quantal solution (Brenig, Kasai & Muller, 1985).

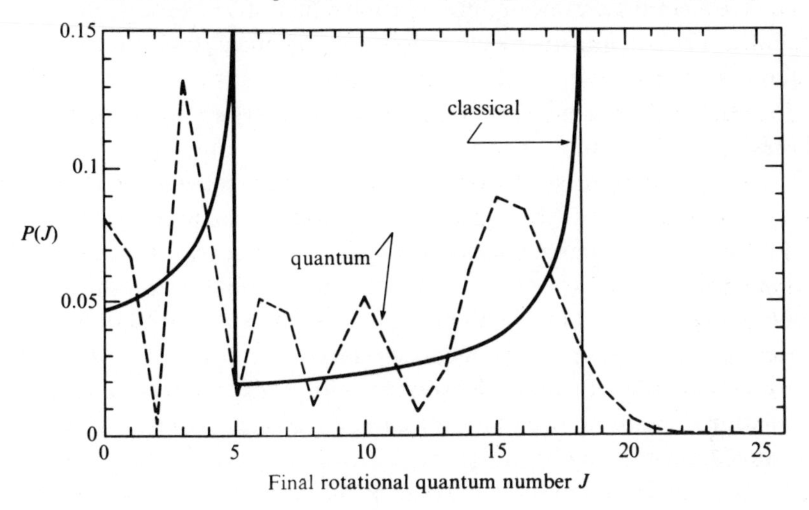

Fig. 13.16. Classical trajectories for scattering from a spherically symmetric potential as a function of impact parameter. A singular trajectory scatters into the rainbow angle θ_R (Pauly, 1979).

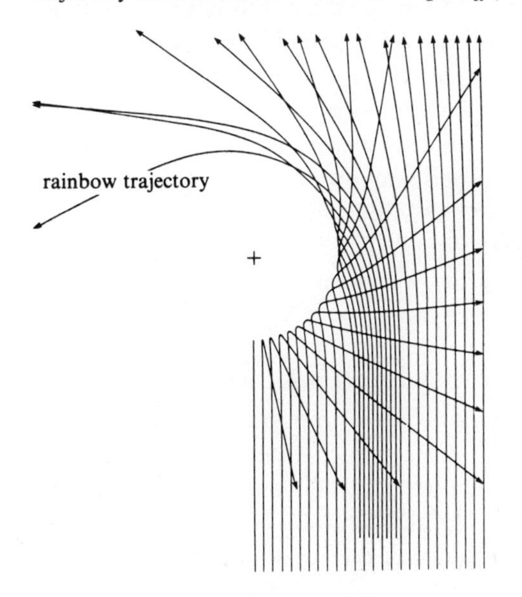

inelastic rotational case, scattering from incoming molecules with both large and small orientation angles γ produces momentum transfers which converge toward one particular value of induced torque. There is a large cross section at the corresponding rotational final state (see, e.g., Bowman & Park, 1982). Non-spherical scattering potentials typically lead to multiple rainbows.

The quantum mechanical rotational state distribution of Fig. 13.15 (dashed curve), obtained by solution of (13.26), plainly shows marked differences from the classical result. First, the rainbow singularity is reduced to a relative maximum. Second, interference among scattering paths leads to oscillations about the classical distribution. Third, energy transfer occurs to high rotational states in the classically inaccessible region above the second singularity. These predictions (particularly the oscillations) are obvious candidates for experimental scrutiny.

We conclude our theoretical overview with a few remarks concerning translational to vibrational ($T \rightarrow V$) energy transfer – a mode of energy exchange unmentioned to this point. Classical trajectory studies indicate that vibrational excitation following impact scattering is an inefficient process; rotational excitation dominates unless a (relatively rare) head-on collision occurs. On the other hand, one can imagine another mechanism – known as harpooning – which is familiar from gas phase collision studies. The idea is simple (Gadzuk, 1985) and follows from our by-now familiar notion that an electron hops from substrate to adsorbate if the latter's affinity level drops below the substrate Fermi level (Fig. 13.8).

Consider a molecule with a low-lying anti-bonding affinity level which scatters from a metal surface. At a certain critical distance of approach, the substrate 'throws out' an electron to the molecule and 'reels in' its catch via the Coulomb interaction of the ion with its electrical image. But, a molecule in this intermediate state certainly stretches apart in response to the filling of its anti-bonding orbital. Therefore, the backscattered neutral projectile can emerge vibrationally 'hot' if the particle does not adiabatically return to its equilibrium bond length when the affinity level rises back above the substrate Fermi level. The relative importance of this mechanism is unknown at present.

Molecular beams and state selection

The most interesting and important surface energy transfer processes probably occur under the non-equilibrium conditions of heterogeneous catalysis: a high temperature/high pressure gas comes into contact with a solid at relatively low temperature. Needless to say, these are not the ideal conditions for study of the basic physical processes of energy transfer. One typically resorts to model systems (but see Chapter 15), focuses on one (or a few) of the processes discussed above and then transfers the intuition so gained to the more complex problem. The same philosophy is used in gas phase studies of reaction dynamics (Levine & Bernstein, 1974). A particularly fruitful approach to our problem has been through the use of *state selected spectroscopy*. The idea is to experimentally determine the quantum state of a single species as fully as possible both before (if possible) and (definitely) after a gas–surface interaction. For the problems of concern to us here, the model system of choice is a molecular* beam scattered from a single crystal surface.

Fig. 13.17 illustrates a generic molecular beam arrangement. The particles emerge from a so-called supersonic nozzle beam source. Therein, one initially maintains the molecules of interest within a high pressure oven. As usual, the Maxwellian temperature of the gas T_0 characterizes the width of the velocity distribution. This completely changes when the gas rapidly expands from the reservoir into an evacuated sample chamber through a very small ($\sim$ 50 μm) conical nozzle. The speed v of the resulting beam is much greater than the local velocity of sound c – hence the use of the term supersonic. More importantly, it is easy to show that the molecules of the beam possess a velocity spread that is *much* narrower than that of the molecules in the oven (Kantrowitz & Grey, 1951). More precisely, the effective translational temperature of the beam T_{eff} is given by

$$\frac{T_{\text{eff}}}{T_0} = \frac{1}{1 + \frac{1}{2}(\gamma - 1)M}, \qquad (13.28)$$

Fig. 13.17. Schematic view of a molecular beam scattering apparatus capable of final translational, rotational and vibrational state selection. Standard equipment for surface cleaning (Ar^+ sputter gun) and composition (Auger) and structure (LEED) analysis is shown as well (D'Evelyn & Madix, 1984; Barker & Auerbach, 1984).

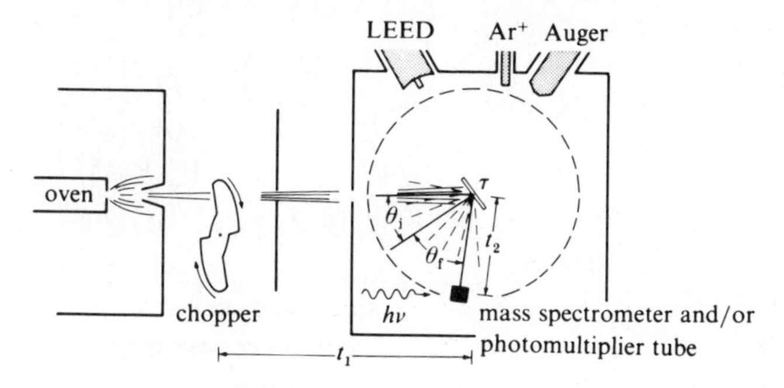

* We use the term 'molecular' to stand for both atomic and molecular beams.

where M is the Mach number (v/c) and $\gamma = C_p/C_V$ is the specific heat ratio of the beam molecules. The collisional mean free path within the beam is very short so that the streaming molecules rapidly reach local thermodynamic equilibrium. This means that one can have surface scattering with a large flux of molecules that are very cold both vibrationally and rotationally.

The beam particles typically pass a rotating chopper which permits both measurement of the arrival time t_1 and low-noise AC detection. They then scatter from the surface (possibly after some 'dwell' time τ) and arrive at a detector after time t_2. With a rotatable mass spectrometer as the detector, a time-of-flight analysis yields the final velocity distribution as a function of scattering angle. More elaborate methods are used to extract the final rotational and vibrational state distributions. Two popular schemes, laser excited fluorescence (LEF) and multi-photon ionization (MPI), involve an interrogation of the scattered beam with a tuned laser. The basic idea is simple but requires a detailed knowledge of the free molecule optical spectrum. Given the latter, it is clear that a specified laser frequency excites only those molecules with one particular set of rotational and vibrational quantum numbers (J, v) into a higher-lying roto-vibrational level (J', v'). From there, the molecules either fluoresce or they might be ionized by a second laser photon. One then records either the intensity of fluorescent

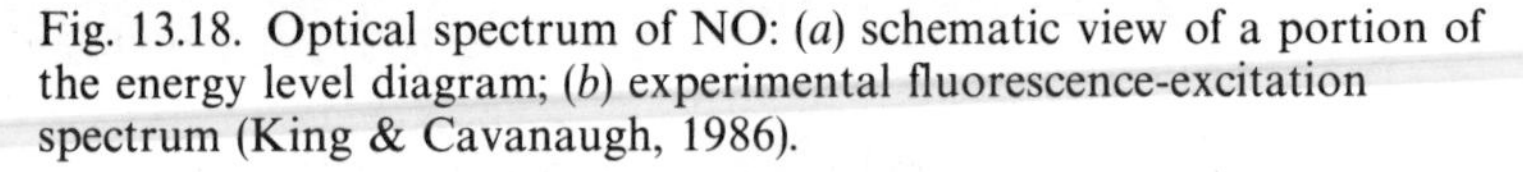

Fig. 13.18. Optical spectrum of NO: (*a*) schematic view of a portion of the energy level diagram; (*b*) experimental fluorescence-excitation spectrum (King & Cavanaugh, 1986).

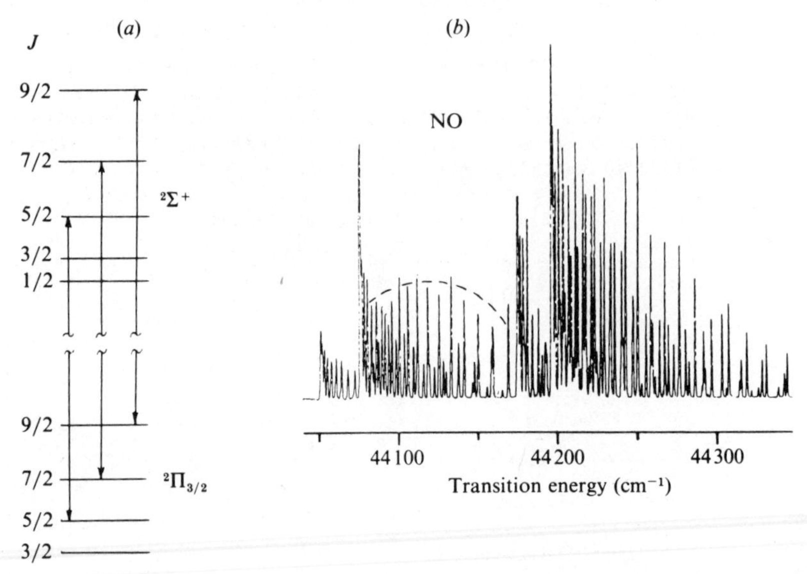

emission from the (J', v') level (LEF) or the flux of ions produced by electron ejection from the (J', v') level (MPI). In either case, the signal strength is proportional to the fraction of molecules originally scattered into the (J, v) quantum state. Hence, by sweeping the laser frequency, one maps out the distribution of final molecular states.

An LEF spectrum for nitric oxide (along with a portion of the NO energy level scheme) serves to illustrate the sensitivity of this class of techniques (Fig. 13.18). Focus attention on the spectral features beneath the hemispherical envelope sketched around 44125 wave numbers. It is easy to pick out a sequence of high intensity lines interlaced with a sequence of low intensity lines. The intense lines correspond to $\Delta J = 0$ transitions from the excited $^2\Sigma^+$ electronic state manifold to the $^2\Pi_{3/2}$ ground manifold with $J = 5/2, 7/2, \ldots, 27/2$. Clearly significant state selection is possible.

It is time to confront the theoretical ideas outlined earlier with detailed molecular beam scattering results for energy transfer between atoms or diatomic molecules and clean crystal surfaces. At present, we restrict ourselves to cases where the 'dwell' time is zero, i.e., where the gas–surface interaction occurs in a single inelastic scattering event.

Let us begin with a check of (13.11) and (13.13), the simplest predictions of the naive classical model. In fact, they describe reality rather well (Fig. 13.19). Time-of-flight measurements show that the energy lost by incident I_2 molecules scattered from MgO(100) is indeed roughly linear in the incident kinetic energy – at least at the higher energies. Perhaps more striking is the case of argon atoms scattered from a polycrystalline

Fig. 13.19. Tests of the Baule model of energy loss to bulk phonons: (*a*) energy loss vs. incident kinetic energy for I_2/MgO(100) (Kolodney, Amirav, Elber & Gerber, 1984); (*b*) exit kinetic energy vs. incident kinetic energy (both scaled to the surface temperature) for Ar/W (Janda *et al.*, 1980). Solid lines are guides to the eye.

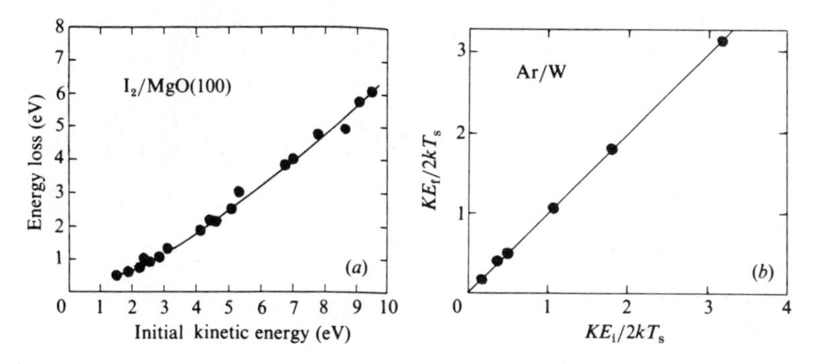

tungsten surface. The data show scaling behavior with the substrate temperature just as expected.

The well-studied* NO/Ag(111) system provides an opportunity to compare a variety of different theoretical approaches to rotational excitation data. Fig. 13.20 illustrates the final rotational state distribution obtained by LEF (filled circles) for an incident beam of molecules with a rotational temperature of about 40 K. The kinetic energy of the beam is 0.75 eV but the same curve results as long as the 'normal' kinetic energy $E_n = E_i \cos^2 \theta_i$ (where θ_i is the incidence angle measured from the surface

Fig. 13.20. Boltzmann plot of the final state rotational distribution of NO molecules scattered from Ag(111) at $E_n = 0.7$ eV. The data (filled circles) are shown along with a quantum mechanical calculation for a smooth, rigid substrate (open circles), a classical Langevin simulation for a non-rigid substrate (dashed curve) and a phenomenological analysis (solid curve). See text for details (Kleyn, Luntz & Auerbach, 1981; Voges & Schinke, 1983; Muhlhausen, Williams & Tully, 1985; Zamir & Levine, 1984).

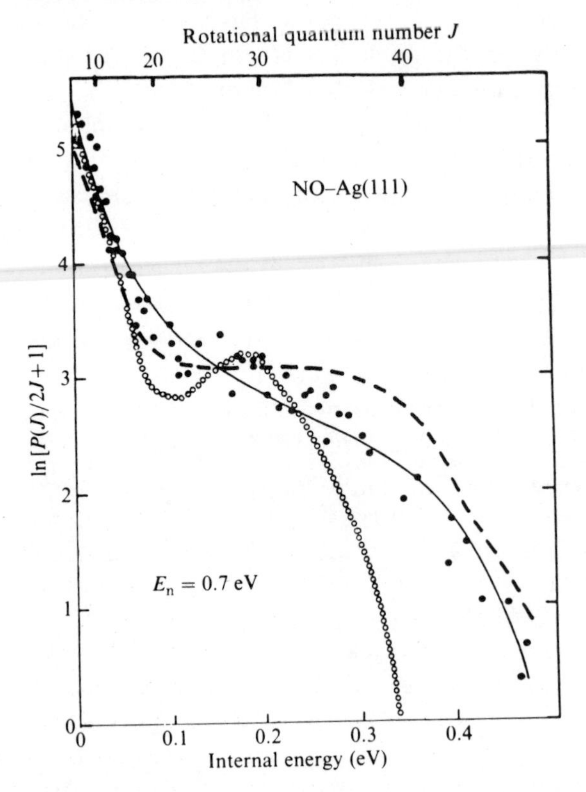

* Because the relevant excited states of NO are accessible conveniently by visible laser photons.

normal) is held fixed. This is consistent with the idea that a collision with a smooth surface conserves a particle's momentum parallel to the substrate. The vertical axis displays the *logarithm* of the probability for excitation into state J. In this way, a thermal Boltzmann distribution at temperature T_{rot} appears as a straight line with a slope of $-1/kT_{rot}$. We observe seemingly Boltzmann behavior at low final state energy ($J < 20$) with a distinct shoulder at higher values of energy. However, the final rotational temperature extracted from the data matches neither the incident beam temperature nor the surface temperature!

Fig. 13.20 compares the experimental rotational state data with three completely different calculations. The first is based on a quantum scattering theory solution to (13.26) with a simple potential function similar to (13.27). An anisotropy is built into $A(\gamma)$ so that the two ends of the heteronuclear NO molecule scatter from the surface somewhat differently. The results (open circles) identify the high-J plateau in the data with a rotational rainbow. But, in addition, a second, intense scattering feature appears in the calculation at very low J. According to this analysis, it is the second rainbow, combined with thermal averaging over the intial states (which wipes out any oscillatory behavior (cf. Fig. 13.15)), which mimics Boltzmann behavior at low energy.

The dashed curve in Fig. 13.20 is a completely classical Langevin simulation based on solution to (13.14) for the scattering process. The NO–Ag(111) scattering potential adopted for this study is rather different from the one chosen in the quantal calculation. Nevertheless, one again finds an excellent account of the data. Analysis of the classical trajectories also attributes the high energy shoulder to rainbow scattering. However, this calculation (which includes the influence of substrate lattice vibrations) suggests that many complex trajectories, including multiple hits, contribute to the exponential behavior at low energy. No single low-J rainbow is identified unambiguously.

Finally, the solid curve in Fig. 13.20 follows from a theoretical study which is not based on any kind of scattering theory. Nor is an interaction potential specified. It is a purely phenomenological analysis in which one presumes that the final states are occupied completely statistically (to maximize entropy) subject to certain constraints, e.g., that the average energy transfer (taken from experiment) be correctly reproduced. The fit to the data so obtained does not lead to much microscopic insight but does implicate at most two principal transfer mechanisms. Although unspecified in detail, these might be taken as single and multiple encounters with the surface to be consistent with the Langevin analysis.

What can we conclude from all of this? Disappointingly, thermal averaging not only dominates quantal effects but renders the experiment insensitive to the details of the gas–surface interaction potential. We are back in a familiar situation: reliable surface science information requires a consistent interpretation of a battery of experimental results. What is needed is more state selection and experiments definitely are moving in this (difficult) direction. We conclude with a prototype study – again for the NO/Ag(111) system. It involves the *interplay* between energy transfer to two different channels: rotations of the projectile molecule and vibrations of substrate target.

Fig. 13.21 shows the measured average final NO translational energy plotted against the measured average final NO rotational energy. Data are shown for a number of different initial kinetic energies E_i with a very cold beam ($T_R < 5$ K). To extract the message in this data, partition the total energy of the system as

$$E_i = E_f + E_{rot} + E_{phonon}. \tag{13.29}$$

The collision transfers substantial energy to the substrate ($\sim 30\%$) when

Fig. 13.21. Variation of the mean translational energy of NO molecules scattered from Ag(111) plotted as a function of their final rotational energies. Data (symbols) are shown at several different beam kinetic energies (0.9–0.1 eV from top to bottom). Dashed lines are the results of Langevin molecular dynamics simulations (Kimman *et al.*, 1986).

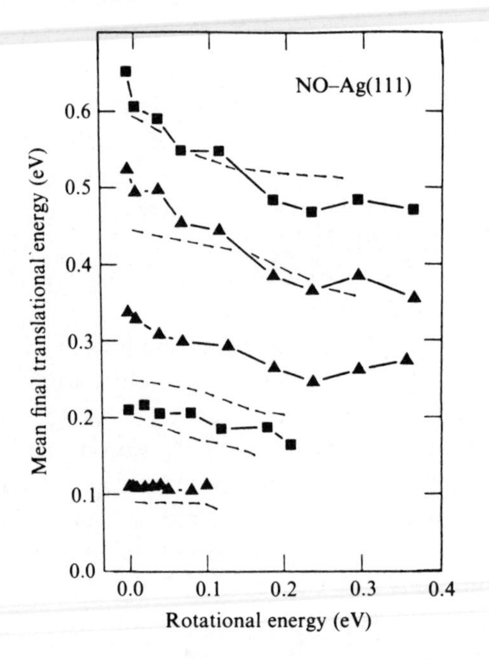

the molecules emerge unrotating. The data would lie on a line of slope -1 if phonons extracted precisely this same amount of energy (on average) regardless of the final rotational state. Instead, the slope is less negative than -1. We conclude that energy transfer to the substrate *decreases* as the final rotational energy *increases*. This is an entirely classical effect which is easy to understand on the basis of energy conservation. Basically, energy deposited in rotation reduces the energy (and hence the momentum) available for transfer to the substrate through a Baule two-body collision process.

Suppose we adopt the classical molecular dynamics approach to analyze this experiment. In this context, it is sensible to adjust the NO–metal surface interaction potential to produce a best fit to the data without, of course, destroying agreement with Fig. 13.20. The dashed lines in Fig. 13.21 are the results obtained from thousands of Langevin trajectory simulations for a potential adjusted in this way. Although the results are hardly perfect, they serve to underscore an important point reminiscent of a comment made much earlier regarding LEED (Chapter 3): the veracity and relevance of a theoretical approach to intimate energy transfer at surfaces increases in proportion to the number of independent experiments it can reproduce quantitatively.

General references

Electronic and vibrational states

Avouris, P. & Persson, B.N.J. (1984). Excited States at Metal Surfaces and Their Non-Radiative Relaxation. *Journal of Physical Chemistry* **88**, 837–48.

Translational and rotational states

Tully, J.C. (1980). Theories of the Dynamics of Inelastic and Reactive Processes at Surfaces. *Annual Reviews of Physical Chemistry* **31**, 319–43.

Molecular beams and state selection

Barker, J.A. & Auerbach, D.J. (1984). Gas–Surface Interactions and Dynamics; Thermal Energy Atomic and Molecular Beam Studies. *Surface Science Reports* **4**, 1–100.

14

KINETICS AND DYNAMICS

Introduction

Among many other seminal contributions to surface science, Langmuir championed the idea that the encounters of a gaseous species with a solid surface constitute a class of chemical reactions. Thus, if we denote a surface site by S and an adsorbate atom by A,

$$A(\text{gas}) + S \rightarrow S{:}A(\text{ads}) \quad \text{and} \quad 2S{:}A(\text{ads}) \rightarrow 2S + A_2(\text{gas}) \qquad (14.1)$$

represent the processes of atomic adsorption and recombinative molecular desorption, respectively. This notion is useful because one then can bring to bear on the problem all the machinery developed to study conventional chemical reactions. The analogy takes us in two directions. First there is the question of *kinetics*, that is, the influence of external macroscopic variables on the overall reaction rate. Typical control parameters in solution phase kinetic studies are the temperature, pressure and relative concentration of reactants (see, e.g., Gardiner, 1969). Alternatively, one can focus attention on the detailed atomic motions that characterize an elementary act of reaction, i.e., the *dynamics* of the process. Molecular beam and laser techniques are the tools of this trade (see, e.g., Bernstein, 1982).

Most of what we know about the energetics of chemisorption comes from many years' acquisition of surface kinetic data. A typical experiment involves measurement of the rate of gas adsorption or desorption as a function of surface temperature and coverage. The microscopic parameters of the system follow from fitting the data to simple postulated rate laws. Experimental studies of surface reaction dynamics are both more recent and more difficult to perform. These measurements are highly desirable because they illuminate so many details of the gas–surface interaction process. But, as the concluding section of Chapter 13 showed, the interpretation of the data may require considerable theoretical input. In the discussion to follow, we use kinetic studies to establish the range of possible phenomena. In a few selected cases, a dynamical analysis provides a window into the underlying physics.

Suppose a solid is in thermal equilibrium with an ambient gas. What is the fate of a mobile gas particle that approaches the solid's surface? The answer depends on the nature of the gas–surface interaction potential. Fig. 14.1 illustrates two possibilities. High kinetic energy particles probably scatter back into the gas phase from the back of the adsorption well. Even low kinetic energy particles may scatter away if the potential possesses a repulsive barrier outside the well. On the other hand, one (or several) of the energy loss mechanisms detailed in Chapter 13 may reduce the normal kinetic energy to the point where the particle 'traps' in the well. It becomes adsorbed. Strange as it may seem at first, we best can understand both the kinetics and dynamics of this process by consideration of the inverse process: desorption.

Consider a group of adsorbed particles thermally distributed amongst the energy levels ε of the one-dimensional adsorption well in Fig. 14.1(*a*).

Fig. 14.1. Schematic one-dimensional view of the approach of a gas particle to a solid surface: (*a*) a simple adsorption well; (*b*) well with a barrier to adsorption (Tully, 1981).

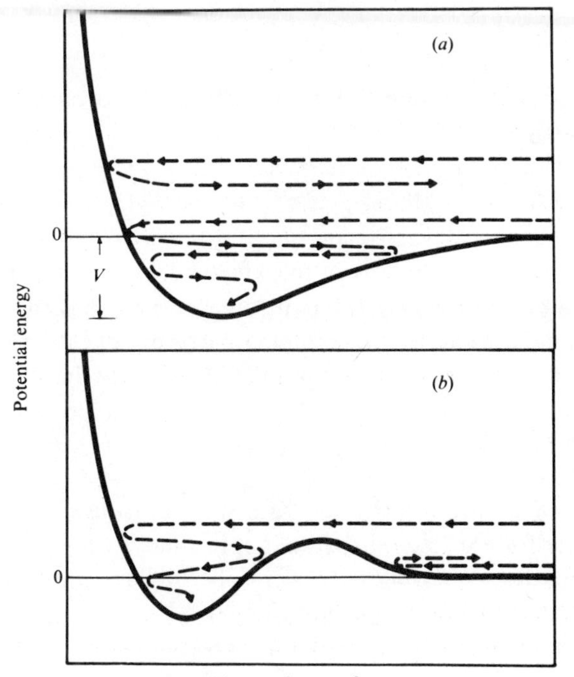

Each particle oscillates to and fro with a frequency $\omega(\varepsilon)$. Hence, in the classical limit ($T \gg \omega(\varepsilon)$), the total number of trapped particles is

$$n_0(T) = \int_{-V}^{0} \mathrm{d}\varepsilon\, g(\varepsilon) f_0(\varepsilon) \tag{14.2}$$

where $g(\varepsilon)$ is the density of oscillator states in the well and

$$f_0(\varepsilon) = Z_{\mathrm{ads}} \mathrm{e}^{-(\varepsilon - \mu)/kT}, \tag{14.3}$$

is the classical occupation factor. The quantity Z_{ads} is the partition function for any internal degrees of freedom which might be present such as adsorbate rotations and μ is the chemical potential. For deep wells, $V \gg T$, the integral (14.2) is dominated by energies within kT of the bottom of the well. Hence, it is sufficient to take $g(\varepsilon) = 1/\hbar\omega(\varepsilon) \cong 1/\hbar\omega_0$ where $\hbar\omega_0$ is the oscillation frequency at the bottom of the well. In that case,

$$n_0(T) = \frac{kT}{\hbar\omega_0} Z_{\mathrm{ads}} \mathrm{e}^{(\mu + V)/kT}. \tag{14.4}$$

Now suppose there is no gas phase above the solid. The trapped particles still are in equilibrium with the surface except for a thin layer near the top of the well. We thus retain (14.4) and consider the net flux of particles J_{out} that leave the well per unit time:

$$J_{\mathrm{out}} = n_0 R. \tag{14.5}$$

This defines the desorption rate R. But the net flux also is calculable from kinetic theory as

$$J_{\mathrm{out}}\, \mathrm{d}t = \int_0^{\infty} \frac{\mathrm{d}x\, \mathrm{d}p}{h} f(\varepsilon) = \frac{kT}{h} Z\, \mathrm{e}^{\mu/kT}\, \mathrm{d}t. \tag{14.6}$$

In this expression, $\varepsilon = p^2/2m$ now is the kinetic energy of the particles, $\mathrm{d}x = p\,\mathrm{d}t/m$ is the distance they travel in time $\mathrm{d}t$, and Z is the internal partition function characteristic of escaping particles near the top of the well. The rate follows from comparison of (14.6) with (14.5):

$$R = R_0 = \frac{\omega_0}{2\pi} \frac{Z}{Z_{\mathrm{ads}}} \mathrm{e}^{-V/kT}. \tag{14.7}$$

The desorption rate has an Arrhenius form with an *activation energy* V. A similar analysis of the *adsorption* rate appropriate to Fig. 14.1(*b*) also leads to Arrhenius behavior (Iche & Nozieres, 1976). The activation energy in this case is simply the barrier height.

There is a certain inconsistency with our derivation of (14.7). Note that we took the thermal equilibrium distribution function $f(\varepsilon)$ to extend to all energies in the integral in (14.6) – despite our explicit comment that this actually is correct only for particles deep within the well. The assumption of equilibrium is the cornerstone of so-called absolute rate theory (Glasstone, Laidler & Eyring, 1941). To correct it, return again to the case where the adsorbed layer actually *is* in global equilibrium with its vapor and use the principle of detailed balance. That is, demand that the net flux leaving the surface J_{out} be balanced by a net flux J_{in} approaching the surface. J_{in} is given by an expression just like (14.5) with an important exception.

Fig. 14.1 shows that some particles that approach the surface from the right bounce back into the gas phase. These cannot be counted in J_{in}. Consequently, we write

$$J_{in} = \int_0^\infty \frac{v\,dp}{h} f(\varepsilon)\, s(\varepsilon), \tag{14.8}$$

where $s(\varepsilon)$ is the probability that a particle incident with energy ε loses enough energy to remain trapped in the well. Equating the fluxes we find

$$R = R_0 \int_0^\infty d\left(\frac{\varepsilon}{kT}\right) e^{-\varepsilon/kT} s(\varepsilon) = s(T) R_0, \tag{14.9}$$

which defines the *thermal sticking coefficient* $s(T)$ as the correction factor to the prediction of absolute rate theory. Evidently, $s(\varepsilon)$ is related to dissipative processes near the surface and, as such, it can be written in terms of the energy loss distribution function $P(\varepsilon)$ defined in Chapter 13:

$$s(\varepsilon) = \int_\varepsilon^\infty P_\varepsilon(\varepsilon')\, d\varepsilon'. \tag{14.10}$$

The subscript on $P_\varepsilon(\varepsilon')$ is a reminder that the loss function depends on the particular forcing function (cf. (13.19)) appropriate to a particle of initial kinetic energy ε.

We now have what we want. Equation (14.9) is a kinetic expression for desorption. Such measurements determine the adsorption energy V and an exponential prefactor. Part of the prefactor comes from equilibrium thermodynamics. But another part – the sticking coefficient – arises from local non-equilibrium processes. The average value $s(T)$ can be found from adsorption kinetics (see below). The constituent quantities, $s(\varepsilon)$ or $P(\varepsilon)$, intimately involve dynamics.

Adsorption

A typical measurement of adsorption kinetics is quite straightforward. One simply exposes the sample to gas at a fixed pressure and monitors the adsorbate coverage as a function of exposure. Data collected in this way are shown in Fig. 14.2 for the case of dissociative chemisorption of oxygen on Rh(111). In this case, Auger spectroscopy is used to monitor the *relative* coverage. The solid curve through the data is a fit to so-called Langmuir kinetics. This model is based on the idea that adsorbate particles randomly occupy the sites of an adsorption checkerboard. Accordingly, the coverage increases at a rate in direct proportion to the product of two factors: the gas phase pressure – which determines the particle impingement rate (2.1) – and the number of available adsorption sites.

By varying the conditions of such an experiment, striking qualitative effects sometimes leap out directly from the raw data. For example, it is immediately evident from Fig. 14.3 that N_2 adsorbs onto a clean Fe(111) surface much more rapidly than onto a clean Fe(100) surface (but vastly less rapidly than O_2/Rh(111) – compare the abscissas). Moreover, we observe that the initial *slopes* of the uptake curves depend sensitively on the surface temperature for the (100) crystal face whereas the corresponding quantity for Fe(111) is temperature independent. The two observations are intimately related. To begin to understand this, note first that the slope of such curves is precisely the ratio of the adsorption rate to the impingement rate; in other words, the sticking coefficient. The data of Figs. 14.1 and 14.2 then directly yield $s(\theta)/s_0$, the ratio of the coverage dependent sticking coefficient to its initial or zero-coverage value. Absolute values for $s(\theta)$ follow from a calibration of the relative coverage scale (see, e.g., Roberts & McKee, 1978).

The physical content of the nitrogen/iron kinetics data emerges most

Fig. 14.2. O_2/Rh(111) adsorption kinetics at 335 K. The unit of exposure is 1 L = 10^{-6} Torr s (Yates, Thiel & Weinberg, 1979).

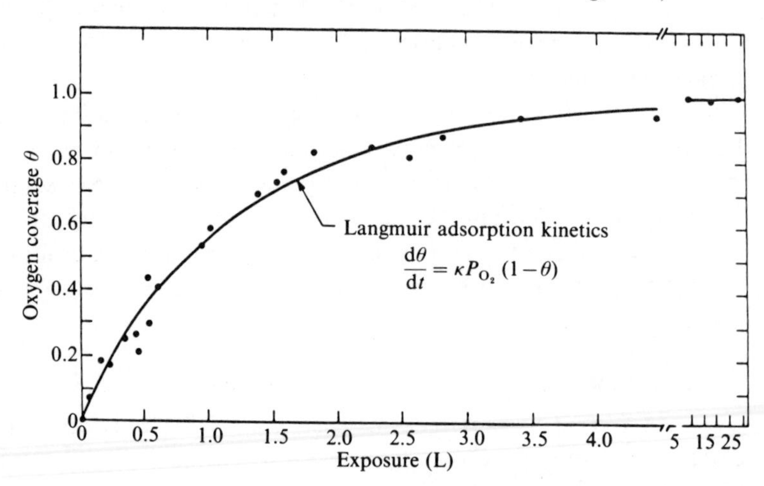

naturally if it is considered in the context of the broader data base available for the nitrogen/tungsten chemisorption system. Here, we turn immediately to the absolute sticking coefficient (Fig. 14.4). Both the initial value of this quantity and its functional dependence on coverage vary substantially from one single crystal face to another. The latter is particularly striking because Langmuir kinetics predicts $s(\theta)/s_0 = 1 - \theta$, which is consistent with at most two of the curves in Fig. 14.4. The explanation of this result, as well as the temperature and surface structure anisotropy of s_0, rests on the fact that the interaction potential between

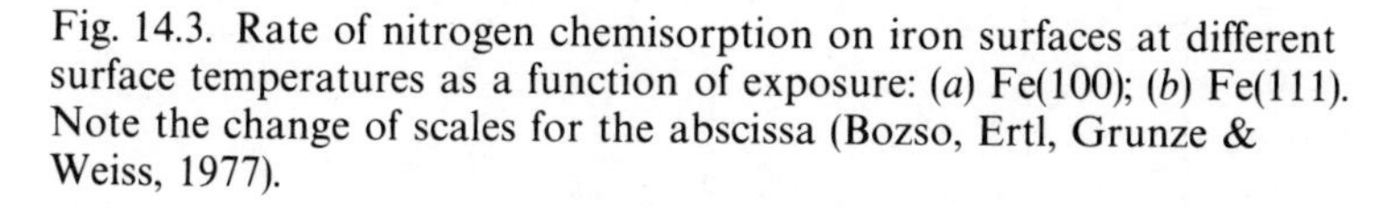

Fig. 14.3. Rate of nitrogen chemisorption on iron surfaces at different surface temperatures as a function of exposure: (*a*) Fe(100); (*b*) Fe(111). Note the change of scales for the abscissa (Bozso, Ertl, Grunze & Weiss, 1977).

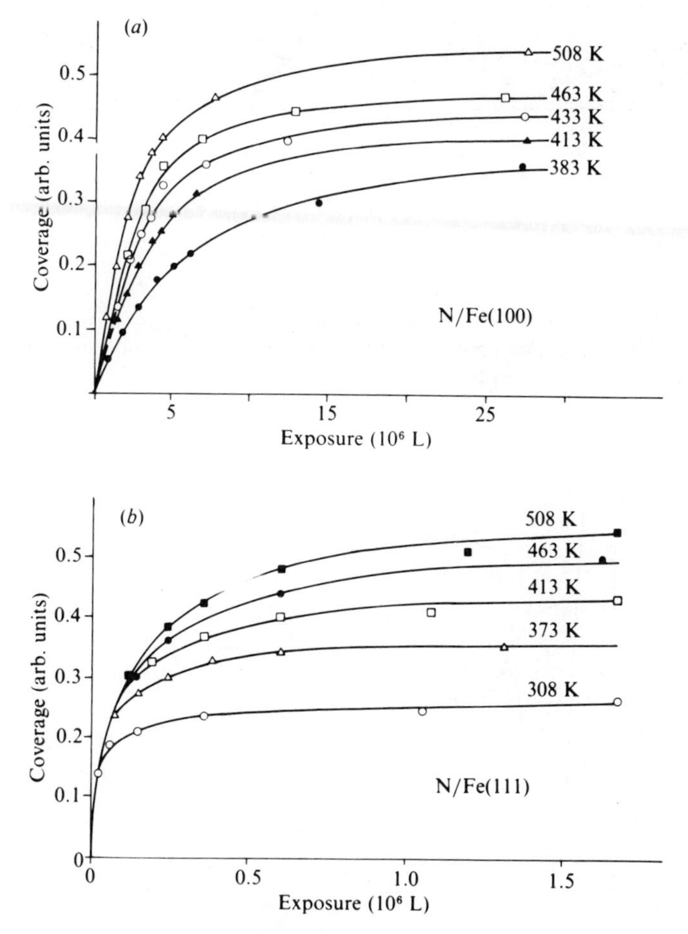

Fig. 14.4. Converge dependence of the absolute sticking probability of N_2 on various single crystal surfaces of tungsten. Gas and surface are both at room temperature (King, 1977).

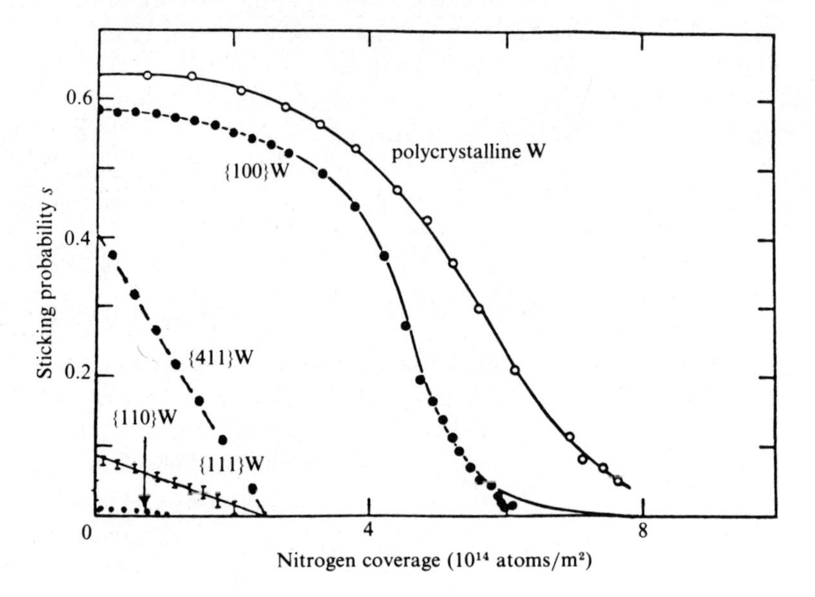

Fig. 14.5. Schematic view of a one-dimensional gas–surface interaction potential which contains both a precursor physisorption well and a deeper chemisorption well.

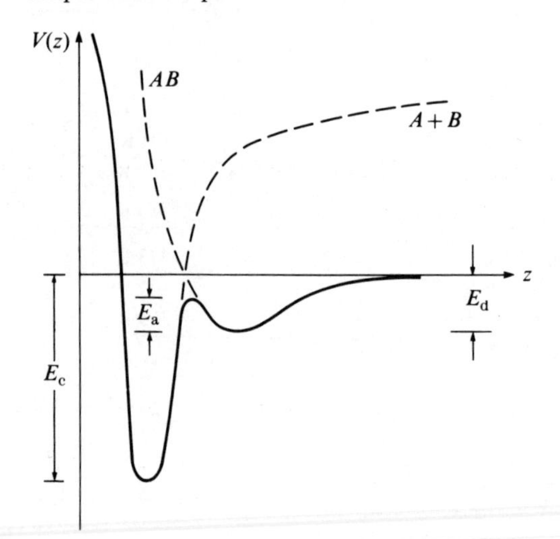

a molecule and a surface generally is much more complex than Fig. 14.1 implies.

The energetic competition between *associative* molecular physisorption and *dissociative* atomic chemisorption is sufficient to produce most of the experimental diversity sketched in the preceding paragraphs. This is so because, recalling the one-dimensional model of Lennard–Jones (1932), the adsorbate adiabatic potential energy surface generally exhibits both a deep chemisorption well and a shallow physisorption well somewhat farther from the surface (Fig. 9.5). Fig. 14.5 shows an expanded view of one conceivable potential function. From this picture it is clear that incoming molecules can trap first in the outer well – called a *precursor state* – and then, perhaps somewhat later, attempt to dissociate and enter the inner well. This is significant because the measured sticking coefficient refers exclusively to dissociated atoms bound in the chemisorption well.

It is simplest to begin with the probable effect of a precursor on the coverage dependence of $s(\theta)$. We make two assumptions. First, every site of the surface can support physisorption, regardless of the occupancy of the chemisorption well directly below. Second, adsorbed molecules can roam across the surface in search of unoccupied chemisorption sites to occupy. To be more quantitative, consider a molecule physisorbed above an 'empty' site. From there, it can either pass into the chemisorption well with probability p_a, desorb back into the gas phase (p_d), or migrate to an adjoining physisorption well (p_m). Only the corresponding probabilities p'_d and p'_m are non-zero for a molecule physisorbed above an 'occupied' site. With these definitions, it is straightforward to verify that the probabilities for chemisorption to, and migration from, the first site visited by a molecule are:

$$\begin{aligned} p_a(1) &= p_a(1-\theta), \\ p_m(1) &= 1 - p_a - p_d + \theta(p_a + p_d - p'_d). \end{aligned} \tag{14.11}$$

Now repeat the calculation for a second site 2 which adjoins 1. One quickly sees that $p_m(2) = p_m^2(1)$ and further that $p_m(3) = p_m^3(1)$, etc. Hence,

$$\begin{aligned} s &= p_a(1-\theta)[1 + p_m(1) + p_m(2) + \cdots] \\ &= \frac{p_a(1-\theta)}{1 - p_m(1)}, \end{aligned} \tag{14.12}$$

which can be rewritten in terms of $s_0 = p_a/(p_a + p_d)$ as

$$s/s_0 = \left(1 + \frac{K\theta}{1-\theta}\right)^{-1}, \tag{14.13}$$

where

$$K = \frac{p'_d}{p_a + p_d}. \tag{14.14}$$

Equation (14.13) defines a family of curves (Fig. 14.6). The data (Fig. 14.4) most often resemble the cases where $K < 1$. This is reasonable because there is one less option open to a molecule physisorbed above an 'occupied' site. But there is a caveat. Dissociative chemisorption of a diatomic molecule such as N_2 requires the presence of at least *two* unoccupied sites. This suggests that (14.11) be replaced by $p_a(1) = p_a(1 - \theta)^2$. However, even that choice is doubtful if there exist strong lateral interactions among the chemisorbed atoms. This is easy to see using our experience gained in Chapter 11.

LEED studies of the dissociative chemisorption of O_2/Ni(100) show that the adsorbed oxygen atoms form islands of p(2 × 2) order for coverages up to $\theta = 0.25$ followed by a transition to c(2 × 2) order up to $\theta = 0.35$. Monte Carlo simulations reproduce this behavior using a lattice gas Hamiltonian (11.2) with nearest neighbor, next-nearest neighbor, and third neighbor interactions which are strongly repulsive, weakly repulsive and weakly attractive, respectively. As suggested above, the number of adjacent empty sites is proportional to $(1 - \theta)^2$ if one randomly throws $N\theta$ atoms onto an adsorption lattice of N total sites.

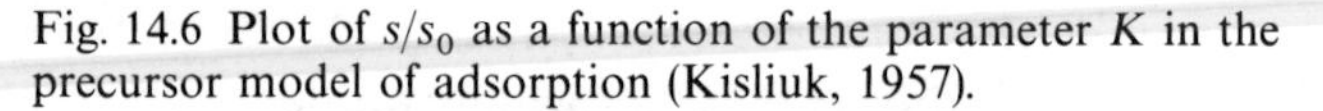

Fig. 14.6 Plot of s/s_0 as a function of the parameter K in the precursor model of adsorption (Kisliuk, 1957).

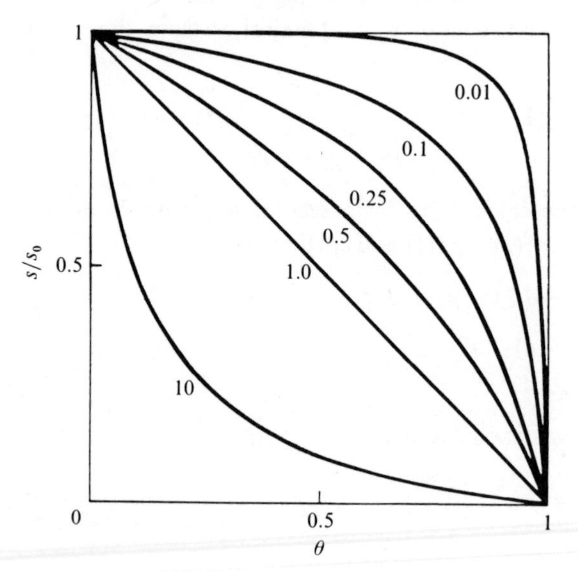

Now use the Monte Carlo method to move the atoms around the lattice so that their final configurations reflect a Boltzmann distribution (at T = 300 K) determined by the interaction Hamiltonian. In this way, we compute the quantity θ_k, the fraction of connected clusters of k sites that are unoccupied. The sticking probability would be proportional to θ_2 if only two adjacent empty sites were required for dissociative chemisorption. Instead, one finds that the measured values of $s(\theta) \propto \theta_8$ (Fig. 14.7). In other words, lateral interactions restrict dissociation of O_2 molecules to portions of the lattice where there exist two adjacent adsorption sites *relative to the p(2 × 2) overlayer structure*. This corresponds to eight adjacent sites of the Ni(100) surface (see inset). Notice that no precursor is required.

The dynamics of sticking resides in the normalization constant in Fig. 14.6. This is s_0, the zero coverage limit of the sticking coefficient. Consider first the dependence of this quantity on the initial kinetic energy

Fig. 14.7. Measured sticking probability of O_2/Ni(100) (solid curve) along with empty site cluster distributions obtained by the lattice gas model. See text for discussion. Inset: two adjacent vacancies in a Ni(100)c(2 × 2)-O adsorption lattice. Open circles are Ni sites, small dots are adsorption sites and large filled circles are O atoms (Brundle, 1985).

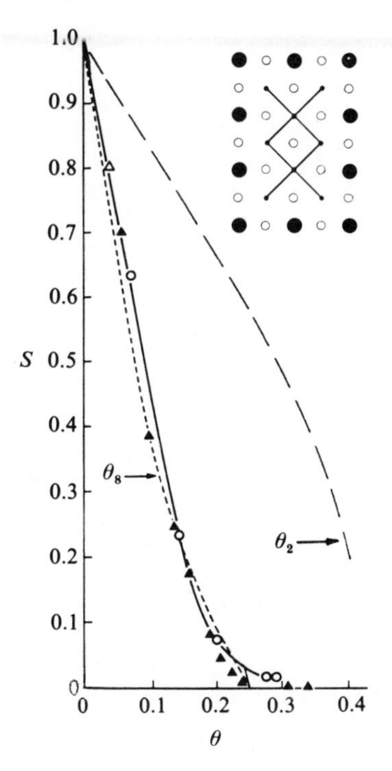

of gas particles seeking adsorption. Fig. 14.8 presents results from both theory and experiment when the gas is in thermal equilibrium with the substrate and when the gas is directed at the surface with a fixed translational kinetic energy/particle. The calculations are from classical Langevin trajectory simulations (see Chapter 13) for unreactive argon and xenon atoms which exchange energy with the phonons of a Pt(111) surface. The gas–surface interaction potential is not unlike Fig. 14.1(*a*). We see that the sticking coefficient rapidly falls as the normal kinetic energy of the particles increases. This is in accord with the simple intuition noted earlier. However, molecular beam measurements exhibit the same general behavior for the substantially more complex N_2/W(100) system, for which a potential energy curve like Fig. 14.5 is more appropriate. One way to distinguish the two is by study of the *surface* temperature dependence of s_0.

Suppose that molecules incident from the gas phase trap and come to thermal equilibrium in the physisorption well. Then, according to Fig. 14.5, the particles see a barrier to desorption of magnitude E_d and a barrier to chemisorption of magnitude E_a. Applying the results of (14.6),

$$\begin{aligned} p_a &= \nu_a \exp(-E_a/kT), \\ p_d &= \nu_d \exp(-E_d/kT), \end{aligned} \tag{14.15}$$

Fig. 14.8. Variation of the initial sticking coefficient as a function of: (*a*) equilibrium gas–surface temperature; (*b*) kinetic energy of the incoming gas particle at $T_s \cong 275$ K. Results are shown for computer simulations of Ar/Pt(111) and Xe/Pt(111) (closed circles) and data for N_2/W(100) (open circles) (Tully, 1981; King & Wells, 1974 and courtesy of C.T. Rettner, IBM Almaden Research Center).

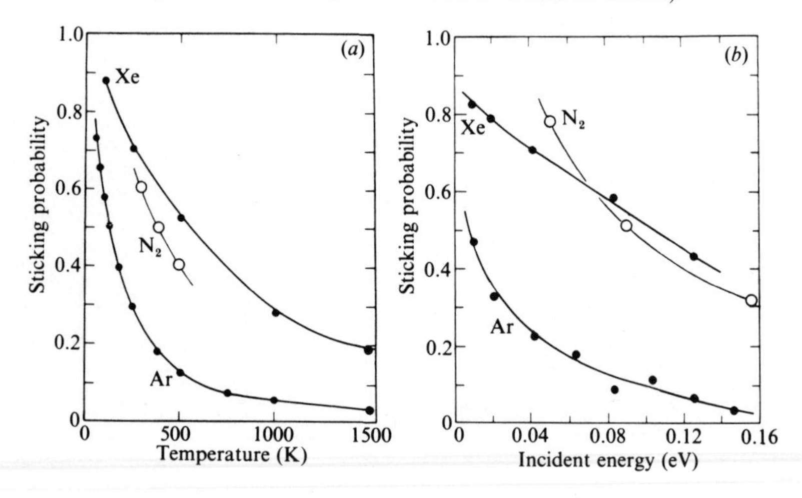

where ν_a and ν_d are (generally unknown) pre-exponential factors, we re-express the initial sticking coefficient as

$$s_0 = \frac{p_a}{p_a + p_d} = [1 + \nu_d/\nu_a e^{-(E_d - E_a)/kT}]^{-1}. \tag{14.16}$$

Consequently, s_0 decreases as the surface temperature increases for the case sketched in Fig. 14.5.

The scenario changes when the crossing between the di-atom and molecular potential curves (cf. Fig. 9.5(*b*)) occurs *above* the zero of energy, i.e., $E_a > E_d$. In that case, adsorption is an activated process and s_0 increases as the surface temperature increases. This is a consistent interpretation of the N_2/Fe(100) data of Fig. 14.3(*a*). Moreover, it should be obvious that the result of a beam scattering experiment designed to probe $s(\varepsilon)$ for a case of activated adsorption will differ markedly from the behavior seen in Fig. 14.8(*b*). Very little sticking can occur until the normal component of the kinetic energy is sufficient to overcome the barrier. The dissociative chemisorption of H_2 on Cu(100) is an example (Fig. 14.9).

The role to be played by theory in this discussion is clear. First, we expect electronic structure calculations to provide a microscopic rationale for the myriad of potential energy curves inferred from experiments on different adsorbate/substrate combinations. Second, dynamics calculations

Fig. 14.9. Molecular beam data for the sticking probability of hydrogen on Cu(100) as a function of the 'normal' component of the incident translational kinetic energy (Balooch, Cardillo, Miller & Stickney, 1974).

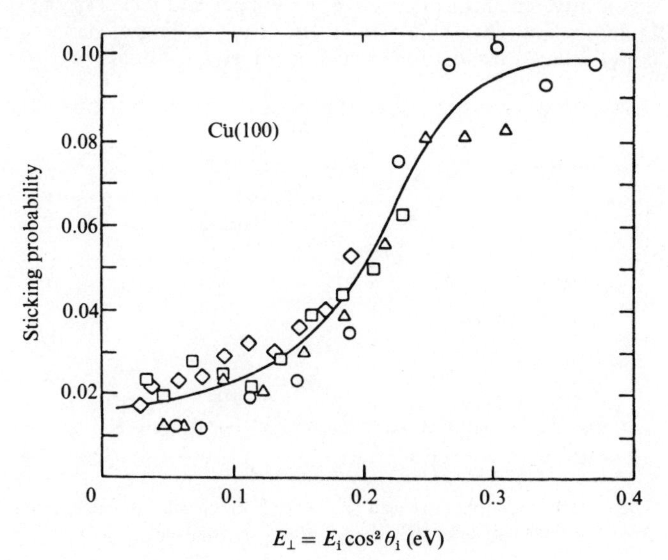

can address energy transfer processes which occur as an adsorbate moves along the gas–surface potential energy surface. Both tasks are formidable and little detailed information is available.

The question of whether, say, H_2 adsorption on a metal surface is activated or non-activated (or dissociative or associative) depends on the precise nature of the energetics associated with pushing the molecule close into the surface. As we saw in Chapter 8, there is always an energy gain at large distances (h) due to van der Waals attraction. At closer distance, Pauli repulsion guarantees that the energy goes back up as the closed-shell molecule samples regions of increasing surface charge density (cf. the He immersion energy curve in Fig. 4.7). Of course, we know from experiment that it ultimately pays to break the molecular bond and form metal–H chemisorption bonds with the constituent atoms. Let us look at this process from the surface molecule point of view.

Fig. 14.10 illustrates schematic energy level diagrams extracted from LDA electronic structure calculations of the interaction of H_2 with Cu_2 and Ni_2 in three different geometrical configurations. Consider the copper case first (Fig. 14.10(*a*)). Far above the surface, the $1s^2$ and $4s^2$ bonding orbitals of H_2 and Cu_2 are filled and the corresponding anti-bonding orbitals are empty.* As the adsorbate approaches the 'surface', the occupied metal level rises – this is the Pauli repulsion barrier – but the $(1s^2)^*$ level of the adsorbate drops. The latter is consistent with our previous results (cf. Fig. 13.8). Moreover, this level begins to acquire Cu–H bonding

Fig. 14.10. Schematic energy level diagrams appropriate to the dissociative chemisorption of H_2 on copper and nickel in the surface molecule limit: (*a*) H_2Cu_2; (*b*) H_2Ni_2. The labelling of states is not conventional. See text for discussion (Harris & Andersson, 1985).

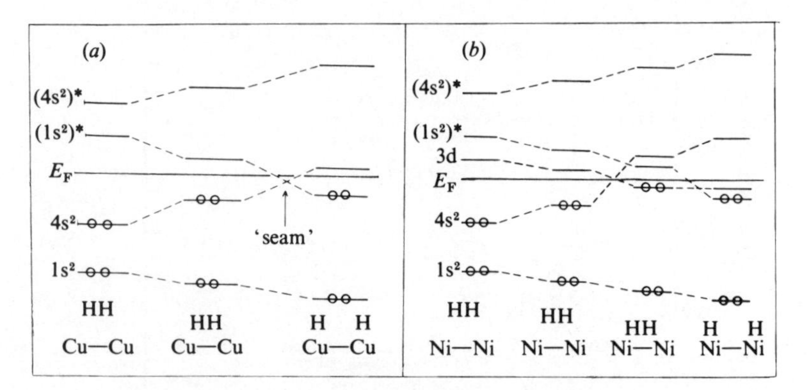

* The 3d levels of copper are well below the Fermi level. For clarity, we omit them from Fig. 14.10 (*a*) since they do not affect our analysis.

character. The change in wave function character is crucial because at smaller values of h and greater H–H separation (d) this level crosses the $4s^2$ level and becomes occupied. Thus, the lower energy of the chemisorbed geometry results from an electronic configurational switch to a surface bonding state of dissociated hydrogen atoms. The switching point (usually called the 'seam') is actually a line in the two-dimensional (h, d) plane and is the generalization of the curve crossing points in Fig. 9.5.

It is instructive to compare the H_2/Cu results with a situation where there exist empty 3d states in the immediate vicinity of E_F. This is the case for H_2/Ni. Here again, an adsorption barrier begins to form as the adsorbate approaches the surface. But now it is possible to reduce the maximum value of the Pauli repulsion (and hence the adsorption barrier) by transferring s-electrons into more compact d-orbitals. This accounts for the experimental fact that the sticking coefficient for dissociative chemisorption of H_2 on nickel is virtually independent of surface temperature (Robota *et al.*, 1985), i.e., $E_a \cong E_d$ in Fig. 14.5. Ultimately, a second seam develops in our model although, in reality, it is likely that d-electrons participate actively in the H–Ni chemisorption bond (see Chapter 12).

It is impractical to compute the complete adsorption potential energy surface for any problem beyond the most trivial. This is true even for the case of a diatomic molecule (six degrees of freedom) interacting with a static substrate. Consequently, one typically freezes most of the coordinates

Fig. 14.11. Total energy contour plot for N_2 adsorption on Fe(111) calculated with the tight-binding method. The dashed line is the path of minimum energy between the molecular precursor and the chemisorbed state (Tomanek & Bennemann, 1985b).

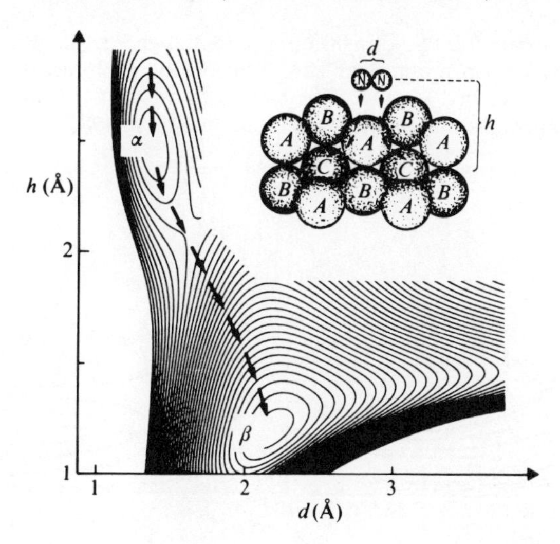

and considers only certain slices through this multi-dimensional space. For example, Fig. 14.11 shows a contour plot of the total gas–surface interaction energy for N_2/Fe(111) in the same (h, d) variable space as the previous example. A twelve atom cluster takes the place of the semi-infinite solid. The results (calculated with a parameterized tight-binding scheme) reveal the anticipated physisorption precursor state (α) and a dissociated chemisorption state (β) much closer to the surface. Both the computed barrier to chemisorption, $E_a = 0.4$ eV, and the depth of the chemisorption well, $E_c = 2.5$ eV, are in reasonable agreement with experiment.

Potential energy surface diagrams similar to that of Fig. 14.11 appear commonly in discussions of gas phase chemical reaction dynamics (Bernstein, 1982). In that context, one identifies two characteristically different reaction scenarios. Fig. 14.12 illustrates the main point for the present case of dissociative chemisorption. Focus attention on the relative position of the point (often called the 'transition state') at which the seam crosses the path of minimum energy. Here one sits atop the adsorption barrier at a saddle point of the potential energy surface. Suppose first that this point lies in the so-called 'entrance channel'. That is, it is encountered as an unstretched molecule quits its physisorption well and moves closer to the surface. This is the case for H_2/Cu(100). Increased translational kinetic energy directed toward the surface promotes dissociation.

Fig. 14.12(*b*) suggests a different means to promote dissociative chemisorption. In this case, a molecular adsorbed state exists at an adsorbate–substrate separation not far from the eventual dissociated atom chemisorption bond length. The rate-limiting barrier to dissociation now

Fig. 14.12. Schematic two-dimensional potential energy diagrams for dissociative chemisorption. Both situations exhibit a physisorbed state (α) and a chemisorbed state (β). The barrier to adsorption can occur in the (*a*) entrance channel or the (*b*) exit channel (Ertl, 1982).

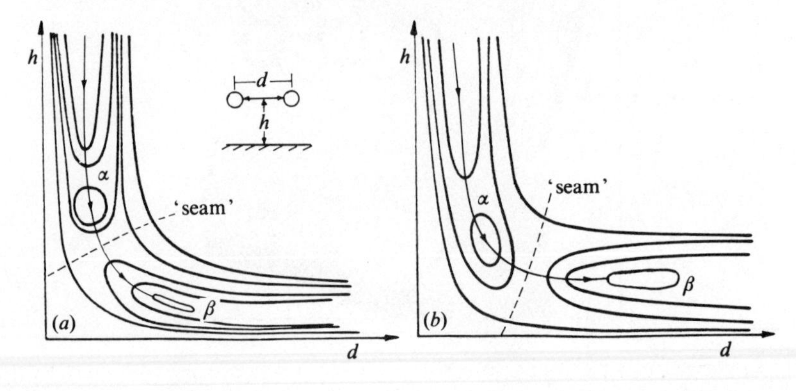

occurs only as the intramolecular bond stretches apart. One says that the transition state lies in the 'exit channel'. Increasing the kinetic energy of the incident molecule does little to promote dissociation. Instead, it pays to prepare the molecule in a high state of vibrational excitation. This is vibrationally assisted sticking.

Diffusion

Successful adsorption into a chemisorbed state places an atom or molecule at the bottom of a potential well in thermal equilibrium with the underlying solid. According to (14.7), thermal fluctuations tend to drive the particle back into the gas phase at a rate proportional to $\exp(-E_c/kT)$. This rate is quite small since typical values of the heat of adsorption E_c are 1–5 eV (cf. Fig. 9.4). But there is another possibility. The adsorbate also can jump *laterally* along the surface from one well to the next (Fig. 14.13). The energy barrier to this motion, E_m, is bound to be less than E_c because the particle always is in intimate contact with the substrate. In other words, the adsorbate (by definition a reactive species) climbs a lower barrier because it samples portions of its immersion energy curve (Fig. 4.7) away from the minimum but still far from the zero of energy (which corresponds to desorption).

It is clear from Fig. 14.13 that there is no *a priori* bias for the particle to jump to the right or jump to the left. The direction of each jump is completely random and uncorrelated from jump to jump. This kind of behavior is called a *random walk*. Generalize now to the case where such a walker is confined to a two-dimensional square lattice with lattice constant l. Let $P(x, y, t)$ be the probability that the particle is at lattice site (x, y) at time t. If each jump occurs after a time τ, it must be the case that

$$\begin{aligned} P(x, y, t+\tau) = {} & \tfrac{1}{4}P(x+l, y, t) + \tfrac{1}{4}P(x-l, y, t) \\ & + \tfrac{1}{4}P(x, y+l, t) + \tfrac{1}{4}P(x, y-l, t), \end{aligned} \tag{14.17}$$

which can be rewritten

$$\begin{aligned} & P(x, y, t+\tau) - P(x, y, t) \\ & \quad = \frac{l^2}{4l^2}[P(x+l, y, t) - 2P(x, y, t) + P(x-l, y, t)] \\ & \quad + \frac{l^2}{4l^2}[P(x, y+l, t) - 2P(x, y, t) + P(x, y-l, t)]. \end{aligned} \tag{14.18}$$

Now divide (4.18) by τ and let both l and τ approach zero in such a way that

$D = l^2/4\tau$ remains finite. The probability function then satisfies the familiar *diffusion* equation:

$$\frac{\partial P}{\partial t} = D\nabla^2 P. \tag{14.19}$$

The properties of the diffusion constant D are best understood if one returns to the discrete jump model. Suppose that the random walker takes N steps in time t. The total displacement of the particle is $\Delta\mathbf{r} = \mathbf{r}(t) - \mathbf{r}(0) = \sum \mathbf{r}_i$ so that

$$\langle|\Delta\mathbf{r}|^2\rangle = \sum_{i=1}^{N} \langle r_i^2 \rangle = \frac{t}{\tau} l^2 = 4Dt. \tag{14.20}$$

This suggests a general definition of the diffusion constant which is independent of the details of the particle dynamics:

$$D = \lim_{t\to\infty} \frac{\langle[\mathbf{r}(t) - \mathbf{r}(0)]^2\rangle}{4t}. \tag{14.21}$$

Another useful expression for D follows if one inserts

$$\mathbf{r}(t) = \mathbf{r}(0) + \int_0^t \mathbf{v}(t')\,\mathrm{d}t', \tag{14.22}$$

into the numerator of (14.21) and notes that $\langle\mathbf{v}\rangle = 0$ for a random walk:

$$\langle|\Delta\mathbf{r}|^2\rangle = \int_0^t \mathrm{d}t' \int_0^t \mathrm{d}t'' \langle\mathbf{v}(t')\cdot\mathbf{v}(t'')\rangle. \tag{14.23}$$

To make progress, we use the fact that the fluctuations of a system in equilibrium with a thermal bath are described by a so-called 'stationary' random process (see e.g., van Kampen (1981)). In which case,

$$\langle\mathbf{v}(t')\cdot\mathbf{v}(t'')\rangle = \langle v^2\rangle f(t' - t'') \tag{14.24}$$

Fig. 14.13. Schematic view of the adsorbate–substrate interaction potential as a function of distance along the surface.

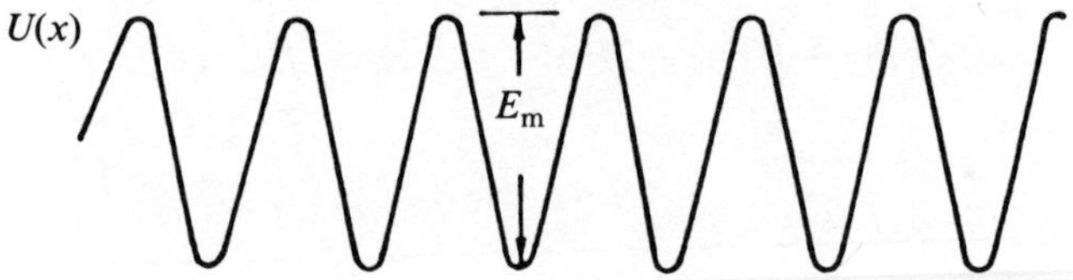

U(x)
E_m

where $f(t)=f(-t)$. Therefore,

$$\langle|\Delta\mathbf{r}|^2\rangle=\langle v^2\rangle\int_{-t}^{t}\mathrm{d}\tau\int_{\tau/2}^{t-\tau/2}\mathrm{d}\theta f(\tau)$$

$$=2\langle v^2\rangle\int_0^t(t-\tau)f(\tau)\,\mathrm{d}\tau$$

$$\cong 2\langle v^2\rangle t\int_0^\infty f(\tau)\,\mathrm{d}\tau. \tag{14.25}$$

The last equation is valid in the limit as $t\to\infty$. Finally, comparing (14.25) with (14.20) we obtain

$$D=\tfrac{1}{2}\langle v^2\rangle\int_0^\infty \mathrm{d}\tau f(\tau)=\tfrac{1}{2}\int_0^\infty\langle\mathbf{v}(\tau)\cdot\mathbf{v}(0)\rangle\,\mathrm{d}\tau. \tag{14.26}$$

This representation of the diffusion constant is useful in a number of respects. First, it helps to clarify the point that D as defined above is *not* the same as the *chemical* diffusion constant D_c which occurs in the diffusion equation for the adsorbate particle density $n(\mathbf{r})$ at finite coverage (Fick's law):

$$\frac{\partial n}{\partial t}=D_c\nabla^2 n. \tag{14.27}$$

This differs from (14.19) because $n(\mathbf{r})$ is not identical to $P(\mathbf{r})$. The latter describes the probability distribution function for diffusion of a single *isolated* particle. By constant, the chemical diffusion constant refers to transport of the total flux of M particles,

$$\mathbf{J}=\sum_{i=1}^{M}\mathbf{v}_i \tag{14.28}$$

and can be expressed in terms of a corresponding correlation function (Kubo, Toda & Hashitsume, 1985):

$$D_c=\frac{1}{2\kappa\langle n\rangle^2}\frac{1}{kTA}\int_0^\infty\langle\mathbf{J}(\tau)\cdot\mathbf{J}(0)\rangle\,\mathrm{d}\tau, \tag{14.29}$$

where κ is the isothermal compressibility and A is the surface area. In fact, one can show (Mazenko, Banavar & Gomer, 1981) that the two are related by

$$D_c=\frac{D}{kT\langle n\rangle\kappa}=D\frac{\partial(\mu/kT)}{\partial(\ln\langle n\rangle)}, \tag{14.30}$$

Fig. 14.14. Field ion microscope images of the diffusion of rhenium atoms on W(211) at $T = 327$ K. Successive images are separated by 60 second intervals (Ehrlich, 1982).

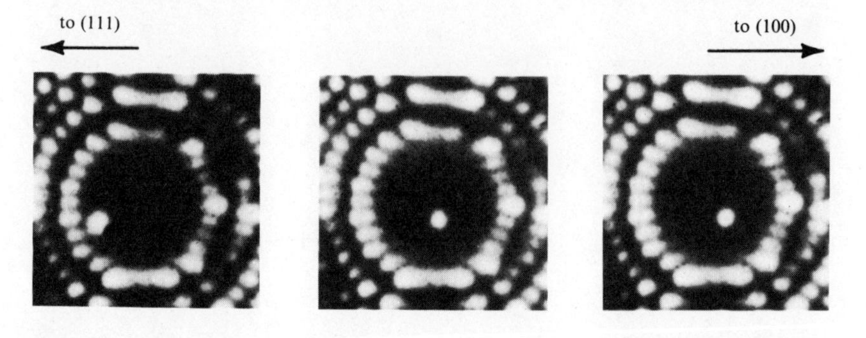

Fig. 14.15. Arrhenius plots of single-particle diffusion constants as determined by field ion microscope studies: (*a*) rhenium atoms and rhenium *dimers* on W(211); (*b*) rhodium atoms on various single crystal surfaces of rhodium (Stolt, Graham & Ehrlich, 1976; Ayrault & Ehrlich, 1974).

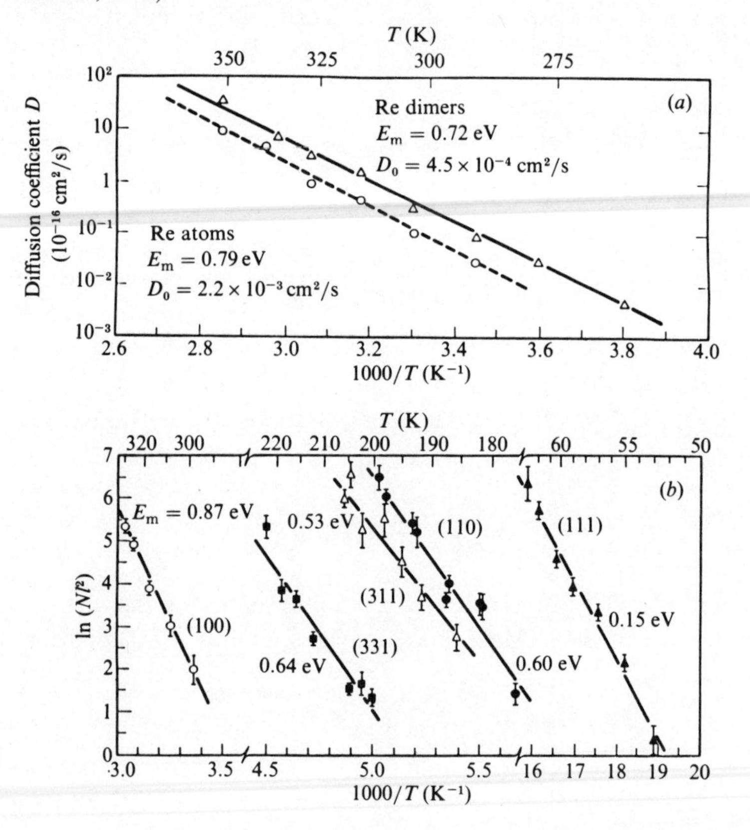

where μ is the chemical potential. The two diffusion constants agrec only when the velocities of different particles are totally uncorrelated – a situation rarely encountered in the presence of adsorbate–adsorbate interactions.

Experimental measurements of surface diffusion constants for adsorbates on refractory metal surfaces employ both (14.21) and (14.29). For the former, the field ion microscope (see Chapter 3) directly images the diffusive motion of individual adatoms deposited onto an otherwise clean crystal surface. Fig. 14.14 shows three successive images of the diffusion of a rhenium atom across a single facet of W(211). The tungsten atoms belonging to {211} planes appear dark in these photographs. Many trials at temperature T are used to compute the thermal average $\langle[\mathbf{r}(t)-\mathbf{r}(0)]^2\rangle$ from the measured particle excursion distances. In accordance with expectation, the values of D obtained in this way fit well to an Arrhenius form (Fig. 14.15(*a*)):

$$D = D_0 \mathrm{e}^{-E_{\mathrm{m}}/kT}. \tag{14.31}$$

The diffusion constant of Re/W(211) exhibits two characteristics common to similar measurements on a wide variety of gases on high-melting point transition metal surfaces. The activation barrier E_{m} is of the order of 5–20% of the chemisorption energy and the pre-exponential factor D_0 is of the order of 10^{-3} cm^2/s. The first observation is consistent with the qualitative discussion given at the beginning of this section. As to the second, suppose we write $D_0 = \nu l^2/4$ as in the simplest random walk model. Then, with the choice of $l = 3$ Å as a typical nearest-neighbor jump distance, we obtain $\nu \cong 10^{13}$/s – a perfectly reasonable estimate of the oscillation frequency at the bottom of a chemisorption well. However, one should not be surprised by examples which exhibit 'anomalous' values of D_0 that differ substantially from this value. Equations (14.7) and (14.9) show that unusual contributions to the relevant partition functions or dynamical factors* can easily renormalize this elementary estimate.

Fig. 14.15(*b*) displays the results of a trend study of rhodium atom diffusion on various crystal surfaces of rhodium metal. Diffusion proceeds rapidly on the close-packed (111) surface and rather more slowly on the (atomically) rougher surfaces of the crystal (cf. Fig. 3.2). This type of data is directly relevant to the question of whether a small chunk of elemental

* Suppose an adsorbed particle gains enough energy from a sequence of small thermal kicks to surmount the diffusion barrier and begin its descent into an adjacent chemisorption well. The analog of the sticking coefficient $s(T)$ for this problem is the probability that the particle receives a kick which sends it back over the barrier into its original well.

material can achieve its equilibrium crystal shape (see Chapter 1). This is so because surface diffusion is the rate-limiting step to the dissolution of high Miller index crystal faces into facets of low index planes with low surface tension. The same issue arises (with considerably more commercial import) in the context of the growth of high quality semiconductor materials by deposition of the constituent atoms from the vapor phase (see, e.g., Chernov (1984)). In this case, a good single crystal results only if surface diffusion permits each atomic plane to completely fill in before the plane above begins to grow (see Chapter 16).

Surface diffusion in crystal growth and adsorbate reaction processes involves the motion of surface species in the presence of other mobile particles. This brings us to measurements of the chemical diffusion constant D_c. Volmer & Estermann (1921) and Langmuir & Taylor (1932) pioneered such studies many years ago. Modern experiments focus on the fluctuations in adsorbate number density reflected in the correlation function (14.29). These in turn produce fluctuations in the current produced by field emission – a process whereby a strong external electric field strips electrons from the outer shells of adsorbate particles (see, e.g., Plummer, Gadzuk & Penn (1975)).

Fig. 14.16. Surface chemical diffusion parameters for O/W(110) at finite coverages obtained from field emission measurements (Chen & Gomer, 1979).

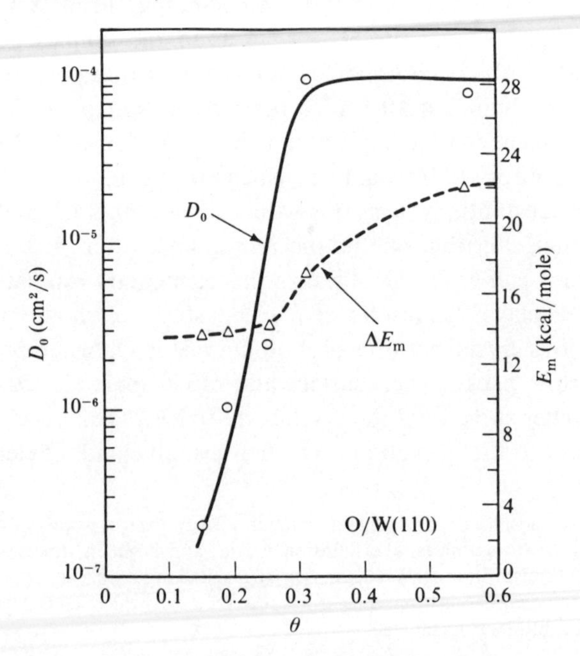

Diffusion data collected in this way again display Arrhenius behavior. Unfortunately, the presence of inter-adsorbate interactions generally precludes a simple interpretation in terms of an activation barrier to single atom jumps. Nevertheless, one can extract values for the coverage dependence of an *effective* energy barrier and pre-exponential factor to concerted atom motion. Fig. 14.16 illustrates an example for the case of oxygen diffusion on W(110). The variations in these quantities around $\theta = 0.25$ are associated with the onset of p(2 × 1) ordering. For example, the increase in E_m is explicable because the system loses condensation energy when particles which belong to islands of the ordered phase attempt to diffuse.

Theoretical study of surface diffusion is possible with either molecular dynamics or Monte Carlo simulation techniques. Of course, both methods require a suitable adsorbate–substrate interaction potential. The case of rhodium atom diffusion on Rh(100) is particularly appealing in this regard because it is reasonable to suppose that a phenomenological potential

Fig. 14.17. Plot of $\langle[\mathbf{r}(t) - \mathbf{r}(0)]^2\rangle$ calculated by a molecular dynamics simulation of Rh/Rh(100) self-diffusion at 1750 K (McDowell & Doll, 1983).

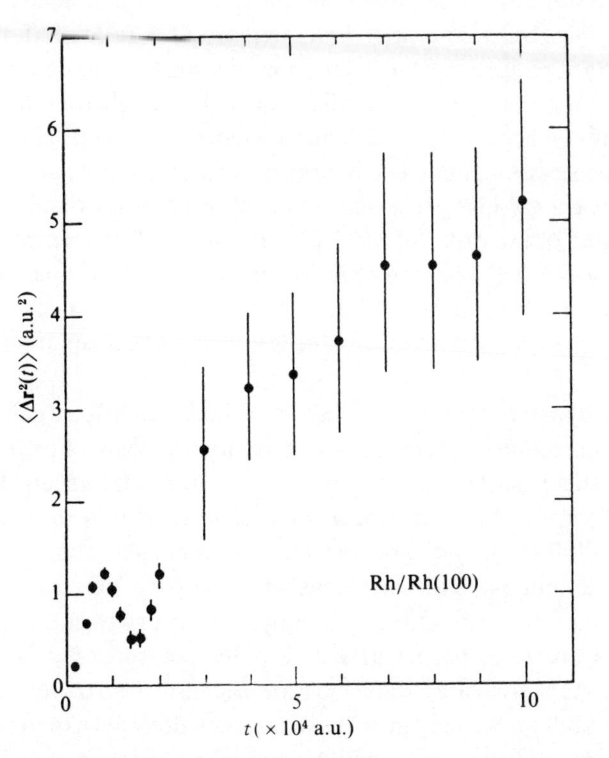

obtained by fitting to bulk properties might well be adequate for surface studies. We focus on the molecular dynamics approach because the analysis of computer-generated surface atom trajectories in that case is particularly simple. One applies (14.21) directly, just as in the case of field ion microscope data.

Fig. 14.17 shows the time dependence of the mean square displacement of a rhodium surface atom on Rh(100) as a function of time. Notice that this quantity is directly proportional to the elapsed time only in the limit of long times – a result in accord with the algebra displayed in (14.25). The slope of the line gives D, and many trials at different temperatures do indeed exhibit activated behavior. The computed E_m differs from the experimental value by about 3% whereas the pre-exponential factor is about 1/4 the observed size. Presumably these discrepancies reflect inadequacies in the assumed Lennard–Jones potential.

The short time behavior in Fig. 14.17 is also of interest. To understand this, let us write a simplified version of the Langevin equation (13.14) for our diffusion problem in terms of the adsorbate velocity $\mathbf{v}$:

$$M\dot{\mathbf{v}} = -\nabla U(\mathbf{r}) - \lambda(\mathbf{r})\mathbf{v} + \xi(t). \tag{14.32}$$

In this expression, M is the adsorbate mass, $U(\mathbf{r})$ is the potential energy function of Fig. 14.13 and $\xi(t)$ and $\lambda(\mathbf{r})$ are the random force and friction constant required by the fluctuation–dissipation theorem to ensure that the diffusing particle is in equilibrium with the substrate heat bath. It is important to observe that this equation correctly reflects the fact that dissipative processes cannot occur arbitrarily rapidly. In particular, we presume that energy loss processes to substrate phonons require (on the average) a time of the order of $M/|\lambda(\mathbf{r})|$. This means that a particle at the bottom of a well oscillates harmonically for times short compared to this characteristic dissipation time. These oscillations are reflected in the correlation function $\langle \mathbf{v}(t) \cdot \mathbf{v}(0) \rangle$ and, hence, in the mean square displacement itself.

Molecular dynamics simulations also are valuable to help us gain insight into the microscopic nature of the diffusion process. For example, Fig. 14.18 displays surface atom trajectories for an adatom on the (100) surface of a generic Lennard–Jones solid. The temperature is approximately $E_m/3$. Most of the time the adatom oscillates about substrate potential well minima with an occasional jump to adjacent minima. However, notice that sometimes a long jump occurs to a distant well. These events are rare (particularly at low temperature) but contribute noticeably to the calculated diffusion rate because they traverse a large distance. To date, such 'long jumps' have not been detected experimentally.

We close with an example for which no classical approach to adsorbate

Fig. 14.18. A single molecular dynamics trajectory (2×10^4 time steps) of adatom motion on the (100) surface of a Lennard-Jones solid at $kT = E_m/3$. Solid circles denote the thermally displaced position of the substrate atoms at the beginning of the run. Periodic boundary conditions are used (DeLorenzi, Jacucci & Pontikis, 1982).

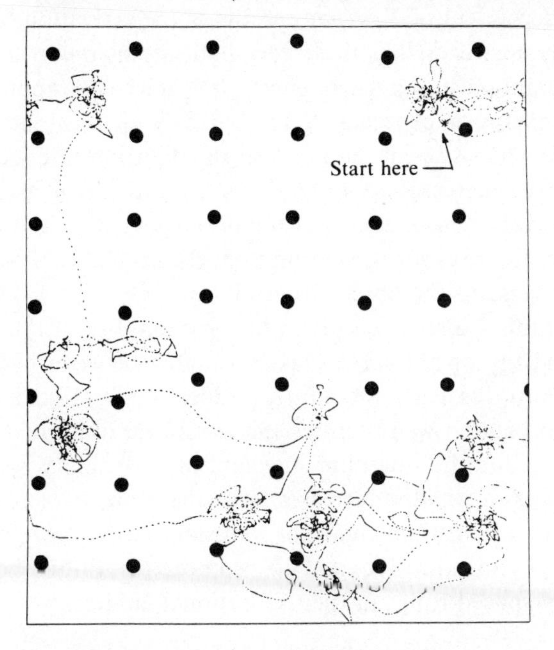

Fig. 14.19. Low-coverage diffusion constant data for hydrogen isotopes on W(110) as a function of substrate temperature (DiFoggio & Gomer, 1982; Wang & Gomer, 1985).

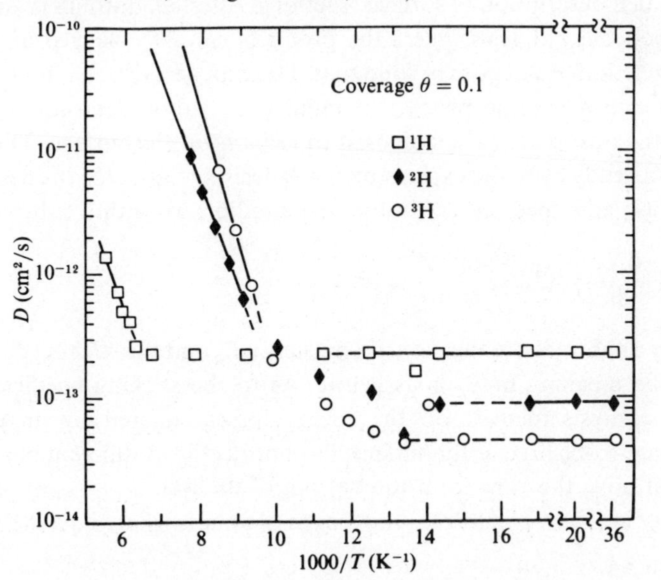

dynamics can be sufficient. Specifically, Fig. 14.19 shows the results of field emission measurements of $D(T)$ for hydrogen and its isotopes on W(110) at low coverage. The high temperature data apparently exhibit conventional Arrhenius behavior. However, the diffusion constant at low temperature is independent of T. What became of the activation barrier? The only plausible answer is that these very light atoms *tunnel* through the barrier – a quantum mechanical effect for which no appreciable temperature dependence is expected. Moreover, the 'classical' activated regime is not quite what is seems. Notice that the diffusion rate increases as the adsorbate mass increases although the activation energy itself does not change appreciably. This means that the pre-exponential factor must be responsible. But the relevant oscillation frequencies (cf. (14.7)) scale as $M^{-1/2}$ – exactly inverse to the observed trend!

It turns out that this is also a quantum phenomenon (Freed, 1985). The effect depends crucially on the large mass mismatch between hydrogen and tungsten. Due to this, each step up the ladder of vibrational energy levels in the chemisorption well requires the assistance of a great many substrate phonons. But the quantum mechanical probability for such multi-phonon processes rapidly decreases with the number of phonons involved. Deuterium and tritium climb the diffusion barrier more rapidly because fewer W phonons are required to promote transitions between the more narrowly spaced rungs of their vibrational ladders.

Desorption

Desorption is the final elementary surface process that must enter any unified description of surface reactions. After all, nothing is gained by heterogeneous catalysis unless the products can be collected at the end of the reaction. But it goes beyond that. Thermal desorption is unavoidable even if one has some process in mind (e.g., vapor deposition growth) where the adsorbates are supposed to remain on the surface. This much is clear already from the expression (14.9) derived earlier for the desorption rate of N_a adsorbed species in thermal equilibrium with a substrate:

$$\frac{dN_a}{dt} = s(T)N_a \frac{\omega_0}{2\pi} \frac{Z}{Z_{ads}} e^{-E_d/kT}. \qquad (14.33)$$

The key quantities are the desorption energy E_d and the sticking coefficient $s(T)$. The meaning of E_d is as before. As to the sticking coefficient, our earlier analysis focused on the energy losses suffered by an external adsorbate-to-be fired at the surface like a projectile. A different perspective is useful now: the view from the bottom of the well.

Under ambient conditions, the escape of an adsorbate into the vacuum occurs because the particle receives kicks from substrate atoms as the latter perform random thermal motion about their equilibrium positions. This is the same process invoked in the previous section to explain surface diffusion. Therefore, it is correct to apply the phenomenological Langevin approach to the desorption problem as well. The goal is to study (14.32) where the potential energy $U(\mathbf{r})$ now looks like Fig. 14.1(*a*). This problem was solved by Kramers (1940) with the result that the escape rate always has the form of (14.33). This is no surprise. But Kramers also established a connection between the sticking coefficient and the friction coefficient λ. There are three regimes:

$$\begin{aligned} &\lambda \gg M\omega_0 \frac{kT}{E_\mathrm{d}} \quad s(T) \propto \frac{1}{\lambda}, \\ &\lambda \sim M\omega_0 \frac{kT}{E_\mathrm{d}} \quad s(T) \sim 1, \\ &\lambda \ll M\omega_0 \frac{kT}{E_\mathrm{d}} \quad s(T) \propto \lambda. \end{aligned} \tag{14.34}$$

The high friction limit is intuitively sensible. No particle in a highly viscous medium can travel far. The low friction limit is more subtle. Why does the desorption rate fall below the prediction of absolute rate theory if the damping is very small? The fluctuation–dissipation theorem provides the answer. If thermodynamic equilibrium is to be maintained, inefficient energy transfer processes from adsorbate to substrate must be accompanied by inefficient energy transfer in the opposite direction. As a result, at low friction, only weak random forces from the substrate are present to drive the particle over the desorption barrier. The escape rate drops concomitantly. It is well worth pondering the connection between this point of view and the reasoning which lead to (14.9).

Return now to (14.33). Conventional discussions (Glasstone, Laidler & Eyring, 1941) quote a slightly different formula for the desorption rate. To make the connection, recall that Z is the partition function for the adsorbed species in its 'transition state' at the top of the barrier to desorption. Similarly, Z_ads is the corresponding internal partition function for the adsorbate except for the one vibrational degree of freedom explicitly considered in our derivation. Nothing is changed if we multiply and divide (14.33) by $Z_\mathrm{vib} = kT/h\omega_0$:

$$\begin{aligned} \frac{\mathrm{d}N_\mathrm{a}}{\mathrm{d}t} &= s(T) N_\mathrm{a} \frac{kT}{h} \frac{Z}{Z_\mathrm{vib} Z_\mathrm{ads}} \mathrm{e}^{-E_\mathrm{d}/kT} \\ &= s(T) N_\mathrm{a} \frac{kT}{h} \mathrm{e}^{\Delta S/k} \mathrm{e}^{-\Delta H/kT}. \end{aligned} \tag{14.35}$$

The last line follows from application of the law of mass action to the equilibrium reaction: adsorption state $\rightleftharpoons$ transition state. In particular, one associates E_d with the *enthalpy* change between these two states and the ratio of partition functions with the exponential of the *entropy* change between the states. This representation is useful because, if the adsorption process is not activated, the top of the desorption barrier occurs well into the gas phase (cf. Fig. 14.1) and, from (8.17), E_d may be identified with the isosteric heat of adsorption. In fact, most of the values of q_{st} plotted in Figs. 9.3 and 9.4 derive from measurements of E_d.

The activated nature of desorption implies that E_d can be found most simply from the slope of an Arrhenius plot of isothermal desorption rate data. As it happens, technical simplicity leads most workers to adopt the method of temperature programmed desorption (TPD). Therein, one varies the substrate temperature in a controlled fashion (e.g., $T(t) = T_0 + \beta t$) and records the temporal change in some quantity connected to the coverage in a known way. The pressure rise in the surrounding vacuum chamber is a common choice. Other possibilities introduced in earlier chapters are a (calibrated) change in the sample work function or changes in the adsorbate Auger signal. In any event, by inserting the temperature 'program' $T(t)$ into (14.33) it is easy to show (Redhead, 1962) that there is a simple relation between E_d and the maximum in the rate curve (they are not equal!). This is illustrated in Fig. 14.20(*a*) for the case of Xe/W(111). Notice that the derived desorption energy is quite similar to the values obtained from isotherm data for Xe/Pd(810) at similar coverages (Fig. 8.9).

The desorption spectrum of CO/Ru(100) exhibits a new feature not seen in the Xe/W(111) data; a second peak grows in as the initial coverage rises above $\theta \cong 1/3$ (Fig. 14.20(*b*)). Generally speaking, bumps and distortions of TPD spectra can arise from a variety of sources: 'second-order' kinetic effects where two atoms combine to desorb as a molecule, adsorbate–adsorbate interactions, precursor and dynamical effects. etc.* However, in the simplest view, a clear second peak corresponds to desorption from a second, inequivalent adsorption site which only becomes occupied after all the primary binding sites are filled. As usual we need corroboration from other surface science probes. In the present case, LEED studies show that a $\sqrt{3} \times \sqrt{3}$ ordered overlayer structure grows in to saturation at $\theta = 1/3$. From our experience in Chapter 11, we conclude that adsorbed CO molecules probably interact via short range repulsive and slightly longer range attractive forces. If this is correct,

* See Menzel (1982) and Yates (1985) for a detailed discussion and critique of the TPD method.

additional molecules forced onto the surface must occupy unfavorable 'interstitial' sites of the $\sqrt{3} \times \sqrt{3}$ lattice. They are, in addition, subject to strong repulsion from their immediate neighbors. A lower desorption energy is inevitable.

Desorption measurements yield equilibrium adsorption information beyond the heat of adsorption. For example, one can construct a plot of coverage versus exposure by integrating the total area under desorption curves such as those in Fig. 14.20. The slope of the resulting curve is the sticking coefficient $s(\theta, T)$. Of course, it is not sufficient merely to collect disconnected bits of kinetic data for any particular adsorption system. The *raison d'être* for these experiments is to combine the results with spectroscopic data in order to construct a reasonably complete picture of the equilibrium adsorption/desorption process. As an illustration, we return to the example of dissociative chemisorption of N_2/Fe(111).

The left panel of Fig. 14.21 displays a one-dimensional representation of structural and energetic information for this system gleaned from several different experiments. Evidently, there is at least one more molecular precursor to adsorption than Fig. 14.11 suggests. Well depths and barrier

Fig. 14.20. Temperature programmed desorption spectra obtained by monitoring pressure rise as a function of initial adsorbate coverage: (*a*) Xe/W(111). Analysis of the lineshape with (14.33) yields $E_d = 0.37$ eV; (*b*) Co/Ru(100). The derived heats of adsorption for the two binding sites are 1.6 eV and 1.1 eV (cf. Fig. 9.4) (Dresser, Madey & Yates, 1974; Pfnur, Feulner, Engelhardt & Menzel, 1978).

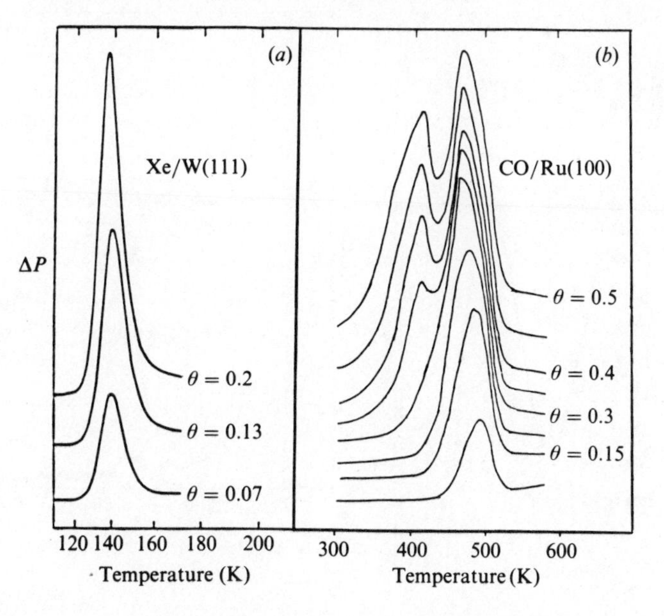

heights come from TPD kinetic measurements. XPS provides state specificity and the orientations are inferred from EELS (right panel). The latter argument (typical of its genre) goes as follows. The N–N stretch frequency of γ-N_2 is only slightly down-shifted from its gas phase value. It presumably interacts very weakly with the substrate. By contrast, α-N_2 exhibits an N–N loss feature a full 85 meV lower still. The 'lying down' geometry is suggested by gas phase experience where similar large frequency shifts occur in complexes where both nitrogen atoms interact with metal atoms. The intra-molecular bond weakens because electrons flow into the normally unoccupied anti-bonding 2π level (N_2 is iso-electronic to CO, cf. Fig. 9.6).

A complementary, real-space view of an adsorption–diffusion–desorption reaction on a 'real', i.e., stepped, surface is depicted in Fig. 14.22. Suppose that the sticking coefficient is high. In that case, most approaching particles adsorb onto terrace sites. However, since $E_m < E_d$, adsorbates at low coverage rapidly diffuse to step sites to gain greater binding energy (cf. Fig. 8.9). The surface career of such a particle consists mainly of many

Fig. 14.21. Interpretive summary of the physics of N_2/Fe(111): (*a*) suggested structure and one-dimensional potential energy (eV) diagram showing two molecular precursor states and the dissociated atomic state; (*b*) electron energy loss data for adsorbate as-deposited at 74 K (γ), after heating to 110 K (α) and after heating to 160 K (β). The final spectrum reveals only the Fe–N stretching mode (Grunze *et al.*, 1984; Whitman *et al.*, 1986).

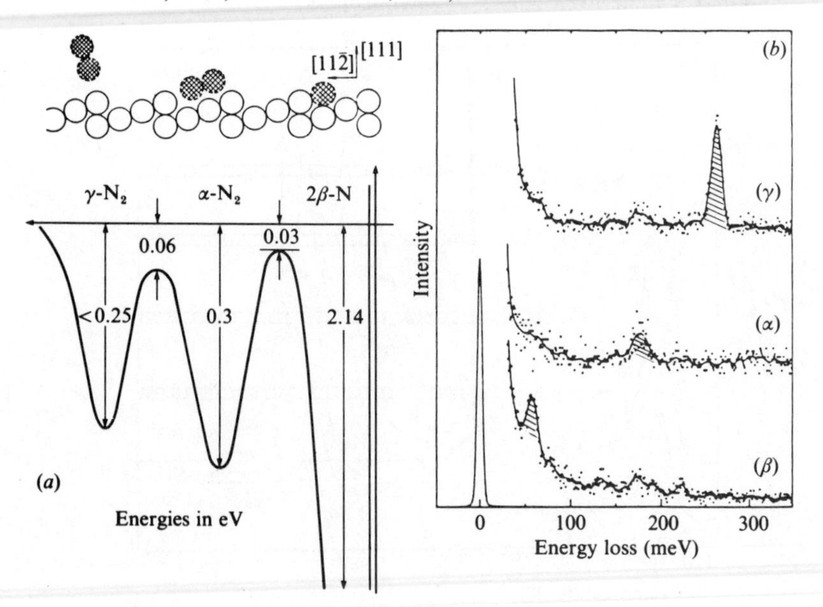

diffusive excursions between various step (or defect) sites. But detailed balance requires that the majority of desorption events occur from the terraces. Hence, the adsorbate quits the surface during one of its sojourns between strong binding sites. We conclude that the surface residence time τ is strongly determined by the density of steps.

The preceding scenario was developed to account for desorption rate constant data obtained by molecular beam methods. This sounds peculiar at first glance. How can one study thermal desorption by scattering a molecule from a surface? Fig. 14.23 answers the question. The diagram shows the angular distribution for different projectile particles scattered from a Pd(111) surface. Helium is a simple case. The incident beam diffracts elastically with most of the intensity going into the specular beam. Oxygen molecules also scatter primarily into the specular direction. However, inelastic scattering events of the sort discussed in Chapter 13 are also present. These tend to broaden the distribution in a characteristic fashion. Finally, we see that CO scattered from Pd(111) exhibits a nearly isotropic angular distribution. It is almost as if the scattered particles had lost memory of their initial beam direction. In fact, this is just what has occurred. To reproduce this behavior we need only imagine that particles incident on the surface trap into adsorption wells, come to thermal equilibrium with the surface, and then desorb back into the gas phase. The mass spectrometer cannot tell the difference between directly scattered

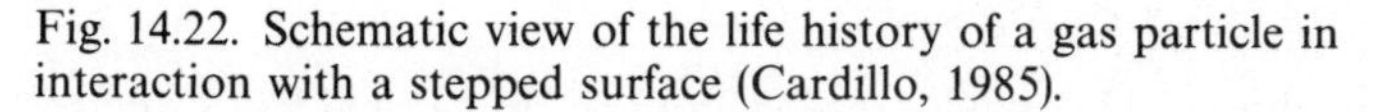
Fig. 14.22. Schematic view of the life history of a gas particle in interaction with a stepped surface (Cardillo, 1985).

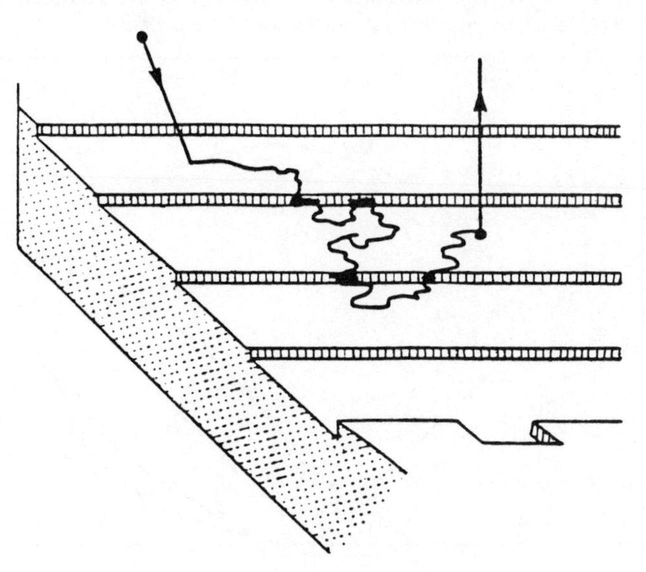

particles and trapped/desorbed particles. The angular distribution is the signature of the trapping channel.

In favorable cases, the technique of molecular beam relaxation spectroscopy (MBRS) directly measures both the desorption rate and the sticking coefficient. If the molecular beam source delivers an incident flux of molecules $I(t)$, the rate of change of the concentration of an adsorbed species $N_a(t)$ is given by

$$\frac{dN_a}{dt} = sI(t) - R_d N_a(t). \tag{14.36}$$

As usual, s is the sticking coefficient and R_d is the desorption rate. In a typical experiment, one chops (cf. Fig. 13.17) the beam flux so that it is a periodic function of time. In that case, the scattered (desorbed) flux $I_d(t)$ follows immediately upon Fourier transformation of (14.36):

$$I_d(\omega) = R_d N_a(\omega) = \frac{sI(\omega)}{1 + i\omega/R_d}. \tag{14.37}$$

Measurement of the amplitude and phase of this complex number is sufficient to extract the two quantities of interest. The desorption energy and pre-exponential factor then follow from an Arrhenius plot of R_d obtained at a sequence of substrate temperatures (Fig. 14.24). Bear in mind however that a more complicated analysis is required if the outgoing flux

Fig. 14.23. Angular distributions of He, O_2 and CO scattered from a Pd(111) surface. See text for discussion (Engel, 1978).

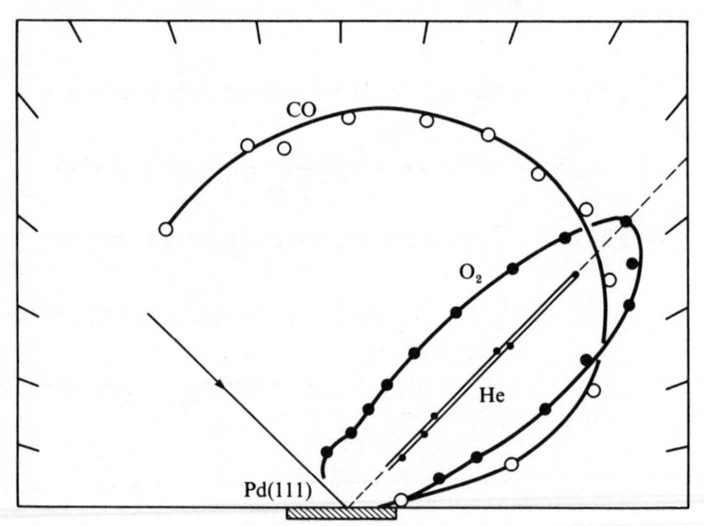

term differs from that assumed in (14.36) (D'Evelyn & Madix, 1984). For example, a term proportional to N_a^2 is appropriate for the case of recombinative desorption of a diatomic molecule. We return to this point in Chapter 15.

We pass from the macroscopic to the microscopic with a change of subject from desorption kinetics to desorption dynamics. For the case of energy transfer, laser spectroscopy opened a window to the internal state distribution of molecules which had suffered inelastic collisions with a solid surface (Chapter 13). It is obvious that this technique can be extended to the case of molecules which thermally desorb or trap/desorb. But another possibility presents itself. Energy flow between adsorbate and substrate is amenable to study if one feeds energy into the complex in a controlled fashion, say, with photons. Stimulated desorption occurs if the flow is such as to rupture a chemisorption bond and the particle escapes before a new bond forms. Both types of experiments require attention – but the results are not always as one might guess.

Imagine a collection of adsorbed molecules in thermal equilibrium with their substrate at temperature T_s. Under these conditions a steady state flux of particles spontaneously desorbs and enters the gas phase. Surely it is reasonable to expect that the translational velocity distribution of these molecules exhibits a Maxwell distribution with a width determined by the surface temperature. Right? Not necessarily. Look at Fig. 14.1(*a*) and construct an imaginary dividing plane parallel to the surface a few Ångströms into the gas phase. According to detailed balance, a Maxwell distribution of particles passes through the plane in both directions. But,

Fig. 14.24. Arrhenius plot of desorption rate constant data for CO/Pd(111) obtained by MBRS (Engel, 1978).

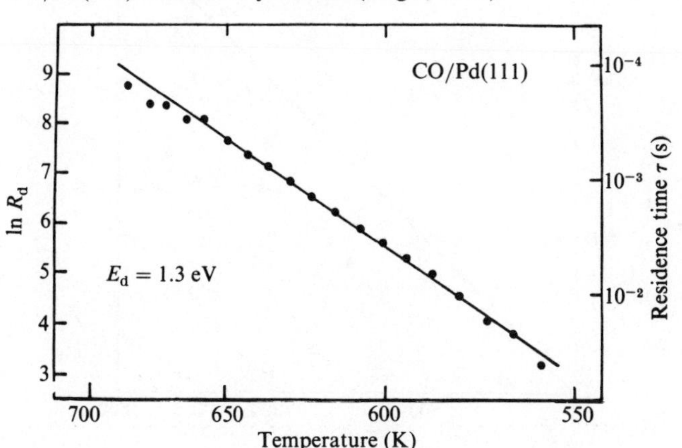

if the sticking coefficient is not unity (a purely dynamical effect), high velocity particles which impinge on the surface bounce back and contribute to the flux of particles passing through the plane from left to right. Therefore, a perfect Maxwell distribution can be maintained only if fewer particles desorb with high velocity. Hence, if one monitors only desorbing species, the apparent temperature of the particles will be *less* than T_s. Stochastic trajectory simulations confirm this statistical requirement precisely (Fig. 14.25).

What about rotations and vibrations? Generally speaking, final state vibrational distributions reflect the surface temperature more faithfully than the corresponding rotational distributions. This is illustrated by state-selected measurements of NO molecules which trap and desorb from Pt(111) (Fig. 14.26). In the simplest case, a detailed balance argument suggests that deviations from T_s Boltzmann behavior must occur for sufficiently high surface temperature. At lower temperature we invoke the results of Chapter 14. Recall that, at the microscopic level, the translational energy of a diatomic molecule converts to rotational energy much more efficiently than it does to vibrational energy. The reverse must be true as

Fig. 14.25. Mean translational energy (expressed as a temperature) of NO molecules thermally desorbed from Ag(111) as a function of surface temperature as calculated by classical Langevin molecular dynamics simulations. The normal (solid circles) and parallel (open squares) components of kinetic energy are plotted separately. The curves would fall on the diagonal solid line if the sticking coefficient were unity (Muhlhausen, Williams & Tully, 1985).

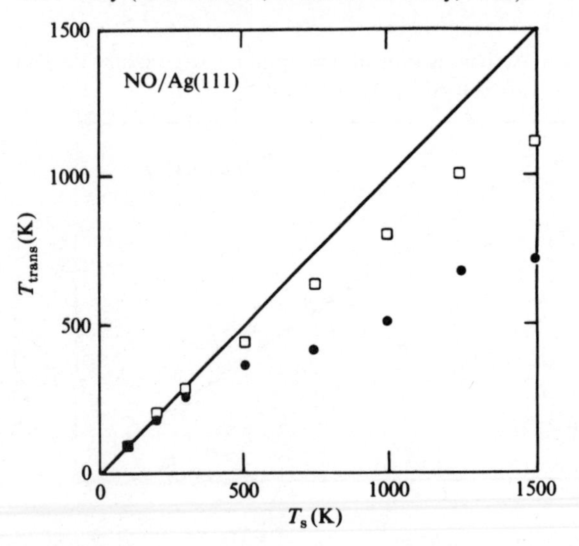

well. Consequently, nascent desorbed molecules can acquire some of the translational energy required for escape from (frustrated) rotations in the chemisorbed state. This depletes the high energy tail of the Boltzmann distribution and produces apparent rotational cooling.

It is important to point out that the foregoing example is representative only of adsorbate–substrate combinations whose interaction potential more or less resembles Fig. 14.1(*a*). One easily can imagine situations where adsorbate final state distributions would look rather different from Fig. 14.26. Consider a case of dissociative chemisorption where the

Fig. 14.26. Effective Boltzmann temperature of NO scattered from Pt(111) in the trapping/desorption regime: (*a*) vibrational states probed by multi-photon ionization; (*b*) rotational states probed by laser-excited fluorescence (Asscher, Guthrie, Lin & Somorjai, 1983; Segner *et al.*, 1983).

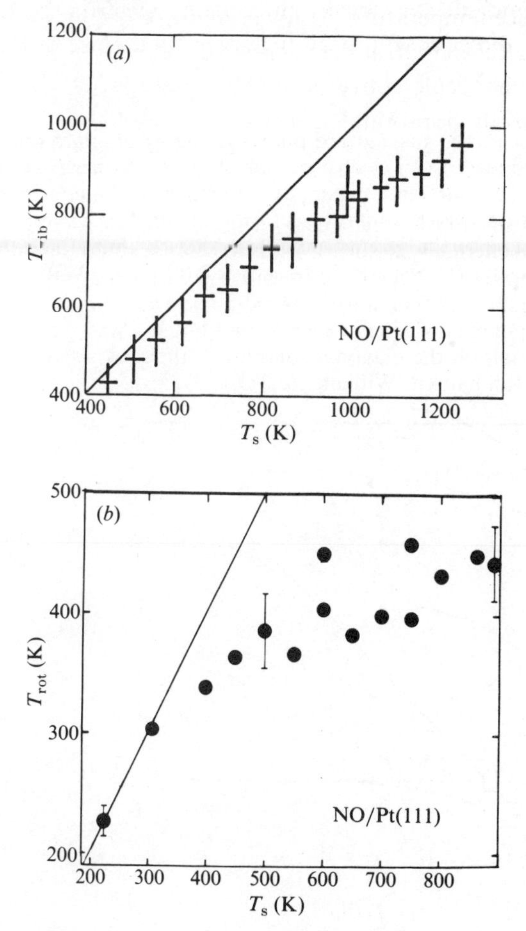

potential energy surface has the form of Fig. 14.12(*b*). The barrier to adsorption occurs in the exit channel. Under these conditions, vibrationally hot molecules stick with greater probability than vibrationally cold molecules. Our familiar detailed balance argument then predicts that molecules which desorb under equilibrium conditions emerge with an effective vibrational temperature *greater* than T_s. Evidently, this technique shows great promise as a means to probe various cuts through the potential energy surface – particularly in combination with trajectory calculations.

State-selected spectroscopy examines the microscopic state of particles ejected from a solid surface by a random process: thermal desorption. An alternative approach to surface dynamics takes just the opposite point of view. We try to manipulate the microscopics in the initial state by state-selected desorption and (at least at first) not worry too much about the details in the final state. This is the idea behind photon stimulated desorption (PSD) and its historical antecedent, electron stimulated desorption (ESD). The reader will recall (Chapter 10) the use of ESD ion

Fig. 14.27. Schematic gas–surface potential energy diagram relevant to desorption induced by an electronic transition. High energy copies (G' and G'') of the ground state curve (G) are members of a continuous manifold of states which contain electron–hole pair excitations of the substrate relative to the ground state. Ions desorb along the repulsive excited state curve (I). See text for discussion (Gomer, 1983).

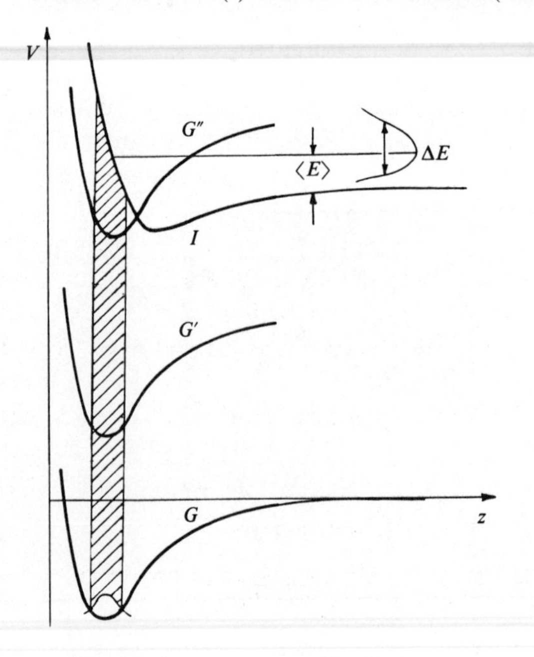

distribution patterns for adsorbate crystallography. Here we have a different use in mind.

Adsorbates become *desorbates* when the chemical bond breaks between surface species and substrate. One way to achieve this result is by direct removal of electrons from electronic bonding orbitals, say, by ionization. The basic idea is sketched in Fig. 14.27. Begin in the ground state where the adsorbate oscillates at the bottom of the adsorbate potential energy curve G. If electron or photon bombardment removes an electron from a bonding orbital, the particle suddenly finds itself (by a Franck–Condon transition) on the repulsive part of an *ionic* potential energy surface I. It accelerates down the potential hill and, in the absence of competing processes, enters the gas phase as an ion A^+. But, an important competing process *does* exist.

The curves marked G' and G'' in Fig. 14.27 represent excited states of the complex which contain an electron–hole pair in the substrate. This type of excitation leaves the basic gas–surface interaction unchanged. Clearly, there are an infinite number of such curves, each a copy of the ground state potential, stacked one atop the next in order of increasing energy. Some of these, G'' for instance, intersect the ionic desorption curve. Barring symmetry restrictions, there is a non-zero probability for the desorbing particle to hop from I to G' at the crossing point. This is a reneutralization process. An electron tunnels from the substrate into the adsorbate state emptied by the original excitation step. The substrate is left in an excited state but the chemisorption bond reforms and desorption is arrested.*

The most common stimulated desorption measurements focus on the threshold for desorption and the desorbate kinetic energy distribution (Fig. 14.28). As Fig. 14.27 shows, this information helps one identify the relevant ionic final state and probes the magnitude and slope of the repulsive potential around the equilibrium adsorption bond length. The interesting dynamical aspects of this problem are buried in the curve-crossing reneutralization probability function. Unfortunately, current theoretical methods are unable to extract this quantity from the data with much reliability. This difficulty has led to an attack on a different, yet related, problem, which has significant bearing on surface chemical dynamics.

Suppose that one knows the dynamics of *photofragmentation* for some

* The desorbate kinetic energy will be large enough to avoid recapture if the curve crossing occurs beyond some critical distance from the surface. The result is *neutral* particle desorption. See, e.g. Gomer (1983).

polyatomic gas phase species. This implies that translational, rotational and vibrational distributions for the final state fragments have been measured and that the appropriate ground state and excited state potential energy curves are known. Now adsorb this molecule on a solid and ask: how does the surface affect the dynamics of the photolysis process? Fig. 14.29 presents a partial answer for the case of laser stimulated C–Br bond breaking in CH_3Br/LiF(100) at submonolayer coverage. Gas phase studies exhibit a kinetic energy distribution for the CH_3 fragment which shows two narrow peaks (the final electronic state is spin-orbit split). The corresponding distribution for fragments produced by laser excitation (at the same frequency) of the adsorbed species is broader and noticeably shifted to lower kinetic energy. Obviously, energy assigned to translation in the gas phase process goes elsewhere when the bromine atom is anchored to a surface. One possibility is that the fragment comes off vibrationally hot (inset). Or perhaps the energy goes into phonons. The details are unknown – even for this simple prototype.

Fig. 14.28. H^+ ion yield versus incident electron energy for ESD from H/Ni(111). Inset shows the ion kinetic energy distribution at 35 eV incident beam energy (Melius, Stulen & Noell, 1982).

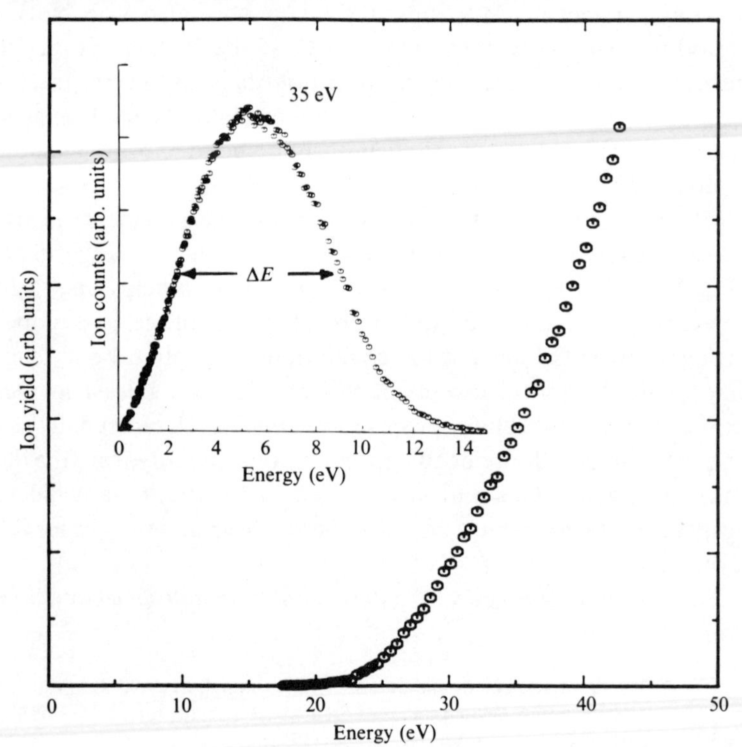

Somewhat more is known about PSD events at both higher and lower photon excitation energy than the foregoing UV example. For example, there is a good understanding of desorption processes associated with core level ionization thresholds (Knotek, 1984). Photo-generated deep core holes do not, in themselves, break chemisorption bonds. But, an interatomic Auger process (Fig. 2.3) can do so if valence bonding levels participate in the radiationless decay process. Reneutralization questions dominate the dynamics as in the aforementioned valence excitation case. By contrast, new issues arise if one examines desorption induced by radiation in the visible and infrared.

Consider the case of molecular adsorption on a clean semiconductor surface such as Si(100). The laser excited photodesorption yield is shown in Fig. 14.30. One observes that desorption turns on at a distinct threshold and increases dramatically as the laser frequency sweeps through the optical gap (vertical arrow). Moreover, the experimental signal is practically independent of the chemical nature of the adsorbate. The most plausible explanation is just the inverse of the reneutralization process discussed earlier. Incident photons create electron–hole pairs in the surface region. Inevitably, some of the diffusing holes encounter adsorption sites and

Fig. 14.29. Measured kinetic energy distribution of CH_3 fragements produced by photolysis of CH_3Br both in the gas phase (right side) and when adsorbed on LiF(100) (left side) (Van Veen, Baller & De Vries, 1985; Bourdon *et al.*, 1984 and courtesy of J.C. Polanyi, University of Toronto).

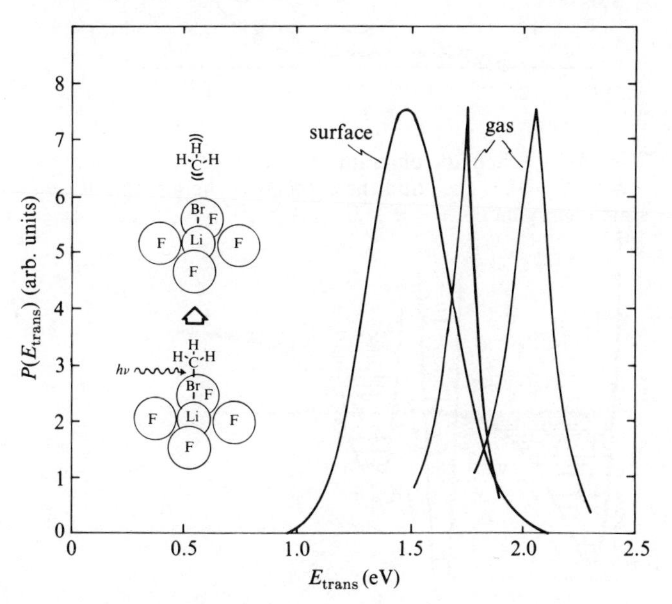

annihilate an electron in a bonding orbital. This process has clear implications for photochemistry at ionic surfaces (see, e.g., Morrison (1977)). In the present case, the threshold correlates well with the energy position of an occupied surface state (below the conduction band edge) in the Si(100) fundamental gap (cf. Fig. 4.36).

Finally, we draw attention to the host of physical processes which can attend irradiation of an adsorbate–substrate complex at infrared

Fig. 14.30. Photodesorption intensity vs. incident laser frequency for NO, CO_2 and CO adsorbed on Si(100) (Ekwelundu & Ignatiev, 1986).

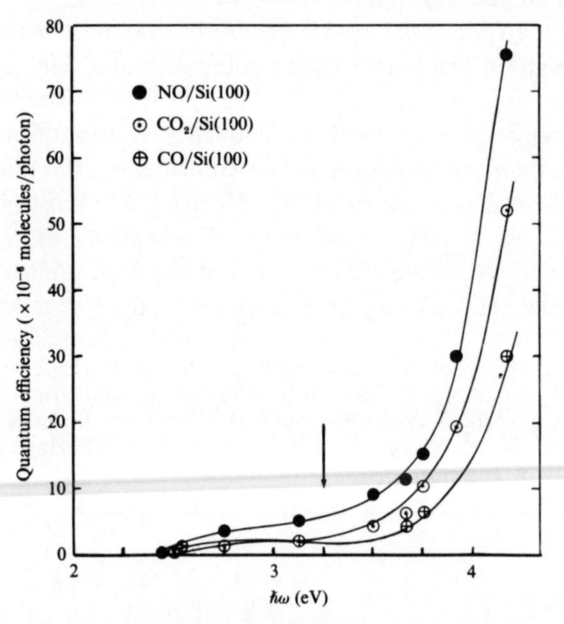

Fig. 14.31. Schematic diagram of an adsorbate vibrational level scheme ($v = 0, 1, 2, \ldots$) and the associated chemisorption well bound state energy ladder ($i = 0, 1, 2, \ldots$) (Gortel, Kreuzer, Piercy & Teshima, 1983).

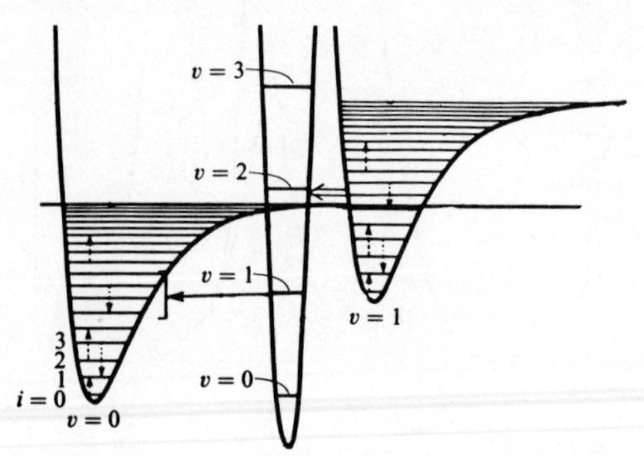

frequencies. To set the energy scales for this problem, it is simplest to consider a diatomic molecule AB bound to a surface by the potential energy curve labeled $v = 0$ in Fig. 14.31. The bound states of this well ($i = 1, 2, 3, \ldots$) correspond to stretching excitations of the molecule–surface bond. These quanta are intermediate in energy between the substrate phonons (which extend down to zero frequency) and typical intramolecular vibrations such as the A–B stretching mode ($v = 1$). Desorption results if energy can be delivered to the adsorption bond in sufficient quantity to exceed the continuum limit (horizontal solid line). We examine three distinct means to this end – all triggered by infrared irradiation.

Our goal is to populate states near the top of the adsorption well. Perhaps one can directly climb the bound state ladder by tuning a laser to the surface bond stretching frequency. Unfortunately, as Fig. 14.31 shows, this generally requires a great many quanta – delivered either coherently (multiphoton absorption) or incoherently (single photon absorption). The probability for such a process is very small except at sample-destroying laser intensities. More promising is an indirect mechanism whereby one pumps the higher energy intramolecular mode with a different laser and relies upon intramolecular relaxation to transfer the energy to the surface bond. The difficulty here is that another pathway for energy flow exists. The excitation can drain away into substrate phonons before desorption can occur. As a result, it turns out that one can interpret the majority of experiments to date (Chuang, 1985) in terms of a third, rather prosaic mechanism: simple thermal desorption due to laser heating of the substrate. Perhaps future experiments will isolate other competing mechanisms.

General references

Adsorption

King, D.A. (1977). Kinetics of Adsorption, Desorption and Migration at Single Crystal Surfaces. *CRC Critical Reviews of Solid State and Materials Sciences* 7, 167–208.

Ertl, G. (1982). Chemical Dynamics in Surface Reactions. *Berichte Bunsen-Gesellschaft fur Physikalische Chemie* **86**, pp. 425–32.

Diffusion

Naumovets A.G. & Vedula, Y.S. (1985). Surface Diffusion of Adsorbates. *Surface Science Reports* **4**, pp. 365–434.

Desorption

Menzel, D. (1975). Desorption Phenomena. In *Interactions on Metal Surfaces* (ed. Gomer), pp. 101–42. Berlin: Springer-Verlag.

Chuang, T.J. (1983). Laser-Induced Gas–Surface Interactions. *Surface Science Reports* **3**, pp. 1–106.

15

SURFACE REACTIONS

Introduction

Our penultimate chapter is concerned with the chemical physics of surface reactions. More precisely, the consistent microscopic perspective adopted throughout this book demands that we ask the following question: can the experimental and theoretical methods of surface physics provide a useful account of real-life surface reaction processes? Let us emphasize the word 'useful'. It is one thing to construct a *post facto* analysis which faithfully reproduces some set of observations. It is quite another to formulate general principles which provide qualitative insight and lead to quantitative predictive power. The selected examples below are intended to demonstrate that this discipline is just now passing from the former perspective to the latter.

Surface reactions are complex events which come in many guises for many purposes. Often (but not always!) one begins with some combination of species in the gas phase: the reactants. In heterogeneous catalysis, the purpose of the surface is to confine the reactants to a two-dimensional space in order to increase the probability for collision and reaction. The sought-after reaction products desorb for collection with no material change in the surface itself. Compare this to metalorganic chemical vapor deposition (MOCVD) growth of compound semiconductors (Dupuis, 1984). Therein, the surface stimulates a decomposition reaction. Unwanted species desorb and the desired species incorporate themselves into the solid. Chemical etching is different again – the inverse of the crystal growth problem in some sense – and it is easy to think of others. Do this; but also, keep the litany in mind so that the specific ways and means of surface analysis we detail do not become too firmly associated with any one particular class of reactions.

Catalysis

'Build a better catalyst and the world will beat a path to your door'. This statement is the driving force behind much of what we know about chemical reactions on metal surfaces. Unfortunately, the profit motive alone does not directly produce answers to some of the remarkably simple questions one can pose. For example, consider a reaction $A + B \rightarrow C + D$ known to be catalyzed by a solid surface. Does the accelerated reaction occur between two chemisorbed species or between one chemisorbed species and an impinging gas phase particle? This is the distinction between a so-called Langmuir–Hinshelwood (LH) process and an Eley–Rideal (ER) process, respectively, although, in fact, they more correctly represent two limiting cases. One easily imagines an intermediate-type reaction between a chemisorbed species and a mobile physisorbed (precursor) species. Nonetheless, it remains a fact that traditional kinetics measurements do not clearly distinguish between the two.

The technique of molecular beam relaxation spectroscopy is well suited to this purpose. Consider the oxidation of carbon monoxide on Pt(111). The alternative reaction schemes are:

$$\begin{array}{ll} \text{Langmuir–Hinshelwood:} & \text{Eley–Rideal:} \\ CO \rightarrow CO\,(\text{ads}), & O_2 \rightarrow 2O\,(\text{ads}), \\ O_2 \rightarrow 2O\,(\text{ads}), & O(\text{ads}) + CO \rightarrow CO_2. \\ CO(\text{ads}) + O(\text{ads}) \rightarrow CO_2. & \end{array} \tag{15.1}$$

In our previous application of MBRS to trapping/desorption, the surface residence time $\tau = R_d^{-1}$ was extracted from the phase lag (imaginary part) of the scattered beam intensity with respect to the modulated incident beam intensity (14.37). Here, we direct a CO molecular beam at a surface previously saturated with dissociated oxygen. But, rather than collect the scattered CO intensity, measure instead the modulated signal from desorbing CO_2 molecules. The two possibilities in (15.1) make rather different predictions for the expected surface residence time. The ER mechanism involves no intrinsic surface processes so that $\tau = 0$. By contrast, a straightforward analysis (Campbell, Ertl, Kuipers & Segner, 1980) shows that τ for the LH process involves both the CO desorption rate and the surface reaction rate. The larger of the two determines the observed phase lag.

Fig. 15.1 illustrates the time dependence of three relevant quantities from the moment the CO beam is turned on: the surface residence time, the instantaneous oxygen coverage and the rate of CO_2 production. Observe that $\tau \cong 6 \times 10^{-4}$ s is constant as long as there is appreciable oxygen left on the sample. When the concentration of O(ads) becomes too low, CO desorption overtakes reaction as the primary CO depletion process. CO_2 production declines and τ rapidly rises towards the clean surface value of R_d^{-1} (~ 10 s) for CO/Pt(111). This is direct evidence for the LH surface reaction mechanism.

The same CO oxidation process catalyzed by a *non-metallic* substrate furnishes a clear example of the fact that many surface reactions involve transient adsorbed species which are of negligible importance during the adsorption and desorption steps. Suppose one prepares a stoichiometric mixture of CO and O_2 over a polycrystalline, indium-doped sample of ZnO. As the reaction proceeds, these gases disappear and CO_2 replaces them (Fig. 15.2(*a*)). Moreover, we know from Chapter 9 that oxygen draws free carriers from the bulk conduction band of this material and chemisorbs as O_2^-. Measurements of the Hall voltage confirm a rapid decrease in the concentration of bulk carriers (Fig. 15.2(*b*)). It is possible that a carbon monoxide molecule (gas phase or adsorbed) collides with this complex leading to $CO_2 + O^-$(ads). However, CO disappears from the gas phase at about the same rate that the carrier concentration returns to its clean surface value.

It turns out that a quantitative account of the kinetics sketched in

Fig. 15.1. Molecular beam study of CO oxidation on Pt(111) at 442 K. The curves show the time dependence of the surface residence time (open circles), the CO_2 production rate (closed circles) and the oxygen coverage (dashed). The initial oxygen coverage is $\theta = 0.25$ (Campbell *et al.*, 1980).

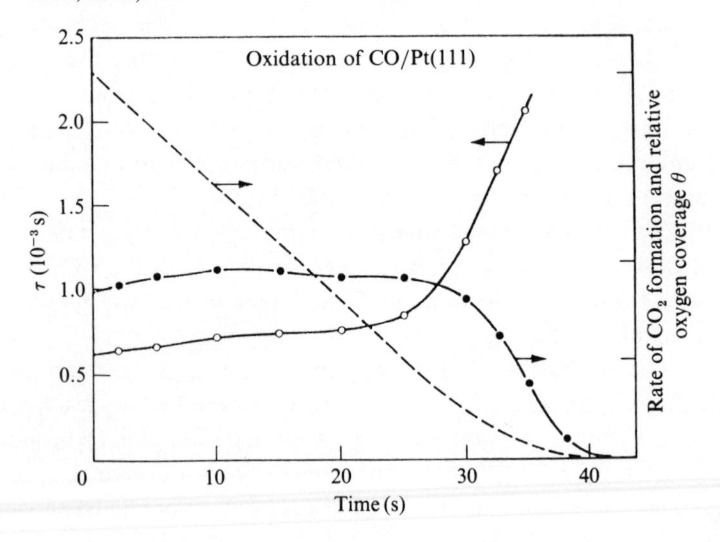

Fig. 15.2 is possible if one assumes the following ER reaction sequence:

$$
\begin{aligned}
&O_2 + e \rightarrow O_2^-(\text{ads}),\\
&O_2^-(\text{ads}) + e \rightarrow 2O^-(\text{ads}),\\
&O^-(\text{ads}) + CO \rightarrow CO_2 + e.
\end{aligned}
\tag{15.2}
$$

This is not to say that (15.2) uniquely describes the precise microscopics. Perhaps (for example) chemisorbed CO plays some role. Nevertheless, the overall kinetics do appear to require that each CO react with the charged transient species O^-. This is not too surprising; atomic oxygen adsorbed onto an insulator doubtless draws free charge to itself. Our point is simply that the presence of this species is special to the surface-catalyzed reaction. The oxidation of carbon monoxide in the gas phase requires no such intermediate.

Also special to *heterogeneous* catalysis is the notion of an 'active site'. The idea is that chemical reaction occurs preferentially (or exclusively) at one set of adsorption sites as opposed to others. Fig. 14.22 already makes clear that terrace edge sites are good candidates. Surface defects are another possibility. Normally this means a structural anomaly such as a vacancy

Fig. 15.2. Kinetic data for CO oxidation to CO_2 on polycrystalline ZnO at 623 K: (*a*) change in the ambient gas partial pressure as the reaction proceeds; (*b*) change in the electron concentration in the conduction band of the substrate (Chon & Prater, 1966).

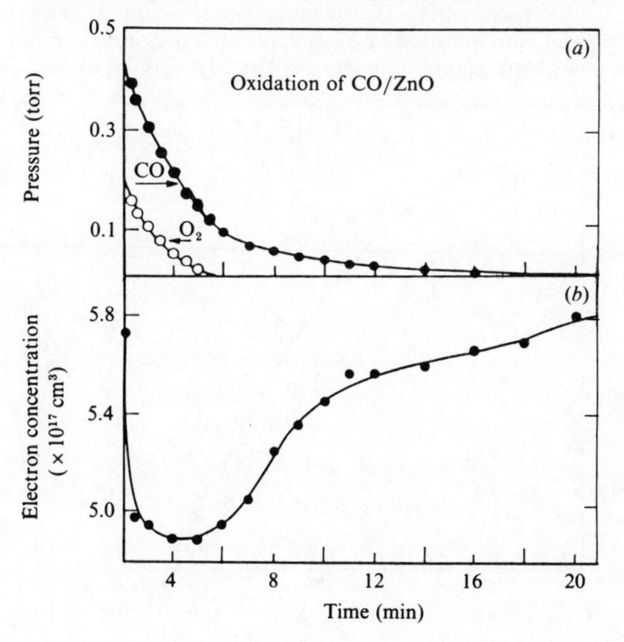

in the adsorption lattice. But, there is good evidence that 'electronic' defects stimulate certain surface reactions as well. Fig. 15.3 illustrates an example of the latter. The process of interest is an H_2–D_2 isotopic exchange reaction which involves dissociative chemisorption of the products and recombination and desorption of HD. Polycrystalline MgO efficiently catalyzes this exchange at 78 K.

Electron spin resonance experiments establish a solid correlation between the observed reaction rate and the concentration of V_k color centers in the substrate. One controls the latter by pre-treatment of the sample at high temperature. In the bulk, this type of defect is associated with a positive carrier localized near adjacent negative ions (see, e.g., Ashcroft & Mermin (1976)). Here, we imagine a hole bound to a complex of three oxygen surface ions at an exposed MgO(111) plane. It is not known precisely how this site catalyzes H_2–D_2 exchange.

The three examples outlined above identify a few typical features of numerous catalytic surface processes: specific reaction mechanisms, transient intermediates and active sites. However, to obtain a more global understanding of heterogeneous catalysis we must identify trends from which to extrapolate and build models. Moreover, it is necessary to establish a link between commercial catalysts and the single crystal clean surfaces of surface science. Three figures of merit are in common use. The

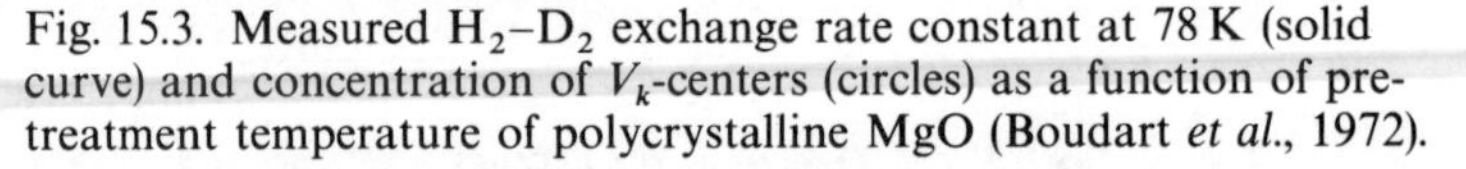

Fig. 15.3. Measured H_2–D_2 exchange rate constant at 78 K (solid curve) and concentration of V_k-centers (circles) as a function of pre-treatment temperature of polycrystalline MgO (Boudart *et al.*, 1972).

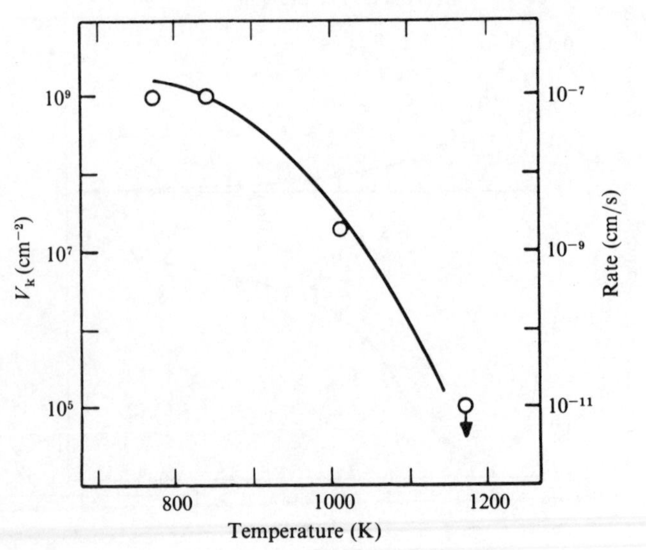

first is called the *activity* of the catalyst. This quantity indexes the efficiency of a surface to accelerate a reaction. We speak of a catalyst's 'turnover number' – defined as the number of product molecules per surface site per unit time. The second concept of interest is called *selectivity*. It arises from the fact that unwanted competing reactions often accompany any particular desirable reaction one seeks to catalyze. In the best situation, we seek a catalyst which promotes – or selects – only the primary reaction and suppresses all others. Finally, the action and efficiency of commercial catalysts (which operate under non-equilibrium steady state conditions) can depend quite sensitively on the presence of *poisons* and/or *promoters*. These are chemical additives that retard or promote specific reactions, respectively, relative to the behavior of the pristine catalyst. We examine each of these with an eye toward the connection to microscopics.

The experimental literature contains a number of correlations which purport to show a connection between absolute catalytic activity and simple physical quantities. The relationship between the rate constant for methane synthesis ($3H_2 + CO \rightarrow CH_4 + H_2O$) over various transition metal supports and the CO heat of adsorption on these surfaces is a typical example (Fig. 15.4(*a*)). This so-called 'volcano plot' suggests that low turnover is associated with either very weak binding (low equilibrium

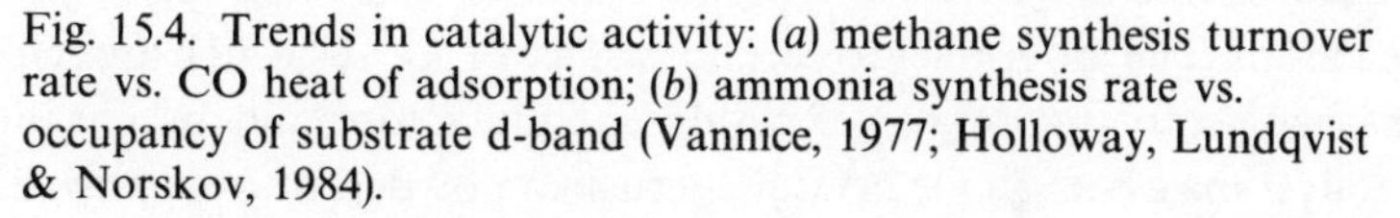

Fig. 15.4. Trends in catalytic activity: (*a*) methane synthesis turnover rate vs. CO heat of adsorption; (*b*) ammonia synthesis rate vs. occupancy of substrate d-band (Vannice, 1977; Holloway, Lundqvist & Norskov, 1984).

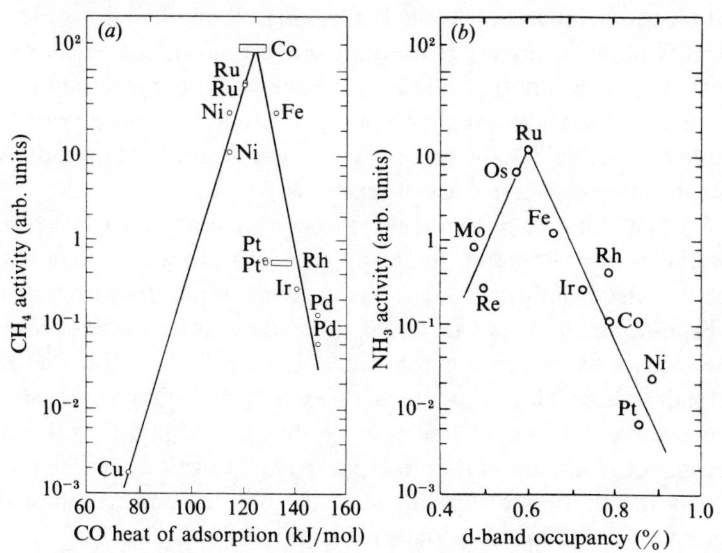

concentration of adsorbed CO) or very strong binding (saturation coverage of CO blocks hydrogen adsorption sites). However, this organization of the data might be misleading because, for example, it is silent on the question of the CO dissociation probability – a crucial step in the synthesis.

Suppose we arrange activity data for another important surface reaction, ammonia synthesis ($N_2 + 3H_2 \rightarrow 2NH_3$), in a slightly different way. Fig. 15.4(*b*) is again a volcano plot but now the abscissa represents the number of electrons in the d-band of the substrate. Invoking our general approach to dissociation, we examine the occupancy of the *intra-molecular anti-bonding* orbital of N_2 (or CO, H_2, etc.) as it hybridizes with substrate d-electrons near the Fermi level. Explicit model calculations using the surface molecule limit of the resonant level model (Chapter 9) show that this quantity steadily decreases as one traverses Fig. 15.4(*b*) from left to right (Holloway, Lundqvist & Norskov, 1984). In other words, molecules adsorbed onto the surface of metals on the right hand side of a transition row are relatively more stable against dissociation (and subsequent reaction) than their counterparts near the middle of the row. The moral of the story is: the overall activity of a reaction depends upon a number of (possibly competing) microscopic steps. Always evaluate macroscopic correlations in the light of available microscopic information.

The possible selectivity of a given catalyst is discussed best in the context of the distinction between 'structure sensitive' and 'structure insensitive' reactions. The latter refer to certain surface processes which are remarkably insensitive to the details of catalyst manufacture and preparation. The catalyst may be a single crystal, a crushed powder, or otherwise dispersed within another material. Roughly speaking, all that matters is the chemical identity of the catalytic agent. Many interesting surface reactions fall into this category. By contrast, the activity of certain other reactions is observed to vary dramatically even as one moves from one single crystal surface plane to another. We begin our survey with some examples drawn from this structure sensitive class of reactions.

Consider first a reaction like the conversion of linear hydrocarbon molecules into aromatic ring compounds – a process of some interest to the petroleum industry. A specific example is the dehydrocyclization of n-heptane into toluene. Fig. 15.5(*a*) shows the dramatic difference in activity that occurs when the reaction proceeds over Pt(111) and Pt(100) single crystal surfaces. These experiments are conducted at high ambient reactant pressures as one might find in a commercial application. Perhaps the hexagonal symmetry of the close-packed (111) surface accommodates (i.e., lowers the barrier to) formation of the product ring compounds more readily than the square surface net of the (100) surface.

Equally striking is the selectivity of various platinum crystal faces for dehydrocyclization compared to a competing process (hydrogenolysis) which simply breaks down the long chain alkanes into shorter linear chain alkanes (Fig. 15.5(*b*)). Note that the greatest selectivity for the desired process occurs on the high-index (557) plane of platinum. This surface consists of very regular close-packed terraces which are five atom rows in width. The selectivity results support the suggestion that symmetry alone may distinguish an active site from a non-active site. The presence of steps apparently retards simple alkane cracking. Perhaps C–C bond scission is too costly at these sites.

A microscopic interpretation of selectivity seems particularly apt for the case of two other hydrocarbon reactions carried out over Cu–Ni *alloy*

Fig. 15.5. Catalytic performance of platinum surfaces for hydrocarbon decomposition: (*a*) activity of Pt(111) and Pt(100) for conversion of n-heptane to toluene at 573 K; (*b*) relative activity (selectivity) of two competing reactions on various stepped surfaces (Gillespie, Herz, Petersen & Somorjai, 1981).

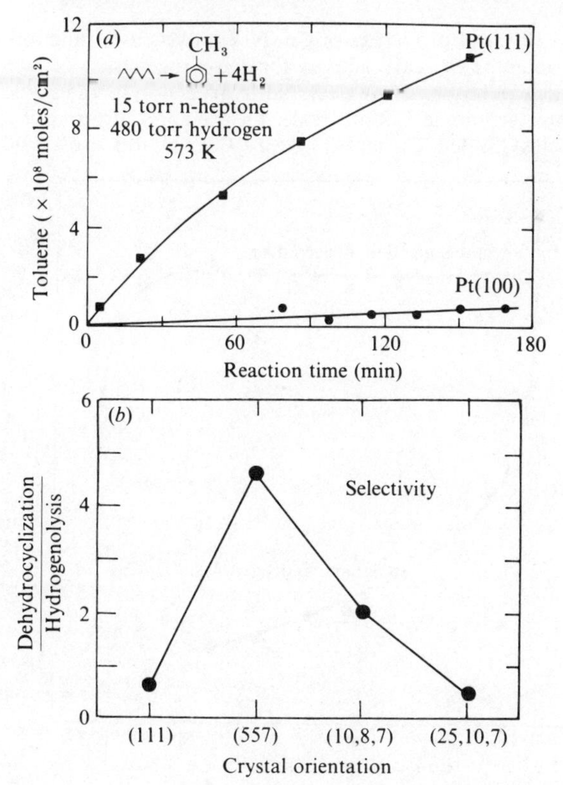

catalysts. Ethane to methane conversion and cyclohexane to benzene conversion are the processes of interest:

$$\begin{aligned} &C_2H_6 + H_2 \rightarrow 2CH_4 \quad \text{(hydrogenolysis)},\\ &C_6H_{12} \rightarrow C_6H_6 + 3H_2 \quad \text{(dehydrogenation)}. \end{aligned} \tag{15.3}$$

Fig. 15.6 shows the measured activity of these reactions as a function of the relative concentration of copper in the bulk catalyst. Evidently, one is strongly selected over the other for a wide range of alloy compositions. Note particularly the precipitous drop in methane (benzene) production for very low (high) alloy copper content. The clue to the origin of this behavior is the measured concentration of copper at the surface of CuNi(111) as obtained by a field ion technique (open circles). The surface composition is copper-rich (nearly) independent of the bulk composition. This segregation phenomenon is in complete accord with the arguments advanced in Chapter 4 (cf. Fig. 4.30). With this information, we argue that reactant adsorption and product desorption, respectively, are the rate-limiting steps for hydrogenolysis and dehydrogenation.

Fig. 15.6. Physical properties of Cu–Ni catalysts as a function of bulk copper content. Left scale: activity for dehydrogenation of cyclohexane to benzene (squares) and hydrogenolysis of ethane to methane (closed circles). Right scale: surface concentration of copper (open circles) (Sinfelt, Carter & Yates, 1972; Sakurai *et al.*, 1985).

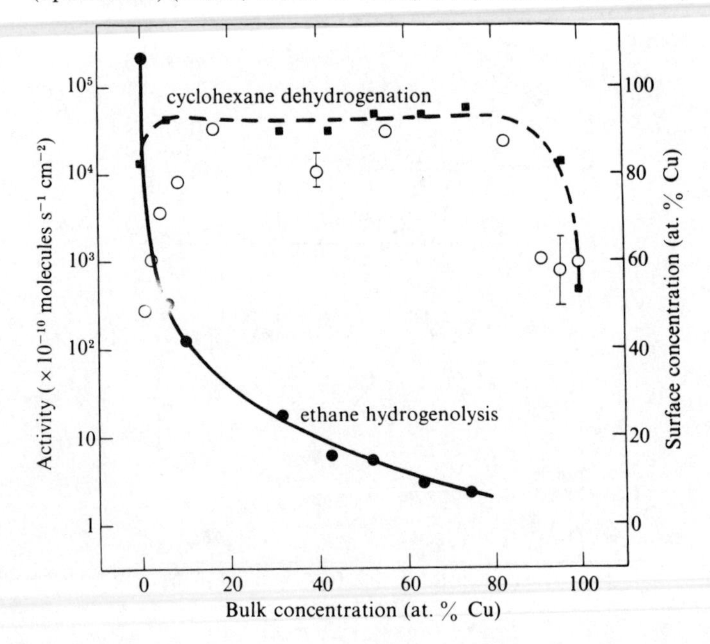

Equation (15.3) shows that dissociative chemisorption of H_2 is essential for ethane conversion to methane. The observed activity follows immediately if hydrogen dissociation occurs readily at nickel surfaces but not at copper surfaces. Evidence that this is indeed the case has appeared repeatedly in this volume. For example, we argued explicitly in the preceding chapter that the observed barrier to hydrogen sticking on Cu(100) will be much reduced (if not eliminated) on nickel surfaces. Moreover, the presence or absence of 4d density of states at the Fermi level of Pd/Nb(110) correlated precisely with the dissociative chemisorption or non-sticking of hydrogen to that surface, respectively (cf. discussion following Fig. 12.13). To complete the story we must argue that a nickel surface retards cyclohexane dehydrogenation to benzene relative to copper. The general trends in chemisorption heats of adsorption (Fig. 9.4) guarantee that this will be the case if desorption of the benzene product is the rate-limiting step.

The reasoning applied to the conversion reactions of (15.3) depended only on the chemical identity of the majority species at the catalyst surface. This suggests that they are structure insensitive reactions. If so, experiments performed on small-area single crystals and large-area dispersed particles of the same metal ought to yield identical catalytic reaction parameters. The Arrhenius plot in Fig. 15.7 illustrates that methane synthesis (Fig. 15.4(*a*)) over nickel substrates is structure insensitive in just this manner. Indeed, the insensitivity of both the turnover rate activation energy and pre-exponential factor raises the possibility that UHV experiments on single crystals might permit one to *predict* the behavior of 'real' catalysts which operate at vastly different temperatures and pressures. To investigate this possibility, return again to the ammonia synthesis reaction of Fig. 15.4(*b*).

The Haber–Bosch process for the synthesis of ammonia from nitrogen and hydrogen gas has remained essentially unchanged since the first commercial plant began operation in 1913 (Topham, 1985). A typical catalyst for this reaction is a porous, high surface area structure consisting of small (~ 250 Å) Fe particles (with partially reduced K_2O adsorbed to submonolayer coverage) interspersed with Al_2O_3. The chemical identities of various reaction intermediates are known from XPS and extensive kinetics studies indicate that the overall potential energy diagram is not unlike Fig. 15.8. Kinetic measurements also implicate the dissociative chemisorption of nitrogen as the rate-limiting step. The latter is related to the observed very small sticking coefficient ($\sim 10^{-6}$) of N_2/Fe rather than to the energy barrier E^*, whose magnitude anyway varies with surface structure and coverage (cf. Fig. 14.21).

Fig. 15.7. Comparison of the rate of methane synthesis over single-crystal nickel surfaces and supported Ni/Al_2O_3 catalysts at 120 Torr total reactant pressure (Kelley & Goodman, 1982).

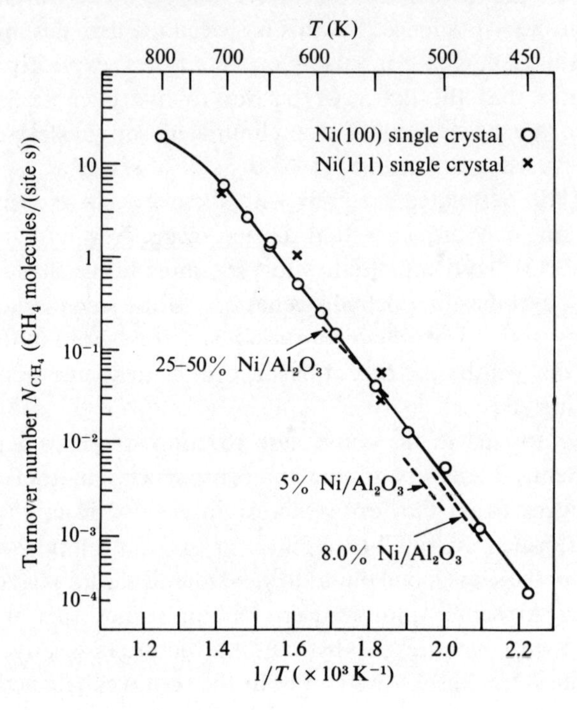

Fig. 15.8. Schematic potential energy diagram indicating six principal steps in the synthesis of ammonia over iron. (Ertl, 1983).

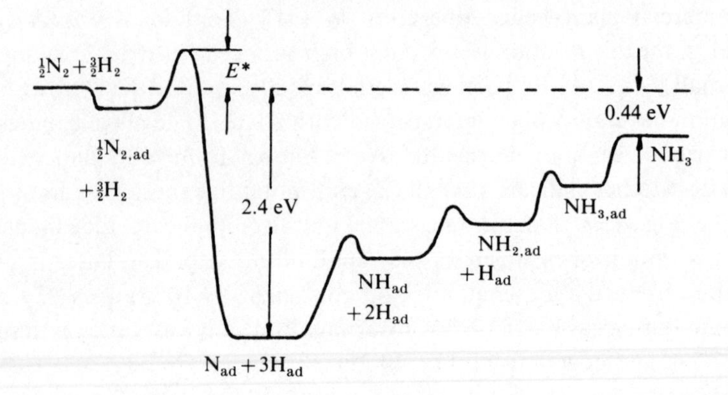

Our goal is to calculate the overall rate of this reaction as a function of reactant temperature and pressure. The simplest thing one can do is to assume that each of the six elementary reaction steps indicated in Fig. 15.8 is in equilibrium except the second: dissociative chemisorption. In that case, it is easy to see that the law of mass action quickly leads to an expression for the output concentration of NH_3 in terms of the input concentration of reactants, the individual sub-reaction equilibrium rate constants and the adsorption and desorption rates appropriate to $N_2(ads) \rightleftharpoons 2N(ads)$. The strategy is to use *gas phase* data for the various equilibrium constants (corrected by appropriate partition function ratios to account for the adsorbed state of the species involved) and surface physics measurements of nitrogen on *single crystal* surfaces to model the rate-limiting step. No information about ammonia synthesis itself or of the physical nature of the catalyst (save its surface area) need be considered. The result of this exercise – using only the statistical mechanics of non-interacting adsorbates – is in remarkable accord with activity measurements for a commercial catalyst over a wide range of reactant pressures and temperatures (Fig. 15.9).

The alert reader will notice an inconsistency in our treatment of surface catalyzed ammonia synthesis. This was supposedly a structure insensitive

Fig. 15.9. Comparison of calculated and measured NH_3 mole fraction output from an Fe-based catalytic reactor (Stoltze & Norskov, 1985).

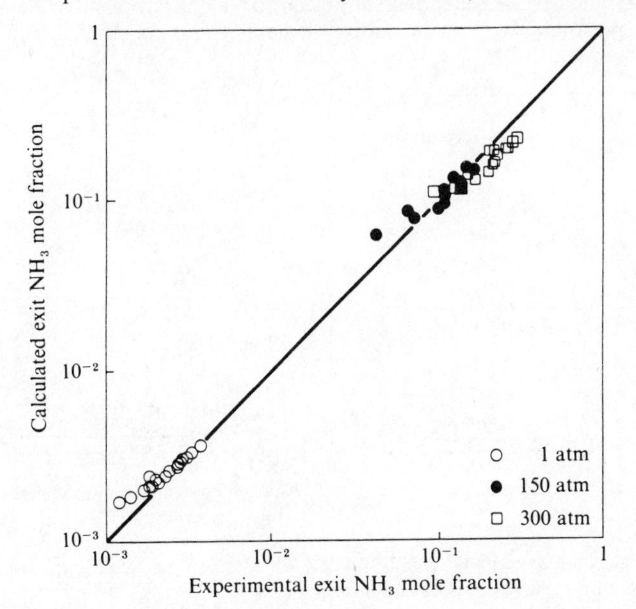

reaction, yet, Fig. 14.3 illustrates clearly that the sticking coefficient for N_2/Fe(111) differs markedly from that of N_2/Fe(100). The resolution of this problem is to be found in the *potassium* content of the catalyst. In the presence of pre-adsorbed K atoms, the sticking coefficient of nitrogen on iron is, in fact, structure insensitive.* Moreover, potassium is a promoter for ammonia synthesis. This is not merely a convenience, for, without its presence, large scale production of NH_3 at acceptable costs is not possible. Accordingly, a search for the microscopic origins of catalytic promotion and poisoning is a particularly pressing goal of modern surface physics. To date, only the first steps along this path have been taken.

Fig. 15.10 illustrates the striking effect that foreign additives can have on catalytic activity. The promotion of ammonia synthesis over a Ru catalyst by alkali metal adsorbates (left panel) and the poisoning of methane synthesis over a Ni catalyst (right panel) by sulfur and phosphorus (right panel) are characteristic of many other surface reactions. The relative position of these additives in the periodic table strongly suggests that their action is related to their electronegativity (relative to the substrate) and hence is electrostatic in nature. This idea is made more precise below. However, it is important to remind oneself that some kind of change in local electronic structure invariably accompanies chemisorption. Given this, we must expect variations in surface reaction activity if these changes

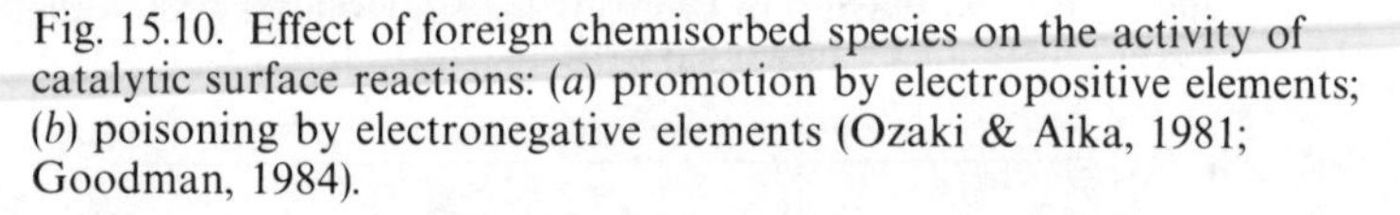

Fig. 15.10. Effect of foreign chemisorbed species on the activity of catalytic surface reactions: (*a*) promotion by electropositive elements; (*b*) poisoning by electronegative elements (Ozaki & Aika, 1981; Goodman, 1984).

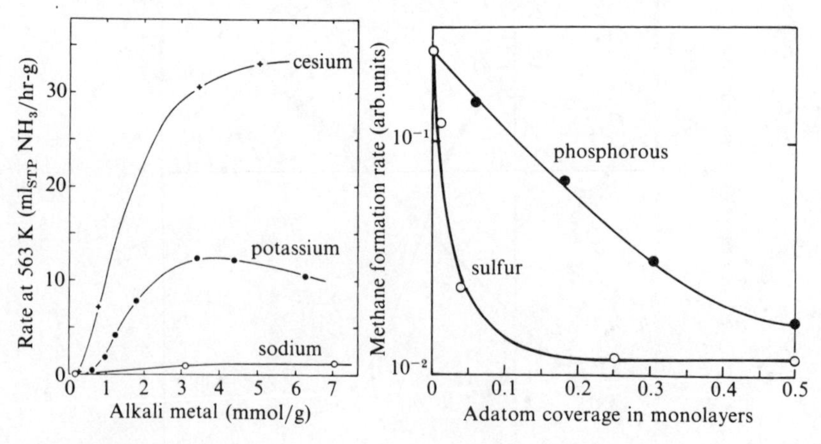

* In general, we cannot reliably predict whether any particular reaction will be structure sensitive or structure insensitive.

occur in the electron states in the immediate vicinity of the Fermi level. After all, it is just the filling and unfilling of bonding and anti-bonding orbitals which dictate the magnitude of molecular adsorption energies and activation barriers. This is especially true of metal surfaces where the continuum of electronic states broadens sharp molecular levels into resonances and facilitates overlaps in energy space for curve crossings, resonant tunnelling, etc. The distinction between 'structural' factors and 'electronic' factors in catalysis is largely a misnomer.

The basic physics of catalytic promotion and poisoning by electropositive and electronegative adsorbates clearly appears in the atom-on-jellium model introduced in Chapters 8 and 9. This self-consistent technique properly describes chemisorption-induced charge rearrangements and the accompanying changes in electrostatic potential. To wit, Fig. 15.11 shows the change in the potential seen by an electron as a function of distance from the surface in the immediate neighborhood of various adsorbates. Note that a significant change occurs only at distances in *excess* of a typical atomic chemisorption bond length. Rather, one

Fig. 15.11. Change in the electrostatic potential as a function of distance in the vicinity of foreign atoms adsorbed at their equilibrium distance on a jellium suface ($r_s = 2$). The curves are drawn in a plane at a lateral distance of 3.5 a.u. and 5 a.u. from the electronegative and electropositive species, respectively (Lang, Holloway & Norskov, 1985).

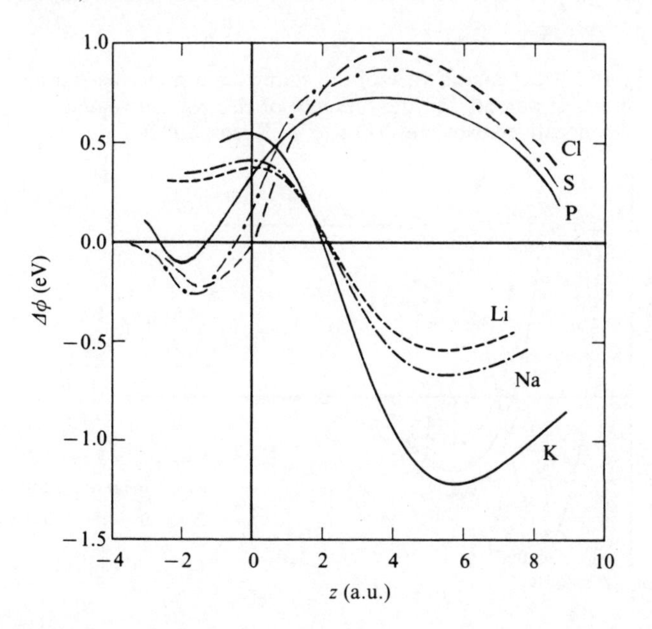

expects molecular adsorption wells at these distances. The net electrostatic effect on the full gas–surface interaction potential is shown in Fig. 15.12. In the notation of Fig. 14.5, the desorption energy E_d increases (decreases) in the presence of a nearby electropositive (electronegative) adsorbate. The value of E_a changes by somewhat less and E_c is unchanged. Consequently, we interpret the effect of the additive as a promotion (poisoning) of the dissociative adsorption step since (14.16) predicts an increase (decrease) of the corresponding sticking probability. The same result holds if the maximum change in potential occurs in the vicinity of the barrier. In that case, we associate promotion or poisoning with acceleration or retardation of the dissociation step.

Full-scale LDA surface electronic structure calculations for CO/Ni(100) with co-adsorbed K or S largely bear out the qualitative picture abstracted from the atom-on-jellium model. Fig. 15.13 is a charge density difference contour map analogous to that of Fig. 12.14 except that now a potassium atom (left panel) or a sulfur atom (right panel) occupies an adsorption site adjacent to the CO molecule. Potassium does indeed induce a large drop (65%) in the work function of the overlayer and electron transfer into the anti-bonding 2π orbital is evident (cf. Fig. 12.14(*a*)). The situation is less clear cut for the sulfur case. The gross effect is surely opposite to that of K. However, the work function increases only negligibly (3%) and the behavior of both additives in the vicinity of the oxygen atom is quite similar. Evidently, electrostatics is not the whole story. For example, we have seen that adsorption commonly alters the local density of states and

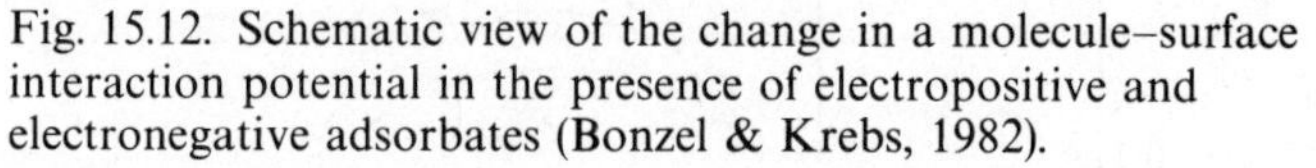

Fig. 15.12. Schematic view of the change in a molecule–surface interaction potential in the presence of electropositive and electronegative adsorbates (Bonzel & Krebs, 1982).

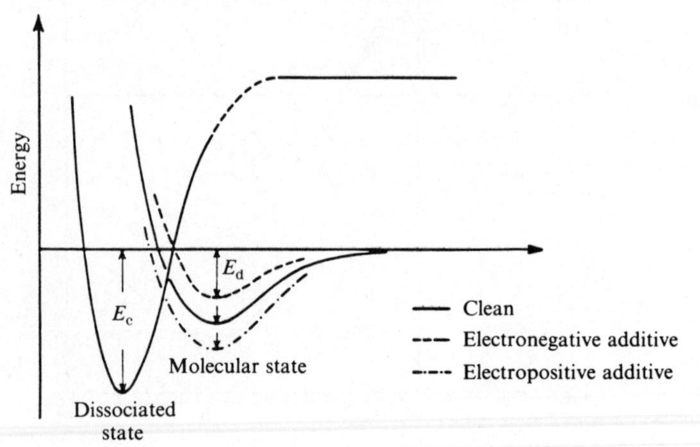

wave function character of electrons in the immediate vicinity of the Fermi level. Since it is these electrons that are the 'reactive' ones, it is evident that a complete description of promotion and poisoning must take account of these effects as well. Theoretical trend studies will be valuable in this regard.

Crystal growth

Surface reactions play a crucial role in modern methods of crystal growth used in the fabrication of materials for electronic devices. Unlike growth from the melt, molecular beam epitaxy (MBE) and the MOCVD technique introduced above involve transport of fresh material from afar (Fig. 15.14). MBE growth occurs in an ultra-high vacuum environment.

Fig. 15.13. Contour plot of the charge density difference: (CO + K)/Ni(100) – CO/Ni(100) (left panel) and (CO + S)/Ni(100) – CO/Ni(100) (right panel) based on LDA surface band structure calculations. Solid (dashed) lines indicate a gain (loss) of electronic charge (Wimmer, Fu & Freeman, 1985).

For say, III–V compound semiconductor growth, an atomic beam of group III atoms and a molecular (typically dimer or tetramer) beam of group V material impinge on the sample.* The growth kinetics is determined by the dissociation rate of the molecules and the relative sticking coefficients of the two species. By contrast, transport of reactant species in MOCVD occurs via hydrodynamic flow of organometallic molecules in a carrier gas such as H_2. Diffusion brings these species into intimate contact with the surface. The growth rate may be limited either by the diffusive step or by the kinetics of cracking apart the organometallic reactants.

From the point of view of fundamental physics, microscopic studies of molecular beam epitaxy are the most advanced. For example, a good deal is known about the kinetics of GaAs growth from beam scattering studies. These experiments show that atomic Ga and As_2 molecules adsorb onto GaAs surfaces rather differently. The sticking coefficient of Ga is

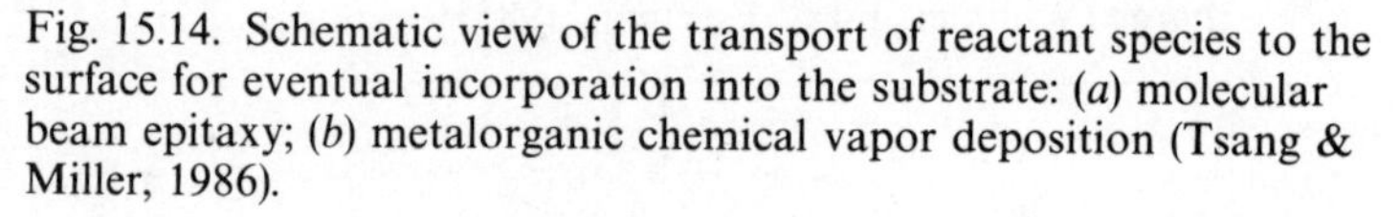

Fig. 15.14. Schematic view of the transport of reactant species to the surface for eventual incorporation into the substrate: (*a*) molecular beam epitaxy; (*b*) metalorganic chemical vapor deposition (Tsang & Miller, 1986).

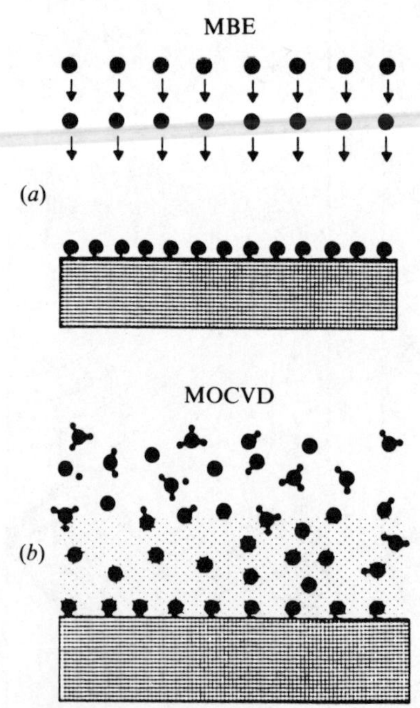

* The beams are typically of the effusive Knudsen variety as opposed to the supersonic beams employed in most reaction dynamics studies.

unity for temperatures up to nearly 10^3 K. The sticking coefficient for arsenic apparently depends quite sensitively on the ambient coverage of gallium. To illustrate this, Fig. 15.15 reproduces an experimental trace of the time-dependent flux of As_2 scattered from GaAs(111). The clean surface reflects essentially all molecules incident upon it. But, if the surface is pre-dosed by a gallium beam, the As_2 sticking coefficient jumps to unity and declines to zero only as the Ga population declines (presumably by formation of GaAs). A plausible growth model based on these and other experiments is presented in Fig. 15.16. Arsenic dimers initially adsorb into mobile precursor states. Stoichiometric gallium arsenide compound formation follows dissociative chemisorption above free gallium sites. Excess As_2 desorbs or associates with other dimers to desorb as As_4 molecules.

The surface reaction physics of MOCVD is basically unknown. There are two reasons for this. First, it is simply a newer technique. Second, it operates under conditions which are far from compatible with UHV experiments. Presumably the future will bring experimental methods which bridge the 'pressure gap' in this case as already has occurred for more conventional catalytic processes (e.g. Fig. 15.7). In the meantime, one must be content with information gleaned somewhat indirectly. Kinetic studies of the growth of the II–VI semiconductor compound CdTe provide an example.

The reaction of interest involves the decomposition of dimethylcadmium

Fig. 15.15. Time dependence of the intensity of As_2 molecules scattered from GaAs(111) with and without pre-dosing by a Ga beam. The incident beam is chopped is a manner indicated by the vertical arrows (Arthur, 1969).

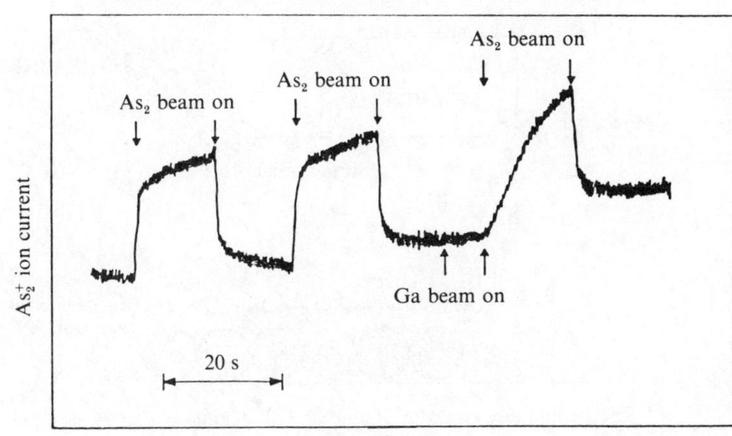

and triethyltellurium in the presence of H_2:

$$(CH_3)_2Cd + (C_2H_5)_2Te + 2H_2 \rightarrow CdTe + 2CH_4 + 2C_2H_6 \quad (15.4)$$

Of course, this equation expresses nothing more than the net stoichiometry. An understanding of the growth mechanism requires knowledge of the individual intermediate reactions which sum to (15.4). A clue comes from measurements of the temperature dependence of the rate of CdTe growth (Fig. 15.17). The process obeys an Arrhenius law with an activation energy of above 0.8 eV. This is substantially lower than the gas phase barrier to the first step in the dissolution of either of the reactant organometallics. However, it is quite close to the observed energy to deposit Cd alone (in the absence of $(C_2H_5)_2Te$) and also to the energy required to detach the *second* methyl group from gaseous dimethylcadmium. This suggests the following tentative Langmuir–Hinshelwood reaction sequence:

$$\begin{array}{ll} \text{(i)} & \begin{cases} (CH_3)_2Cd \rightarrow (CH_3)_2Cd(ads), \\ (C_2H_5)_2Te \rightarrow (C_2H_5)_2Te(ads), \end{cases} \\ \text{(ii)} & (CH_3)_2Cd(ads) + \tfrac{1}{2}H_2 \rightarrow (CH_3)Cd(ads) + CH_4, \qquad (15.5) \\ \text{(iii)} & (CH_3)Cd(ads) + \tfrac{1}{2}H_2 \rightarrow Cd(ads) + CH_4, \\ \text{(iv)} & Cd(ads) + (C_2H_5)_2Te(ads) + H_2 \rightarrow CdTe(ads) + 2C_2H_6. \end{array}$$

The idea is that the growing CdTe surface heterogeneously catalyzes the

Fig. 15.16. Schematic model of GaAs crystal growth via Ga and As_2 molecular beam deposition (Foxon & Joyce, 1981).

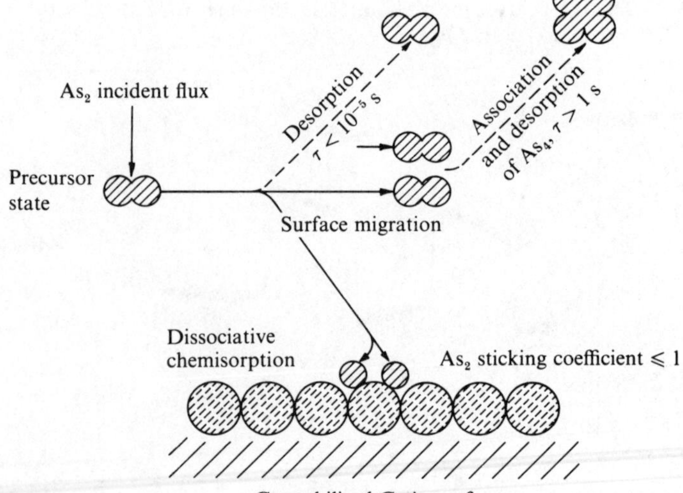

decomposition steps (ii) and (iv) to the point where step (iii) becomes rate limiting. Spectroscopic surface chemical analysis is a minimal step required to determine the veracity of (15.5).

It is appropriate to close with a few remarks about the probable role of surface diffusion in reactions of the sort examined in this chapter. Doubtless there are cases where this process is the rate-limiting step. In general, both species diffuse and reaction occurs when their mutual separation reaches some small value R_0. For simplicity, imagine reactant A as fixed at the origin with B reactants diffusing towards it with diffusion constant D. In *steady state*, the diffusion equation (14.19) reduces to Laplace's equation $\nabla^2 P(\mathbf{r}) = 0$ for the probability that a B molecule be found at position $\mathbf{r}$. The boundary conditions (assuming spherical symmetry) are:

$$\left.\begin{aligned} P(R_0) &= 0, \\ P(R) &= c_0. \end{aligned}\right\} \tag{15.6}$$

The first of these directs us to remove the random walker from the problem after the reaction occurs. The second maintains the steady state by fixing the concentration of B molecules at some large radial distance R from the origin. With a solution in hand, we take the limit $R \to \infty$ and calculate

Fig. 15.17. Arrhenius plot of the observed rate of CdTe growth as a function of substrate temperature in an MOCVD reactor (Bhat, Taskar & Ghandi, 1987).

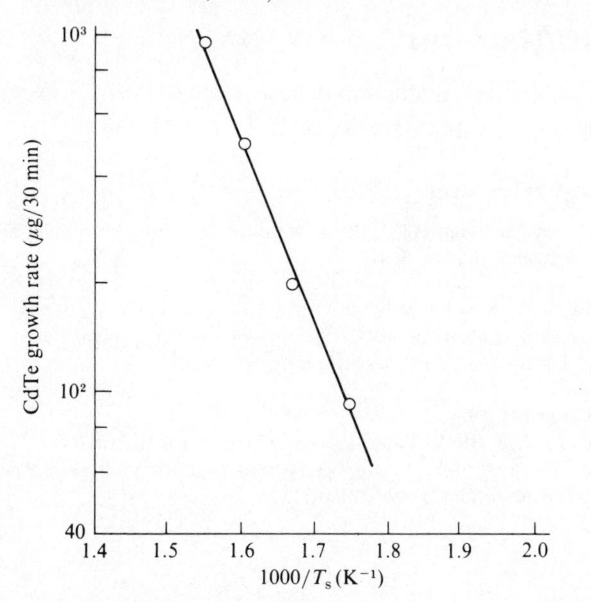

the diffusion-controlled reaction rate k_D from the requirement that the flux of particles injected at R equal the flux of particles which disappear at R_0:

$$c_0 k_D = D \left.\frac{dP}{dr}\right|_{r=R_0} \cdot \begin{cases} 4\pi R_0^2 & \text{3-D,} \\ 2\pi R_0 & \text{2-D.} \end{cases} \tag{15.7}$$

In three dimensions one readily verifies that $P(r) = c_0(1 - R_0/r)$ so that $k_D = 4\pi R_0 D$. However, in two dimensions the solution to

$$\frac{d^2P}{dr^2} + \frac{1}{r}\frac{dP}{dr} = 0 \tag{15.8}$$

which satisfies (15.6) is

$$P(r) = c_0 \frac{\ln(r/R_0)}{\ln(R/R_0)}, \tag{15.9}$$

which *diverges* as R is taken to infinity! This implies a vanishing rate constant – clearly an unphysical result. It turns out that the problem is *not* that we have confined the diffusion to two dimensions. Instead, it reflects our neglect of any interactions amongst the B molecules and/or the fact that the reactive species have a finite surface lifetime τ due to desorption. It is possible to correct for these effects at various levels of sophistication (Keizer, 1982; Freeman & Doll, 1983). Here, we merely quote the result. Ignoring inter-adsorbate interactions, the reaction rate may be written

$$k_D = 4\pi D\Phi(R_0/\sqrt{D\tau}), \tag{15.10}$$

where $\Phi(x)$ is a well-behaved function of its argument. Perhaps experiments can test (at least) the proportionality of k_D to D.

General references

D'Evelyn, M.P. & Madix, R.J. (1984). Reactive Scattering from Solid Surfaces. *Surface Science Reports* **3**, 413–98.

Catalysis

Somorjai, G.A. & Zaera, F. (1982). Heterogeneous Catalysis on a Molecular Scale. *The Journal of Physical Chemistry* **86**, 3070–8.

Crystal growth

Stringfellow, G.B. (1985). Vapor Phase Epitaxial Growth of III/V Semiconductors. In *Crystal Growth of Electronic Materials* (ed. Kaldis), pp. 247–67. Amsterdam: North-Holland.

16

EPITAXY

Introduction

Our account of adsorption to this point has been restricted largely to one particular, albeit important, special case: the situation where adsorbate–substrate interactions dominate adsorbate–adsorbate interactions. This is sufficient for discussion of the vast majority of interesting chemical processes that occur at surfaces. Important inter-adsorbate forces surely come into play – there would be no surface reactions otherwise – but what counts the most is just the fact that the species do in fact find themselves on a surface. This is what we mean by heterogeneous catalysis.

The rules of the game change somewhat when we consider the other major driving force for research into our subject: the microelectronics industry. Here, surface physics *per se* is not so crucial as the closely related field of *interface* physics. The interfaces in question typically involve the junction of two micron-sized wafers of metal, semiconductor, ceramic, etc. Since these junctions break translational invariance, it is unsurprising that certain ideas (such as interface localized electronic and vibrational states) reappear almost unchanged. But a great many new features enter which would carry us far outside the intended scope of this book. Luckily, there is one aspect of the problem which does fall within our purview: the concept of *epitaxy* and epitaxial growth. The purpose of this brief chapter is merely to introduce the subject and prepare the reader for further exploration elsewhere.

The term epitaxy (from *ε'πι*, 'on' and *τάξις*, 'arrangement') refers explicitly to a situation where the structural integrity of the overlayer material (taken as an independent whole) is of at least equal energetic importance when compared to adsorbate–substrate bonding across the interface. Evidently, this occurs only when the separation between adatoms is quite small. In this limit (which is opposite to that considered above), the role of adsorbate–adsorbate interactions is not merely to order atoms into particular arrangements on an adsorption checkerboard determined by the substrate (see Chapter 11). Now these interactions determine a 'natural' lattice constant for the overlayer material. In the simplest case, we seek the lowest energy 'arrangement' of such a lattice forced into contact with a substrate which exhibits a different lattice constant. Interesting complications arise if one permits one (or both) lattices of this *bicrystal* to be elastically non-rigid. As we shall see, these considerations have important implications for the growth of artificial materials.

Orientation and strain

Early experimental studies of crystal growth showed quite clearly that there exist preferred *orientational relationships* between dissimilar crystal lattices when they are forced into intimate contact (see. e.g., Seifert (1953)). It turns out that one can rationalize the observations on the basis of purely geometrical 'row-matching' considerations. Consider the case of a bicrystal formed by the placement of a close-packed FCC(111) monolayer (nearest neighbor distance *a*) atop the close-packed (110) surface of a BCC substrate (nearest neighbor distance *b*). For the orientation depicted in Fig. 16.1(*a*), it is easy to see that consecutive atomic rows parallel to the

Fig. 16.1. Overlay of an FCC(111) monolayer (filled circles) onto a BCC(110) substrate surface (open circles): (*a*) FCC $[0\bar{1}1]$ parallel to BCC [001]; (*b*) 5.26° rotation relative to (*a*). The lattice constants of the two crystals are chosen to produce row-matching in the rotated case (Dahmen, 1982).

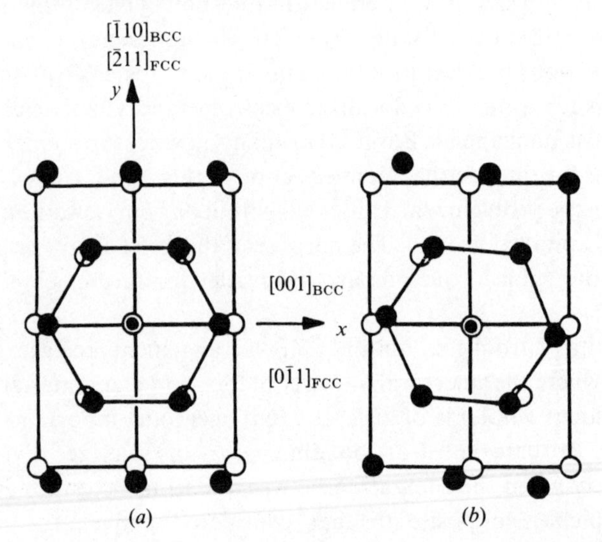

$y(x)$ axis of the two lattices match if the overlayer lattice is slightly expanded (contracted) along the $x(y)$ axis. Moreover, for one particular value of $r = a/b = 1.0887$ one can achieve matching along the most close-packed row of these lattices by rotating the overlayer through 5.26° (Fig. 16.1(*b*)).

It seems reasonable that a row-matching condition must have something to do with a requirement that many overlayer atoms sit (on average) in minima of the substrate corrugation potential. This turns out to be true – despite the fact that individual atoms along matching rows generally are not coincident. To be more precise, suppose that each pair of atoms of the overlayer and substrate interact with one another via a conventional Lennard–Jones 6–12 potential. Now evaluate the total energy of the foregoing FCC(111)/BCC(110) epitaxial system as a function of the parameter r (defined above) and the orientation angle θ at the lattices' equilibrium separation. The energy turns out to be independent of r for all angles except $\theta = 0$ and $\theta = 5.26°$ (Fig. 16.2). The deepest minimum corresponds precisely to the close-packed row-matching condition. One says that the lattices exhibit a Kurdjumov–Sachs (KS) orientational relationship.* The principal minimum for $\theta = 0$, which corresponds to a so-called Nishiyama–Wasserman (NW) orientational relationship, is simply row-matching parallel to the x-axis of Fig. 16.1. The Lennard–Jones potential predicts no distinct minimum for row-matching along the y-axis.

Fig. 16.2. Model calculation of the total adsorbate–substrate interaction energy for rigid lattice FCC(111)/BCC(110) epitaxy as a function of the nearest neighbor distance ratio a/b for two angles of orientation relative to Fig. 16.1(*a*) (Ramirez, Rahman & Schuller, 1984).

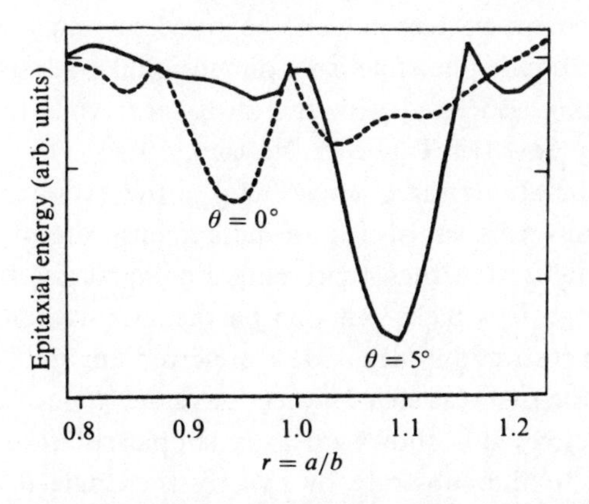

* The names given to various orientational relationships derive from the metallurgical literature of structural phase transitions. See, e.g., Nishiyama (1978).

The KS and NW orientational relationships are observed for many metal–metal adsorbate–substrate combinations (Bauer, 1982). Of course, this does not mean that the natural lattice constants are in precisely the ratios predicted by Fig. 16.2. Generally there is some non-zero 'misfit' $f = (a - b)/a$. Epitaxy occurs because the overlayer material distorts somewhat to achieve row-matching. The reader will recall that this terminology arose earlier in connection with our discussion of the commensurate–incommensurate transition in physisorbed overlayers (Chapter 11). In that context, we saw that for small values of misfit, the overlayer strains to match itself to the rigid lattice constant of the substrate. Domain wall defects appear for larger values of f (Fig. 11.10). The same occurs here except that a new feature – KS rotational epitaxy – enters by virtue of the two-dimensionality of the problem. As a matter of fact, rotational epitaxy occurs for the case of incommensurate physisorbed overlayers as well. It is instructive to examine the connection between the two.

The essential ingredient is the interaction energy between a rigid substrate (reciprocal lattice vectors **G**) and a non-rigid overlayer lattice (reciprocal lattice vectors **g**). By 'non-rigid' we mean that the atoms of the overlayer are permitted to strain away from their equilibrium positions. For small excursions, an approximate energy expression is:

$$\frac{E}{N} = \sum_{\mathbf{G},\mathbf{g}} V_{\mathbf{G}} \delta_{\mathbf{G},\mathbf{g}} - \frac{1}{2} \sum_{\mathbf{G},\mathbf{g}} V_{\mathbf{G}} \mathbf{u}(\mathbf{q}) \cdot \mathbf{G} \delta_{\mathbf{G},\mathbf{q}+\mathbf{g}} + \frac{1}{2} \sum_{\mathbf{G},\mathbf{g}} \hbar\omega(\mathbf{q}) \delta_{\mathbf{G},\mathbf{q}+\mathbf{g}}. \tag{16.1}$$

The first term we have seen before (cf. (11.7) and (11.8)). It is present whether strain is present or not and, in effect, is all that enters the Lennard-Jones calculations discussed above. Now let $\mathbf{u}(\mathbf{q})$ denote the Fourier components of some static displacement pattern which the overlayer may wish to adopt. The second term (made plausible from dimensional analysis alone) describes the energy gain associated with this distortion while the third term counts the energy cost (McTague & Novaco, 1979). The latter is expressed in terms of the phonon frequencies $\omega(\mathbf{q})$ of the overlayer.

By definition, the first terms vanish for an incommensurate overlayer. The lowest energy configuration then represents a compromise between the remaining two terms. This picks out one particular **q**-vector which, from the delta function restriction, picks out a preferred angle θ between the vectors **G** and **g**. If the two real-space lattices have the same symmetry (as in our examples below) this shows up as a simple rotation of the overlayer with respect to the substrate by exactly the angle θ. LEED experiments directly test such predictions (Fig. 16.3) since one can vary the natural lattice constant of the overlayer (and hence the misfit) simply by changing the coverage. Note that the theory appears to account for data from not only a bona fide incommensurate system – physisorbed Ar/graphite – but also for a strong *chemisorption* system – Na/Ru(100).

For commensurate systems, it is necessary to consider explicitly the first term in (16.1). At 1/3 coverage, this 'lock-in' energy stabilizes the ($\sqrt{3} \times \sqrt{3}$–30°) structure of Na/Ru(100). But, as the coverage increases, more sodium atoms must occupy chemisorption sites. Where do the atoms go? It is easy to convince oneself that (at any coverage) it is always possible to find a commensurate overlayer structure (with unit cell axes rotated from the substrate axes by some angle) with a **g**-vector that matches to some **G**-vector of the substrate. Of course, the requisite overlayer unit cell might be very large. Nonetheless, one can always gain *some* lock-in energy. On the other hand, glance back at Fig. 11.15 and recall our discussion of the commensurate–incommensurate transition. The key idea there was that a commensurate solid interspersed with extra atoms ('domain walls') can be regarded equally well as an incommensurate solid. Consequently, the foregoing analysis will remain correct for high-order commensurate structures if the strain terms in (16.1) successfully compete with the lock-in term.

Return now to the FCC(111)/BCC(110) epitaxy problem where lock-in

Fig. 16.3. Rotational epitaxy of Ar/graphite (triangles) and Na/Ru(100) (circles) as a function of overlayer lattice misfit. Solid curve is the prediction of the last two terms of (16.1) (Shaw, Fain & Chinn, 1978); Doering & Semancik, 1984).

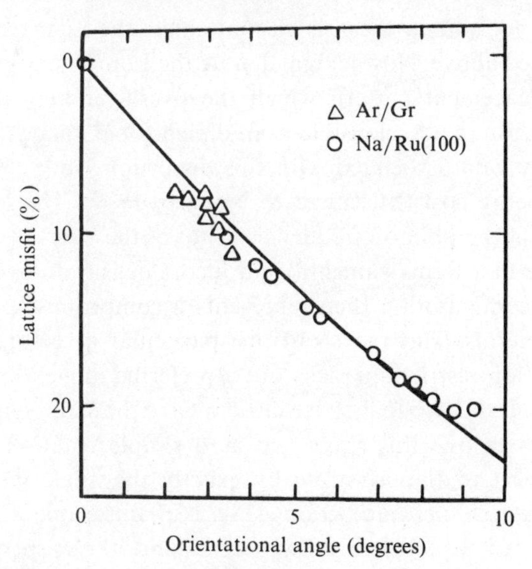

dominates the problem and NW and KS orientational relationships are the rule. This does not imply that strain effects are insignificant. In fact, a vast richness opens up with just the slightest allowance for atomic relaxation. Fig. 16.4 presents a structural phase diagram for this problem where only a small number of possible strain patterns have been considered. The parameter λ is a measure of the ratio of intralayer coupling strength to interlayer coupling strength. Hence, near $\lambda = 0$, the overlayer is strained into commensurability (or 'coherence') with the substrate regardless of the lattice mismatch. Unstrained KS or NW behavior occurs nowhere. In general, the film achieves row-matching only in an average sense. Coherence is lost through a complicated interplay of domain walls along one direction and non-uniform strains in the other. In analogy with the bulk, the former are called 'misfit dislocations' in this context because they appear as added rows of atoms arranged in periodic arrays (Fig. 16.5).

The results of Fig. 16.4 notwithstanding, it is imperative to minimize

Fig. 16.4. Structural phase diagram of an FCC(111)/BCC(110) bicrystal as a function of geometrical (r) and energetic (λ) parameters of the system. Dark shading denotes regions of one-dimensional coherence. Light shading denotes regions where all coherence with the substrate is lost. See text for discussion (Stoop & Van der Merwe, 1982).

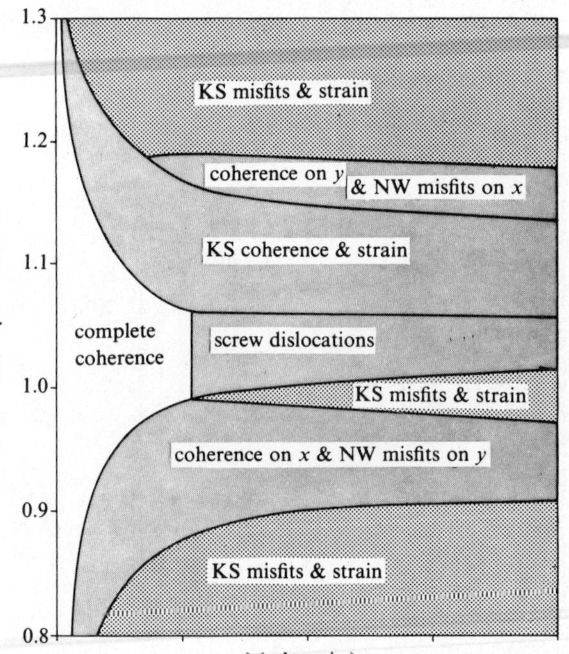

the density of misfit dislocations present at a bicrystal interface if issues of electrical transport are important. Dislocations scatter electrons and hence reduce mobility. For this reason, artificial semiconductor materials generally are constructed from constituents which (*a*) have the same bulk crystal structure and (*b*) are as nearly lattice-matched as one can arrange (see, e.g., Bean (1985)). But this may not always be an option. Suppose one desires a heterostructure fabricated from two materials which differ markedly in both crystal structure and lattice constant. Usually, this leads to a mess, i.e., no epitaxy. However, in certain cases, nature has arranged an elegant solution.

Pseudomorphy refers to a situation where the overlayer material adopts a crystal structure and lattice constant which differ from their normal bulk manifestations but which match coherently to the underlying substrate. This is the normal state of affairs in submonolayer chemisorption but is highly non-trivial to arrange when the atoms of the deposit material are within a few Ångströms of one another. For example, elemental tin adopts its familiar body-centered tetragonal 'white tin' structure at room temperature. It is a metal with a lattice constant of 5.83 Å. However, the same material crystallizes into the 'grey tin' diamond structure ($a = 6.49$ Å) when deposited onto (100) surfaces of InSb and CdTe ($a = 6.48$ Å). There is essentially no misfit at the epitaxial interface and, since grey tin is a

Fig. 16.5. A misfit dislocation (MD) in the epitaxy of an FCC(111) monolayer on a BCC(110) substrate in NW orientation (Bauer & Van der Merwe, 1986).

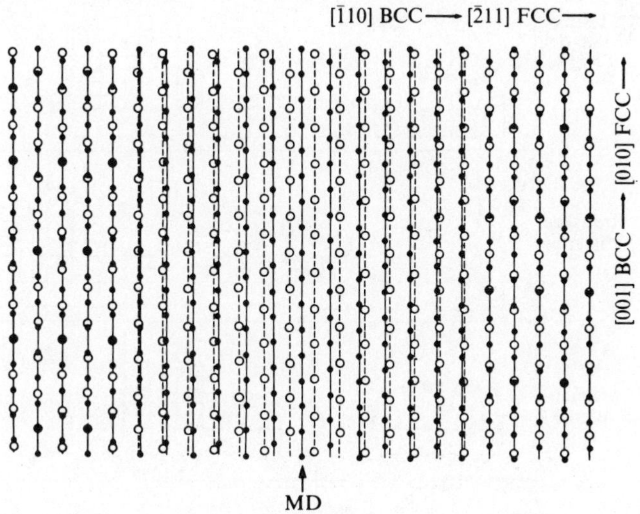

semiconductor, one has fabricated a novel heterojunction material (Farrow, 1983). Up to a point that is. The overlayer grows as one adds more material until eventually at some thickness ($\sim 0.5\,\mu$m in this case) the tin transforms to its bulk stable phase – as it must.

Crystal growth

The growth of a perfect semiconducting tin crystal at 300 K as imagined above contains a crucial hidden assumption. It is presumed that growth proceeds in a two-dimensional fashion, one monolayer after the next, up to some desired thickness of overlayer material. As it happens, this actually appears to be the case for Sn/InSb(100) and Sn/CdTe(100). However, it is not the usual situation observed for either the growth of metals on metals (Vook, 1982) or for the growth of metals on semiconductors (Ludeke, 1984). Instead, one often finds that the deposited material 'balls up' into three-dimensional clumps which only later coalesce into a thick polycrystalline film. In fact, extensive experimental results point to the existence of three distinct growth modes, each named after investigators associated with their initial description: Frank–Van der Merwe (FV) growth, Stranski–Krastanov (SK) growth and Volmer–Weber (VW) growth.

Fig. 16.6 is a schematic representation of the common modes of crystal

Fig. 16.6. Schematic view of the three topologically distinct epitaxial growth modes (Kern, Le Lay & Metois, 1979).

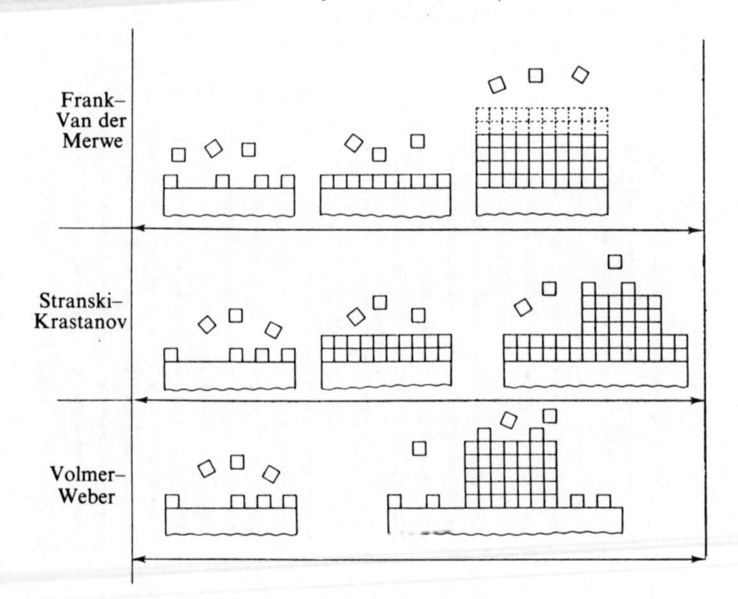

growth one observes under typical deposition conditions (MBE, MOCVD, etc.). FV growth follows the layer-by-layer scenario outlined above. VW growth is just the opposite. Three-dimensional crystallites nucleate immediately upon contact and the overlayer may not completely cover the exposed substrate surface until a great many atoms have been deposited. SK growth lies in between: a few monolayers adsorb in layer-by-layer fashion before three-dimensional clumps begin to form. The obvious question is: how does one know what sort of growth one is dealing with?

Fig. 16.7. Growth of GaAs(100) by MBE. Intensity of the RHEED specular beam as a function of time (top panel). The slow decay of peak intensity reflects a gradual increase in surface roughness. The lower panel illustrates a model of monolayer growth in the Frank–Van der Merwe scenario (Neave, Joyce, Dobson & Norton, 1983).

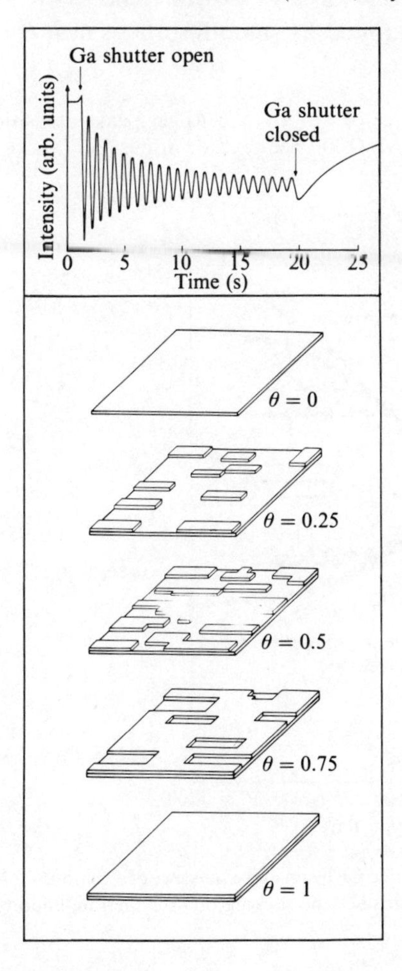

High-quality layer-by-layer growth is essential for the production of electronic materials by molecular beam epitaxy. The *in situ* UHV technique of RHEED (reflection high energy electron diffraction) appears to be a reasonably reliable monitor of the presence or absence of FV growth. In RHEED, a 5–50 keV electron beam is directed towards the sample at extreme grazing incidence. Electrons scattered through small angles sample only the top 1–2 atomic layers of the crystal under these conditions. The characterization experiment is straightforward: one simply monitors the variation of the specular beam intensity as a function of time. An example is reproduced as the top panel of Fig. 16.7 for the case of GaAs growth with MBE.* Remarkably, the signal exhibits extremely regular oscillations whose period exactly corresponds to the growth rate of a single layer of GaAs (as determined independently). The lower panel of Fig. 16.7 suggests a simple interpretation in terms of FV growth: reflectivity maxima correspond to scattering from atomically smooth surfaces near $\theta = 0$ and

Fig. 16.8. Time dependence of Mo and Cu Auger peak intensities as Cu grows epitaxially on Mo(100). See text for discussion (Soria & Poppa, 1980).

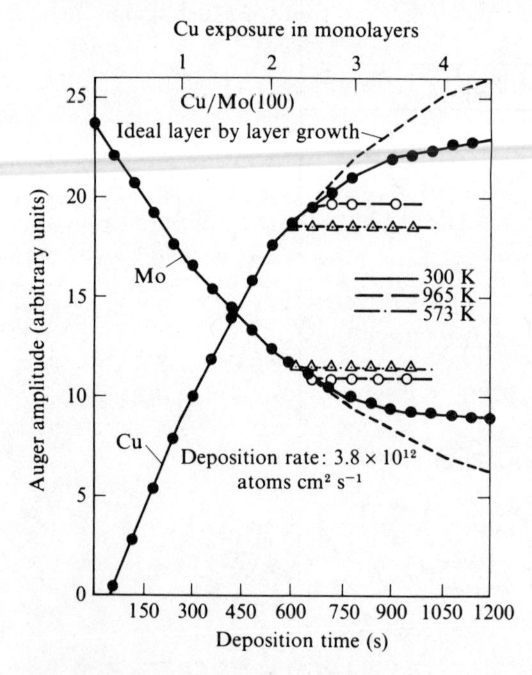

* We need only control the atomic Ga beam in the presence of a continuous supply of arsenic to study this problem since the sticking of As is the rate-limiting step in GaAs growth (Chapter 15).

$\theta = 1$ while reflectivity minima correspond to scattering from maximally disordered surfaces near $\theta = 0.5$. No persistent oscillations are expected for either SK or VW growth.

Auger spectroscopy is another common technique used to identify epitaxial growth modes. Fig. 16.8 illustrates the time dependence of the strength of two peaks in the Auger spectrum collected from a Mo(100) surface during deposition of copper atoms from a vapor source. Notice that both the increasing Cu signal and the decreasing Mo signal consist of a sequence of line segments with uniformly decreasing slope. We certainly expect linear behavior if a single monolayer of adsorbate uniformly covers the layer beneath it. But, since only those electrons within about one escape depth of the surface actually emerge, the slope of the signal must decline as each new layer is added. In the limit of a thick overlayer, the substrate signal vanishes and the adsorbate signal has zero slope.

Based on escape depth data (Fig. 2.1), the dashed curves in Fig. 16.8 are the predicted Auger amplitudes for Cu/Mo(100) if FV growth is operative. Evidently, the data follow these predictions for at most three layers. The observed behavior is indicative of the Stranski–Krastanov growth mode. It is easy to see, as well, that the Auger technique is sensitive to the difference between SK growth and VW growth. However, sometimes direct inspection is sufficient. The electron micrograph of lead crystallites adsorbed onto a single crystal graphite surface discussed first in Chapter 1 (Fig. 1.7) clearly shows that these solid 'droplets' do not 'wet' the substrate surface.

For obvious reasons, a good deal of effort has been expended in pursuit of a tractable theory of crystal growth. We encountered some of this work previously in our analysis of the roughening transition (Chapter 1). This example is typical in the sense that, in order to simplify the statistical mechanics, one adopts extremely simple models which hopefully admit analytic solutions. On the other hand, we have seen throughout this book that certain delicate matters of principle can depend quite sensitively on

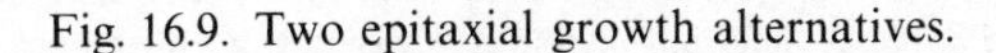
Fig. 16.9. Two epitaxial growth alternatives.

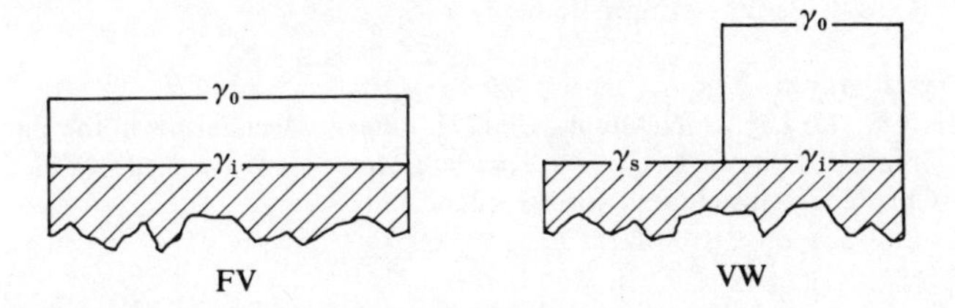

details of the systems under investigation. This observation leads to the prediction that the emphasis in theoretical analysis of epitaxial growth is likely to shift toward molecular dynamics simulations as supercomputer availability increases.

We conclude with an example which illustrates both the power of simple analysis and a clear source of the sensitivity to parameters which bedevils our subject. Let us try to estimate when any one of the three growth modes examined above is likely to occur. To do so, consider the energy difference between the two epitaxial overlayer arrangements depicted in Fig. 16.9. The inputs to this calculation are three macroscopic surface tensions: γ_0, γ_i and γ_s – the free energy/unit area at the overlayer–vacuum interface, the overlayer–substrate interface and the substrate–vacuum interface, respectively. We may assume, without loss of generality, that the Volmer–Weber cluster occupies half the available surface area (A). It follows immediately that

$$\begin{aligned}\Delta E &= E_{\mathrm{FV}} - E_{\mathrm{VW}} \\ &= (\gamma_0 + \gamma_i)A - \tfrac{1}{2}(\gamma_0 + \gamma_i + \gamma_s)A.\end{aligned} \tag{16.2}$$

Therefore, we expect complete wetting (FV growth) when $\Delta\gamma = \gamma_0 + \gamma_i - \gamma_s < 0$, VW growth when $\Delta\gamma > 0$ and SK growth when $\Delta\gamma \cong 0$.

It is important to bear in mind that this estimate is only qualitatively useful because we have completely neglected any effects which might arise from the anisotropy of the surface tensions (Fig. 1.8). Even then, it is not entirely clear what one should choose for γ_i for any particular epitaxial pair. Existing methods of measurement (see, e.g., Eustathopoulos & Joud, 1980) rely on bulk grain boundary data which are not obviously transferable to the epitaxial situation of interest where additional contributions from misfit dislocations, strain, etc., might be significant. A truly microscopic replacement for (16.2) is yet another challenge for the future of surface physics!

General references

Orientation and strain

Van der Merwe, J.H. (1984). Recent Development in the Theory of Epitaxy. In *Chemistry and Physics of Solid Surfaces* (eds. Vanselow & Howe), vol. 5, pp. 365–401. Berlin: Springer-Verlag.

Crystal growth

Kern, R., Le Lay, G. & Metois, J.J. (1979). Basic Mechanisms in the Early Stages of Epitaxy. In *Current Topics in Materials Science* (ed. Kaldis), vol. 3, Chapter 3. Amsterdam: North-Holland.

REFERENCES

Abraham, F.F. (1984). *J. Vac. Sci. Tech. B***2**, 534.
Adelman, S.A. & Doll, J.D. (1976). *J. Chem. Phys.* **64**, 2375.
Agnolet, G. (1983). Ph.D. Thesis, Cornell University (unpublished).
Agranovich, V.M. & Mills, D.L. (1982). *Surface Polaritons.* Amsterdam: North-Holland.
Alivisatos, A.P., Waldeck, D.H. & Harris, C.B. (1985). *J. Chem. Phys.* **82**, 541.
Allan, D.C. & Mele, E.J. (1984). *Phys. Rev. Lett.* **53**, 826.
Allan, G. (1981). *Surf. Sci. Rep.* **1**, 121.
Allen, F.J. & Gobeli, G.W. (1962). *Phys. Rev.* **127**, 152.
Almbladh, C.-O. & von Barth, U. (1985). *Phys. Rev. B***31**, 3231.
Alvarado, S., Campagna, M. & Hopster, H. (1982). *Phys. Rev. Lett.* **48**, 51.
Anderson, P.W. (1963). *Concepts in Solids.* Reading: Benjamin.
Andersson, S. (1975). In *Surface Science*, vol. II, pp. 113–44. Vienna: IAEC.
Andersson, S. & Harris, J. (1983). *Phys. Rev. B***27**, 9.
Andreussi, F. & Gurtin, M. (1977). *J. Appl. Phys.* **48**, 3798.
Andronikashvili, E.L. (1946). *Zh. Eksp. Teor. Fiz.* **16**, 780.
Aono, M., Hou, Y., Souda, R., Oshima, C., Otani, S. & Ishizawa, Y. (1983). *Phys. Rev. Lett.* **50**, 1293.
Apell, P., Ljungbert, A. & Lundqvist, S. (1984). *Phys. Scr.* **30**, 367.
Appelbaum, J.A. & Hamann, D.R. (1972). *Phys. Rev. B***6**, 2166.
Appelbaum, J.A. & Hamann, D.R. (1974). *Phys. Rev. B***10**, 4973.
Arfken, G. (1970). *Mathematical Methods for Physicists*, 2nd edn, pp. 662–6. New York: Academic.
Arthur, J.R. (1969). In *The Structure and Chemistry of Solid Surfaces* (ed. Somorjai), pp. 46:1–46:17. New York: Wiley.
Aruga, T., Tochihara, H. & Murata, Y. (1984). *Phys. Rev. Lett.* **53**, 372.
Ashcroft, N.A. & Mermin, N.D. (1976). *Solid State Physics.* Philadelphia: Saunders.
Asada, H. & Matsui, T. (1982). *Jpn. J. Appl. Phys.* **21**, 259.
Asonen, H., Lindroos, M., Pessa, M., Prasad, R., Rao, R.S. & Bansil, A. (1982). *Phys. Rev. B***25**, 7080.
Asscher, M., Guthrie, W.L., Lin, T.-H. & Somorjai, G.A. (1983). *J. Chem. Phys.* **78**, 6992.
Atkins, P.W. (1983). *Molecular Quantum Mechanics*, 2nd edn. Oxford: Oxford University Press.
Avouris, P. & Persson, B.N.J. (1984). *J. Phys. Chem.* **88**, 837.
Avron, J.E., Balfour, L.S., Kuper, C.G., Landau, J., Lipson, S.G. & Schulman, L.S. (1980). *Phys. Rev. Lett.* **45**, 814.
Ayrault, G. & Ehrlich, G. (1974). *J. Chem. Phys.* **60**, 281.

Bachelet, G.B. & Schluter, M. (1983). *J. Vac. Sci. Tech. B***1**, 726.
Bachrach, R.Z., Bauer, R.S., Chiaradia, P. & Hansson, G.V. (1981). *J. Vac. Sci. Tech.* **19**, 335.
Bader, R.F.W., Henneker, W.H. & Cade, P.E. (1967). *J. Chem. Phys.* **46**, 3341.
Bak, P. (1979). *Solid State Commun.* **32**, 581.
Bak, P. (1982). *Rep. Prog. Phys.* **45**, 587.
Bak, P. (1984). In *Chemistry and Physics of Solid Surfaces* (eds. Vanselow & Howe), vol. 5, pp. 317–37. Berlin: Springer-Verlag.
Bak, P. & Bruinsma, R. (1982). *Phys. Rev. Lett.* **49**, 249.
Balooch, M., Cardillo, M.J., Miller, D.R. & Stickney, R.E. (1974). *Surf. Sci.* **46**, 358.
Barash, Y.S. & Ginzburg, V.L. (1984). *Sov. Phys. Usp.* **27**, 467.
Bardeen, J. (1936). *Phys. Rev.* **49**, 653.
Bardeen, J. (1947). *Phys. Rev.* **71**, 717.
Bardeen, J. & Brattain, W.H. (1949). *Phys. Rev.* **75**, 1208.
Barker, J.A. & Auerbach, D.J. (1984). *Surf. Sci. Rep.* **4**, 1.
Barnes, M.R. & Willis, R.F. (1978). *Phys. Rev. Lett.* **41**, 1729.
Batra, I.P. & Ciraci, S. (1986). *Phys. Rev. B***33**, 4312.
Bauer, C.E., Speiser, R. & Hirth, J.P. (1976). *Met. Trans. A***7**, 75.
Bauer, E. (1982). *Appl. Surf. Sci.* **11/12**, 479.
Bauer, E. & Van der Merwe, J.H. (1986). *Phys. Rev. B***33**, 3657.
Baule, B. (1914). *Ann. der Physik* **44**, 145.
Baym, G. (1974). *Lectures on Quantum Mechanics.* Reading: Benjamin.
Bean, J.C. (1985). *Science* **230**, 127.
Beasley, M.R., Mooij, J.E. & Orlando, T.P. (1979). *Phys. Rev. Lett.* **42**, 1165.
Becker, R.S., Golovchenko, J.A., McRae, E.G. & Swartzentruber, B.S. (1985). *Phys. Rev. Lett.* **55**, 2028.
Benninghoven, A. (1973). *Surf. Sci.* **35**, 427.
Bernstein, R.B. (1982). *Chemical Dynamics via Molecular Beam and Laser Techniques.* Oxford: Clarendon.
Besocke, K., Krahl-Urban, B. & Wagner, H. (1977). *Surf. Sci.* **68**, 39.
Bethge, H. (1982). In *Interfacial Aspects of Phase Transformations* (ed. Mutaftschiev), pp. 669–96. Dordrect: Reidel.
Bhat, I.B., Taskar, N.R. & Ghandi, S.K. (1987). *J. Electrochem. Soc.* **134**, 195.
Binder, K. (1983). In *Phase Transitions and Critical Phenomena* (eds. Domb & Lebowitz), vol. 8, pp. 1–144. London: Academic.
Binder, K. & Landau, D.P. (1981). *Surf. Sci.* **108**, 503.
Binnig, G. & Rohrer, H. (1986). *IBM J. Res. Develop.* **30**, 355.
Binnig, G., Rohrer, H., Gerber, Ch. & Weibel, E. (1982). *Phys. Rev. Lett.* **49**, 57.
Binnig, G., Rohrer, H., Gerber, Ch. & Weibel, E. (1983). *Surf. Sci.* **131**, L379.
Birgeneau, R.J., Hammons, E.M., Heiney, P., Stephens, P.W. & Horn, P.M. (1980). In *Ordering in Two Dimensions* (ed. Sinha), pp. 29–38. New York: North Holland.
Bishop, D.J. & Reppy, J.D. (1980). *Phys. Rev. B***22**, 5171.
Bohr, J., Feidenhans'l, R., Nielsen, M., Toney, M., Johnson, R.L. & Robinson, I.K. (1985). *Phys. Rev. Lett.* **54**, 1275.
Bonzel, H.P. & Krebs, H.J. (1982). *Surf. Sci.* **117**, 639.
Boudart, M., Delbouille, A., Derouane, E.G., Indovina, V. & Walters, A.B. (1972). *J. Am. Chem. Soc.* **94**, 6622.
Bourdon, E.B.D., Cowin, J.P., Harrison, I., Polanyi, J.C., Stanners, C.D. & Young, P.A. (1984). *J. Phys. Chem.* **88**, 6100.
Bowman, J.M. & Park, S.C. (1982). *J. Chem. Phys.* **77**, 5441.
Bozso, F., Ertl, G., Grunze, M. & Weiss, M. (1977). *J. Catal.* **49**, 18.
Bradshaw, A.M. (1982). *Surf. Sci.* **11/12**, 712.
Brako, R. (1982). *Surf. Sci.* **123**, 439.

Brenig, W., Kasai, H. & Muller, H. (1985). *Surf. Sci.* **161**, 608.
Brennan, S. (1985). *Surf. Sci.* **152/153**, 1.
Brinkman, W.F., Fisher, D.S. & Moncton, D.E. (1982). *Science* **217**, 693.
Brongersma, H.H. & Buck, T.M. (1978). *Nuc. Inst. Meth.* **149**, 569.
Brown, F., Parks, R.E. & Sleeper, A.M. (1965). *Phys. Rev. Lett.* **14**, 1533.
Brundle, C.R. (1974). *J. Vac. Sci. Technol.* **11**, 212.
Brundle, C.R. (1985). *J. Vac. Sci. Technol.* *A*3, 1468.
Brusdeylins, G., Rechsteiner, R., Skofronick, J.G., Toennies, J.P., Benedek, G. & Miglio, L. (1985). *Phys. Rev. Lett.* **54**, 466.
Buck, T.M. (1975). In *Methods of Surface Analysis* (ed. Czanderna), vol. 1, pp. 75–102. Amsterdam: Elsevier.
Buck, T.M., Stensgaard, I., Wheatley, G.H. & Marchut, L. (1980). *Nucl. Inst. Meth.* **170**, 519
Bullett, D.W. (1980). *Surf. Sci.* **93**, 213.
Bullett, D.W. (1982). *Solid State Commun.* **43**, 491.
Burns, M.J., Lince, J.R., Williams, R.S. & Chaikin, P.M. (1984). *Solid State Commun.* **51**, 865.
Burton, W.K. & Cabrera, N. (1949). *Disc. Farad. Soc.* **5**, 33.
Butler, D.M., Litzinger, J.A. & Stewart, G.A. (1980). *Phys. Rev. Lett.* **44**, 466.
Calandra, C., Catellani, A. & Beatrice, C. (1985). *Surf. Sci.* **152/153**, 814.
Callen, H.B. (1985). *Thermodynamics*, 2nd edn. New York: Wiley.
Camley, R.E. & Mills, D.L. (1982). *Phys. Rev. B*26, 1280.
Campbell, C.T., Ertl, G., Kuipers, H. & Segner, J. (1980). *J. Chem. Phys.* **73**, 5862.
Campion, A., Brown, J.K. & Grizzle, V.M. (1982). *Surf. Sci.* **115**, L153.
Campuzano, J.C., King, D.A., Somerton, C. & Inglesfield, J.E. (1980). *Phys. Rev. Lett.* **45**, 1649.
Campuzano, J.C., Foster, M.S., Jennings, G., Willis, R.F. & Unertl, W. (1985). *Phys. Rev. Lett.* **54**, 2684.
Cardillo, M.J. (1985). *Langmuir* **1**, 410.
Carlson, T.A. (1975). *Photoelectron and Auger Spectroscopy*. New York: Plenum.
Caruthers, E., Kleinman, L. & Alldredge, G.P. (1973). *Phys. Rev. B*8, 4570.
Casula, F., Ossicini, S. & Selloni, A. (1979). *Solid State Commun.* **30**, 309.
Ceperley, D.M. (1978). *Phys. Rev. B*18, 3126.
Ceperley, D.M. & Alder, B.J. (1980). *Phys. Rev. Lett.* **45**, 566.
Cerny, S. (1983). In *The Chemical Physics of Solid Surfaces and Heterogeneous Catalysis* (eds. King & Woodruff), vol. 2, pp. 1–57. Amsterdam: Elsevier.
Cerri, A., Mauri, D. & Landolt, M. (1983). *Phys. Rev. B*27, 6526.
Chabal, Y.J. (1985). *Phys. Rev. Lett.* **55**, 845.
Chakraborty, B., Holloway, S. & Norskov, J.K. (1985). *Surf. Sci.* **152/153**, 660.
Chan, E.M., Buckingham, M.J. & Robins, J.L. (1977). *Surf. Sci.* **67**, 285.
Chance, R.R., Prock, A. & Silbey, R. (1978). *Adv. Chem. Phys.* **37**, 1.
Chelikowsky, J.R. & Cohen, M.L. (1979). *Phys. Rev. B*20, 4150.
Chelikowsky, J.R., Schluter, M., Louie, S.G. & Cohen, M.L. (1975). *Solid State Commun.* **17**, 1103.
Chemla, D.S. (1985). *Physics Today*, May 1985, pp. 56–64.
Chen, J.R. & Gomer, R. (1979). *Surf. Sci.* **79**, 413.
Chen, T.S., de Wette, F.W. & Alldredge, G.P. (1977). *Phys. Rev. B*15, 1167.
Chen, Y.C., Cunningham, J.E. & Flynn, C.P. (1984). *Phys. Rev. B*30, 7317.
Chernov, A.A. (1984). *Modern Crystallography* III. Berlin: Springer-Verlag.
Chiaradia, P., Katnani, A.D., Sang, H.W. & Bauer, R.S. (1984). *Phys. Rev. Lett.* **52**, 1246.
Chinn, M.D. & Fain, S.C. (1977). *Phys. Rev. Lett.* **39**, 146.
Chon, H. & Prater, C.D. (1966). *Disc. Farad. Soc.* **41**, 380.
Chuang, T.J. (1983). *Surf. Sci. Rep.* **3**, 1.

Chuang, T.J. (1985). *J. Vac. Sci. Tech. B*3, 1408.
Citrin, P.H., Rowe, J.E. & Eisenberger, P. (1983). *Phys. Rev. B*28, 2299.
Citrin, P.H. & Wertheim, G.K. (1983). *Phys. Rev.* **27**, 3176.
Clarke, L.J. (1985). *Surface Crystallography*. New York: Wiley.
Cohen, M.L. (1985). In *Highlights of Condensed Matter Theory* (eds. Bassani, Fumi & Tosi), pp. 16–58. Bologna: Società Italiana di Fisica.
Compton, A.H. (1923). *Phil. Mag.* **45**, 1121.
Cotton, F.A. & Wilkinson, G. (1962). *Advanced Inorganic Chemistry*. New York: Wiley.
Cowley, J.M. (1986). *Prog. Surf. Sci.* **21**, 209.
Cremaschi, P. & Whitten, J.L. (1981). *Surf. Sci.* **112**, 343.
Dahmen, U. (1982). *Acta Metall.* **30**, 63.
Damon, R.W. & Eshbach, J.R. (1961). *J. Phys. Chem. Solids* **19**, 308.
Dash, J.G. (1975). *Films on Solid Surfaces*. New York: Academic.
Davenport, J. W. & Estrup, P.J. (1987). In *The Chemical Physics of Surfaces and Heterogeneous Catalysis* (eds. King & Woodruff), vol. 3*A*, pp. 27–89.
Davis, H.L. & Noonan, J.R. (1982). *J. Vac. Sci. Tech.* **20**, 842.
Davisson, C.J. & Germer, L.H. (1927). *Phys. Rev.* **30**, 705.
Davydov, B. (1939). *J. Phys. USSR*, **1**, 167.
DeLorenzi, G., Jacucci, G. & Pontikis, V. (1982). *Surf. Sci.* **116**, 391.
Del Sole, R. & Selloni, A. (1984). *Phys. Rev. B*30, 883.
Demuth, J.E., Sanda, P.N., Warlaumont, J.M., Tsang, J.C. & Christman, K. (1982). In *Vibrations at Surfaces* (eds. Caudano, Gilles & Lucas), pp. 391–411. New York: Plenum.
Derry, G., Wesner, D., Krishnaswamy, S.V. & Frankl, D.R. (1978). *Surf. Sci.* **74**, 245.
Desjonqueres, M.C. & Cyrot-Lackmann, F. (1977). *J. Phys. F*7, 61.
D'Evelyn, M.P. & Madix, R.J. (1984). *Surf. Sci. Rep.* **3**, 413.
Devreese, J.T. (1983). *Ab Initio Calculation of Phonon Spectra*. New York: Plenum.
DiFoggio, R. & Gomer, R. (1982). *Phys. Rev. B*25, 3490.
Doak, R.B. & Toennies, J.P. (1982). *Surf. Sci.* **117**, 1.
Doak, R.B., Harten, U. & Toennies, J.P. (1983). *Phys. Rev. Lett.* **51**, 578.
Doering, D.L. & Semancik, S. (1984). *Phys. Rev. Lett.* **53**, 66.
Domany, E., Schick, M., Walker, J.S. & Griffiths, R.B. (1978). *Phys. Rev. B*18, 2209.
Dorn, R., Luth, H. & Buchel, M. (1977). *Phys. Rev. B*16, 4675.
Dose, V. (1985). *Appl. Surf. Sci.* **22/23**, 338.
Douglas, B.E., McDaniel, D.H. & Alexander, J.J. (1983). *Concepts and Models of Inorganic Chemistry*, 2nd edn. New York: Wiley.
Dow, J.D., Allen, R.E. & Sankey, O.F. (1984). In *Chemistry and Physics of Solid Surfaces* (eds. Vanselow & Howe), vol. 5, pp. 483–500. Berlin: Springer-Verlag.
Dresser, M.J., Madey, T.E. & Yates, J.T. (1974). *Surf. Sci.* **42**, 533.
Drude, P. (1890). *Ann. Phys. Chem.* **39**, 481.
Duke, C.B. (1984). *J. Vac. Sci. Technol. A*2, 139.
Dunlap, B.I., Yu, H.L. & Antoniewicz, P.R. (1982). *Phys. Rev. A*25, 7.
Dupuis, R.D. (1984). *Science* **226**, 623.
Eastman, D.E., Himpsel, F.J. & van der Veen, J.F. (1982). *J. Vac. Sci. Tech.* **20**, 609.
Eberhardt, W. & Himpsel, F.J. (1979). *Phys. Rev. Lett.* **42**, 1375.
Ecke, R.E. & Dash, J.G. (1983). *Phys. Rev. B*28, 3738.
Egelhoff, W.F. (1983). *Phys. Rev. Lett.* **50**, 587.
Ehrenreich, H. & Cohen, M.H. (1959). *Phys. Rev.* **115**, 786.
Ehrlich, G. (1982). *CRC Crit. Rev. Sol. State Mat. Sci.* **10**, 391.
Einstein, T.L. (1978). *CRC Crit. Rev. Sol. State Mat. Sci.* **7**, 261.
Ekwelundu, E. & Ignatiev, A. (1986). *Surf. Sci.* **179**, 119.
El-Batanouny, M., Strongin, M., Williams, G.P. & Colbert, J. (1981). *Phys. Rev. Lett.* **46**, 269.
Elgin, R.L., Greif, J.M. & Goodstein, D.L. (1978). *Phys. Rev. Lett.* **41**, 1723.

Engel, T. (1978). *J. Chem. Phys.* **69**, 373.
Engel, T. & Rieder, K.H. (1982). In *Structural Studies of Surfaces* (ed. Hohler), pp. 55–180. Berlin: Springer-Verlag.
Ertl, G. (1982). *Ber. Bunsenges. Phys. Chem.* **86**, 425.
Ertl, G. (1983). *J. Vac. Sci. Tech. A***1**, 1247.
Ertl, G. & Koch, J. (1970). *Z. Naturforsch* **25***A*, 1906.
Estrup, P.J. (1970). In *Modern Diffraction and Imaging Techniques in Materials Science* (eds. Amelinckx, Gevers, Remaut & Van Landuyt), pp. 377–406. Amsterdam: North-Holland.
Euceda, A., Bylander, D.M. & Kleinman, L. (1983). *Phys. Rev. B***28**, 528.
Eustathopoulos, N. & Joud, J.-C. (1980). In *Current Topics in Materials Science* (ed. Kaldis), vol. 4, pp. 281–360. Amsterdam: North-Holland.
Evans, E. & Mills, D.L. (1973). *Phys. Rev. B***8**, 4004.
Falicov, L.M., Victoria, R.H. & Tersoff, J. (1985). In *The Structure of Surfaces* (eds. Van Hove & Tong), pp. 12–17. Berlin: Springer.
Fano, U. (1941). *J. Opt. Soc. Am.* **31**, 213.
Farnsworth, H.E., Schlier, R.E., George, T.H. & Burger, R.M. (1958). *J. Appl. Phys.* **29**, 1150.
Farrell, H.H. & Somorjai, G.A. (1971). *Adv. Chem. Phys.* **20**, 215.
Farrow, R.F.C. (1983). *J. Vac. Sci. Tech. B***1**, 222.
Faulkner, J.S. (1982). *Prog. Mat. Sci.* **27**, 1.
Feder, R. & Pleyer, H. (1982). *Surf. Sci.* **117**, 285.
Feibelman, P.J. (1982). *Prog. Surf. Sci.* **12**, 287.
Feibelman, P.J., Hamann, D.R. & Himpsel, F.J. (1980). *Phys. Rev. B***22**, 1734.
Feldman, L.C. (1980). In *Surface Science: Recent Progress and Perspectives* (ed. Vanselow). Cleveland: CRC.
Feller, W. (1968). *An Introduction to Probability Theory and Its Applications*, 3rd edn, vol. 1. New York: Wiley.
Ferrell, T.L., Callcott, T.A. & Warmack, R.J. (1985). *Am. Sci.* **73**, 344.
Feuerbacher, B., Fitton, B. & Willis, R.F. (1978). *Photoemission and the Electronic Properties of Surfaces.* Chichester: Wiley.
Feynman, R.F. (1972). *Statistical Mechanics.* Reading: Benjamin.
Finnis, M.W. & Heine, V. (1974). *J. Phys. F***4**, L37.
Finzel, H.U., Frank, H., Hoinkes, H., Luschka, M., Nahr, H., Wilsch, H. & Wonka, U. (1975). *Surf. Sci.* **49**, 577.
Forstmann, F. (1970). *Z. Physik* **235**, 69.
Forstmann, F. (1978). In *Photoemission and the Electronic Properties of Surfaces* (eds. Feuerbacher, Fitton & Willis), pp. 193–226. Chichester: Wiley.
Fowler, R.H. (1936). *Proc. Camb. Phil. Soc.* **32**, 144.
Foxon, C.T. & Joyce, B.A. (1981). In *Current Topics in Materials Science* (ed. Kaldis), vol. 7, pp. 1–68. Amsterdam: North-Holland.
Frankel, D.J., Anderson, J.R. & Lapeyre, G.J. (1983). *J. Vac. Sci. Tech. B***1**, 763.
Freed, K.F. (1985). *J. Chem. Phys.* **82**, 5264.
Freeman, D.L. & Doll, J.D. (1983). *J. Chem. Phys.* **78**, 6002.
Frenkel', J. & Kontorova, T. (1939). *J. Phys. USSR* **1**, 137.
Frenken, J.W.M. & van der Veen, J.F. (1985). *Phys. Rev. Lett.* **54**, 134.
Friedel, J. (1964). *Dislocations.* Oxford: Pergamon.
Frisch, R.O. & Stern, O. (1933). *Z. Physik* **84**, 430.
Fu, C.L., Freeman, A.J., Wimmer, E. & Weinert, M. (1985). *Phys. Rev. Lett.* **54**, 2261.
Fuwa, K., Fujima, K., Adachi, H. & Osaka, T. (1984). *Surf. Sci.* **148**, L659.
Gadzuk, J.W. (1974). *Surf. Sci.* **43**, 44.
Gadzuk, J.W. (1985). *Comments At. Mol. Phys.* **16**, 219.
Gallet, F., Nozières, P., Balibar, S. & Rolley, E. (1986). *Europhys. Let.* **2**, 701.

Garcia-Moliner, F. & Flores, F. (1979). *Introduction to the Theory of Solid Surfaces.* Cambridge: Cambridge University Press.
Gardiner, W.C. (1969). *Rates and Mechanisms of Chemical Reactions.* Reading: Benjamin/Cummings.
Gavrilyuk, V.M. & Medvedev, V.K. (1966). *Sov. Phys. Sol. State* **8**, 1439.
Gerken, F., Flodstrom, A.S., Barth, J., Johansson, L.I. & Kunz, C. (1985). *Phys. Scr.* **32**, 43.
Gersten, J.I. & Nitzan, A. (1982). In *Surface Enhanced Raman Scattering* (eds. Chang & Furtak), pp. 89–107. Plenum: New York.
Gibbs, H.M. (1985). *Optical Bistability: Controlling Light with Light.* Orlando: Academic.
Gibbs, J.W. (1948). *Collected Works*, vol. 1, New Haven: Yale.
Gibson, W.M. (1984). In *Chemistry and Physics of Solid Surfaces* (eds. Vanselow & Howe), vol. 5, pp. 427–53. Berlin: Springer-Verlag.
Gillespie, W.D., Herz, R.K., Petersen, E.E. & Somorjai, G.A. (1981). *J. Catal.* **70**, 147.
Glasstone, S., Laidler, K.J. & Eyring, H. (1941). *The Theory of Rate Processes.* New York: McGraw-Hill.
Goddard, P.J. & Lambert, R.M. (1977). *Surf. Sci.* **67**, 180.
Gomer, R. (1983). In *Desorption Induced by Electronic Transitions* (eds. Tolk, Traum, Tully & Madey), pp. 40–52. Berlin: Springer-Verlag.
Goodman, D.W. (1984). *Appl. Surf. Sci.* **19**, 1.
Goodwin, E.T. (1939). *Proc. Camb. Phil. Soc.* **35**, 221.
Gortel, Z.W., Kreuzer, H.J., Piercy, P. & Teshima, R. (1983). *Phys. Rev.* **27**, 5066.
Graybeal, J.M. & Beasley, M.R. (1984). *Phys. Rev. B***29**, 4167.
Grazis, D.C., Herman, R. & Wallis, R.F. (1960). *Phys. Rev.* **119**, 5336.
Grimes, C.C. & Adams, G. (1979). *Phys. Rev. Lett.* **42**, 795.
Grunberg, P.A. (1985). *Prog. Surf. Sci.* **18**, 1.
Grunze, M., Golze, M., Fuhler, J., Neumann, M. & Schwarz, E. (1984). In *Proceedings of the Eighth International Congress on Catalysis*, pp. IV:133–IV:143. Weinheim: Verlag-Chemie.
Gunnarsson, O. & Schonhammer, K. (1982). In *Chemistry and Physics of Solid Surfaces* (eds. Vanselow & Howe), vol. 4, pp. 363–88. Berlin: Springer-Verlag.
Gurney, R.W. (1935). *Phys. Rev.* **47**, 479.
Guyot-Sionnest, P., Chen, W. & Shen, Y.R. (1986). *Phys. Rev. B***33**, 8254.
Habraken, F.H.P.M., Gijzeman, O.L.J. & Bootsma, G.A. (1980). *Surf. Sci.* **96**, 482.
Haight, R. & Bokor, J. (1986). *Phys. Rev. Lett.* **56**, 2846.
Halliday, D. & Resnick, R. (1966). *Physics.* New York: Wiley.
Hamers, R.J., Tromp, R.M. & Demuth, J.E. (1986). *Phys. Rev. Lett.* **56**, 1972.
Hammersley, J.M. & Handscomb, D.C. (1964). *Monte Carlo Methods.* London: Methuen.
Hanke, W. & Wu, C.H. (1977). In *Transition Metals, 1977* (eds. Lee, Perz & Fawcett), pp. 337–40. Bristol: Institute of Physics.
Hansson, G.V. & Flodstrom, S.A. (1978). *Phys. Rev. B***18**, 1562.
Harris, J. & Andersson, S. (1985). *Phys. Rev. Lett.* **55**, 1583.
Harris, L. (1974). *J. Vac. Sci. Technol.* **11**, 23.
Harrison, W.A. (1980). *Electronic Structure and the Properties of Solids.* San Fransisco: W.H. Freeman.
Haydock, R. & Kelly, M.J. (1973). *Surf. Sci.* **38**, 139.
Hecht, E. & Zajac A. (1974). *Optics.* Reading: Addison-Wesley.
Heimann, P., Hermanson, J. Miosga, H. & Neddermeyer, H. (1979). *Phys. Rev. B***20**, 3059.
Heine, V. (1960). *Group Theory in Quantum Mechanics.* Oxford: Pergamon.
Heine, V. (1965). *Phys. Rev. A***138**, 1689.
Heiney, P.A., Stephens, P.W., Birgeneau, R.J., Horn, P.M. & Moncton, D.E. (1983). *Phys. Rev. B***28**, 6416.

Heinz, K., Schmidt, G., Hammer, L. & Muller, K. (1985). *Phys. Rev. B***32**, 6214.
Hellsing, B. & Persson, M. (1984). *Phys. Scr.* **29**, 360.
Henrich, V.E. (1985). *Rep. Prog. Phys.* **48**, 1481.
Henrich, V.E., Dresselhaus, G. & Zeiger, H.J. (1980). *Phys. Rev. B***22**, 4764.
Henzler, M. (1982). *Appl. Surf. Sci.* **11/12**, 450.
Herring, C. (1951a). In *Physics of Powder Metallurgy* (ed. Kingston), pp. 143–79. New York: McGraw-Hill.
Herring, C. (1951b). *Phys. Rev.* **82**, 87.
Herring, C. (1953). In *Structure and Properties of Solid Surfaces* (eds. Gomer & Smith), pp. 5–72. Chicago: University Press.
Heyraud J.C. & Metois, J.J. (1983). *Surf. Sci.* **128**, 334.
Hirabayashi, I., Koda, T., Tokura, Y., Murata, J. & Kaneko, Y. (1976). *J. Phys. Soc. Jpn.* **40**, 1215.
Hoddeson, L. (1981). *Hist. Stud. Phys. Sci.* **12**, 41.
Holland, B.W. & Woodruff, D.P. (1973). *Surf. Sci.* **36**, 488.
Holloway, S., Lundqvist, B.I. & Norskov, J.K. (1984). In *Proceedings of the 8th International Congress on Catalysis*, vol. 4, pp. 85–95. Berlin: BRD.
Holmes, M.I. & Gustafsson, T. (1981). *Phys. Rev. Lett.* **47**, 443.
Holzl, J. & Schulte, F.K. (1979). In *Springer Tracts in Modern Physics* (ed. Hohler), vol. 85, pp. 1–150. Berlin: Springer-Verlag.
Horlacher Smith, A., Barker, R.A. & Estrup, P.J. (1984). *Surf. Sci.* **136**, 327.
Hosler, W., Behm, R.J. & Ritter, E. (1986). *IBM J. Res. Develop.* **30**, 403.
Hulbert, S.L., Johnson, P.D., Stoffel, N.G., Royer, W.A. & Smith, N.V. (1985). *Phys. Rev.* **31**, 6815.
Ibach, H.J. (1972). *J. Vac. Sci. Tech.* **9**, 713.
Ibach, H. & Mills, D.L. (1982). *Electron Energy Loss Spectroscopy and Surface Vibrations.* New York: Academic.
Iche, G. & Nozieres, P. (1976). *J. Phys. (Paris)* **37**, 1313.
Imbihl, R., Behm, R.J., Christmann, K., Ertl, G. & Matsushima, T. (1982). *Surf. Sci.* **117**, 257.
Inglesfield, J.E. (1982). *Rep. Prog. Phys.* **45**, 223.
Inglesfield, J.E. (1985). *Prog. Surf. Sci.* **20**, 105.
Ivanov, I., Mazur, A. & Pollmann, J. (1980). *Surf. Sci.* **92**, 365.
Jahn, H.A. & Teller, E. (1937). *Proc. Roy. Soc. A***161**, 220.
Jancovici, B. (1967). *Phys. Rev. Lett.* **19**, 20.
Janda, K.C., Hurst, J.E., Becker, C.A., Cowin, J.P., Auerbach, D.J. & Wharton, N. (1980). *J. Chem. Phys.* **72**, 2403.
Johansson, P.K. (1981). *Surf. Sci.* **104**, 510.
Jones, G.J.R. & Holland, B.W. (1985). *Solid State Commun.* **53**, 45.
Kahn, A. (1983). *Surf. Sci. Rep.* **3**, 193.
Kalkstein, D. & Soven, P. (1971). *Surf. Sci.* **26**, 85.
Kanash, O.V., Naumovets, A.G. & Fedorus, A.G. (1975). *Sov. Phys. JETP* **40**, 903.
Kantrowitz, A. & Grey, J. (1951). *Rev. Sci. Instr.* **22**, 328.
Katnani, A.D. & Margaritondo, G. (1983). *Phys. Rev. B***28**, 1944.
Keizer, J. (1982). *J. Phys. Chem.* **86**, 5052.
Kelley, R.D. & Goodman, D.W. (1982). *Surf. Sci.* **123**, L743.
Kern, R., Le Lay, G. & Metois, J.J. (1979). In *Current Topics in Materials Science* (ed. Kaldis), vol. 3, Chapter 3. Amsterdam: North-Holland.
Kevan, S.D. (1983). *Phys. Rev. Lett.* **50**, 526.
Khachaturyan, A.G. (1983). *Theory of Structural Transformations in Solids.* New York: Wiley.
Kiejna, A. & Wojciechowski, K.F. (1981). *Prog. Surf. Sci.* **11**, 293.

Kim, H.K. & Chan, M.H.W. (1984). *Phys. Rev. Lett.* **53**, 170.

Kimman, J., Rettner, C.T., Auerbach, D.J., Barker, J.A. & Tully, J.C. (1986). *Phys. Rev. Lett.* **57**, 2053.

King, D.A. (1977). *CRC Crit. Rev. Sol. State Mat. Sci.* **7**, 167.

King, D.A. & Wells, M.G. (1974). *Proc. Roy. Soc. Lond. A***339**, 245.

King, D.S. & Cavanaugh, R.R. (1986). In *Chemistry and Structure at Interfaces: New Laser and Optical Techniques* (eds. Hall & Ellis), pp. 25–83. Deerfield Beach: VCH.

Kingdom, K.H. & Langmuir, I. (1923). *Phys. Rev.* **21**, 380.

Kiselev, V.F. & Krylov, O.V. (1985). *Adsorption Processes on Semiconductor and Dielectric Surfaces.* Berlin: Springer.

Kisker, E., Schroder, K., Gudat, W. & Campagna, M. (1985). *Phys. Rev. B***31**, 329.

Kisliuk, P. (1957). *J. Phys. Chem. Solids* **3**, 95.

Kittel, C. (1966). *Introduction to Solid State Physics*, 3rd edn. New York: Wiley.

Kleyn, A.W., Luntz, A.C. & Auerbach, D.J. (1981). *Phys. Rev. Lett.* **47**, 1169.

Knotek, M.L. (1984). *Rep. Prog. Phys.* **47**, 1499.

Knox, R.S. (1983). In *Collective Excitations* (ed. DiBartolo), pp. 183–245. New York: Plenum.

Knudsen, M. (1934). *The Kinetic Theory of Gases.* London: Methuen.

Koch, S.W., Rudge, W.E. & Abraham, F.F. (1984). *Surf. Sci.* **145**, 329.

Koelling, D.D. (1981). *Rep. Prog. Phys.* **44**, 139.

Kolaczkiewicz, J. & Bauer, E. (1984). *Phys. Rev. Lett.* **53**, 485.

Kolodney, E., Amirav, A., Elber, R. & Gerber, R.A. (1984). *Chem. Phys. Lett.* **111**, 366.

Kono, S., Fadley, C.S., Hall N.F.T. & Hussain, Z. (1978). *Phys. Rev. Lett.* **41**, 117.

Kosterlitz, J.M. & Thouless, D.J. (1973). *J. Phys. C***6**, 1181.

Kramers, H.A. (1940). *Physica* **7**, 284.

Krane, K.J. & Raether, H. (1976). *Phys. Rev. Lett.* **37**, 1355.

Kubo, R., Toda, M. & Hashitsume, N. (1985). *Statistical Physics* II. Berlin: Springer-Verlag.

Kuhn, H. (1983). *Thin Solid Films* **99**, 1.

Kuk, Y. & Feldman, L.C. (1984). *Phys. Rev. B***30**, 5811.

Kumar, P. (1974). *Phys. Rev. B***10**, 2928.

Kumar, V. & Bennemann, K.H. (1983). *Phys. Rev. B***28**, 3138.

Lagally, M.G. (1982). In *Chemistry and Physics of Solid Surfaces* (eds. Vanselow & Howe), vol. 4, pp. 281–313. Berlin: Springer-Verlag.

Lagally, M.G., Lu, T.-M. & Wang, G.-C. (1980). In *Ordering in Two Dimensions* (ed. Sinha), pp. 113–21.

Lagendijk, E. & Huiskamp, W.J. (1972). *Physica* **62**, 444.

Lagois, J. & Fischer, B. (1978). In *Festkorperprobleme* XVIII (ed. Treusch), pp. 197–216. Braunschweig: Vieweg.

Landau, L.D. (1935). *Phys. Z. Sowjet.* **8**, 489.

Landau, L.D. (1965). In *Collected Papers*, pp. 540–5. Oxford: Pergamon.

Landau, L.D. & Lifshitz, E.M. (1969). *Statistical Physics*, 2nd edn. Reading: Addison-Wesley.

Landau, L.D. & Lifshitz, E.M. (1970). *Theory of Elasticity*, 2nd edn. Oxford: Pergamon.

Landgren, G., Ludeke, R., Jugnet, Y., Morar, J.F. & Himpsel, F.J. (1984). *J. Vac. Sci. Tech.* **8**, 351.

Landman, U. & Kleiman, G.G. (1977). In *Surface and Defect Properties of Solids*, vol. 6, pp. 1–105. London: The Chemical Society.

Lang, N.D. (1973). In *Solid State Physics* (eds. Seitz, Turnbull & Ehrenreich), vol. 28, pp. 225–300. New York: Academic.

Lang, N.D. (1981). *Phys. Rev. Lett.* **46**, 842.

Lang, N.D. & Kohn, W. (1970). *Phys. Rev. B***1**, 4555.

Lang, N.D. & Williams, A.R. (1977). *Phys. Rev. B***16**, 2408.

Lang, N.D. & Williams, A.R. (1978). *Phys. Rev. B***18**, 616.
Lang, N.D. & Williams, A.R. (1982). *Phys. Rev. B***25**, 2940.
Lang, N.D., Holloway, S. & Norskov, J.K. (1985). *Surf. Sci.* **150**, 24.
Langmuir, I. (1916). *J. Am. Chem. Soc.* **38**, 2221.
Langmuir, I. & Taylor, J.B. (1932). *Phys. Rev.* **40**, 463.
Langreth, D.C. (1985). *Phys. Rev. Lett.* **54**, 126.
Larher, Y. (1979). *Mol. Phys.* **38**, 789.
Lau, K.H. & Kohn, W. (1977). *Surf. Sci.* **65**, 607.
Lee, P.A. & Ramakrishnan, T.V. (1985). *Rev. Mod. Phys.* **57**, 287.
Lee, P.A., Citrin, P.H., Eisenberger, P. & Kincaid, B.M. (1981). *Rev. Mod. Phys.* **53**, 769.
Lennard-Jones, J.E. (1932). *Trans. Farad. Soc.* **28**, 333.
Lennard-Jones, J.E. & Devonshire, A.F. (1936). *Nature* **137**, 1069.
Levine, J.D. & Mark, P. (1966). *Phys. Rev.* **144**, 751.
Levine, R.D. & Bernstein, R.B. (1974). *Molecular Reaction Dynamics.* Oxford: Clarendon.
Liao, P.F., Bergman, J.G., Chemla, D.S., Wokaun, A., Melngailis, J., Hawryluk, A.M. & Economou, N.P. (1981). *Chem. Phys. Lett.* **82**, 355.
Lieske, N.P. (1984). *J. Phys. Chem. Solids* **45**, 821.
Lindford, R.G. (1973). In *Solid State Surface Science* (ed. Green), vol. 2, pp. 1–152. New York: Marcel Dekker.
Lipowsky, R. & Speth, W. (1983). *Phys. Rev. B***28**, 3983.
Louie, S.G., Chelikowsky, J.R. & Cohen, M.L. (1977). *Phys. Rev. B***15**, 2154.
Lubensky, T.C. & Rubin, M.H. (1975). *Phys. Rev. B***12**, 3885.
Ludeke, R. (1984). *J. Vac. Sci. Tech. B***2**, 400.
Lundqvist, B.I. (1984). In *Many-Body Phenomena at Surfaces* (eds. Langreth & Suhl), pp. 93–144. Orlando: Academic.
Lundqvist, S. & March, N.H. (1983). *Theory of the Inhomogeneous Electron Gas.* New York: Plenum.
Lunsford, J.H. (1974). *Cat. Rev.* **8**, 135.
Luth, H., Rubloff, G.W. & Grobman, W.D. (1976). *Solid State Commun.* **18**, 1427.
Ma, S. (1976). *Modern Theory of Critical Phenomena.* Reading: Benjamin.
Ma, S. (1985). *Statistical Mechanics.* Singapore: World Scientific.
Madelung, E. (1918). *Physik. Z.* **19**, 524.
Madey, T.E., Doering, D.L., Bertel, E. & Stockbauer R. (1983). *Ultramicroscopy* **11**, 187.
Madey, T.E. & Netzer, F.P. (1980). *Surf. Sci.* **117**, 549.
Maekawa, S. & Fukuyama, H. (1981). *J. Phys. Soc. Jpn.* **51**, 1380.
Many, A., Goldstein, Y. & Grover, N.B. (1965). *Semiconductor Surfaces.* Amsterdam: North Holland.
March, N.H. & Parrinello, M. (1982). *Collective Effects in Solids and Liquids.* Bristol: Adam Hilger.
Margaritondo, G. (1983). *Solid State Elect.* **26**, 499.
Margaritondo, G. (1985). *Phys. Rev. B***31**, 2526.
Margoninski, Y. (1986). *Cont. Phys.* **27**, 203.
Marks, L.D., Heine, V. & Smith, D.J. (1984). *Phys. Rev. Lett.* **52**, 656.
Marschall, N., Fischer, B. & Queisser, H.J. (1971). *Phys. Rev. Lett.* **27**, 95.
Martinot, P., Koster, A., Laval, S. & Carvalho, W. (1982). *Appl. Phys.* **29**, 172.
Marx, R. (1985). *Phys. Repts.* **125**, 1.
Mathon, J. (1983). *Phys. Rev. B***27**, 1916.
Mattheiss, L.F. & Hamann, D.R. (1984). *Phys. Rev. B***29**, 5372.
Maue, A.W. (1935). *Z. Physik.* **94**, 717.
Mazenko, G., Banavar, J.R. & Gomer, R. (1981). *Surf. Sci.* **107**, 459.
McDowell, H.K. & Doll, J.D. (1983). *J. Chem. Phys.* **78**, 3219.
McRae, E.G. (1984). *Surf. Sci.* **147**, 663.
McTague, J.P. & Novaco, A.D. (1979). *Phys. Rev. B***19**, 5299.

McTague, J.P., Als-Nielsen, J., Bohr, J. & Nielsen, M. (1982). *Phys. Rev. B***25**, 7765.
Mehl, M.J. & Schaich, W.L. (1975). *Phys. Rev. A***16**, 921.
Mehta, M. & Fadley, C.S. (1979). *Phys. Rev. B***20**, 2280.
Melius, C.F., Stulen, R.H. & Noell, J.O. (1982). *Phys. Rev. Lett.* **48**, 1429.
Menzel, D. (1975). In *Interactions on Metal Surfaces* (ed. Gomer), pp. 101–42. Berlin: Springer-Verlag.
Menzel, D. (1982). In *Chemistry and Physics of Solid Surfaces* (eds. Vanselow & Howe), vol. 4, pp. 389–406. Berlin: Springer-Verlag.
Mermin, N.D. (1968). *Phys. Rev.* **176**, 250.
Meyer, R.J., Duke, C.B., Paton, A., Kahn, A., So, E., Yeh, J.L. & Mark, P. (1979). *Phys. Rev. B***19**, 5194.
Miedema, A.R. (1978). *Z. Metallkunde* **69**, 455.
Migone, A.D., Li, Z.R. & Chan, M.H.W. (1984). *Phys. Rev. Lett.* **53**, 810.
Mills, D.L. (1984). In *Surface Excitations* (eds. Agranovich & Loudon), pp. 379–439. Amsterdam: North-Holland.
Milnes, A.G. & Feucht, D.L. (1972). *Heterojunctions and Metal–Semiconductor Junctions.* New York: Academic.
Miranda, R., Daiser, S., Wandelt, K. & Ertl, G. (1983). *Surf. Sci.* **131**, 61.
Monch, W. (1979). *Surf. Sci.* **86**, 672.
Monch, W. (1983). *Thin Solid Films* **104**, 285.
Morf, R.H. (1979). *Phys. Rev. Lett.* **43**, 931.
Morishige, K., Kittaka, S. & Morimoto, T. (1984). *Surf. Sci.* **148**, 401.
Morrison, S.R. (1977). *The Chemical Physics of Surfaces.* New York: Plenum.
Moskovits, M. (1985). *Rev. Mod. Phys.* **57**, 783.
Mott, N.F. (1938). *Proc. Camb. Phil. Soc.* **34**, 221.
Muhlhausen, C.W., Williams, L.R. & Tully, J.C. (1985). *J. Chem. Phys.* **83**, 2594.
Mukamel, D. & Krinsky, S. (1976). *Phys. Rev. B***13**, 5065.
Muller, E.W. (1951). *Z. Physik* **131**, 136.
Muller, E.W. (1977). In *Chemistry and Physics of Solid Surfaces* (eds. Vanselow & Tong), pp. 1–25. Cleveland: CRC.
Muller-Krumbhaar, H. (1978). In *Current Topics in Materials Science* (ed. Kaldis), vol. 1, pp. 1–47. Amsterdam: North-Holland.
Muscat, J.-P. (1984). *Surf. Sci.* **139**, 491.
Naumovets, A.G. & Vedula, Y.S. (1985). *Surf. Sci. Rep.* **4**, 365.
Neave, J.H., Joyce, B.A., Dobson, P.J. and Norton, N. (1983). *Appl. Phys. A***31**, 1.
Nelson, D.R. & Halperin, B.I. (1979). *Phys. Rev.* **19**, 2457.
Newns, D.M. (1969). *Phys. Rev.* **178**, 1123.
Nicholas, J.F. (1965). *An Atlas of Models of Crystal Surfaces.* New York: Gordon & Breach.
Nishiyama, Z. (1978). *Martensitic Transformation.* New York: Academic.
Nkoma, J., Loudon, R. & Tilley, D.R. (1974). *J. Phys. C***7**, 3547.
Norskov, J.K. & Lang, N.D. (1980). *Phys. Rev. B***21**, 2136.
Northrup, J.E. & Cohen, M.L. (1982). *Phys. Rev. Lett.* **49**, 1349.
Nosker, R.W., Mark, P. & Levine, J.D. (1970). *Surf. Sci.* **19**, 291.
Ohnishi, S., Freeman, A.J. & Weinert, M. (1983). *Phys. Rev. B***28**, 6741.
O'Keeffe, M. & Navrotsky, A. (1981). *Structure and Bonding in Crystals.* New York: Academic.
Osakabe, N., Tanishiro, Y., Yagi, K. & Honjo, G. (1981). *Surf. Sci.* **109**, 353.
Otto, A. (1968). *Z. Phys.* **216**, 398.
Ozaki, A. & Aika, K. (1981). In *Catalysis: Science and Technology* (eds. Anderson & Boudart), vol. 1, pp. 87–167. Berlin: Springer-Verlag.
Pandey, K.C. (1976). *Phys. Rev. B***14**, 1557.

Pandey, K.C. (1981). *Phys. Rev. Lett.* **47**, 1913.
Park, R.L. & den Boer, M.L. (1977). In *Chemistry and Physics of Solid Surfaces* (eds. Vanselow & Tong), pp. 191–203. Cleveland: CRC.
Park, R.L. & Lagally, M.G. (1985). *Solid State Physics: Surfaces.* Orlando: Academic.
Pauling, L. (1960). *The Nature of the Chemical Bond*, 3rd edn. Ithaca: Cornell.
Pauly, H. (1979). In *Atom–Molecule Collision Theory* (ed. Bernstein), pp. 111–99. New York: Plenum.
Pendry, J.B. (1974). *Low Energy Electron Diffraction.* New York: Wiley.
Penn, D.R. (1976). *Phys. Rev. B*13, 5248.
Persson, B.N.J. & Andersson, S. (1984). *Phys. Rev. B*29, 4382.
Pfnur, H., Feulner, P., Engelhardt, H.A. & Menzel, D. (1978). *Chem. Phys. Lett.* **59**, 481.
Philpott, M.R. & Turlet, J.M. (1976). *J. Chem. Phys.* **64**, 3852.
Pierce, D.T. & Meier, F. (1976). *Phys. Rev. B*13, 5484.
Pierce, D.T., Celotta, R.J., Unguris, J. & Siegmann, H.C. (1982). *Phys. Rev. B*26, 2566.
Plummer, E.W. (1985). *Surf. Sci.* **152/153**, 162.
Plummer, E.W. & Eberhardt, W. (1979). *Phys. Rev. B*20, 1444.
Plummer, E.W. & Eberhardt, W. (1982). *Adv. Chem. Phys.* **49**, 533.
Plummer, E.W., Gadzuk, J.W. & Penn, D.R. (1975). *Physics Today*, April, pp. 63–71.
Plummer, E.W., Salaneck, W.R. & Miller, J.S. (1978). *Phys. Rev. B*18, 1673.
Pokrovsky, V.L. & Talapov, A.L. (1984). *Theory of Incommensurate Crystals.* London: Harwood.
Posternak, M., Krakauer, H., Freeman, A.J. & Koelling, D.D. (1980). *Phys. Rev. B*21, 5601.
Prutton, M. (1983). *Surface Physics.* Oxford: Clarendon.
Raatz, F. & Salahub, D.R. (1984). *Surf. Sci.* **146**, 1609.
Raether, H. (1982). In *Surface Polaritons* (eds. Agranovich & Mills), pp. 331–403. Amsterdam: North-Holland.
Rahman, T.S., Mills, D.L., Black, J.E., Szeftel, J.M., Lehwald, S. & Ibach, H. (1984) *Phys Rev. B*30, 589.
Ramirez, R., Rahman, A. & Schuller, I.K. (1984). *Phys. Rev. B*30, 6208.
Rau, C. (1982). *J. Magn. Magn. Mat.* **30**, 141.
Rayleigh, Lord (1885). *Proc. London Math. Soc.* **17**, 4.
Redhead, P.A. (1962). *Vacuum* **12**, 203.
Rhodin, T.N. & Adams, D.L. (1976). In *Treatise on Solid State Chemistry* (ed. Hannay), vol. 6*A*, pp. 343–484. New York: Plenum.
Rhodin, T.N. & Gadzuk, J.W. (1979). In *The Nature of the Surface Chemical Bond* (eds. Rhodin & Ertl), pp. 113–273. Amsterdam: North-Holland.
Rice, J. (1936). In *Commentary on the Scientific Writings of J. Williard Gibbs* (eds. Donnan & Haas), vol. 1, pp. 505–708. New Haven: Yale.
Rieder, K.H. (1983). *Phys. Rev. B*27, 7799.
Rieder, K.H. & Engel, T. (1980). *Phys. Rev. Lett.* **45**, 824.
Rikvold, P.A. (1985). *Phys. Rev. B*32, 4756.
Ritchie, R.H. (1973). *Surf. Sci.* **34**, 1.
Roberts, M.W. & McKee, C.S. (1978). *Chemistry of the Metal–Gas Interface.* Oxford: Clarendon.
Robinson, I.K., Waskiewicz, W.K., Fuoss, P.H., Stark, J.B. & Bennett, P.A. (1986). *Phys. Rev. B*33, 7013.
Robota, H.J., Verheij, L.K., Liu, M.C., Segner, J. & Ertl, G. (1985). *Surf. Sci.* **155**, 101.
Roelofs, L.D. (1982). In *Chemistry and Physics of Solid Surfaces* (eds. Vanselow & Howe), vol. 4, pp. 219–50. Berlin: Springer-Verlag.
Roelofs, L.D. & Estrup, P.J. (1983). *Surf. Sci.* **125**, 51.
Roelofs, L.D., Chung, J.W., Ying, S.C. & Estrup, P.J. (1986). *Phys. Rev. B*33, 6537.

Rosenbaum, T.F., Nagler, S.E., Horn, P.M. & Clarke, R. (1983). *Phys. Rev. Lett.* **50**, 1791.
Rosenfeld, A. (1962). In *The Collected Works of Irving Langmuir* (ed. Suits), vol. 12, pp. 5–229. New York: Pergamon.
Rosengren, A. & Johansson, B. (1981). *Phys. Rev.* **23**, 3852.
Sadiq, A. & Binder, K. (1984). *J. Stat. Phys.* **35**, 517.
Saile, V., Skibowski, M., Steinmann, W., Gurtler, P., Koch, E.E. & Kozevnikov, A. (1970). *Phys. Rev. Lett.* **37**, 305.
Sakurai, T. & Hagstrum, H.D. (1975). *Phys. Rev. B***12**, 5349.
Sakurai, T., Hashizume, T., Jimbo, A., Sakai, A. & Hyodo, S. (1985). *Phys. Rev. Lett.* **55**, 514.
Salahub, D.R., Roche, M. & Messmer, R.P. (1978). *Phys. Rev. B***18**, 6495.
Sandercock, J.R. & Wettling, W. (1979). *J. Appl. Phys.* **50**, 7784.
Scheffler, M. & Bradshaw, A.M. (1983). In *The Chemical Physics of Solid Surfaces and Heterogeneous Catalysis* (eds. King & Woodruff), vol. 2, pp. 165–257. Amsterdam: Elsevier.
Scheffler, M., Kambe, K. & Forstmann, F. (1978). *Solid State Commun.* **25**, 93.
Scheibner, E.J., Germer, L.H. & Hartman, C.D. (1960). *Rev. Sci. Inst.* **31**, 112.
Schick, M. (1981). *Prog. Surf. Sci.* **11**, 245.
Schluter, M. & Sham, L.J. (1982). *Physics Today* **35**, no. 2, 36.
Schluter, M., Chelikowsky, J.R., Louie, S.G. & Cohen, M.L. (1985). *Phys. Rev. B***12**, 4200.
Schmit, J.N. (1974). Ph.D. Thesis, University of Liege (unpublished).
Schottky, W. (1939). *Z. Physik*, **113**, 367.
Schulz, H.J. (1985). *J. Phys. (Paris)* **46**, 257.
Scott, R.Q. & Mills, D.L. (1977). *Phys. Rev. B***15**, 3545.
Seeger, R.J. (1973). *Benjamin Franklin: New World Physicist.* Oxford: Pergamon.
Segner, J., Robota, H., Vielhaber, W., Ertl, G., Frenkel, F., Hager, J., Krieger, W. & Walther, H. (1983). *Surf. Sci.* **131**, 273.
Seifert, H. (1953). In *Structure and Properties of Solid Surfaces* (eds. Gomer & Smith), pp. 318–83. Chicago: University Press.
Seiler, A., Feigerle, C.S., Pena, J.L., Celotta, R.J. & Pierce, D.T. (1985). *Phys. Rev. B***32**, 7776.
Seitz, F. (1940). *The Modern Theory of Solids.* New York: McGraw-Hill.
Selke, W., Binder, K. & Kinzel, W. (1983). *Surf. Sci.* **125**, 74.
Sette, F., Stohr, J. & Hitchcock, A.P. (1984). *J. Chem. Phys.* **81**, 4906.
Shaw, C.G., Fain, S.C. & Chinn, M.D. (1978). *Phys. Rev. Lett.* **41**, 955.
Shen, Y.R. (1984). *The Principles of Non-Linear Optics.* New York: Wiley.
Shockley, W. (1939). *Phys. Rev.* **56**, 317.
Shih, A. & Parsegian, V.A. (1975). *Phys. Rev. A***12**, 835.
Siegbahn, K., Nordling, C., Fahlman, A., Nordberg, R., Hamrin, K., Hedman, J., Johannson, G., Belgmark, T., Karlsson, S.E., Lindgren, I. & Lindberg, B. (1967). *ESCA – Atomic, Molecular and Solid State Structure Studied by Means of Electron Spectroscopy.* Uppsala: Almquist & Wiksell.
Siegmann, H.C., Meier, F., Erbudak, M. & Landolt, M. (1984). *Adv. Elec. Electron Phys.* **62**, 1.
Simon, H.J., Mitchell, D.E. & Watson, J.G. (1974). *Phys. Rev. Lett.* **33**, 1531.
Sinfelt, J.H., Carter, J.L. & Yates, D.J.C. (1972). *J. Catal.* **24**, 283.
Smith, C.S. (1948). *Trans. AIME*, **175**, 15.
Smoluchowski, R. (1941). *Phys. Rev.* **60**, 661.
Sokolov, J., Jona, F. & Marcus, P.M. (1984). *Solid State Commun.* **49**, 307.
Sommerfeld, A. (1909). *Ann. Physik* **28**, 665.
Somorjai, G.A. (1972). *Principles of Surface Chemistry.* Englewood Cliffs: Prentice-Hall.
Somorjai, G.A. (1981). *Chemistry in Two Dimensions: Surfaces.* Ithaca: Cornell.
Somorjai, G.A. & Zaera, F. (1982). *J. Phys. Chem.* **86**, 3070.

Soria, F. & Poppa, H. (1980). *J. Vac. Sci. Tech.* **17**, 449.

Spanjaard, D. & Desjonqueres, M.C. (1984). *Phys. Rev. B***30**, 4822.

Sparks, M. (1970). *Phys. Rev. B***1**, 4439.

Specht, E.D., Sutton, M., Birgeneau, R.J., Moncton, D.E. & Horn, P.M. (1984). *Phys. Rev. B***30**, 1589.

Spicer, W.E., Chye, P.W., Skeath, P.R., Su, C.Y. & Lindau, I. (1979). *J. Vac. Sci. Tech.* **16**, 1422.

Stanley, H.E. (1971). *Introduction to Phase Transitions and Critical Phenomena.* New York: Oxford.

Steele, W.A. (1974). *The Interaction of Gases with Solid Surfaces.* Oxford: Pergamon.

Stensgaard, I., Feldman, L.C. & Silverman, P.J. (1978). *Surf. Sci.* **77**, 513.

Stohr, J. (1984). In *Chemistry and Physics of Solid Surfaces* (eds. Vanselow & Howe), vol. 5, pp. 231–55. Berlin: Springer-Verlag.

Stohr, J. & Jaeger, R. (1982). *Phys. Rev. B***26**, 4111.

Stohr, J., Jaeger, R. & Brennan, S. (1982). *Surf. Sci.* **117**, 503.

Stolt, K., Graham, W.R. & Ehrlich, G. (1976). *J. Chem. Phys.* **65**, 3206.

Stoltze, P. & Norskov, J.K. (1985). *Phys. Rev. Lett.* **55**, 2502.

Stoop, L.C.A. & Van der Merwe, J.H. (1982). *Thin Solid Films* **94**, 341.

Stott, M.J. & Zaremba, E. (1980). *Phys. Rev. B***22**, 1564.

Stringfellow, G.B. (1985). In *Crystal Growth of Electronic Materials* (ed. Kaldis), pp. 247–67. Amsterdam: North-Holland.

Sturge, M.D. (1967). In *Solid State Physics* (eds. Seitz, Turnbull & Ehrenreich), vol. 20, pp. 91–211. New York: Academic.

Sunjic, M. & Lucas, A.A. (1971). *Phys. Rev. B***3**, 719.

Susskind, C. (1980). In *Advances in Electronics and Electron Physics* (eds Marton & Marton), vol. 50, pp. 241–60. New York: Academic.

Tabor, D. (1980). *J. Colloid Interface Sci.* **75**, 240.

Takayanagi, K., Tanishiro, Y., Takahashi, M. & Takahashi, S. (1985). *Surf. Sci.* **164**, 367.

Tamm, I. (1932). *Phys. Z. Soviet Union*, **1**, 733.

Tersoff, J. (1984a). *Phys. Rev. Lett.* **52**, 465.

Tersoff, J. (1984b). *Phys. Rev. B***30**, 4874.

Thomas, L.B. (1967). In *Rarefied Gas Dynamics*, vol. 1, pp. 155–62. New York: Academic.

Thompson, M.D. & Huntington, H.B. (1982). *Surf. Sci.* **116**, 522.

Thomy, A., Duval, X. & Regnier, J. (1981). *Surf. Sci. Rep.* **1**, 1.

Toennies, J.P. (1984). *J. Vac. Sci. Tech. A***2**, 1055.

Tomanek, D. & Bennemann, K.H. (1985a). *Surf. Sci.* **163**, 503.

Tomanek, D. & Bennemann, K.H. (1985b). *Phys. Rev. B***31**, 2488.

Topham, S. (1985). In *Catalysis: Science and Technology* (eds. Anderson & Boudart), vol. 7, pp. 1–50. Berlin: Springer-Verlag.

Toyoshima, I. & Somorjai, G.A. (1979). *Cat. Rev. Sci. Eng.* **19**, 105.

Trilling, L. (1970). *Surf. Sci.* **21**, 337.

Tringides, M., Wu, P.K., Moritz, W. & Lagally, M.G. (1986). *Ber. Bunsenges. Phys. Chem.* **90**, 277.

Tsang, W.T. & Miller, R.C. (1986). *J. Cryst. Growth* **77**, 55.

Tsong, T.T. (1980). *Prog. Surf. Sci.* **10**, 165.

Tsong, T.T. & Casanova, R. (1981). *Phys. Rev. B***24**, 3063.

Tsong, T.T. & Sweeney, J. (1979). *Solid State Commun.* **30**, 767.

Tsukada, M. & Hoshino, T. (1982). *J. Phys. Soc. Jpn.* **51**, 2562.

Tuck, R.A. (1983). *Vacuum* **33**, 715.

Tully, J.C. (1980). *Ann. Rev. Phys. Chem.* **31**, 319.

Tully, J.C. (1981). *Surf. Sci.* **111**, 461.

Tully, J.C., Chabal, Y.J., Raghavachari, K., Bowman, J.M. & Lucchese, R.R. (1985). *Phys. Rev.* **31**, 1184.

Van Kampen, N.G. (1981). *Stochastic Processes in Physics and Chemistry.* Amsterdam: North-Holland.
Van der Merwe, J.H. (1984). In *Chemistry and Physics of Solid Surfaces* (eds. Vanselow & Howe), vol. 5, pp. 365–401. Berlin: Springer-Verlag.
Van der Veen, J.F. (1985). *Surf. Sci. Rep.* **5**, 199.
Van der Veen, J.F., Tromp, R.M., Smeenck, R.G. & Saris, F.W. (1979). *Surf. Sci.* **82**, 468.
Van Veen, G.N.A., Baller, T. & De Vries, A.E. (1985). *Chem. Phys.* **92**, 59.
Vannice, M.A. (1977). *J. Catal.* **50**, 228.
Villain, J. (1980). In *Ordering in Strongly Fluctuating Condensed Matter Systems* (ed. Riste), pp. 221–60. New York: Plenum.
Villain, J. & Gordon, M.B. (1983). *Surf. Sci.* **125**, 1.
Voges, H. & Schinke, R. (1983). *Chem. Phys. Lett.* **100**, 245.
Volmer, M. & Estermann, J. (1921). *Z. Phys.* **7**, 13.
Vook, R.W. (1982). *Int. Metals Rev.* **27**, 209.
Wagner, C.D., Riggs, W.M., Davis, L.E., Moulder, J.F. & Muilenberg, G.E. (1978). *Handbook of X-ray Photoelectron Spectroscopy.* Eden Prairie: Perkin-Elmer.
Wagner, L.F. & Spicer, W.E. (1974). *Phys. Rev. B***9**, 1512.
Wang, D.-S., Freeman, A.J. & Krakauer, H.K. (1981). *Phys. Rev. B***24**, 3092.
Wang, G.-C. & Lu, T.-M. (1985). *Phys. Rev. B***31**, 5918.
Wang, S.C. & Gomer, R. (1985). *J. Chem. Phys.* **83**, 4193.
Watson, R.E., Davenport, J.W., Perlman, M.L. & Sham, T.K. (1981). *Phys. Rev.* **24**, 1791.
Weeks, J.D. (1980). In *Ordering in Strongly Fluctuating Condensed Matter Systems* (ed. Riste), pp. 293–317. New York: Plenum.
Weinert, M. & Davenport, J.W. (1985). *Phys. Rev. Lett.* **54**, 1547.
Weller, D., Alvarado, S.F., Gudat, W., Schroder, K. & Campagna, M. (1985). *Phys. Rev. Lett.* **54**, 1555.
Wendelken, J.F. & Wang, G.C. (1985). *Phys. Rev.* **32**, 7542.
Whitman, L.J., Bartosch, C.E., Ho, W.H., Strasser, G. & Grunze, M. (1986). *Phys. Rev. Lett.* **56**, 1984.
Whitten, J.L. & Pakkanen, T.A. (1980). *Phys. Rev. B***21**, 4357.
Wigner, E.P. (1934). *Phys. Rev.* **46**, 1002.
Wille, L.T. & Durham, P.J. (1985). *Surf. Sci.* **164**, 19.
Williams, A.R. & von Barth, U. (1983). In *Theory of the Inhomogeneous Electron Gas* (eds. Lundqvist & March), pp. 189–308. New York: Plenum.
Williams, L.P. (1965). *Michael Faraday.* London: Chapman & Hall.
Williams, R.H. & McGovern, I.T. (1984). In *The Chemical Physics of Solid Surfaces and Heterogeneous Catalysis* (eds. King & Woodruff), vol. 3, pp. 267–309. Amsterdam: Elsevier.
Williams, R.S. & Yarmoff, J.A. (1983). *Nucl. Inst. Meth.* **218**, 235.
Willis, R.F. (1985). In *Dynamical Phenomena at Surfaces, Interfaces and Superlattices* (eds. Nizzoli, Rieder & Willis), pp. 126–47.
Willis, R.F., Lucas, A.A. & Mahan, G.D. (1983). In *The Chemical Physics of Solid Surfaces and Heterogeneous Catalysis* (eds. King & Woodruff), vol. 2, pp. 59–163. Amsterdam: Elsevier.
Wilson, K.G. (1979). *Sci. Am.* **241**, no. 2, 158.
Wimmer, E., Freeman, A.J., Hiskes, J.R. & Karo, A.M. (1983). *Phys. Rev. B***28**, 3074.
Wimmer, E., Fu, C.L. & Freeman, A.J. (1985). *Phys. Rev. Lett.* **55**, 2618.
Winnick, H. & Doniach, S. (1980). *Synchrotron Radiation Research.* New York: Plenum.
Wolfram, T. & DeWames, R.E. (1972). *Prog. Surf. Sci.* **2**, 233.
Wood, E. (1964). *J. Appl. Phys.* **35**, 1306.
Woodruff, D.P. & Delchar, T.A. (1986). *Modern Techniques of Surface Science.* Cambridge: Cambridge University Press.
Wulff, G. (1901). *Z. Kristallog.* **34**, 449.

Yakovlev, V.A. & Zhizhin, G.N. (1975). *JETP Lett.* **19**, 189.
Yanagihara, T. & Yamaguchi, H. (1984). *Jpn. J. Appl. Phys.* **23**, 529.
Yates, J.T. (1985). In *Methods of Experimental Physics* (eds. Celotta & Levine), vol. 22, pp. 425–64. Orlando: Academic.
Yates, J.T., Thiel, P.A. & Weinberg, W.H. (1979). *Surf. Sci.* **82**, 45.
Zamir, E. & Levine, R.D. (1984). *Chem. Phys. Lett.* **104**, 143.
Zaremba, E. & Kohn, W. (1976). *Phys. Rev. B***13**, 2270.
Zaremba, E. & Kohn, W. (1977). *Phys. Rev. B***15**, 1769.
Zehner, D.M., Noonan, J.R., Davis, H.L. & White, C.W. (1981). *J. Vac. Sci. Tech.* **18**, 852.
Zhang, S.B., Cohen, M.L. & Louie, S.G. (1986). *Phys. Rev. B***34**, 768.
Zhu, X., Huang, H. & Hermanson, J. (1984). *Phys. Rev. B***29**, 3009.
Ziman, J.M. (1972). *Principles of the Theory of Solids.* Cambridge: Cambridge University Press.

INDEX